**Contributions to Management Science**

The series *Contributions to Management Science* contains research publications in all fields of business and management science. These publications are primarily monographs and multiple author works containing new research results, and also feature selected conference-based publications are also considered. The focus of the series lies in presenting the development of latest theoretical and empirical research across different viewpoints.

This book series is indexed in Scopus.

Guy Fournier • Adrian Boos • Dimitri Konstantas •
Danielle Attias
Editors

# Automated Vehicles as a Game Changer for Sustainable Mobility

## Learnings and Solutions

Springer

*Editors*
Guy Fournier
Pforzheim University
Pforzheim, Germany

Adrian Boos
Pforzheim University
Pforzheim, Germany

Dimitri Konstantas
Geneva School of Economics
and Management
University of Geneva
Geneva, Switzerland

Danielle Attias
University of Paris-Saclay
Gif-sur-Yvette, France

ISSN 1431-1941 ISSN 2197-716X (electronic)
Contributions to Management Science
ISBN 978-3-031-61680-8 ISBN 978-3-031-61681-5 (eBook)
https://doi.org/10.1007/978-3-031-61681-5

European Union Horizon 2020 research and innovation programme grant agreement 769033 - AVENUE project



This Springer imprint is published by the registered company Springer Nature Switzerland AG
The registered company address is: Gewerbestrasse 11, 6330 Cham, Switzerland

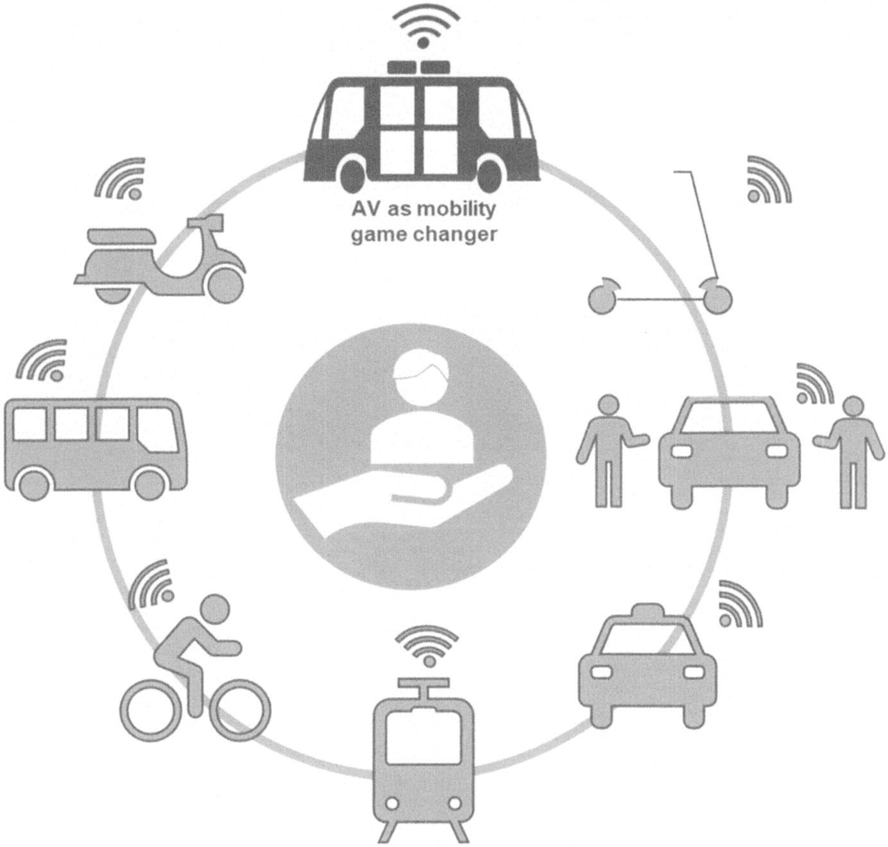


Own design, Poster with more information can be found at the following link: AVENUE

# Foreword

Dear Reader,

In discussions on how we can steer the future direction of our planet in the right way, mobility soon takes a central spot. Today, mobility generates approximately one third of all urban emissions, an impact that will only grow as millions more people move to cities, increasing passenger demand by 79% by 2050.

The desired outcome of this discussion is clear: we need to encourage a shift from the private car towards more sustainable, higher occupancy modes such as public transport. To make this shift, the latter must provide a convenient, reliable, and inclusive alternative.

At UITP (the International Association of Public Transport), we believe that automation can be a key solution here. Automated mobility, autonomous vehicles, self-driving shuttles: in their many forms, they can hugely enrich the mobility mix, providing a kind of flexibility that has the power to fill gaps in terms of times (late at night), location (rural areas), and people's needs.

But, taking an innovative spin on a famous quote, we can strongly argue that alike happiness, automation is best when shared. Only when deployed as shared vehicles that are integrated into an effective public transport network, automated vehicles (AVs) can be a real solution to create healthier and more competitive and sustainable cities.

In order to be a credible alternative to (automated) car ownership, the different sustainable modes need to be coordinated, planned, and delivered in an integrated way not only from a physical perspective (coordinated network planning, stations, urban planning, and algorithmic optimisation of automated fleets) but also from an information perspective whereby services are integrated into a passenger-centric MaaS system. Here, the question of data also remains important: there is a need for a huge collection of data, as well as the regulatory framework to utilise it safely and cooperatively.

The question is thus not (anymore) the technology, but how to involve all stakeholders in the best way and how to create business models and services that work and make passengers want to use AVs. Public transport authorities and operators are experts in organising urban mobility solutions: we should allow them to lead the transition and take the lead in the coordination of tomorrow's mobility.

Research & innovation projects like AVENUE are key to address such questions and to be able to maximise the benefits of automation for our society. UITP is proud to be involved in various EU projects on CCAM, including as Coordinator in SHOW (SHared automation Operating models for Worldwide adoption). Also in ULTIMO, which will build on the learnings of SHOW and AVENUE, UITP is proud to work together with CCAM colleagues to look at AV deployment that considers all elements in a cross-sector business environment and truly wants to unlock the integration of AVs into cities.

This AVENUE book shows the results of a unique collaboration and provides invaluable knowledge on how to ensure automated mobility is deployed as real solution for some of today's greatest challenges. We are confident it will inspire many.

Head of 3rd Party funded projects and the CCAM Colleagues, UITP,
the International Association of Public Transport
Brussels, Belgium

Umberto Guida

# Foreword

Dear Reader

Digitalisation is having a far-reaching impact on us and our lives—also in the way we get around. We cannot say conclusively today what this will look like in 50 or even 100 years' time. What is certain is that even in a hundred years' time, a society's prosperity, welfare, health, education, and culture will still be directly linked to the wide availability of sustainable, safe, and reliable transport. Automated vehicles are likely to play a major role in that respect. They can make more efficient use of existing roads, improve traffic flow, and significantly increase road safety—an opportunity that we, as specialist authorities and road traffic agencies, must not pass up if we want to make our roads fit for the future.

That is why the Federal Roads Office (FEDRO)[1] has made smart mobility and automated and connected transport a focal point of its official activities. In 2015, an automated vehicle travelled on public roads in Switzerland for the first time as part of a pilot scheme. In the meantime, 16 pilot schemes have been conducted throughout Switzerland. This also included the operation of three on-demand shuttle buses in Geneva. As part of this pilot project, authorised and supervised by FEDRO, an automated bus operated by the Geneva transport authority (transports publics genevois, tpg) carried passengers on the semi-public roads of a hospital campus. The routes were not predetermined, but were orientated to the needs of the customers. Similar pilot applications were run in three other cities (Lyon, Copenhagen, and Luxembourg).

These four pilot applications were part of the major EU-funded AVENUE project, in which FEDRO was represented on the Advisory Board. In FEDRO's view, the project has been positive. AVENUE has shown how teams from different cultures across Europe can work together to successfully find solutions in a highly complex and innovative field such as automated mobility. Mutual exchange, a willingness to learn, and cooperation between the private and public sectors are essential. FEDRO's participation on the advisory board contributed to this

[1] Switzerland's federal authority responsible for road infrastructure and private road transport.

constructive cooperation. At the same time, it has gained valuable experience in the field of automated transport. This meant that both sides were able to benefit from FEDRO's involvement in the major AVENUE project.

Today's pilot applications will form an integral part of tomorrow's transport solutions. By 2050, around a third of all vehicles on the roads could be (partially) automated. As one of our studies also shows, driverless shuttles, buses, and trains could cover up to 70% of our transport needs. Traditional motorised private and public transport would lose their dominance. The boundaries between these forms of transport will shift towards 'public individual transport' or 'individual public transport' services. These forms of transport will be characterised by their efficiency: automated vehicles can be on the move continuously, making one journey after another. In future, it will be almost inconceivable for vehicles to stand stationary for more than 90% of the time as they do today.

Switzerland has taken an important step towards smarter mobility: The revision of the Road Traffic Act and the Ordinance on Automated Transport demonstrate a clear commitment to automated transport on the part of the Swiss parliament and government. The potential of conditional and highly automated vehicles (SAE automation levels 3 and 4) will be deployed on Swiss roads as early as 2025.

In addition to the political process, it will be important that we as a society prepare ourselves for the 'automation' megatrend. As specialised authorities, we can ensure that the new technologies are supported in terms of legislation and infrastructure, as the potential in terms of efficiency, costs, safety, and availability is simply too promising to ignore. But there must be broad social acceptance. Projects like AVENUE can play an important part by literally taking users on a journey and incorporating their experiences into improving the technology.

I am confident that this approach will be successful. And I look forward to seeing the upcoming changes in mobility lead us towards greater safety, improved compatibility, and availability at lower costs.

Director of the Federal Roads Office (FEDRO) Jürg Röthlisberger
Ittigen, Switzerland

Federal Roads Office (FEDRO) & Advisory Benno Nager
Board Member AVENUE
Ittigen, Switzerland

# Foreword

## The Five Ts: Transparency, Trust, Teamwork, Try, and Transform

### *We Need a Climate for Change*

People and the ecosystems we life in, in this Digital Age, have great capabilities to improve and sustain the quality of life for all, if we interact and team up. As we face and urgently need to deal with many societal challenges, we need a climate for change.

Societal challenges and related Sustainable Development Goals are a challenging problem set. There is no one solution. There is no one group with the answer. There is no one technical fixture. This is all about boots on the ground, roll up our sleeves, start walking, and work together. This is not about the talk. This is all about walking the talk and achieving outcomes. Transformation and other continuous change is a team sport.

Mobility is one of our main societal challenges locally and globally. Whether it is about getting persons of any demography and in any phase of life, safely, securely, timely, comfortably and affordably while preserving human integrity, or otherwise. Mobility is an integral part of almost every aspect in life. It intertwines with many other societal challenges. So addressing the mobility challenges will also support, co-address, and improve those other societal challenges, whether local, regional, national, or global.

However, thinking that we can address this big societal challenge and other intertwined challenges by doing more of the same and with that we will get a different outcome is a low probability success strategy. We have got to get out of our comfort zones. We have to be willing to try new, alternative approaches to provide these essential services to the public, while improving and sustaining the quality of life and planet.

## *Horizon 2020 Project AVENUE*

With Horizon 2020 Project AVENUE, an essential and great start has been made to do exactly that: trying new alternatives. For once, these included (A) finding the optimal balance between EV and AV in order to provide useful, affordable, viable, and sustainable multi- and inter-modal services to the public—in a people-centric manner, (B) pivoting from fixed stop to on demand—but still—public transportation, while (C) getting the right stakeholders onboard, comfortable, confident, and courageous enough to actively support the mission, and while (D) aiming to create multi-layered yet integrated data sharing ecosystems in this Digital Age for multiple stakeholders, including the traveller, the public transport operator, the platform provider and its supply chain ecosystems, as well as first responders, relevant authorities, other stakeholders, and last but not least: citizens, businesses, and society at large.

The latter—data, information, and knowledge sharing—is essential to unlock data, create transparency, and spur innovation in an accountable manner, and to support the AVENUE mission including to build, achieve, and sustain the appropriate levels of trust that are necessary to be truly able to trigger the envisioned symbiotic transformation of future-proof mobility, green, digital, and societal transition. Step by step.

## *Europe Fit for the Digital Decade*

I want Europe to strive for more by grasping the opportunities from the digital age within safe and ethical boundaries. When advocating its mission called 'A Europe Fit for the Digital Age', the Commission well-phrased the importance of the symbiosis between opportunities, challenges, risks, and ethics. Obviously, that should generally be nothing new to anyone. However, connecting, inter-connecting and even hyper-connecting, and creating new physical, cyber-physical and digital ecosystems and other combinatoric possibilities and innovations to empower people and organisations, public and private, and guide Europe's green and digital transformations requires each of us to get out of our comfort zone, rethink, team up, and most of all, to act and try.

In line with the EU Fit for the Digital Age mission, together with the member states the Commission has developed and started to deliver on the so-called Path to the Digital Decade 2030. It consists of various strategies, common objectives and targets, where mobility is one of the main essential components thereof to support the necessary transformations. Said targets are to be met by 2030.

## *The Five Ts: Transparency, Trust, Teamwork, Try, and Transform*

In the above paragraphs you have already been able to identify the Five Ts. One needs to be willing to meet the right levels of transparency, as transparency leads to trust. Trust can be seen as the equation of consistency through time, where teamwork and the willingness to try, fail, pivot, and try again will be the strategy with the highest probability of success.

Project AVENUE has basically taken these Five Ts in, also by demonstrating that the challenges at hand can be converted with our combined creativity and determination to transform. Meanwhile, obviously many new challenges, big and small have arisen during the project. This, as one will only see and is able to identify and address the true challenges if one starts walking, teams up, tries, and pivots while focusing on transformative outcomes—not on efforts. It is good to see and know that this project has already been succeeded by new collaborations, and namely the Horizon Europe project ULTIMO.

This publication gives you a rich amount and variety of valuable insights and oversights, lessons learned and good practices for you to consider, further, improve, and optimise. It is a great example of the willingness to share and provide transparency.

And with that it also makes available a valuable amount of trust principles and components that can help you, your organisation, and your stakeholders to join forces to addressing the societal challenge of mobility, and the many intertwined challenges—and opportunities. On the latter, I am certain that if one is able to convert those challenges into such opportunities, any person in any part of the mobility market in any place on this planet can benefit from it.

I welcome you to enjoy reading this publication, use its valuable content, and join an ever-increasing community that is determined to foster the climate for change and focus on making the difference, for the short, mid, and long term, whether local, regional, global, or otherwise.

Director Arthur's Legal, Strategies & Systems
& Advisory Board Member AVENUE
Amsterdam, The Netherlands

Arthur van der Wees

# Acknowledgements

This project has received funding from the European Union's Horizon 2020 research and innovation program under grant agreement No. 769033.

The research work developed in this book received funding from the European Union's Horizon 2020 research and innovation program under grant agreement No. 769033, for the AVENUE project (Autonomous Vehicle to Evolve to a New Urban Experience).

We would like to express our sincere thanks to all the authors and related institutions who contributed to this book. Participating in this adventure required considerable effort, commitment, and flexibility, as many of our authors generously shared their knowledge and insights at all hours of the day and throughout the week. Their willingness to stay involved and contribute even after the project was completed is something to be praised and we are very grateful for. We are especially thankful for their enthusiasm to explore and illuminate the potential of automated vehicles for the common good. Your contributions have been invaluable for the team and we greatly appreciate the time, expertise and dedication you have brought to this project. Thank you for being an integral part of this journey and helping to create a thoughtful and informative discourse on such an important topic. Your input for the book will surely be a starting point for further meaningful discussions.

*Acknowledgment for support in artwork, layout, and typesetting:*

Nicole van den Boom, Antonia Bucksch, Charly Beye, Sally Krüger, and Niklas Liebhauser

# About This Book

This book deals with the possibilities of a new, more sustainable form of integrated mobility on the basis of the Horizon 2020 Project AVENUE-Autonomous Vehicles to Evolve to a New Urban Experience. AVENUE designed and carried out full-scale demonstrations of urban transport automation by deploying fleets of automated minibuses in low to medium demand areas of four European demonstrator cities (Geneva, Lyon, Copenhagen, and Luxembourg) and two replicator cities (Esch-sur-Alzette in Luxembourg and Sion in Switzerland). The AVENUE vision for future public transport in urban and suburban areas is that automated vehicles will ensure safe, rapid, economic, sustainable, and personalised transport of passengers. AVENUE introduces disruptive public transportation paradigms based on on-demand, door-to-door services, aiming to set up a new model of public transportation, by revisiting the offered public transportation services, and aiming to suppress prescheduled fixed bus itineraries. In Part I, the book explains how automated vehicles were implemented on the different sites and which progress were achieved on the vehicles, the vehicle safety, in-vehicle services, cybersecurity and on the requirements for Persons with Reduced Mobility. The economic, environmental, and social impact of the implementation of automated vehicles for companies, citizens, and cities is assessed in Part II. One aim of this book is finally to demonstrate that automated vehicles integrated in a MaaS (Mobility-as-a-Service)/ITS (Intelligent Transport System) can become a game changer towards sustainable mobility and a fair share of the created values between the stakeholders. The transport system could become more efficient, flexible, resilient, and citizen centric avoiding thus a coercive transformation policy.

# Contents

# Contributors

**Fabio Antonialli** PPA Business School, Paris, France
École Centrale Lyon, Écully, France

**Danielle Attias** University of Paris-Saclay, Gif-sur-Yvette, France

**Meriem Benyahya** Centre Universitaire d'Informatique, Geneva School of Economics and Management, University of Geneva, Carouge, Switzerland

**Jeroen A. Beukers** tpg, Geneva, Switzerland

**Charly Beye** Pforzheim University, Pforzheim, Germany
CentraleSupélec, Université Paris-Saclay, Gif-sur-Yvette, France

**Lionel Binz** Cantonal Transports Office in Geneva (Etat de Genève), Petit-Lancy, Switzerland

**Nicole van den Boom** Pforzheim University, Pforzheim, Germany
CentraleSupélec, LGI, Gif-sur-Yvette, France

**Adrian Boos** Pforzheim University, Pforzheim, Germany

**Julie Bulteau** Université de Versailles Saint-Quentin-en-Yvelines, Versailles, France

**Pierre Chehwan** NAVYA Group, Courbevoie, France
beti - Experts Mobilité Automatisée, Saint-Donat-sur-l'Herbasse, France

**Philippe Chrétien** Centre Européen d'Etudes de Sécurité et d'Analyse des Risques, CEESAR, Nanterre, France

**Anastasija Collen** Centre Universitaire d'Informatique, Geneva School of Economics and Management, University of Geneva, Carouge, Switzerland

**Minas Dasygenis** Department of Electrical and Computer Engineering, University of Western Macedonia, Kozani, Greece

**Markus Dubielzig** Siemens Technology, Accessibility Competence Center, Paderborn, Germany

**Dorien Duffner-Korbee** Fraunhofer ISI, Karlsruhe, Germany

Pforzheim University, Pforzheim, Germany

**Nanna May Felhaus** Holo, København N, Denmark

**Guy Fournier** Pforzheim University, Pforzheim, Germany

CentraleSupélec, Université Paris-Saclay, Gif-sur-Yvette, France

**Georges Hilbert** SALES-LENTZ (S.L.A. S.A.), Bascharage, Luxembourg

**Albert Hoffmann** SALES-LENTZ (S.L.A. S.A.), Bascharage, Luxembourg

**Ines Jaroudi** Fraunhofer ISI, Karlsruhe, Germany

Pforzheim University, Pforzheim, Germany

CentraleSupélec, LGI, Gif-sur-Yvette, France

**Daniel Kaeding** SALES-LENTZ (S.L.A. S.A.), Bascharage, Luxembourg

**Johannes Klarmann** Pforzheim University, Pforzheim, Germany

**Dimitri Konstantas** Geneva School of Economics and Management, University of Geneva, Geneva, Switzerland

**Antonios Lalas** Centre for Research and Technology Hellas, Information Technologies Institute, Thessaloniki, Greece

**Niklas Liebhauser** Pforzheim University, Pforzheim, Germany

**Matthias Lindemann** Siemens Technology, Accessibility Competence Center, Paderborn, Germany

**Linda Mathé** Siemens Technology, User Experience Design, Munich, Germany

**Sylvie Mira-Bonnardel** École Centrale Lyon, Écully, France

**Maher Ben Moussa** University of Geneva, Carouge, Switzerland

**Gabriele Naderer** Pforzheim University, Pforzheim, Germany

**Eliane Horschutz Nemoto** Pforzheim University, Pforzheim, Germany

CentraleSupélec, LGI, Gif-sur-Yvette, France

ICLEI Europe, Freiburg im Breisgau, Germany

**Niels A. Nijdam** Centre Universitaire d'Informatique, Geneva School of Economics and Management, University of Geneva, Carouge, Switzerland

**Athanasios Papadakis** Centre for Research and Technology Hellas, Information Technologies Institute, Thessaloniki, Greece

Department of Applied Informatics, University of Macedonia, Thessaloniki, Greece

**Sofia Petridou** Department of Applied Informatics, University of Macedonia, Thessaloniki, Greece

**Kevin Salvi** Mobile Thinking, Geneva, Switzerland

**Michael Thalhofer** Software-driven Ecosystems and Market Places, Siemens Technology, Munich, Germany

**Dimitrios Tsiktsiris** Centre for Research and Technology Hellas, Information Technologies Institute, Thessaloniki, Greece

Department of Electrical and Computer Engineering, University of Western Macedonia, Kozani, Greece

**Dimitrios Tzovaras** Centre for Research and Technology Hellas, Information Technologies Institute, Thessaloniki, Greece

**Anastasios Vafeiadis** Centre for Research and Technology Hellas, Information Technologies Institute, Thessaloniki, Greece

**Tobias Viere** Pforzheim University, Pforzheim, Germany

**Konstantinos Votis** Centre for Research and Technology Hellas, Information Technologies Institute, Thessaloniki, Greece

**Christian Zinckernagel** Holo, København N, Denmark

**Quentin Zuttre** Keolis, Lyon, France

# Acronyms

| | |
|---|---|
| ADS | Automated Driving Systems |
| AI | Artificial Intelligence |
| API | Application Protocol Interface |
| AV | Automated Vehicle |
| BMM | Business Modelling Manager |
| CAV | Connected and Automated Vehicles |
| DMP | Data Management Plan |
| DTU test track | Technical University of Denmark test track |
| EC | European Commission |
| ECSEL | Electronic Components and Systems for European Leadership |
| EU | European Union |
| EUCAD | European Conference on Connected and Automated Driving |
| F2F | Face-to-face meeting |
| GDPR | General Data Protection Regulation |
| GNSS | Global Navigation Satellite System |
| HARA | Hazard Analysis and Risk Assessment |
| IPR | Intellectual Property Rights |
| IT | Information Technology |
| ITU | International Telecommunications Union |
| LaaS | Logistics as a Service |
| LIDAR | Light Detection and Ranging |
| MaaS | Mobility as a Service |
| ODD | Operational Domain Design |
| OEDR | Object and Event Detection and Response |
| PRM | Persons with Reduced Mobility |
| PTA | Public Transport Authority |
| PTO | Public Transportation Operator |
| PTS | Public Transportation Services |
| SAE Level | Society of Automotive Engineers Level (Vehicle Autonomy Level) |
| SDK | Software Development Kit |
| SLA | Sales Lentz Autocars |

| | |
|---|---|
| SoA | State of the Art |
| SOTIF | Safety Of The Intended Functionality |
| SWOT | Strengths, Weaknesses, Opportunities, and Threats. |
| TPG | Transports Publics Genevois |
| UITP | Union Internationale des Transports Publics (International Transport Union) |
| V2I | Vehicle-to-Infrastructure communication |

# Chapter 1
# Introduction

**Guy Fournier, Adrian Boos, Dimitri Konstantas, and Danielle Attias**

**Abstract** The AVENUE project, part of European Horizon 2020, investigated the integration of automated vehicles (AVs) into public transport across Europe, focusing on improving mobility through innovative, sustainable solutions. From 2018 to 2022, AVENUE research demonstrated the potential of automated minibuses for safe, efficient and environmentally friendly transport in urban environments. This introduction explains the project's approach to deploy AVs for the first time globally in cities, the impact assessment and the potential for the future. Small summaries of the content of the book parts introduce to the book to give support and guidance to the reader.

## 1.1 Automated Vehicles (AVs) for a New Mobility

Since Adam Smith (1776) and David Ricardo (1817), we know that mobility of people and goods is the basis for the functioning and growth of economies based on the division of labour. In modern economies, in particular, goods can be produced in one place and made available for consumption in another. People can further move long distances to work, consume goods from all over the world and participate in education and social life outside their immediate environment. Mobility is therefore a key factor for growth and prosperity.

G. Fournier (✉)
Pforzheim University, Pforzheim, Germany

CentraleSupélec, Université Paris-Saclay, Gif-sur-Yvette, France
e-mail: guy.fournier@hs-pforzheim.de

A. Boos
Pforzheim University, Pforzheim, Germany

D. Konstantas
Geneva School of Economics and Management, University of Geneva, Geneva, Switzerland

D. Attias
University of Paris-Saclay, Gif-sur-Yvette, France



G. Fournier et al. (eds.), *Automated Vehicles as a Game Changer for Sustainable Mobility*, Contributions to Management Science,
https://doi.org/10.1007/978-3-031-61681-5_1

However, mobility is also criticised for the environmental and social costs it can cause. There are various strategies to deal with this. One option is to accept and continue the current paradigm, accepting the associated costs. Another option is to be more sober and reduce mobility (sufficiency strategy), but this strategy is not very popular and would as a consequence reduce and focus activity, purchasing power and the production on a local level to lower environmental impacts (Matusiewicz, 2024). However, with the new digital and automated technologies, it seems possible to avoid these negative scenarios and (a) keep or even increase current prosperity levels while (b) better meeting people's mobility needs at an affordable price and (c) additionally addressing environmental issues.

The division of labour can indeed be improved and even automatised through digital technologies. Transaction and information costs can be significantly reduced in this way (Rifkin, 2015). As a consequence, the control/management of sophisticated technical systems, such as automated vehicles, can be streamlined and simplified. At a higher organisational level, such as a city, the coordination among stakeholders within the entire mobility/transport ecosystems can further be enhanced and even partly automatised as well. Collaboration could thus optimise the use of material and energy,[1] reduce waste and improve the mobility services. Meeting the mobility needs of citizens by aggregating services and combining them into a seamless and comfortable journey could accordingly be better fulfilled. A transformation towards sustainable mobility could so be the choice of the citizen and be fulfilled without constraints. Beyond that, new opportunities to improve the efficiency and flexibility of the transport system arise from the integration of data generated/made available by the mobility/transport ecosystem, combined with the capabilities of artificial intelligence (AI). As a consequence, these developments support the disappearance of the traditional distinction between individual transport and the transport services provided/offered by public (Haan et al., 2020) and/or private transport operators. Consequently, it appears feasible to revolutionise urban mobility/transport and enable sustainable mobility without a coercive transport policy. This book summarises the learnings from the European Horizon 2020 project AVENUE and would like to contribute how automated vehicles (AVs)[2] could be a game changer to realise this transformation.

[1] The general systems theory is an epistemological approach to solve complex problems, distinguish information, matter (material/resources) and energy as basis of all systems. Information can be exchanged through collaboration to improve the (transport) system as a whole. In contrast to information, data can be objectified because it can be coded, stored, transformed and, above all, transported. Information is "by its very nature processual, knowledge-changing, subject-related and cannot be objectified" (Fournier, 1994; Kesic, 2024; Rifkin, 2015).

[2] According to SAE Mobilus (2018, 2021), the word "autonomous" is a misnomer as applied to automated driving technology, because even the most advanced ADSs are not "self-governing". We thus will follow in the book the recommendation of SAE Mobilus to characterise driving automation and use "automation" instead of the popular term "autonomous".

AVENUE was a European Horizon 2020 project spanning 4.5 years from May 2018 to October 2022. The primary objective of AVENUE was to validate the suitability and efficiency of deploying small- and medium-sized AVs into Europe's public transportation system (Konstantas, 2021). AVENUE aims were to design and carry out full-scale demonstrations of urban transport automation by deploying, for the first time worldwide, fleets of automated minibuses in low- to medium-demand areas of four European demonstrator cities: Geneva (Meyrin, Belle-Idée; Switzerland), Lyon (Parc OL; France), Copenhagen (Nordhavn, Slagelse; Denmark) and Luxembourg (Pfaffenthal, Contern; Luxembourg) and two replicator cities (Uvrier in Switzerland and Esch-sur-Alzette in Luxembourg). The AVENUE vision for future public transport in urban and suburban areas was that AVs will ensure safe, rapid, economic, sustainable and personalised transport of passengers. AVENUE introduced disruptive public transportation paradigms on the basis of on-demand, door-to-door services, aiming to set up a new model of public transportation, by revisiting the offered public transportation services, and aiming to suppress prescheduled fixed bus itineraries.

The progress and findings achieved with the AVENUE project provided evidence supporting the possible role of AVs not only in public transport but also as a possible transformative improvement for the entire transport system of a city (see part 3). The in-depth research by the AVENUE project presented in this book is structured in three parts. Part 1 describes and analyses the complexities of such projects in detail. Part 2 presented the impact assessment of the AVENUE projects where we use an interdisciplinary sustainable urban mobility plan (SUMP)-based holistic approach making a link between social, economic and environmental dimensions and also addressing technical obstacles and legal issues. Part 3 finally explores how AVs could become a game changer for urban transport systems, driving them towards a (more) sustainable transport/mobility system and also examining how this transformation can be realised effectively. The three parts of the book are briefly summarised below and then presented in more detail for each book chapter in the next section "Contents of the Individual Book Parts".

Part 1 of the book ("AVENUE Project: Implementing Automated Minibuses for Door-to-Door and On-Demand Passenger Transportation") examines the fundamental transformative role of AVs in public transport as a whole and in four key European cities as test sites of the AVENUE project. AVENUE demonstrated the possibilities of automating mobility/transport by deploying fleets of automated minibuses for the first time worldwide in areas with low to medium demand. This first part of the book looks in particular at the technological advances through the project and challenges of AVs, particularly automated minibuses, and how these vehicles are being integrated into public transport systems. The individual sub-chapters address safety, service quality and the impact of in-vehicle services on user experience and security. The development of in-vehicle services is particularly relevant to prepare the transition from an in-vehicle safety operator to a teleoperator, who will initially manage one and subsequently multiple vehicles from an external location. To this end, safety issues, for example, were thus addressed with artificial intelligence to substitute the services provided by the safety driver when level 4 will be achieved. Critical aspects of cyber security, data protection and the application of advanced security solutions such as security information and event management (SIEM) are also covered. The part presents last but not least solutions to support vulnerable travellers and concludes with a stakeholder analysis highlighting the need for collaboration in the introduction of automated minibuses in public transport.

Part 2 of the book ("Impact Assessment of AVENUE") presents a multifaceted analysis of the deployment of AMs in mobility ecosystems. It uses a holistic approach to understand the relationships between technical, economic, social, environmental and legal dimensions of such projects. It begins with an introduction to the sustainable urban mobility plan (SUMP) and its integration with AVENUE, emphasising citizen-centric mobility planning. The sections then delve into various aspects of automated minibuses, including technical challenges and advancements, economic analysis focusing on service costs and externalities and the social impacts on mobility behaviour. The governance impact assessment is also explored, highlighting regulatory challenges and solutions. Lastly, a comprehensive sustainability assessment is provided, evaluating the before-mentioned dimensions of the implications of integrating automated minibuses into urban mobility systems. Overall, this section of the book demonstrates the economic, environmental and social potential of AVs for stakeholders like citizen, companies and public transport operators/authorities.

Part 3 of the book ("Future Vision of AVENUE") delves into the prospective advancements and strategies for integrating automated minibuses in intelligent transport systems (ITS). It begins with a discussion on system innovation in passenger transportation, focusing on a citizen-centric approach and a future vision 2030 for that. The section explores various pathways for integrating AVs into mobility ecosystems, emphasising the transformative role of automated minibuses in mobility-as-a-service (MaaS) systems to provide services that are similar to private cars in terms of flexibility (see above). With the use of data, AVs could in this model be further integrated into an ITS. AV would complement private (car sharing, bike sharing, ride sharing, etc.) and public transport and make the transport system as a whole more efficient and resilient. Instead of a product innovation, the entire transport system could be improved towards sustainable mobility. The emphasis is on leveraging data to enhance the efficiency, flexibility and resilience of the transport system, a concept referred to as ambidexterity, meaning that antinomic economic goals can be achieved at the same time. The final chapter addresses transition planning towards a sustainable mobility ecosystem, outlining a strategic, phased approach for this transition, especially to the proposed AM in MaaS/ITS system. However, it highlights the challenges and opportunities in creating effective, sustainable urban mobility plans.

To assist the reader, you will find a brief summary of the chapters below.

## 1.2 Contents of the Individual Book Parts

### *1.2.1 Part 1: The AVENUE Project: Implementing Automated Minibuses for "Door-to-Door" and "On-Demand" Passenger Transportation in Geneva, Lyon, Luxembourg and Copenhagen*

**Chapter 2: AVENUE Site Demonstrators: Geneva, Lyon, Luxembourg and Copenhagen**

The first sections examine the transformation of public transport by AVs in Geneva, Lyon, Luxembourg and Copenhagen, within the AVENUE project. It addresses objectives, deployment (automation level, on demand, door to door), achievements, key success factors and future developments of integrating AV in public transport in the different demonstration sites. The chapter underscores the importance of real-world testing within AVENUE project case studies and aims to show how AVs can reshape twenty-first-century public transport as well as the entire mobility ecosystem.

**Chapter 3: Automated Minibuses: State of the Art and Improvements Through AVENUE**

This chapter focuses on advances in AV technology, particularly automated minibuses for public transport. It discusses the development of AV technology, detailing the contributions of NAVYA and the significant technological advances achieved through the AVENUE project. The chapter addresses the integration of AVs into public transport systems, the development of door-to-door on-demand services and the challenges of managing the entire AV ecosystem. It also documents the continuous improvements in software and hardware, the legal and regulatory challenges and NAVYA's role in shaping the AV landscape in Europe and highlights the impact of the project on future technological developments in this area.

**Chapter 4: Safety, Security and Service Quality for Automated Minibuses: State of the Art, Technical Requirements and Data Privacy in Case of Incident**

This chapter proposes the critical aspects of safety and service quality in the use of AMs. It discusses the state of the art in safety measures, the need for technical reliability and the challenges of data protection and cyber security. The chapter emphasises the importance of maintaining high safety standards for the successful integration of automated minibuses into public transport and examines strategies to continuously improve safety and service quality, including regulatory compliance and incident management.

### Chapter 5: In-Vehicle Services to Improve the User Experience and Security when Traveling with Automated Minibuses

This chapter deals with the development and implementation of various in-vehicle services designed to improve the user experience and safety in automated minibuses. It addresses the integration of advanced technologies such as artificial intelligence and deep learning models to address the challenges related to passenger safety and comfort. The chapter emphasises the importance of these services in the context of driverless public transport and highlights how they contribute to the overall efficiency and acceptance of AVs in the urban environment.

### Chapter 6: Cybersecurity and Data Privacy: Stakeholders' Stand on Regulations and Standards

This chapter discusses the importance and challenges of ensuring cybersecurity and data protection in the context of connected AVs. It investigates technological and regulatory aspects through outlining the threat landscape within the AV's ecosystem and showcasing how such threats are considered within a case study where AVs are foreseen as a means of public transport. The study results demonstrate these vehicles' resilience from cyber and privacy threats. With an alignment to the core regulations and standards, this chapter also derives the main mitigation strategies and requirements to strengthen the evolving technology in automated transport.

### Chapter 7: Technical Cybersecurity Implementation on Automated Minibuses with Security Information and Event Management (SIEM)

This chapter discusses the integration of security information and event management (SIEM) solutions in automated minibuses. It discusses the importance of SIEM in real-time monitoring and response for cybersecurity and examines its application in the context of automated minibuses. The chapter provides an analysis of different SIEM software solutions and their role in protecting against cyberthreats and highlights the critical need for advanced security measures in the automated transport sector.

### Chapter 8: Persons with Reduced Mobility (PRM): Specific Requirements for Passenger Transportation Services

The use of fully automated vehicles in public transport has the potential to revolutionise the complete landscape of public transport. Fully automated vehicles can provide the necessary frequency even in places with a low passenger volume. In addition, these vehicles can stop even in places which are not yet integrated into the public transport network, thus better meeting the needs of passengers as public transport becomes more and more individual. The absence of a driver, however, also implies that these vehicles must provide more services than conventional ones. This chapter presents the findings on passenger services for fully automated minibuses after 2 years of studies with passengers. These findings were gathered based on observations, workshops, questionnaires and interviews with (disabled) passengers, safety drivers, bus drivers, PTOs and associations of disabled persons.

**Annex 8: AnnexMobile Apps for Blind and Low-Vision Public Transport Travellers**

This chapter focuses on mobile applications designed to help blind and visually impaired people on public transport. It lists various iOS apps, such as GoodMaps Outdoors, BlindSquare, myfinder and Seeing AI and describes their functions such as GPS navigation, obstacle detection and information provision. The chapter emphasises that these apps are complementary aids and not a replacement for conventional aids. It also emphasises the need to equip vehicles with assistive technologies to provide visually impaired users with a comprehensive public transport experience.

**Chapter 9: Stakeholder Analysis and AVENUE Strategies**

This chapter presents a comprehensive stakeholder analysis for the use of automated minibuses in public transport systems. The most significant stakeholder groups and their expectations, needs and impacts are identified using qualitative methods. It discusses the diverse impact different stakeholder groups have on the advancement of automated minibuses and provides strategic recommendations for stakeholders such as governments, public transport operators and citizen groups. It emphasises the importance of stakeholder collaboration, collective action in updating regulatory frameworks and the integration of AM into mobility-as-a-service (MaaS) systems.

### *1.2.2 Part 2: Impact Assessment of AVENUE*

**Chapter 10: Research Approach: Introduction to SUMP and AVENUE Methodology**

The first chapter of this second part of the book outlines the concept of the sustainable urban mobility plan (SUMP) and its integration with the AVENUE project. It describes SUMP as a planning paradigm that moves from motorised road planning to people-centred urban mobility planning and how it incorporates elements such as micromobility, AVs and mobility as a service (MaaS). The chapter describes how the different work packages of the AVENUE project, each focusing on automated minibuses, align with SUMP by improving public transport and integrating innovative mobility solutions. It emphasises the importance of interdisciplinary approach to address sustainable planning strategies which integrates economic environmental and social impact assessments (see chapter about sustainable assessment and the developed indicators) but also the stakeholders to formulate recommendations which satisfy citizen, the original transport operator (OEM), the public transport operator and the city, for example.

### Chapter 11: Technical Impact Assessment: Obstacles and Development of Automated Minibuses for Public Transport

The first section begins by exploring the assessment of technical obstacles and advancements of automated minibuses in public transport, drawing upon insights collected from urban areas like Geneva, Lyon, Luxembourg and Copenhagen within the AVENUE project. This initial exploration provides valuable perspectives on the development of mobility ecosystems. Furthermore, a closer examination unfolds, focusing on specific facets of this evaluation which are critical for the seamless integration of automated minibuses into public transport. Through this study, a spectrum of challenges and progress points observed in AVENUE are described, revealing insights into regulatory compliance, safety demonstration and technological constraints. This chapter emphasises in particular the importance of creating and adopting new open standards for the interoperability of automated technologies. Such standards are central to fostering robust competition between different technological stakeholders, including vehicle manufacturers, fleet management services, V2X technology developers and others, thereby promoting a dynamic and sustainable fair ecosystem in the field of automated transport.

### Chapter 12: Economic Impact Assessment: Local Service Costs of Automated Minibuses for Public Transport

This chapter provides a detailed economic analysis of AVs in public transport. It presents an economic assessment tool, EASI-AV, which was developed to support public policy in the introduction of innovative mobility services. The tool takes into account the total cost of mobility (TCM) and reflects a modern approach to fleet management, including automated electric mobility. Real data from the pilot sites of the AVENUE project confirm the relevance of the tool as a decision-making aid for the introduction of AVs. The focus of the chapter is on understanding the economic feasibility and impact of integrating AVs into public transport networks.

### Chapter 13: Environmental Impact Assessment: Automated Minibuses for Public Transport

This chapter assesses the environmental impacts of integrating automated minibuses into public transport systems. It analyses the balance between energy savings offered by automated minibuses through efficient routing and vehicle operation and the increased energy demand due to their connectivity and automation features. Using life cycle assessment (LCA), the study shows that the magnitude of environmental benefits of automated minibuses depends largely on the electricity mix, passenger occupancy (i.e. how many passengers use the vehicle at the same time) and vehicle utilisation over its lifetime.

Future use cases with renewable energy mixes, high passenger occupancies and overall vehicle utilisation highlight the full potential of automated minibuses to become a game changer towards a more environment-friendly transport system.

**Chapter 14: Economic and Environmental Impact Assessment: Externalities of Automated Vehicles for Public Transports**

This chapter analyses the externalities of deploying automated minibuses in public transportation, particularly focusing on possible reductions of external costs. It utilises scenario planning and an externalities model to assess the impact of automated minibus deployment in AVENUE cities. The chapter explores different deployment scenarios, including substituting buses and private cars with automated minibuses, and their potential negative and positive externalities. It provides a nuanced understanding of how different automated minibuses integration strategies can affect transportation systems' economic and environmental aspects.

**Chapter 15: Social Impact Assessment: Changing Mobility Behaviour by Understanding Customer Needs and Attitudes**

This chapter focuses on the social impact of the use of automated minibuses in the four AVENUE cities. It presents the results of studies on mobility needs, attitudes towards automated minibuses and user experiences in pilot and replicator cities. The results show a generally positive, open-minded attitude of citizens towards automated minibuses and a higher willingness to use automated minibus services when they are on demand, door to door and integrated into public transport systems. The chapter emphasises the influence of real-life experiences on acceptance and risk perception and highlights the potential of automated minibuses to transform urban mobility when flexibly integrated into urban transport systems. The quantitative results on user acceptance, which are presented at the end of the chapter, are especially important for the planning of future mobility ecosystems. In particular, the representative survey revealed that a significant proportion of respondents, especially car owners, are willing to switch to automated minibuses if they prove to be efficient and reliable. This shift in preferences highlights the potential impact of AVs on mobility ecosystems and car ownership trends and reflects openness to new, technologically advanced transport solutions. In particular, the representative survey of 1816 citizens in the four AVENUE cities shows that 45% of drivers are "willing" (22%) or even "very willing" (23%) to give up their car to use automated minibuses to cover the first and last mile, if this becomes possible.

**Chapter 16: Governance Impact Assessment, Regulatory Recommendations and Challenges**

This chapter examines the regulatory and governance challenges posed by AVs, focussing in particular on the changes in legislation and governance models required for AVs. It outlines different governance approaches, including global, governmental, self-directed, co-operative, social, technological and financial governance. The chapter highlights the need for an integrated and collaborative approach to creating regulations, taking into account the unique characteristics of AVs and the importance of harmonising laws at different levels. It discusses in detail the impact of the use of AVs on various legal and societal areas and presents legal recommendations for public authorities, emphasising the need for governance that supports societal benefits and addresses ethical, privacy and liability issues.

**Chapter 17: Sustainability Assessment of the Integration of Automated Minibuses in Urban Mobility Systems: Learnings from the AVENUE Project**

The sustainability assessment provides a comprehensive evaluation of the sustainability of automated minibuses in urban mobility ecosystems based on the pilot tests of the AVENUE project. It takes an interdisciplinary approach, integrating environmental, economic and social assessments with a range of sustainability indicators. The chapter assesses the impact of integrating AMs into urban mobility and discusses related challenges, potential benefits and strategies for their integration into sustainable urban mobility plans. It emphasises the importance of aligning AMs with sustainable development goals and the need for system innovation to achieve public transport-centric mobility as a service (MaaS).

### *1.2.3 Future Vision of AVENUE*

**Chapter 18: System Innovation in Passenger Transportation with Automated Minibuses in ITS: The Citizen-Centric Approach of AVENUE**

The first chapter of this third part of the book discusses the integration of automated minibuses in intelligent transport systems (ITS) with a citizen-centric approach, as part of the AVENUE project. It explores three pathways for incorporating AVs into mobility ecosystems: private AVs, robotaxis and automated minibuses in a mobility-as-a-service (MaaS) system. The chapter emphasises that automated minibuses in MaaS could be a “game changer”, enhancing public transport efficiency and flexibility and filling mobility gaps. The successful implementation of this concept hinges on open data and APIs and the potential of artificial intelligence to create a self-learning transport system. The chapter argues that this integration aligns with SUMP, aiming to meet citizens’ needs more effectively while addressing sustainability challenges.

### Chapter 19: Transition Planning Towards a Sustainable Urban Mobility Ecosystem

The last chapter discusses the process of transitioning towards sustainable mobility, focusing on automated minibuses within MaaS/ITS. It outlines a three-phase process: analysing the status quo, defining a future vision and designing the transition from current scenarios to the envisioned future. The chapter provides a road map for this transition, encompassing critical success factors, stakeholder identification, goal setting and strategic recommendations. It emphasises the need for an integrated approach to create effective transition plans and highlights the challenges and opportunities in realising a sustainable mobility ecosystem.

**Acknowledgements** This project has received funding by the European Union's Horizon 2020 research and innovation program under grant agreement No. 769033.

## References

Fournier, G. (1994). Informationstechnologien in Wirtschaft und Gesellschaft - Sozioökonomische Analyse einer technologischen Herausforderung. *Volkswirtschaftliche Schriften (VWS), 439*, 1–253.

Haan, P. d., Straumann, R., Bianchetti, R., Stetter, A., Oehry, B., & Jermann, J. (2020). Verkehr der Zukunft 2060: Technologischer Wandel und seine Folgen für Mobilität und Verkehr. In *Forschungsprojekt SVI 2017/003 auf Antrag der Schweizerischen Vereinigung der Verkehrsingenieure und Verkehrsexperten (SVI)*. Bundesamt für Strassen (ASTRA). http://www.mobilityplatform.ch

Kesic, S. (2024). Universal complexity science and theory of everything: Challenges and prospects. *Systems, 2024*(12), 29. https://doi.org/10.3390/systems12010029

Konstantas, D. (2021). From demonstrator to public service: The AVENUE experience. In S. Mira-Bonnardel, F. Antonialli, & D. Attias (Eds.), *The robomobility revolution of urban public transport. Transportation research, economics and policy*. Springer. https://doi.org/10.1007/978-3-030-72976-9_5

Matusiewicz, M. (2024). Framework for physical Internet deployment in cities. *Urban, Planning and Transport Research, 12*(1), 2303341. https://www.tandfonline.com/doi/full/10.1080/21650020.2024.2303341

Rifkin, J. (2015). *Zero marginal cost society*. The rise of the collaborative commons and the end of capitalism, Palgrave Macmillan.

SAE International. (2018). Taxonomy and definitions for terms related to driving automation systems for on-road motor vehicles. Ground Vehicle Standard J3016_201806. Published June 15, 2018 by SAE International in United States. Available online at https://saemobilus.sae.org/content/J3016_201806/

SAE International. (2021). SAE Levels of Driving Automation™ Refined for Clarity and International Audience. SAE Blog, Published May 3, 2021 by SAE International in United States. Available online at https://www.sae.org/blog/sae-j3016-update

# Part I
# The AVENUE Project: Implementing Automated Minibuses for "Door-to-Door" and "On-Demand" Passenger Transportation in Geneva, Lyon, Luxembourg and Copenhagen

# Chapter 2
# AVENUE Site Demonstrators: Geneva, Lyon, Luxembourg, and Copenhagen

**Jeroen A. Beukers, Quentin Zuttre, Georges Hilbert, Daniel Kaeding, Albert Hoffmann, Christian Zinckernagel, Nanna May Felhaus, Nicole van den Boom, Adrian Boos, and Dimitri Konstantas**

**Abstract** The AVENUE project, which was carried out in Geneva, Lyon, Luxembourg, and Copenhagen, demonstrates the innovative integration of automated vehicles (AVs) into urban public transport systems. This initiative investigated the use of AVs in different urban environments and achieved remarkable progress in the areas of mobility, environmental sustainability, and accessibility. Through real-world testing, the project identified key success factors such as technological integration, social acceptance, and infrastructural adaptation. Achievements include on-demand services, seamless integration into existing transport networks and the promotion of multimodal transport solutions. The experiences from the AVENUE project provide crucial insights into operational and societal aspects of AV implementation in public transport and emphasise the importance of ongoing innovation and research in the field of automated mobility for future urban transport ecosystems.

---

J. A. Beukers (✉)
tpg, Geneva, Switzerland
e-mail: Beukers.Jeroen@tpg.ch

Q. Zuttre
Keolis, Lyon, France

G. Hilbert · D. Kaeding · A. Hoffmann
SALES-LENTZ (S.L.A. S.A.), Bascharage, Luxembourg

C. Zinckernagel · N. M. Felhaus
Holo, København N, Denmark

N. van den Boom
Pforzheim University, Pforzheim, Germany

CentraleSupélec, LGI, Gif-sur-Yvette, France

A. Boos
Pforzheim University, Pforzheim, Germany

D. Konstantas
Geneva School of Economics and Management, University of Geneva, Geneva, Switzerland



G. Fournier et al. (eds.), *Automated Vehicles as a Game Changer for Sustainable Mobility*, Contributions to Management Science,
https://doi.org/10.1007/978-3-031-61681-5_2

## 2.1 Introduction

As urban populations continue to surge and our cities grapple with ever-increasing traffic congestion and environmental concerns, the quest for innovative and sustainable public transportation solutions has intensified. Among the emerging technologies at the forefront of this transformation is fully automated vehicles. These self-driving vehicles have the potential to revolutionise public transportation, offering eco-friendly, efficient, and convenient mobility options for commuters in urban, suburban, and even rural areas.

Today, automated vehicles are proposed and employed across various use cases, leading to some confusion and unrealistic expectations. To provide clarity on the objectives of the AVENUE project, we categorised them into the following applications of automated[1] vehicles in today's context:

1. Public Transport Fixed Line or Predefined Geofenced Zone Local Transport:
   - Operates within predetermined geofenced areas with established stop points.
   - May or may not require physical infrastructure.
   - On-demand transport is possible, with ride pooling options typically mandatory.
2. Shared Autonomy for Long and Short Distances (VDV, 2015):
   - Involves shared or private transportation using non-privately owned vehicles.
   - These vehicles are capable of traveling and stopping at various locations.
3. Owned Autonomy (ibidem):
   - Refers to private transportation using privately owned vehicles.
   - Offers 100% control over mobility requirements for the vehicle owner.

This categorisation helps define the different levels of automated mobility services, each catering to specific needs and preferences in urban transportation. Public transport, which is the target of the AVENUE project (Konstantas, 2021), targets the first of the above categories. Public transport serves a distinct role that distinguishes it from a taxi service as it is intended to be the most cost-effective mode of non-human-powered travel, remaining cost-effective and accessible, while other options provide varying degrees of flexibility and control to users but at higher costs.

The deployment of automated shuttles for public transportation is a promising development, but it is crucial to emphasise that their success hinges on testing and evaluation under real-world conditions, as the road to a seamless automated future is paved with uncertainties, technical challenges, and regulatory hurdles (Antonialli et al., 2022). Therefore, it is imperative to conduct real tests in actual deployment

[1]According to the taxonomy recommendation of the On-Road Automated Driving (ORAD) Committee of SAE Mobilus related to driving automation systems for on-road motor vehicles, we will use in the following the term "automated" instead of the popular "autonomous" (SAE International, 2023).

scenarios to identify potential issues, correct mistakes, and obtain invaluable feedback from users.

In this chapter, we present the deployments in the different sites of the AVENUE project where fully automated shuttles were (and in some still are) being deployed for shared public transportation services. These case studies serve as compelling examples of how automated shuttle technology is evolving and adapting to the unique challenges and opportunities presented in different urban landscapes. Through these deployments, we were able to understand and concretise the hidden issues, for which we had just some vague idea, and refine this models, technology, and services developed in order to ensure integration into our daily lives and related urban environment.

The significance of this research lies not only in its contribution to the development of automated shuttle technology but also in its potential to reshape the way we envision public transportation in the twenty-first century. By highlighting the challenges faced, lessons learned, and the ultimate goal of enhancing the daily commute for countless individuals, this chapter aims to provide a comprehensive perspective on the deployment of automated shuttles for public transportation.

In the pages that follow, we will delve into the specific case studies, examining the unique contexts and challenges of each deployment site (an overview is provided in the following table). Through these analyses, we hope to shed light on the multifaceted nature of automated shuttle integration and inspire continued research, innovation, and collaboration in the field of automated public transportation (Table 2.1).

**Table 2.1** Summary of AVENUE operating site (+ODD components)

| | (TPG) Geneva | | (Holo) Copenhagen | (Holo) Oslo | (Holo) Copenhagen | (Keolis) Lyon | (Sales-Lentz) Luxembourg | | |
|---|---|---|---|---|---|---|---|---|---|
| Site | Meyrin | Belle-Idée | Nordhavn | Ormøya | Slagelse | Parc OL | Pfaffenthal | Contern | Esch-sur-Alzette |
| Funding | TPG | EU + TPG | EU + Holo + AM + Ruter/Movia | EU + Holo + AM + Ruter/Movia | EU + Holo + AM + Ruter/Movia | EU + Keolis | EU + SLA | EU + SLA | EU + SLA + Ville dÉsch |
| Start date of the project | August 2017 | May 2018 | May 2017 | August 2019 | August 2019 | May 2017 | July 2018 | July 2018 | February 2021 |
| Start date of trial | July 2018 | June 2020 | September 2020 | December 2019 | August 2021 | November 2019 | September 2018 | September 2018 | September 2021 |
| Type of route | Fixed circular line | Area | Fixed circular line | Fixed circular line | Area | Fixed circular line | Fixed circular line | Fixed circular line | Fixed circular line |
| Level of on-demand service* | Fixed route/fixed stops | Flexible route/ on-demand stops | Fixed route/fixed stops | Fixed route/fixed stops | Flexible route/ on-demand stops | Fixed route/ fixed stops | Fixed route/ fixed stops | Fixed route/ fixed stops | Fixed route/fixed stops |
| Route length | 2.1 km | 38 hectares | 1.3 km | 1.6 km | 5 km | 1.3 km | 1.2 km | 2.3 km | 1 km |
| Road environment | Open road | Semiprivate | Open road | Open road | Open road | Open road | Public road | Public road | Public road/ pedestrian zone |
| Type of traffic | Mixed | Mixed | Mixed | Mixed | Mixed | Mixed | Mixed | Mixed | Mixed |
| Speed limit | 30 km/h | 30 km/h | 30 km/h | 30 km/h | 30 km/h | 8–10 km/h | 30–50 km/h | 50 km/h | 20 km/h |
| Roundabouts | Yes | Yes | No | No | No | Yes | No | No | No |
| Traffic lights | No | No | No | No | No | Yes | Yes | Yes | No |
| Type of service | Fixed line | On demand | Fixed line | Fixed line | On demand | Fixed line | Fixed line | Fixed line | Fixed line/on demand |
| Concession | Line (circular) | Area | Line (circular) | Line (circular) | Area | Line (circular) | Line (circular) | Line (circular) | Line (circular) |
| Number of stops | 4 | >35 | 6 | 6 | 6 | 2 | 4 | 2 | 5 (+4 on demand) |
| Type of bus stop | Fixed | Fixed | Fixed | Fixed | Fixed | Fixed | Fixed | Fixed | Fixed (+4 virtual) |
| Bus stop infrastructure | Yes | Sometimes, mostly not | Yes | Yes | Yes | Yes | Yes | Yes | Yes |
| Number of vehicles | 1 | 3–4 | 1 | 2 | 2 | 2 | 2 | 1 | 1 |
| Timetable | Fixed | On demand | Fixed | Fixed | On demand | Fixed | Fixed | Fixed | Fixed/on demand |

| | | | | | | | | | |
|---|---|---|---|---|---|---|---|---|---|
| Operation hours | Monday–Friday (5 days) | Sunday–Saturday (7 days) | Monday–Friday (5 days) | Monday–Sunday (7 days) | Monday–Friday (5 days) | Monday–Saturday (6 days) | Tuesday and Thursday, Saturday, Sunday, and every public holiday | Monday–Friday | Monday–Saturday |
| Timeframe weekdays | 06:30–08:30/ 16:00–18:15 | 07:00–19:00 | 10:00–18:00 | 7:30–21:30 | 07:00–18:00 | 08:30–19:30 | 12:00–16:00, 16:45–20 h00 | 7:00–9:10, 15:50–18:35 | 11:00–18:00, 18:00–21:00 (on demand) |
| Timeframe weekends | No service | 07:00–19:00 | No service | 9:00–18:00 | Only at special events | 08:30–19:30 | 10:00–21:00 | No service | 11:00–18:00, 18:00–21:00 (on demand) |
| Depot | 400 m distance | On-site | 800 m distance | 200 m distance | 200 m distance | On-site | On-site | On-site | On-site |
| Driverless service | No | No | No | No | No | No | No | No | No |
| Drive area type/ODD | B roads | Minor roads/parking | B roads/minor roads | B roads | B roads/parking | B roads | B roads | B roads/parking | B roads |
| Drive area geo/ODD | Straight lines/plane | Straight lines/plane | Straight lines/plane | Curves/slopes | Straight lines/curves | Straight lines/plane | Straight lines/plane | Straight lines/plane | Straight lines/plane |
| Lane specification/ODD | Traffic lane | Traffic lane | Traffic lane | Traffic lane | Traffic lane | Traffic lane | Traffic lane | Traffic lane | Traffic lane |
| Drive area signs/ODD | Regulatory | Regulatory | Regulatory, warning | Regulatory | Regulatory | Regulatory | Regulatory | Regulatory | Regulatory |
| Drive area surface/ODD | Standard surface, speed humps | Standard surface, speed humps | Standard surface, speed humps, roadworks | Frequent ice, snow | Standard surface, snow during winter | Standard surface, potholes | Standard surface | Standard surface | Standard surface |

## 2.2 The Changing Landscape of Mobility

There is a noticeable trend of people leaving rural areas and migrating towards more densely populated cities and suburbs. This mass movement results in higher population concentrations per square mile (United Nations, 2018). The consequence of this increased population density is an upsurge in commuter numbers and a growing demand for urban and suburban transportation services, leading to an increased use of private vehicles for commuting.

In response to the ever-increasing number of private vehicles within cities, a current prevailing trend in Europe is the restriction of traditional motorised individual traffic in city centres. Instead, there is a focus on providing shared public transportation options with the aim of reducing environmental impacts and enhancing the overall quality of life for both citizens and commuters (Nieuwenhuijsen & Khreis, 2016).

The future of transportation hinges on collaborative efforts to create sustainable mobility options, with on-demand automated public transport at the forefront. This represents a transformative shift in public transportation and has the potential to be a catalyst for social change.

In the following sections, we will explore the AVENUE deployments in Europe. Before delving into these specific deployments, we will elucidate the driving factors behind this trend and explore potential solutions that have paved the way for the adoption of automated vehicles in public transportation.

### *2.2.1 Fighting Congestion*

The primary goal of contemporary mobility regulation is to mitigate the environmental impact, specifically stemming from $CO_2$ emissions resulting from various modes of transportation (European Commission, 2020). A fundamental step towards addressing climate protection is the transition from internal combustion engine vehicles to electric ones. However, this shift alone may not be adequate, as it fails to alleviate the overall strain on road infrastructure. Mere replacement of traditional individual vehicles with electric counterparts still results in the same road occupancy and congestion issues.

An alternative approach to combat traffic congestion could involve restricting the entry of single-occupant or dual-occupant vehicles into inner cities. Instead, priority could be given to purpose-designed vehicles capable of efficiently transporting a significant number of passengers or goods while optimising road space utilisation. This approach might encompass larger vehicles for mass transportation along prominent routes, alongside smaller vehicles ready to promptly respond to on-demand requests.

By adopting this strategy, non-essential traffic could be reduced, and curbside parking areas could be repurposed for intelligent mobility solutions, bicycle lanes, and other innovative transportation methods. This transformation would inevitably reshape urban infrastructure and create fresh opportunities for public transportation agencies, provided they can offer more customer-centric transport services (OECD/ITF, 2015, 2016, 2017, 2018).

### 2.2.2 The Transformation of Public Transportation

In response to the growing need to transport larger numbers of people efficiently, especially in urban centres and mobility hubs, the conventional approach has often involved separate ground-level railway and bus lines. These systems serve the majority of commuters effectively but fall short of providing the door-to-door convenience associated with individual car travel.

Public transportation journeys often consist of a collective core, with personalised segments at the start and end to accommodate the individual needs of passengers. To address the growing demand for smooth door-to-door transportation solutions within the public transport system, it is imperative to adopt a multifaceted approach. This approach entails the integration of various transportation modes into a unified and interconnected system, commonly referred to as mobility as a service (MaaS). The central challenge for public transport operators is to create a multimodal transportation solution that not only matches the convenience of private car usage but actually surpasses it in terms of quality.

The transformation of city centres, coupled with the phasing out of private cars, necessitates the introduction of green and health-conscious mobility alternatives or robust public transport systems. While walking and cycling are gaining traction, there will always be a segment of the population reliant on vehicle transit. On-demand public transport with smaller vehicles could address this need within car-free city areas. To integrate the related external costs (competition with public transport in particular mass transit, additional congestion costs, and/or other costs) for the city and the citizen, a different price should be claimed to the users.

On-demand public transport represents enhanced efficiency and sustainability. Passengers can schedule their trips at their convenience within designated service hours, arranging for minibus pickups and drop-offs at agreed-upon locations, all facilitated through online booking or dedicated applications. This flexibility streamlines the public transport experience, making it more convenient for daily commuters (Alonso-González et al., 2018).

Another crucial aspect of this evolution is the development of high-capacity, demand-responsive vehicles designed to maximise passenger capacity and minimise overall travel distance. The primary challenge lies in coordinating passenger collection and drop-off for on-demand travel, which may require either a few larger vehicles or numerous smaller ones, with the latter incurring higher costs due to the need for dedicated drivers for each.

Currently, public transport operators tend to utilise vehicles that are specialised for either bulk transportation or individual travel. However, it can be highly inefficient and impractical to transfer a large group of, for example, 200 people using minibuses designed for just 12 passengers each, as this approach essentially forces the implementation of a monomodal transport solution.

Nevertheless, there are innovative options available. For example, large trolleybuses, specifically designed for fixed main routes equipped with overhead electricity lines, can offer efficient transportation solutions. These vehicles can also serve the first and last mile on an on-demand basis by using an additional battery pack when traveling off the main electrified routes. However, it is important to note that large trolleybuses may come with a higher cost and could present challenges when navigating through smaller neighbourhoods due to their size and overhead line constraints.

A potential solution to overcome these challenges is the utilisation of shared driverless vehicles for public transportation. Automated and connected vehicles have already undergone successful testing, with the primary hurdle being their interaction with older vehicles, bicycles, pedestrians, and existing infrastructure.

### *2.2.3 Readiness to Adopt New Transportation Means*

End users seek a seamless, user-friendly travel experience. They desire an all-encompassing travel assistant that plans their journey, anticipates their travel habits, and provides real-time information on vehicle arrivals, appearance, pickup locations, drop-off points, delays, and essential itinerary changes, all while ensuring an efficient flow of pertinent information.

In the modern transportation landscape, passengers are increasingly open to adopting innovative transportation modes that make use of cutting-edge technologies and alternative means of mobility. A prime example of this is the surging popularity of electric bicycles. These battery-powered models have become a captivating category within the evolving transportation ecosystem, attracting a growing number of enthusiasts. Electric bicycle riders often sidestep the complexities of multimodal transportation chains, as they can journey directly from their starting point, "A", to their destination, "B", without relying on other modes of transport. The introduction of electric bicycles has not only extended the range of cyclists but has also enhanced their capabilities, allowing them to cover even greater distances by combining their pedal power with electric assistance. This trend underscores passengers' readiness to embrace innovative, eco-friendly transportation solutions. Holiday travel stands as a well-established example of the widespread shift towards multimodal transportation. It vividly illustrates the transformation of our travel habits over time. Historically, families used to embark on monomodal journeys, primarily traveling by car directly from their residences to their holiday destinations. However, the landscape of holiday travel has evolved significantly, and contemporary vacations now often involve a more intricate and multimodal approach.

Today, families embark on individualised journeys, departing from their homes to reach airports. Once at the airport, they collectively board planes, and upon arriving at their destination, they are individually dispatched to reach their final holiday spots. Surprisingly, many travellers have come to embrace this somewhat complex multimodal approach, despite its inconveniences. This shift is primarily driven by the numerous advantages it offers, including cost-effectiveness, speed, and the ability to access more remote and distant locations. It underscores how passengers are adapting to and appreciating the benefits of multimodal travel experiences.

### 2.2.4 Challenges for Public Transport Operators (PTOs)

We find ourselves at a critical juncture in public transportation, where a paradigm shift is on the horizon. The necessary technology is readily available, and passengers are eager to embrace a new and improved transportation experience. However, public transport operators (PTOs) face a complex array of challenges that must be addressed to make this transformation a reality. These challenges include V2X communication, fleet orchestration intelligence, GDPR compliance, cybersecurity, data collection, legal considerations, societal changes, economic implications, environmental costs, and more (Davidsson et al., 2016; Konstantas & Fournier, 2023).

In addition to addressing these challenges, achieving passenger and citizen acceptance is a pivotal aspect of the successful transformation of public transportation. This involves not only overcoming technological and regulatory challenges but also re-evaluating the cost structure and providing new services that align with evolving passenger expectations. Here we can mention some expectations of passenger like transparent and affordable pricing, recognising that passengers have diverse needs and preferences and offering personalised services, introducing innovative and convenient services that enhance the passenger experience, creating feedback channels for passengers to express their opinions, and finally effectively communicating the benefits of the new transportation options and services.

By focusing on these aspects, public transport operators can foster passenger acceptance and encourage the adoption of multimodal and innovative transportation solutions. This, in turn, can contribute to the success of the paradigm shift in public transportation towards more sustainable mobility.

In the subsequent sections of this document, we delve into the endeavours of public transport operators within the AVENUE project. They aim to address the questions raised earlier by drawing from their experiences gained through various deployments in diverse European cities.

## 2.3 The Geneva Sites

### 2.3.1 Objectives

Starting 2017, Transport Publics Genevois (TPG), the public transport operator in the Swiss Canton of Geneva, has been testing an automated vehicle on a fixed 2.1-km route with four bus stops. This first experience allowed TPG to be a partner in

**Fig. 2.1** The Belle-Idée with 75 stop points

the European co-funded H2020 AVENUE project and to set up a fully automated transport system on the Belle-Idée estate in Thônex, Geneva, Switzerland, combining all future key elements of on-demand automated public transport (Fig. 2.1).

As part of the AVENUE project, the aim was to demonstrate a comprehensive, shared, on-demand, door-to-door, and dynamically routed public transport service. This service utilised a fleet of automated minibuses operating within a geofenced zone, eliminating the need for fixed routes, set timetables, or traditional bus stops. The system was seamlessly integrated into the existing transport network.

Passengers had the exclusive option to book rides through a dedicated application. The project also explored the feasibility of allowing passengers to board and disembark at system-defined stop points without the need for physical infrastructure like ground markings or information display poles.

*The objective was to develop a 100% automated public transport service in which the passenger is the only human link.*

The following steps showcase how this objective was brought into practice:

1. A user on-site wants to be transported as soon as possible from his current position to the entrance of building 8, named reception.
2. The user takes his smartphone and opens up the tpgFlex on-demand application from TPG.
3. The user only has to submit his final destination. His smartphone already knows the user's current position and, if not changed by the user, assumes that he is alone and wants to travel right away.
4. The moment the user validates one of the proposed offers according to his requirements regarding travel time, service level, and pricing, a door of the on-site depot opens, and one of the induction-charged electric minibuses drives out automatically and heads down to the user's location via the fastest route.

5. Once the vehicle arrives at the pickup location, the user opens the doors of the minibus, gets in, and finds a seat. The vehicle automatically closes its doors and directly hovers to the user's destination by choosing the fastest or most convenient route.
6. If another person makes a reservation at the same time on a very similar itinerary as the first user, the minibus will divert its trajectory to combine both reservations and serve the other user as well.
7. After it arrives at the user's destination, the on-demand system can send the vehicle to another location in order to achieve a new mission, back to the depot to charge batteries, or place the vehicle at a key location since the on-demand system knows from experience that most of the bookings are being made at that bus stop at that time of the day.

The 100% automated system, comprising three to four vehicles, is continuously supervised by an on-site supervisor who is supported by two automated monitoring systems:

1. The vehicle monitoring system supervises the vehicle's road-handling capabilities, such as monitoring harsh braking and object detection.
2. The passenger monitoring system is responsible for ensuring the passengers' well-being while on board. It monitors for situations like passengers falling, experiencing health issues, engaging in altercations, or potential incidents like bag snatching. Additionally, this system checks the available interior space for the on-demand system.

Both systems provide the supervisor with visual and acoustic alerts, enabling him/her to take appropriate action. These actions may include observing monitoring tools to assess the situation, inspecting the interior of the vehicle, stopping the vehicle remotely, opening doors from a distance, or contacting emergency services such as calling an ambulance (Fig. 2.2).

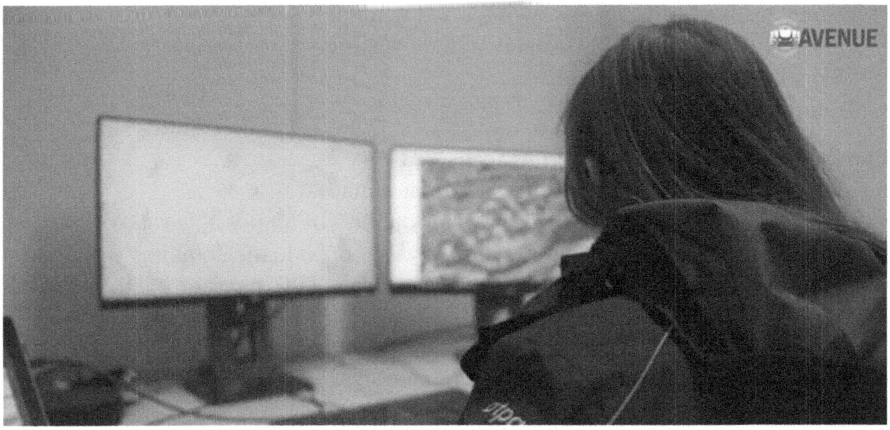


**Fig. 2.2** Belle-Idee on-site supervision office

It is worth noting that, as of the current date, there is no functionality for the supervisor to remotely control or manoeuvre a vehicle at low speeds (<6 km/h) to a safe position. This capability is not included in the project's objectives due to technical limitations.

The on-demand system is fully accessible from the supervision office, allowing the supervisor to dispatch a vehicle to individuals who may not have access to a smartphone and have made reservations by telephone.

As of the date of this document and not included in the project's objectives due to legal restrictions, the possibility of removing the operator (driver) from the vehicle is not yet allowed. However, the primary objective within the AVENUE project was to demonstrate the ability to operate without the need for an onboard operator.

### 2.3.2 Deployment

Due to the complexity of the project and the fact that most partners did not have experience in combining all these technologies, the deployment of the entire project was divided into three phases.

**Phase 1: Between July and September 2020**
- Mapping of the entire Belle-Idée site (summer map)
- Deployment of one vehicle on 45% of all available roads
- Creation of 27 stop points, 4 existing TPG bus stops, and 23 virtual stop points
- First tests with a safety operator and without passengers
- Training of safety operators

**Phase 2: Between October and December 2020**
- Deployment of/increase to two vehicles on 70% of all available roads
- Creation of/increase to 53 stop points, 4 existing TPG bus stops, and 49 virtual stop points
- Vehicle testing with a safety operator and without passengers
- First "fully automated on-demand system" testing.
- First software release 6.0 test
- Training of safety operators

**Phase 3: Between January and March 2021**
- Mapping of the entire Belle-Idée site (winter map)
- Deployment of/increase to three vehicles on 99% of all possible roads
- Tracing of a route to the TPG Seymaz bus stop outside the domain
- Creation of/increase to 75 stop points, 5 existing TPG bus stops and 70 virtual stop points
- Installation of software update to version 6.1 which enabled better road holding and easier on-demand
- Fully automatic depot entrance and exit of vehicles
- Vehicle testing with a safety operator and without passengers
- On-demand system testing with application booking, 100% automated
- Training of safety operators

**Fig. 2.3** Belle-Idee service travel information

At the end of 2021, software update 6.2 was installed which enabled for a dynamic update of an ongoing on-demand mission, hence vehicles already on their way to the next bus stop could be rerouted to being able to pick up or drop off passengers elsewhere (Fig. 2.3).

At the end of 2022, the operating hardware and software of the vehicles was changed from Windows to Linux for better system stability.

During the deployment phases, every safety operator drove numerous kilometres on-site to test and identify issues as well as to get acquainted with the overall traffic after being trained and accompanied by the automated vehicle assistant trainer.

### *2.3.3 Achievements and Key Success Factors*

Due to the willingness to make the project happen and the combined forces from the Swiss Federal and Geneva Cantonal Authorities, the AVENUE project partners, as well as the TPG Managing Board, the project team was able to create and set up the project and start giving demos for select groups of people and testers after 2.5 years.

Unfortunately, the partner in charge of the on-demand vehicle dispatching system went bankrupt soon after and the project was brought to a halt. It then took another 8 months to start for the second time. Since the service started its operation again, various public transport delegations from all over the world who systematically validated the importance of the solution have been welcomed.

At the end of the project, the only issue that kept the vehicles from driving fully automated was obstacles on the streets, such as wrongly parked cars, which the vehicle cannot bypass (yet). When the vehicle detects an object that is not moving, it will stop driving until the object has been removed. When it arrives at an object moving in the same direction as the vehicle, the vehicle will adapt its speed and follow the object from a distance. When an object is moving in another direction as the vehicle, the vehicle comes to a halt and waits for the object to go out of the way, and when the road is free, it starts driving again.

### 2.3.4 Recommendations

**SAE Level 4+ for Public Transport** The development of fully automated vehicles has progressed at a slower pace than initially anticipated and previously announced by international car manufacturers. The most notable achievements currently available on the market can be categorised into two main use cases:

1. SAE Level 3 Automated Highway Lane and Traffic Jam Driving: This represents an advanced driver assistance system that enables automation for highway lane keeping and managing traffic jams.
2. SAE Level 4 Fully Automated Driving: This pertains to fully automated driving on pre-programmed virtual routes, typically at very low speeds.

These advancements illustrate the current status of automated driving technology, where different levels of autonomy and capabilities are achieved. The key distinguishing factors revolve around speed and the complexity of driving conditions. As of the present, achieving SAE Level 5 fully automated driving, as well as SAE Level 4 pre-programmed automated driving at speeds exceeding 30 km/h, remains unattainable. Commercial availability for such advanced capabilities is not anticipated before 2027. This, however, does not mean that the development and testing of fully automated vehicles need to stop. With regard to public transport operators, there is an important use case being created on a political level: thermic-powered vehicles are starting to be banned from city centres, parking spots removed, and speed limits reduced to 30 and even 20 km/h. This means that, as public transport is deployed in a well-defined geofenced area in the city, with all the roads and bus stops known in advance, a highly efficient public transportation service can be provided with SAE Level 4 vehicles, integrated in a MaaS system with the existing main lines of bus, metro, and tram.

As it is not economically feasible for public transport companies to equip every fully automated vehicle with a legally required safety operator, removing the safety operator as in SAE Level 5 is essential. The interim solution could be to create an extra SAE Level, between Levels 4 and 5, for public transport only which aligns with the following requirements:

- 30 km/h speed limit
- Pre-programmed virtual line driving
- Without a safety operator
- Supervision at near distance

The creation of such a "SAE Level 4+ for public transport" or SAE Level 4.5, hence between SAE Level 4 and 5, has already been proposed by the Association of German Transport Companies—VDV—in 2020.[2]

### 2.3.5 *Future Developments*

Due to the maturity of the project and Europe-wide recognition as being an urban public transport solution, TPG will continue gaining experience with new transport modes and technologies within the Horizon Europe Ultimo project. Firstly, TPG will scale up the current Belle-Idée living lab test site with the following steps:

1. Connect the Belle-Idée estate to the nearest train/metro station via a 1.5-km route.
2. Test at least one vehicle in 100% fully automated mode, without an operator on board, with the authorisation to supervise several vehicles simultaneously at a distance.
3. Reduce or eliminate the motorised traffic on the Belle-Idée estate completely.
4. Offer a made-to-measure mobility service to be able to transport everyone on-site.
5. Test the transport of goods in off-peak hours.

The second objective is to transform the existing classic driver-based rural on-demand service tpgFlex of the Champagne/Mandement region in the Swiss Canton of Geneva into an on-demand, door-to-door, fully automated, people and goods transport service integrated in the TPG transport system and later perhaps in a MaaS.

## 2.4 Denmark and Norway

In the AVENUE project, AM was running three test sites:

- Nordhavn, Copenhagen, and Denmark
- Ormøya, Oslo, and Norway
- Slagelse and Denmark

[2] Verband Deutscher Verkehrsunternehmen e. V. (VDV): Eckpunkte zum Rechtsrahmen für einen vollautomatisierten und fahrerlosen Level 4 Betrieb im öffentlichen Verkehr, Positionspapier / September 2020, Köln.

The Ormøya route was originally initiated without being a part of AVENUE, but an agreement has been made to include the site for 5 months, to begin with. The Norwegian site ended in December 2020. A new Danish site in Slagelse Hospital began in September 2021, with a focus on on-demand transportation service.

### *2.4.1 Nordhavn*

The Copenhagen pilot site was situated in an area of the city called Nordhavn. Nordhavn is an active industrial port that is undergoing a transformation—turning into Copenhagen's new international waterfront district offering residential and commercial buildings. When the development of Nordhavn is done, the area will house more than 40,000 residents and 40,000 employees.

Nordhavn aims at being an eco-friendly neighbourhood and contributes to boosting Copenhagen's image as an environmental metropolis. The city should vibrate with life as a versatile urban area with a multitude of activities and a wide range of shops, cultural facilities, and sports facilities. The area is becoming more and more populated, and the need for local transportation is expected to keep growing (Fig. 2.4).

Currently, the Nordhavn area is serviced by a nearby S-train station (approx. 1.1 km away) and bus stops located near the train station. There are, however, no buses or trains running directly in the area—creating a great opportunity for

**Fig. 2.4** The Nordhavn route area seen from above

automated vehicles to function as a new public transport solution, connecting the area much better than it is today. In 2020 two new metro stations have been built—opening in the periphery of the neighbourhoods.

#### 2.4.1.1 Objectives

The main users of the shuttle service have been the residents of Nordhavn (including families, children, and elderly), commuters working in Nordhavn, and visitors to the area. Several usage scenarios can thereby be anticipated:

- Ease mobility within the area for the residents and commuters working in the area.
- Used for the first/last mile from the main road/entry point to the different stops within the area for residents and commuters working there.
- Provide easier access from the main road to, e.g. the harbour pool, restaurants, and cultural facilities for visitors and families.

Planned services provided for the end users:

- The shuttles are free of charge during the pilot project in Denmark, so there is no ticketing yet.
- There are static bus stops providing the position of the bus, relative to the given stop.
- The real-time location of buses can be seen in the mobile application.
- Besides the bus stop signs, users can find information about the pilot project on the AM website and in the AVENUE mobile application.

During the project period, it was the aim to test the services developed through the AVENUE project, e.g. real-time position of the bus, on-demand booking, and accessibility for disabled persons.

#### 2.4.1.2 Deployment

The first route was placed in the area called Århusgadekvarteret. This area was the first one finished, and residents started moving there in 2015. Since then, different squares, the harbour promenade, and a rooftop gym have been evolved and taken into use. Furthermore, special attention has been paid to developing local retail, so today there are supermarkets, cafes, restaurants, and different specialised retailers. There are several shared space areas on the route including a bathing zone.

The first route is a circle line around the area (blue line on the map below), making it easier to get around and to enter the area from outside Nordhavn. Our garage is located on the next peninsula close to Århusgadekvarteret (the red line on the below map) (Fig. 2.5).

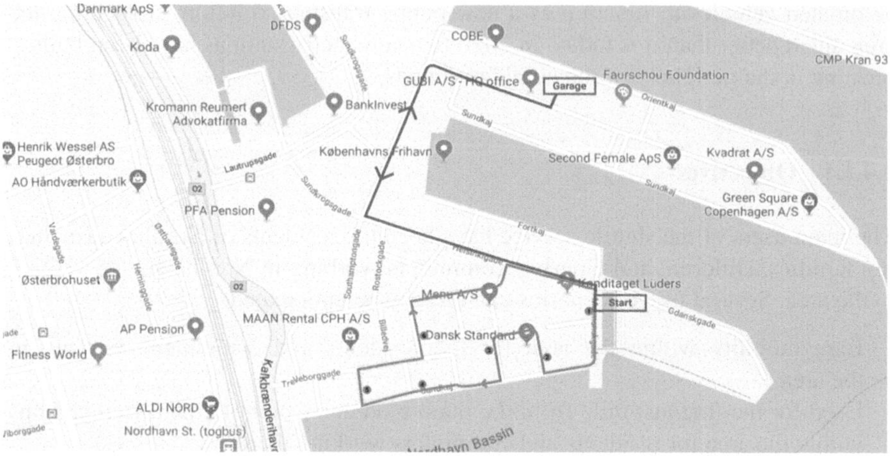


**Fig. 2.5** Map showing Nordhavn route

The pilot route is going to be in mixed traffic with cars, pedestrians, bicycles, etc. The area is, in general, a low-speed area with 20–50 km/h speed limits on the route, and in the 50 km/h limit areas, the recommended speed for cars is 30 km/h.

Operation facts:

- 2 AVs running (initially)
- Mon–Fri 10.00–18.00
- Loop route with six stops

#### 2.4.1.3 Achievements and Key Success Factors

Passengers and Distance Driven

Passengers we counted every day of operation. In total, we have transported 1579 passengers. It is seen that most passengers were transported at the beginning of the pilot, possibly due to the route location close to the harbour popular in the summer time. Another possible explanation could be that locals got used to the shuttle as an attraction and as time went by a sense of novelty disappeared (Fig. 2.6).

A total of 2.417 km has been driven in Nordhavn. The count can be seen increasing in almost linear progress since August 2020 (Fig. 2.7).

Driving Speed and Automated vs. Manual Mode

The driving speed is seen to be increasing slightly over the operational time period from an initial 7 km/h to 7.89 km/h. This is despite an increase in manually driven kilometres primarily related to an increase in parked cars on the route.

In total, close to 992 h have been driven during the pilot. In regard to the navigation mode, 82.6% of overall driving on the route was done in automated mode. The

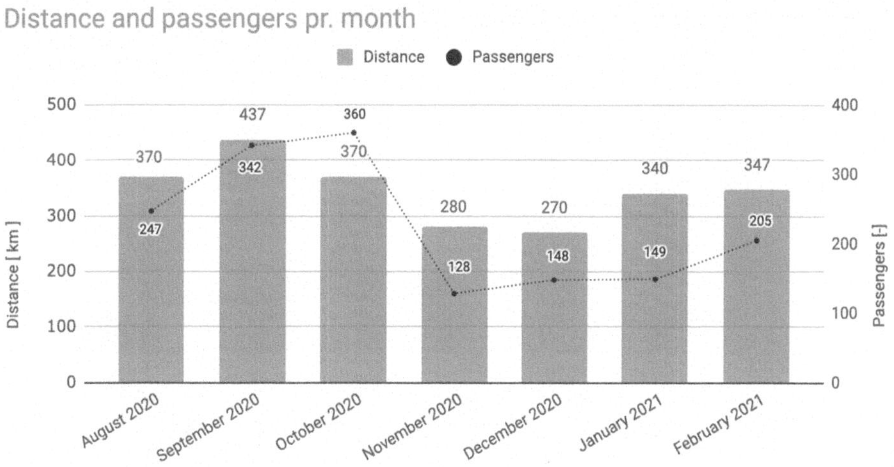


Fig. 2.6 Distance and passengers per month in Nordhavn

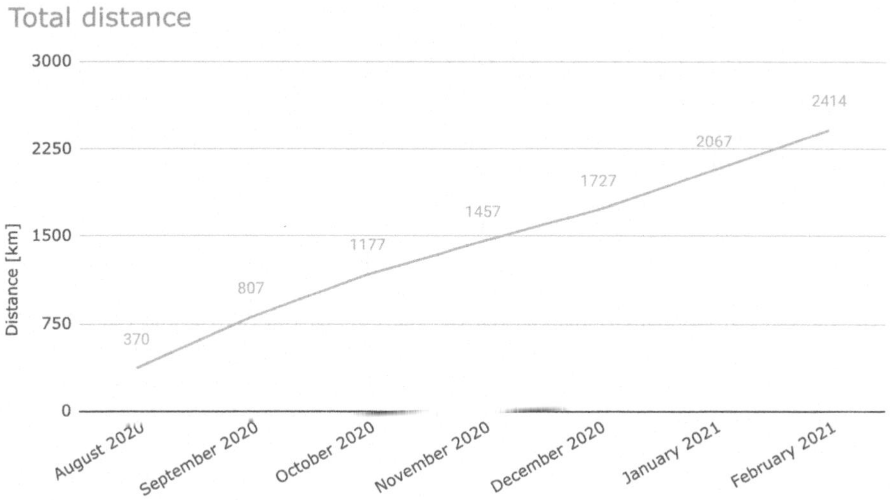


Fig. 2.7 Accumulated distance driven in Nordhavn

rest of 17.4% driven in manual mode was mostly due to a large amount of illegally parked cars on the route and due to roadworks. Driving manual to and from the garage is filtered out.

### Issues Reported on Route

The main issue reported through the safety operator app was parked cars on the road. The parked cars disturb the driving, as the AV is not able to diverge from its planned trajectory, and hence the operator is forced to do a manual takeover in order to pass the parked vehicle. In total, this has been registered 1995 times.

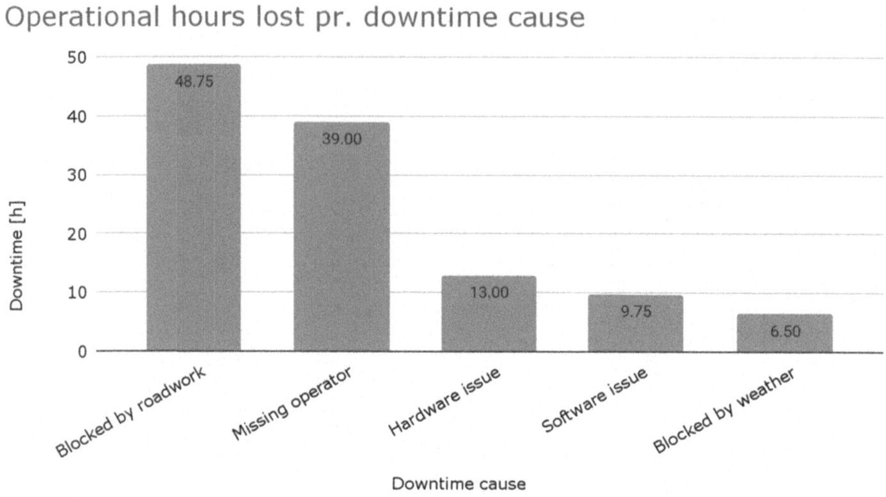


**Fig. 2.8** Causes of downtime in Nordhavn by downtime hours

The issue of parking on the road has been reported in several locations on the route. Mostly around Århusgade, Sandkaj, and Göteborg Plaza. The issues with parked cars in the Nordhavn area were significantly contributing to impeding the operation, finally resulting in its closure.

Downtime and Cancelled Operation

The following figure illustrates the distribution of the five most common reasons for downtime/cancelled operation per downtime hour it caused. The downtime was primarily due to roadwork on the route, which made driving in the designated trajectory impossible and manual takeover necessary (Fig. 2.8).

#### 2.4.1.4 Recommendations

During the deployment of the shuttles in Nordhavn, multiple learnings were achieved. They are described in the following.

Object-Detection Challenges

The Nordhavn area is packed with restaurants, cafes, and shops with outdoor seating and display areas. Larger parts of the Nordhavn route are shared space areas, where pedestrians, bicycles, and vehicles interact in a shared road, rather than separated lanes. This type of city planning causes many situations where the automated shuttles detect obstacles and stop as a consequence. Furniture and other objects cross into the driving lanes, causing the shuttle to stop and requiring the safety operator to move the objects away from the driving lanes or manoeuvre manually around obstacles. For a completely automated operation without a safety operator, the area would probably not be ideal because of these shared space areas, and at least it would require significant technological breakthroughs before an automated vehicle would be able to operate in the Nordhavn area without a safety operator. This is also due to narrow roads, where two cars cannot pass each other without human interaction.

#### Increased Mixed Traffic in High Seasons

During the summer period, the swimming areas at Nordhavn attract many visitors who arrive by bike. Due to the large number of bikes, bikes are sporadically parked outside of the appointed areas for bike parking. This caused difficult driving conditions, and many operational days were shortened as the vehicles were not able to drive due to the many parked bikes, and it was deemed impossible to have the safety operator remove the bikes during every round of driving.

#### Consequences of Construction Work

As the Nordhavn area is under heavy construction, the shuttles have had to operate in an area with many trucks and work-related vehicles being parked illegally and shutting off parts of the route. This has caused delays and cancellations of the operation for shorter periods and sometimes days.

#### Lack of Parking Spots Compared to the Number of Cars

The Nordhavn area is a very busy area with local residents, offices, shops and restaurants, and many daily visitors. Some of these visitors travel by car and can find it very difficult to find a place to park the car during peak hours. This resulted in many illegally parked cars on the route, and the safety operators have been taking over manually multiple times every day to overtake parked cars. Driving in an area that requires manual overtaking does affect the total percentage of kilometres driven in automated mode.

#### Low-Speed Limit

Most streets in Nordhavn have a speed limit of 30 km/h which suits the automated vehicles' speed capabilities. In general, a low gap between the vehicle's top speed and the speed limit provides a more safe operation, with less risky overtaking from other road users. This contributes to Holo's general assessment of the Nordhavn area as a safe environment to drive and test low-speed automated vehicles.

#### 2.4.1.5 Future Developments

Unfortunately, the route in Nordhavn had to be shut down, and AM has investigated several scenarios to still meet the learning objectives for the AVENUE project. In the following the process and reasoning for shutting down the Nordhavn site will be further elaborated.

Complications in Nordhavn

In the fall of 2020 AM started to investigate the potential for expanding the Nordhavn site. AM investigated several routes which could be added to the already active route in Nordhavn. The selection and evaluation focused on last-mile transportation from the two metro stations and to both residential and various business areas of Nordhavn.

In November 2020 all the potential new routes were discarded by By & Havn, as construction plans for the area would interfere with the automated minibuses. The area of Nordhavn is undergoing heavy construction, and due to delays in the construction work, several roads have been unavailable to use for the remaining period of the AVENUE project.

The above challenges in Nordhavn caused AM to seek alternatives for automated operations in the Copenhagen area. AM has evaluated several scenarios for how to still meet the objectives of the AVENUE project and decided to use the AVENUE shuttles on an existing Amobility site. In order to minimise the risk of delaying operations substantially, caused by long approval processes, AM investigated the use of an existing site in Denmark for the remaining months of operations in AVENUE.

AM was granted approval to start operations at Slagelse Hospital (Denmark) in March 2020. However, the project had been paused due to COVID-19. In the fourth quarter of 2020 AM and Movia (PTA) decided to restart the project in August 2021. AM and Movia discussed the possibility of integrating AVENUE as part of the Slagelse Hospital site and agreed to proceed with this possibility.

Even though this alternative only can support operations with Navya vehicles, AM has identified several crucial learnings in relation to on-demand and integration with PTA for ordering.

### *2.4.2 Ormøya*

Due to the delays in launching the Danish demonstration site, it took some time before all four buses were in operation. Therefore, in May 2019 AM agreed with the consortium to include our subsidiary in Norway as a third party, so that two of AM's AVENUE buses could be deployed on a route there. This way AVENUE would still gain useful insights into the operation while awaiting the launch of the Copenhagen site.

Originally the plan was to include the two buses on the route Akershusstranda in central Oslo for 5 months beginning in June 2019. This would be a route with four buses running and the service fully integrated with existing public transport in Oslo. However, due to heavy construction, this route had to be cancelled. It was therefore decided to integrate AM's AVENUE buses on the second route "Ormøya" just outside Oslo centre. Ormøya is an island south of Oslo city centre connected by a bridge to the mainland and a bridge to a second island called Malmøya.

#### 2.4.2.1 Objectives

AM is collaborating with Oslo Municipality, the Norwegian Public Roads Administration, and Ruter[3] on a 3-year self-driving trial project. The project is an important milestone in the process of getting self-driving buses to the Oslo area. Oslo and Akershus wish to have 0% emissions across their public transportation, and this project will test if self-driving buses can support these ambitions for a sustainable public transport system. The end goal is for automated buses to be part of Ruter's regular offer in a few years.

The main purpose of the project was to investigate what self-driving vehicles can mean for everyday logistics in a neighbourhood. By increasing the frequency of public transport by means of small self-driving vehicles, the goal was to reduce the need for private cars in the area. One road leads in and out of the two islands that have a total of around 500 households. The local residents have a 12-m bus service which departs around once an hour for most of the day. On the mainland just off the inland is one of the major thoroughfares going into Oslo from the south, Mosseveien/E18. This main road has frequent express buses going in and out of Oslo. The automated bus service provided a high-frequency last-mile solution for the residents of Ormøya and Malmøya which connected them to the express service on Mosseveien/E18.

#### 2.4.2.2 Deployment

The route is 1.6 km one way (3.2 km round trip) and has six bus stops. It runs from Nedre Bekkelaget bus stop which is located near Mosseveien/E18 where users can access high-frequency express buses to and from Oslo. Also near this end point is the local area public school that kids from Malmøya and Ormøya attend.

The other end point, the Malmøya bus stop, is right on the landing on the island of Malmøya where there is a turning place for the vehicles. This bus stop is also located close to a marina, where lots of Oslo residents keep their recreational boats. The four other bus stops are evenly distributed along the two end points. The bus stop Mailand is also located close to a public beach/swimming area and a marina

[3] The public transport authority for Oslo and Akershus counties.

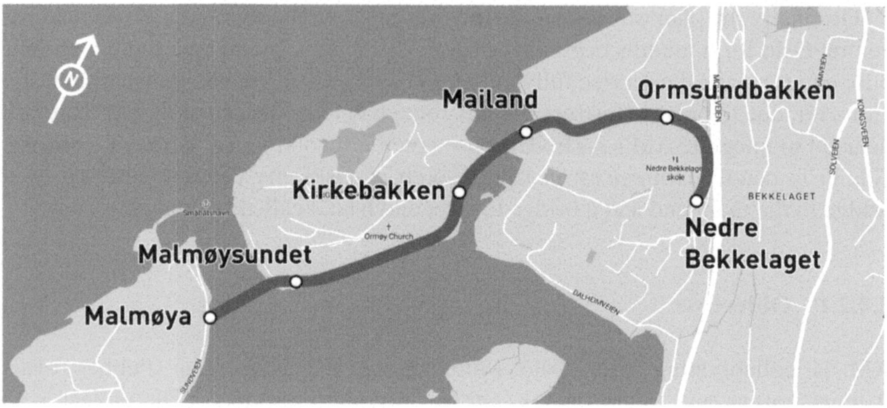


**Fig. 2.9** Map of Ormøya route

which attracts lots of visitors in the summer. The route can be seen below with the stops marked (Fig. 2.9).

The speed limit on the entire route is 30 km/h, and it contains several speed bumps which generally keep the speed in the area low. The condition and build of the road vary quite a bit along the route. Several places are very narrow, just barely wide enough for two vehicles to pass each other, and several stretches have poor asphalt quality. There is also a lot of vegetation close to the route.

In order to be able to offer the inhabitants a valuable self-driving travel service, we had to ensure high operational stability along the stretch. This has been challenging due to several elements, and we have therefore made ongoing adjustments in the offer to explore what it takes to ensure stable and reliable operation. Operational stability will be a success factor in initiating new, more complex self-driving bus lines in the years to come.

#### 2.4.2.3 Achievements and Key Success Factors

Passengers and Distance

In total, 6637 passengers were transported and 22,984 km driven on the Ormøya route. During the 1-year pilot, there were over 5233 h of operation (total number when all three vehicles are added together). That is equal to approximately 395 operational days or 131.67 full operational days with 3 vehicles (1 operational day = 13.25 h) (Figs. 2.10 and 2.11).

The route in Ormoya was severely impacted by operational challenges and COVID-19 lockdowns in some months during the project period. This resulted in some periods with few passengers and less distance driven. In terms of passenger distribution, passengers were using the shuttle service for transport every day of the week. However, there was a somewhat higher number of passengers during the weekdays than the weekend.

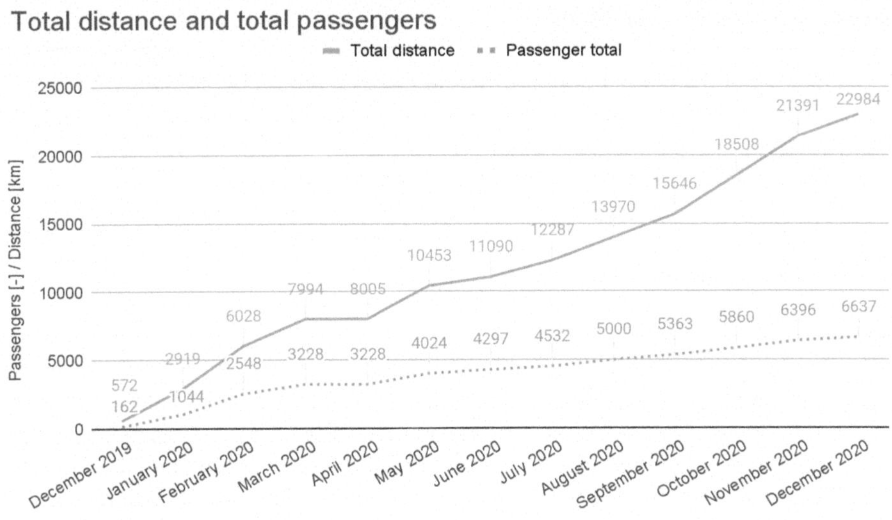


**Fig. 2.10** Accumulated distance and passenger count at Ormøya

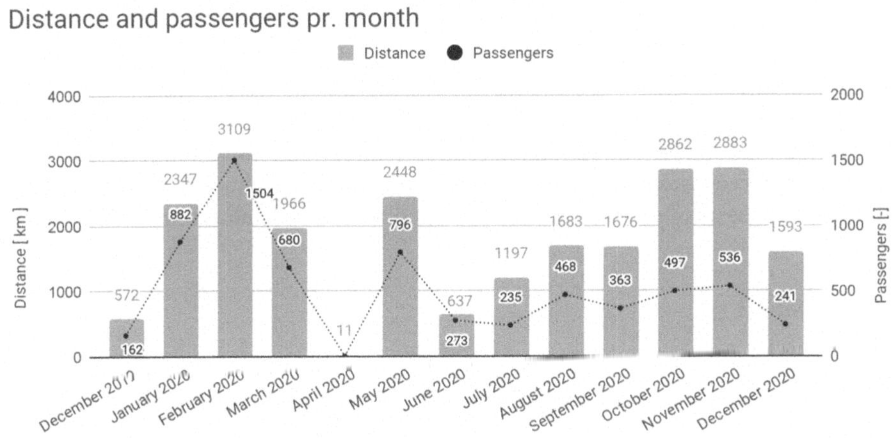


**Fig. 2.11** Distance and passenger count per month at Ormøya

### Automated vs. Manual Driving

In regard to the navigation mode, 93.8% of overall driving on the route was driven in automated mode. The rest of 6.8% driven in manual mode was mostly due to technical issues. Driving to and from the garage is filtered out.

Driving speed on the Ormøya route was between 9.44 and 10.53 km/h. The overall average speed was approximately 10 km/h. By the summer of 2020, a recommissioning was done to improve driving in areas of the route.

Issues Encountered on the Route

When looking at the distribution of different issues reported by the safety operator, again for this route "parking on road" is the most frequent issue occurring with a total occurrence of 1725 times. Operators fall as many as 62 times, signalling a safety issue in the working position of the operators (Fig. 2.12).

#### 2.4.2.4 Recommendations

Public Transport in Oslo

The service at Ormøya functioned as an integrated part of Ruter's public transport offerings in the general Oslo area. This meant that the bus required a standard ticket which gives access to the entire network and that the service was included in Ruter's overall route planning tools for users. Neither Holo nor Ruter performed checks of valid tickets.

By all indications, users did not complain about the ticket requirement but viewed it as natural for service in Ruter's network.

User Experience

During the final quarter of operations, Ruter conducted a user survey around the shuttle. The survey was conducted by interviewing people walking or moving around the general area of Ormøya. In total Ruter collected 107 interviews each lasting

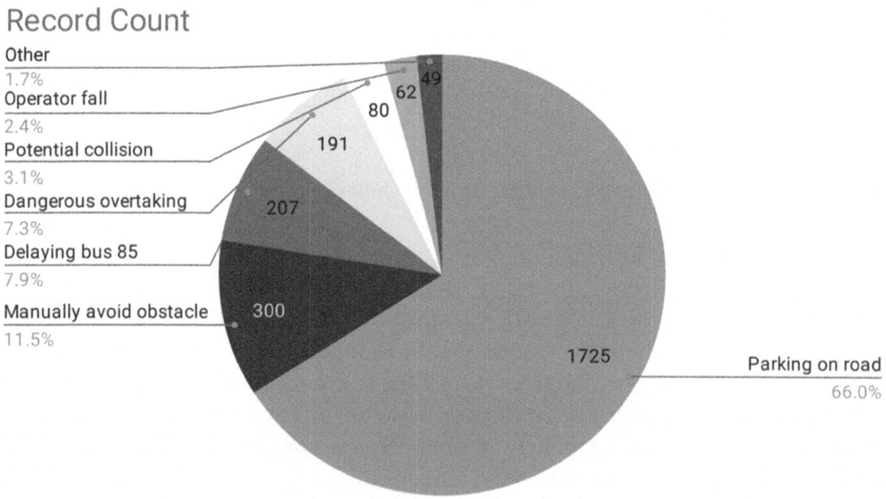


Fig. 2.12 Issue distribution at Ormøya

5–8 min during weeks 38, 39, 41, and 42 in 2020. The main purpose of the survey was to evaluate what users and local residents think of the self-driving bus service.

In general, respondents were positive towards the tests and felt that it was safe. However, the survey also showed clearly that the service was not providing a valuable mobility solution. Approximately 82% of passengers took the bus merely out of curiosity, while only 12% used the service for their daily commute. When local residents at Ormøya and Malmøya were asked why they did not use the automated bus for their daily mobility needs, two of the top answers were the low speed and that the bus did not go where they needed to go. Furthermore, low reliability was also noted as a reason why the service was not used for their daily mobility needs.

### Vegetation and Snow

The narrow road on Ormøya makes this site especially challenging when it comes to vegetation. There is little to no margin for the growth of vegetation. Both vegetation on the sidewalk and overhanging bushes and branches are stretching into the safety zones interfering with smooth operation.

Another issue at Ormøya is the high and low overhanging branches interfering are then detected by the vehicle causing it to brake or slow down (Fig. 2.13).

The rapid growth of vegetation in the summer months has led to constant interference with operations. In periods operations have been halted until the vegetation has been properly cut. In the summer months, April to September, we have seen the need to cut vegetation up to every second week.

**Fig. 2.13** Photos showing the vegetation close to the vehicle trajectory on the route in Ormøya

**Fig. 2.14** Photos showing snow and snow banks in and close to the vehicle's trajectories on the route at Ormøya

Vegetation-related issues have heavily impacted the stability of the operation on this route. Comparing this site to other routes with less vegetation clearly shows a negative impact on the performance of the vehicles. Additionally, sudden braking caused by vegetation is a major safety concern. The sudden braking can cause cyclists, pedestrians, and following cars to potentially collide with the vehicle when it makes unexpected and seemingly irrational decisions like braking hard for a small branch near the trajectory.

In the winter months snow and ice proved a similarly big challenge. The operational design domain (ODD) for the Navya Arma vehicles states that they cannot operate when snow is falling or when the route is covered in snow. Falling snow proved quickly to be unsuitable for operation. With snow in the air, the vehicles make continuous false obstacle detections, and every snowfall resulted in a halt of operations.

The year 2020 did not see very much snow at the Ormøya site, and only on a couple of occasions did snow on the ground cancel operations, mainly because snow clearing was not done well enough to clear the vehicle's trajectory as highlighted in the image below (Fig. 2.14).

Major Safety Issues

Already in the risk analysis of the route, before the operation commenced, the major safety concerns were identified as other road users behaving dangerously around the shuttle due to low speed and slow reaction to clear road or similar. Also, passengers and operators getting hurt inside the shuttle due to sudden and hard braking were highlighted along with intentional disturbances. These safety concerns proved to have been correctly assessed from the beginning, and they were the major safety concerns throughout the operation.

Through the Holo operator app, Holo was able to collect data on the potentially most dangerous situations that occurred on the route. The four most important categories that data was collected on are dangerous overtaking of the vehicle, intentional disturbance, operator falls, and passenger falls. Two major safety concerns that remain with an operation like the one on Ormøya are:

- The operators of Navya Arma vehicles do not have ideal working conditions. Sudden and hard braking has resulted in several mild injuries because of the lack of proper seating and suitable working positions.
- The low speed of the shuttle does cause other road users to act irresponsibly and attempt overtaking in areas that are not suited for overtaking like areas with bad overview or near pedestrian crosswalks.

### *2.4.3 Slagelse*

The Nordhavn area should originally have been finished in 2020, but due to major delays in construction plans, parts of the Nordhavn route (streets) were closed permanently for longer periods of time during the remainder of the AVENUE project in that area. Hence, AM and Copenhagen area PTA Movia agreed on introducing an AVENUE shuttle on the two shuttle project at Slagelse Hospital. Here the main learnings were aimed at on-demand driving and integrations with public transport PTA Movia, their client systems, etc.

The Slagelse site was an interesting case, as the distances between the departments in the hospital are too long for patients to walk between them and a shuttle moving between different departments and connecting parking lots would improve the mobility in the area.

#### 2.4.3.1 Objectives

The goal for the on-demand service is (1) hospital staff can book trips for patients and visitors, and (2) patients and visitors can book their trip. Transportation between parking lots and other entrances makes good sense because of the large distances in the hospital area. The status of the trips booked is updated through the tool used by hospital staff, hereby passengers can get info on when to expect pickup time. Safety stewards will greet passengers when the vehicle arrives at the pickup point. Without any direct input from the safety steward, the vehicle will begin its trip once the doors are closed.

The on-demand service initiated with one vehicle. The second vehicle was included in the on-demand service as soon as possible. The goal was to service both vehicles on demand as much as possible.

At this stage, there were some technical activities that need to be performed in order to prepare the service:

- Update Navya vehicle software to 6.1
- Integrate Holo system with Navya API
- Integrate Holo system with a dispatcher at Movia
- Movia to develop User Interface for booking

There will be a full focus on the technical integration and stability/performance of the service created, in order to gain the full learning experience in trying to service a robust on-demand service.

There will be less focus on the development of apps and screen content, hereby limiting the investigations into the whole automated minibus customer journey. This is in order to favour the technical development and achievement of back-end software—customer interfaces are owned by Movia.

This priority is possible because the user who is booking the trips will be hospital staff and passengers via the Movia interfaces. The passengers will receive the necessary info from the hospital staff and the interfaces. Furthermore, the safety operator still has to be present in the vehicles and will be utilised to give the needed information to the user.

The practical test process will look like this:

- Outline vehicle behaviour in all possible on-demand situations on Holo's test track in Copenhagen. SW version 6.1. Holo internally dispatches missions to vehicles.
- Outline vehicle and integration behaviour in all possible on-demand situations with missions received from Movia. Still on Holo's test track.
- Move to Slagelse and perform similar tests on the real route. Reach a satisfactory level of performance before servicing passengers.

#### 2.4.3.2 Deployment

The shuttle was driving on a 770-m-long stretch on Fælledvej at Slagelse Hospital. Besides driving on Fællesvej, the shuttle was driving in five parking areas, with multiple stops, to turn the shuttle and re-enter the stretch of Fælledvej. The route and the stops can be seen below. Besides the stops placed in the parking areas, the shuttle also stopped at the western part of Fælledvej, in both directions (Fig. 2.15).

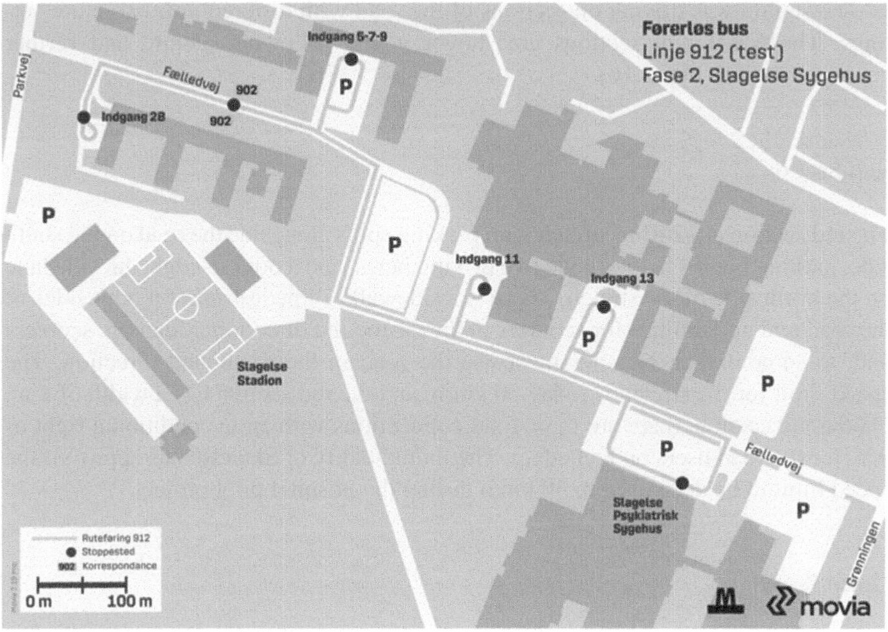


Fig. 2.15 Map showing the route at Slagelse

Fig. 2.16 Map showing the route sections at Slagelse

The shuttle was driving on sections of the road with different infrastructure settings. These different sections are shown in the following picture and further described below (Fig. 2.16).

Red Section

The red section is a 240-m stretch with parking spots alongside the road on the south side, parking booths between the two driving lanes, and a double-edged bicycle lane on the north side of the stretch. There is a sidewalk for pedestrians on both sides of the road and no facilities for scooters in the eastbound direction. Meaning, scooters and, to some extent, bicycles must use the road in the eastbound direction. The speed limit for the stretch is today 50 km/h, and the red section has a width of 6 m. On the red section, there are three road connections with an unconditional right of way for the road users on Fælledvej. The municipality of Slagelse has approved the speed limit to be decreased to 30 km/h during the hospital pilot project.

Green Section

The green section is a 300-m stretch with parking booths between the two driving lanes and a walking path on the west and south sides of the road, as shown in the picture below. There is no sidewalk on the north side of the stretch (meaning that it is assumed that pedestrians may walk on the roadway on the north side). On the green section, there are no implemented facilities for scooters or bicycles, meaning that they will drive on the roadway on the green section. The speed limit for the stretch is today 50 km/h, but it is assumed that the driving speed is lower because of two sharp curves on the stretch, with a small turning radius, which cannot be driven with 50 km/h. The green section has a width of 6 m. On the green section, there are five entry roads for parking facilities. From the parking facilities to Fælledvej drivers have an unconditional obligation to give way. There is a road connection in one of the sharp curves, where drivers have an unconditional obligation to give way to drivers on Fælledvej. The municipality of Slagelse has approved the speed limit to be decreased to 30 km/h during the Hospital Trials.

Blue Section

The blue section is a 230-m stretch with barriers separating the road and newly implemented speed signs, recommending 20 km/h on the stretch. There are sidewalks for pedestrians on both sides of the stretch. There are approximately 35 m of bicycle lanes on both sides by the exit from entrance 11. On the remaining part of the stretch, the road is shared with bicycles and scooters. The driving lane has a width of 3.25 m in both directions, and there are five cross sections with an obligation to give way for the traffic on Fælledvej. Slagelse municipality has approved speed limits of 30 km/h during the Hospital Trials.

A regular Movia bus (line 902) operates on Fælledvej with 30-min intervals during weekdays. The line has two stops (for each direction) on the part of Fælledvej that the shuttles were driving on. At the main entrance of Slagelse Hospital, a patient bus of the same size as Movia's regular bus departed a couple of times during the day. The patient bus had a marked parking area in front of Slagelse Hospital that has been placed outside the self-driving shuttle's route. During entry and exit to and from Slagelse Hospital, the self-driving shuttle and the patient bus could lock. This required the safety driver to manually take over and give way for the patient bus.

Parking Conditions

The area in which the automated minibus had been operating has a high density of parking areas. There are parking facilities in Fælledvej, where drivers dismount their cars directly onto the road. Further, there are marked parking booths on these parking facilities, which the automated minibus had to drive past. Besides parking

**Fig. 2.17** Photos showing parking conditions at Slagelse

areas defined by regulations or signs, there was a high degree of parking outside of these designated areas, as seen below. Based on this, it is uncertain how much effect the parked cars outside of designated areas will have on the operation of the shuttle. Unwanted stops and brakes may occur if parked cars are blocking the route of the shuttle (Fig. 2.17).

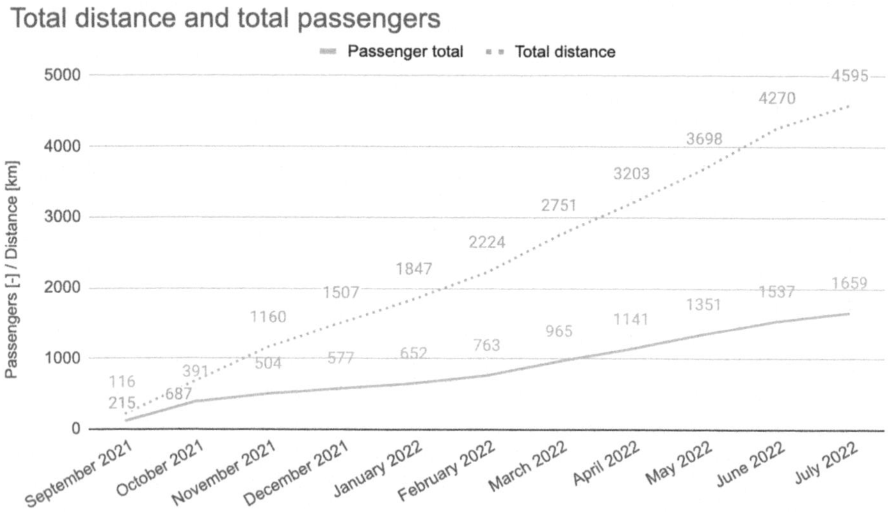


**Fig. 2.18** Total number of passengers and distance driven in the Slagelse project

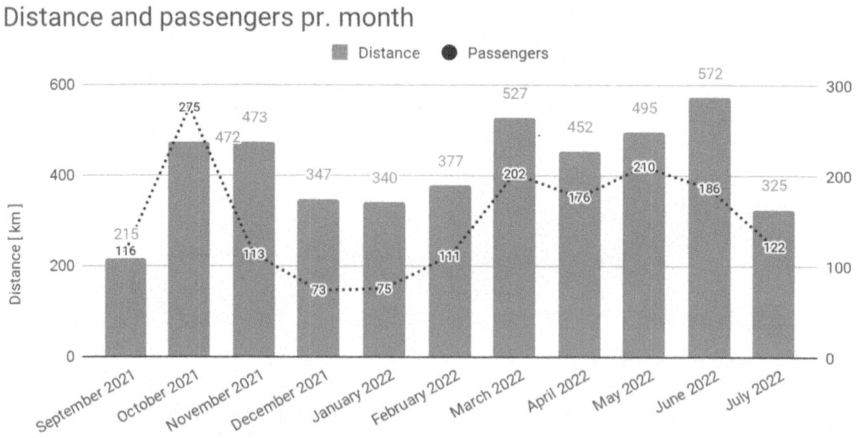


**Fig. 2.19** Distribution of passengers and distance driven per month in the Slagelse project

#### 2.4.3.3 Achievements and Key Success Factors

Distance and Passengers

In total, 1659 passengers were transported and 4595 km driven on the Slagelse route. During the pilot, there were over 2117 h of operation (Figs. 2.18 and 2.19).

Automated Vs. Manual Driving

In regard to the navigation mode, 93.9% of overall driving on the route was driven in automated mode. The rest of 6.1% driven in manual mode was mostly due to technical issues. Driving to and from the garage is filtered out.

#### 2.4.3.4 Recommendations

User Experience Learnings

*Patients*

Moving patients from A to B on a hospital site has proven to be a very good use case for the Navya vehicles, as patients often do not need high speeds but rely on the comfort and the ability to be moved. This means that even if the shuttle's top speed is 18 km/h, the patients still experience high value, as the alternative would be to walk or wait for local flex taxis, etc.

*Relatives/Visitors*

Being a relative at a hospital site is often associated with difficult parking conditions and lots of walking on the hospital sites. With the service provided in the AVENUE project, relatives and visitors have had the opportunity to park in large parking areas away from the hospital entrances but then being carried by the shuttle to the entrances. This means less congestion and car hassle at the entrances of the hospital, resulting in emergency vehicles having more space and less reckless parking from visitors.

*Employees*

As for the relatives and the visitors, the employees have used the shuttle service to get from larger parking areas to the entrances of the different departments—but also to accompany patients from one department to another—cutting off time from walking and waiting on flex taxis.

Performance Learnings

*Low-Speed Environment*

The Slagelse Hospital site has shown to be a very good fit for the deployment of the Navya vehicles. Being the site in Amobility's history with the highest automated uptime percentage, 93.9% of the planned operation has been delivered in automated mode. The environment at Slagelse Hospital is a low-speed (20–30 km/h) zone. Passengers, employees, and relatives visiting the hospital with cars drive slowly and are in general not in a hurry—meaning less dangerous overtaking and reckless driving around the shuttle.

Low Complexity Environment

The roads on the Slagelse Hospital are public with low-speed zones. The roads are wide and have designated lanes for pedestrians, bicycles, and cars. This has provided a good environment and level of complexity for the deployment of the Navya vehicles. Historically Amobility has experienced many issues with bicycles sharing the roads with cars, resulting in many overtaking close to the vehicle causing severe and hard braking of the vehicle—a huge risk for the passengers inside the vehicle. Having this environment, more suited for the Navya vehicle, has had a huge impact on the high uptime in a positive direction.

### *2.4.4 Conclusions*

The main conclusions that can be drawn from the Amobility efforts in the AVENUE project and the implementation of self-driving vehicles in Nordhavn, Copenhagen, Ørmoya, Oslo, and Slagelse, Copenhagen, are summarised in the following points.

- The approval process in Denmark has been a slow start where multiple stakeholders in the approval process had to learn the basics of self-driving vehicles, resulting in an approval process with inspirations from the railway approval systems—hence a documentation level out of the ordinary, seen from the perspective of the self-driving vehicles industry, where the technology is still at a low maturity level. A new approval is on average considered to take between 9 and 14 months in Denmark, and the adjustments and changes to an already given approval are rigid and require a new approval in most cases.
- The entire approval process in Denmark with multiple approvers is very expensive and requires a huge amount of work from Amobility, in terms of documentation, separate approvals, tests, risk assessments, and so forth.
- The approval process in Norway is structured in a very different way, allowing for a more dynamic and communicative approach. It is Amobility's experience that the Norwegian system is more agile and ready to adjust to the rapid develop-

ment of self-driving technologies. A new approval is on average considered to take between 3 and 5 months in Norway, and the adjustments and changes to an already given approval are seen as dynamic and innovative.

- Based on the experiences and knowledge that Amobility has gained so far, a recommendation for the Danish legal framework has been provided, with both direct and indirect changes that can be made to ease up the approval process.
- Amobility has suggested that when testing/driving in SAE Level 3 or lower, only approval from the DRSA, local police, and road owners should be necessary. Furthermore in projects with SAE Level 4–5 operation, AM recommends having either (1) approval from the appointed assessor, DRSA, police, and road owner(s) or (2) an approval by the Danish Road Directorate (taskforce), DRSA, police, and road owner(s). AM also suggests authorities to allow approvals for areas rather than single route approvals. Also, AM recommends that the Danish authorities target public funds more towards AV implementation. Ultimately AM recommends removing the political approval process entirely. For each project to have its own executive order completely diminishes the flexibility and innovation for existing and future pilot projects.
- Inspiration and know-how from the Norwegian approval system have been recommended to the Danish authorities as a comparison, aiming at showcasing the potential loss of innovative projects in Denmark if the approval system does not adapt to the innovative and rapid developments of self-driving technologies.

  - From a technical perspective, Amobility has experienced that the technology and industry in general have developed at a slower pace than anticipated and the objectives of AVENUE have been difficult to meet.
  - The Navya vehicles are not able to drive in SAE Level 4 as there are still many aspects of the safety-related features that need to mature before Amobility can take out the safety operator from the vehicles. As a part of the risk assessment in Denmark, there are certain jobs related to risks in traffic that the safety operator has to perform to verify and support the vehicle in public transport
  - During the project of AVENUE, Amobility deployed Navya vehicles at three different sites in both Norway and Copenhagen, Denmark. Experience has shown that the route in Slagelse offers the best circumstances for operating Navya vehicles with the right amount of complexity still ensuring the opportunity for high uptime. The maintenance cost is historically low in Slagelse, and major breakdowns and issues have not been experienced. At the same time, passengers at Slagelse Hospital have been very happy with the service of moving patients from department to department. This highlights why the use case in Slagelse is better than the other sites, as the patients do not demand speed but comfort and the ability to avoid walking. They are not in a hurry, as Amobility experienced passengers in Nordhavn to be due to the city centre position having both residents, employees, and tourists.
  - To be able to reach higher speeds with the shuttles, the sensory systems have to be improved allowing for a larger safety net of lidars, etc. Given the

Amobility experience in Nordhavn and Ormøya, the average speed is still as low as 7–8 km/h. The top speed of the Navya vehicle is currently 18 km/h. It is Amobility's opinion that the average speed is more important to increase than the speed of vehicles as the most important improvement for the operation is to move the passenger faster from A to B and not at the highest top speed. The objective is to try to raise the average speed to 14–16 km/h.

- Driving in snow and heavy rain in Oslo has turned out to be quite the challenge for the Navya shuttles. The big snowflakes are often confusing the lidars and tricking the system into thinking that there is an object that the vehicle has to avoid, hence usually a severe braking causing the passengers and safety operator to fall.
- The map of the vehicle (from the commissioning process) does not take into account that vegetation changes from season to season, even from the day the commissioning was executed to the day the operation begins. This has caused many problems for the vehicles that have performed many hard and severe braking. Often the branches are seen as obstacles in front of the vehicle in the safety zones. Amobility and the clients have spent both time and resources on keeping the vegetation to a minimum (preferably exactly as it was during the commissioning).

## 2.5 Lyon, France

### *2.5.1 Objectives*

Launching a transport service with automated vehicles open to the public on an open road in the Confluence district in September 2016, Keolis Lyon is an early adapter of this research field. They are part of a local dynamic that is pushing the development of automated vehicles, with a political vision supported by the public transport operators (PTOs), as the means of the future to complete public transport offers. This first experiment allowed the creation of a public transport line serving a 1.6-km-long neighbourhood, thus connecting residents to a tramway line. This helped Keolis Lyon to lay the foundations for the use of automated vehicles as part of a strategy to bring passengers back to other modes of the urban transportation network (tramway, metro, high service level bus) making the public transport network more attractive.

The objective is to test a technology capable of revolutionising public transport, by offering PTOs a new way of developing their transport offer. Keolis Lyon is hoping that automated vehicles could modify the distribution of funding in public transport, which decides whether new public transport lines can be developed.

To understand the context of this project, it is important to acquire knowledge about the economic background of public transport. In Europe, mobility and the development of public transport in urban areas are two important issues. Therefore, the involvement of public authorities in urban transport is essential to ensure the availability of a quality service at affordable prices. The management of a high-quality urban transport network represents an important public service mission (PSM) for the community, as it justifies the granting of subsidies to transport companies.

Indeed, most urban transport networks are not commercially viable. The public authorities must therefore cover all or at least part of the costs of investment in infrastructure and vehicles. Operators also receive subsidies for the operation of public transport, which vary in size from community to community. For example, in 2002, subsidies covered 45% of operational costs in Helsinki, 50% in Stockholm, 51% in Lyon, and 60% in Brussels. To meet the mobility needs of the urban population and the expansion of the areas served, the supply of urban transport has increased considerably over the last two decades. The French networks have become increasingly busy, and operating costs have increased very rapidly. However, the networks have not compensated for the increase in their operating costs with fare increases. Thus, from 2000 to 2015, the average rate of coverage of operating expenses by fare revenues has been steadily decreasing. It fell from 31% to 18% in networks with 50–100,000 inhabitants, from 33% to 18% in those without metro or tramway systems, and from 37% to 32% in those including them (Coppe & Gautier, 2004).

Reviewing these figures, it is important to point out that a significant part of the operating costs is represented by the drivers' wage bill. This is how the idea that the creation of a new public transport line depends mainly on the expected volume of demand can be confirmed. Following this logic, it can be considered that if the operating costs of a new line are fixed according to its frequency (driving costs, investment, and maintenance of vehicles and infrastructure), the share of the territorial authority will decrease with the increase in frequency. Busy lines are therefore more profitable because they benefit from additional revenue. Thus, in an area with insufficient passenger density, the local authority will not be able to support the costs of creating and operating a new line.

In this paradigm, the automated vehicle could change the equation by considerably lowering operating costs, particularly by reducing the cost of labour, which is not negotiable for vehicles with drivers.

The experiment carried out in 2016 was therefore intended to enable a better assessment of the capacities of automated vehicles to accompany public transport networks in this transition.

### 2.5.2 *Deployment*

In 2018, after 2 years of operation of the experiment in the Confluence district, the opportunity to join the AVENUE consortium and to participate in the H2020 project led Keolis Lyon to look for a new and more ambitious field of experimentation. The Parc OL experimentation made it possible to work on several use cases.

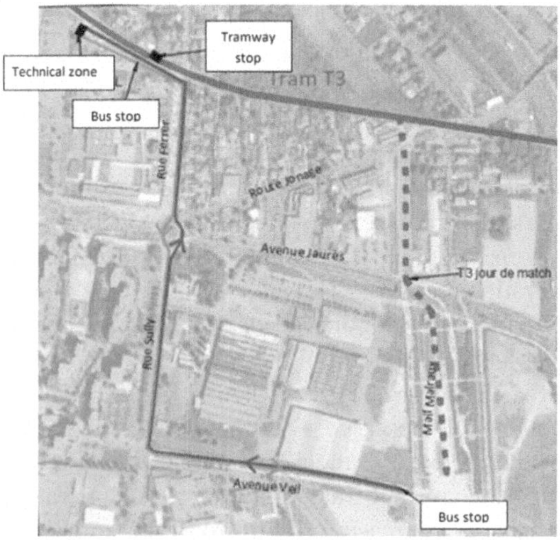


**Description:**
- 2.6 km round trip
- 3 crossroads with V2X
- 1 roundabout with V2X
- High frequentation open road
- 4 schools on the itinerary which cause high pedestrian traffic

**Goals:**
- Full integration on Public Transport
- Support for OL City development (hotel, restaurant, office)
- Fluid interaction between automated vehicles and urban traffic
- Social acceptance of automated vehicles
- Improve passenger's information

**Fig. 2.20** Parc OL route

Economic use case: The Parc OL is the stadium of the professional football team which attracts about 40,000 spectators on event nights. The ambition is to create an economic activity around this stadium, in a project called "OL Vallée". Therefore, there was a certain number of buildings to be built throughout the experiment which would completely change the way people travel in the area (leisure centre, restaurants, office buildings, medical centre and analysis laboratory, hotel). With a progressive increase in the number of passengers per day, the automated shuttles were seen as a way to offer a complementary public transport solution to accompany the economic development of this district. While it was clear that in the long term a mode of transport with more capacity would have to serve the OL Valley area, the automated shuttles were to provide a link during the increase in transport demand.

Technological use case: The first phase of experimentation was to connect the nearest tram stop (T3 Décines-Grand Large) with the Parc OL, by carrying out a route of approximately 2.6 km round trip. The route that the automated shuttles were to take allowed them to be integrated into a large flow of cars, passing crossroads and roundabouts for which the automated shuttle technology alone was insufficient. It was, therefore, necessary to equip four intersections with communicating traffic lights and to develop communications between the automated shuttles and the traffic light intersections. Thus, several insertion levels with or without priority for the automated shuttles were developed.

Social acceptance use case: The route of this experiment required the shuttles to cross a sensitive neighbourhood, in which road incivilities are regularly observed by the police. The question of the appropriation of these new technologies by the local population was therefore a central issue. This project was an opportunity to meet the inhabitants of the district to present the project, the technology used, its operation, and its constraints. This promotion of the project was done through several communication channels and by going to meet the schools of the district. The objective of this phase was to allow residents to take ownership of the project and to integrate the automated shuttles into the urban landscape as a new everyday object (Fig. 2.20).

### 2.5.3 Achievements and Key Success Factors

Unfortunately, after an inauguration of the experiment on 18 November 2019, the crisis of COVID imposed the stop of this experimentation on 16 March 2020. New health regulations have forced French public transport operators to respect social

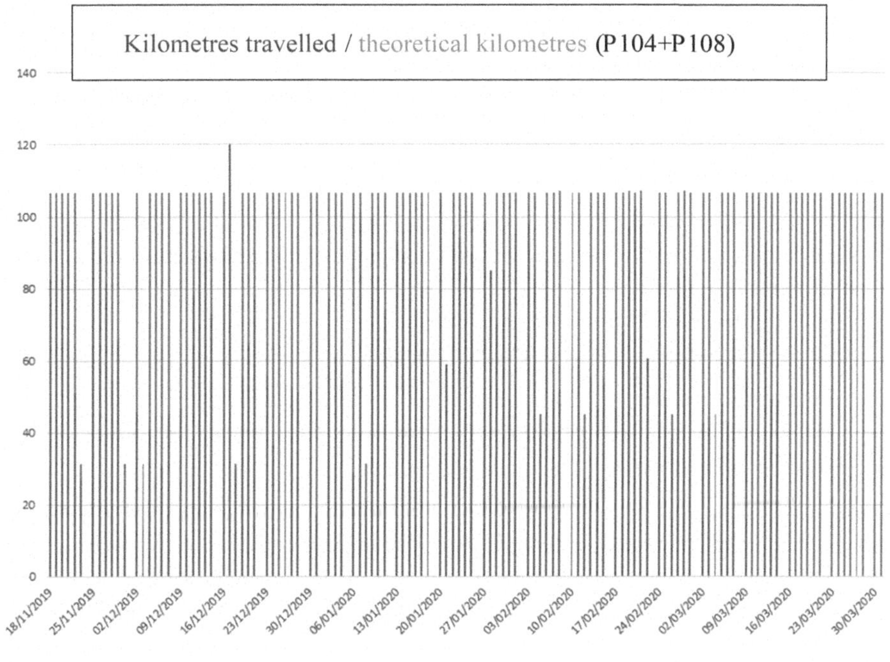


**Fig. 2.21** Kilometres travelled/theoretical kilometres

**Table 2.2** Shuttle data

| Month | P108 % Auto | P108 Average speed (km/h) | P104 % Auto | P104 Average speed (km/h) |
|---|---|---|---|---|
| Nov 2019 (starting from 15 Nov) | 93% | 9.6 | 95% | 9.7 |
| Dec 2019 | 91% | 9.7 | 92% | 9.8 |
| Jan 2020 | 88% | 9.9 | 89% | 10.0 |
| Feb 2020 | 85% | 10.1 | 87% | 10.1 |
| Mar 2020 | 75% | 8.9 | 93% | 10.1 |

distancing in their vehicles. However, given the size of the automated shuttles, respecting these social distances was not possible. Although a resumption was allowed in September 2020, the context had evolved and no longer allowed for qualitative observations of this experiment. The number of passengers had dropped significantly on the whole network, with traffic of less than 70% of the normal attendance. The figures presented below are therefore only for the period before the COVID-19 crisis (Fig. 2.21 and Table 2.2).

Monitoring of the experiment:

- 4000 passengers
- 6400 km
- 1.13 passengers/journey made

Awareness among CT users (650 people from the panel to link with the social impact chapter):

- 46% are aware of the experiments in Lyon.
- 55% have a favourable opinion of automated vehicles.
- 33% would be ready to give up their car if a TAD service with automated vehicles was proposed.

Rate of completion of the service:

- 85% (theoretical kilometre to kilometre to be achieved by a shuttle when it is out of service and kilometre to be achieved during operational stops caused by external reasons).
- 65% if all theoretical kilometres are kept.
- Equipment reliability problem and known cause:
  - Prototype vehicle
  - (1800 km of untested operation + Transpolis tests where the limits of the shuttles were tested)

The average speed was provided by Navya. It represents the average of all operating speeds. The average speed is usually low because it includes slowdowns when approaching stops where passengers get on and off the vehicle, as well as all the slowdowns caused by events on the route (red lights, crossing times at junctions, awkward parking, etc.).

### *2.5.4 Future Development*

While the question of the technical capabilities of automated vehicles seems central to the anticipated developments, it is necessary to put them in perspective with the needs of mobility services. For automated vehicles to be commercially successful in public transport, it is imperative that the service offered is relevant.

By integrating this reflection, it is easily understandable that the principle of a fixed line, with regular and reliable schedules, can be a relevant service. The regularity of the passages could facilitate the habitual travels of the users, as well as their confidence in this service. Nevertheless, this is only true when the frequency of passage of automated vehicles at the stops is very high. This allows the customer to remove the constraint of planning these trips. In the same way as a metro, we do not look at the time of arrival because we know that they run frequently and that the waiting time between two metros is always acceptable. The logic for a fixed line of automated shuttles is therefore the same. Even if this is feasible, it implies a constraint on the size of the vehicle fleet. To be able to guarantee a high-frequency operation, the transport operator will have to exponentially increase the size of their fleet with the increase in the size of the area to be served. This principle may therefore create a strain on investment requirements and operating costs. The increase in fleet size implies more vehicles but also more supervisors and intervention teams, which will still be needed in case of problems and disrupted situations.

In this paradigm, we could therefore deduce that the development of fixed public transport lines operated with automated shuttles is only relevant considering certain characteristics of the site, including size or certain constraints related to the types of routes available for automated shuttles (e.g. industrial site with a single road).

Based on these observations, we can therefore anticipate that the aggregation of automated shuttle technology and vehicle fleet management technologies to offer an on-demand transport service would offer an answer to several constraints.

#### 2.5.4.1 The Constraints of Availability for Users

With an on-demand transport system, the constraints on the availability and regularity of automated shuttles are perceived differently from a fixed route. The automated shuttles will be available through a spontaneous reservation, and waiting times can be anticipated by the users. For example, users of conventional on-demand transport platforms (e.g. Uber) have learnt through use that certain times are to be preferred and others to be avoided to use the services offered.

The second option for the user is to book their journey in advance, giving them more freedom to choose the time they want to be picked up. With this proposal, the transport operator encourages the user to anticipate his trips, which subsequently facilitates the organisation of services and the dispatching of vehicles according to the time slot of the day. The more customers anticipate their journeys, the more efficient the transport operators can be in managing production.

#### 2.5.4.2 Energy Constraints and Battery Capacity

In addition, on-demand transport helps to increase the ratio of passengers to kilometres travelled, thus improving energy efficiency as the energy requirement on the batteries is reduced. While creating a fixed route, the battery capacity is an important factor for the dimensioning of the vehicle fleet. This is less important for a fleet of automated shuttles in demand-responsive transport because the services are composed of breaks during which the automated shuttles can recharge the batteries.

#### 2.5.4.3 Facilitate the Relationship with the User

As on-demand transport services are mostly based on a digital solution that integrates a smartphone interface, the relationship with the user becomes easier. Important top-down information such as the waiting time before the next trip, the proposal to use another more efficient mode in a mobility-as-a-service (MaaS) environment, the monitoring of use and the valorisation of loyalty, and the listening of the customer can be managed via this interface. Thanks to this possibility, the transport operator minimises their investment in infrastructure by creating passenger information.

On the other hand, it is important to note that this vision of the service excludes users without a smartphone, which is notably the case for the oldest and most isolated population. It is therefore necessary to think about an accompanying service if automated shuttles are to fulfil a public interest mission and thus be a fully fledged mode of the public transport network. It is important to not exclude any population from a new service.

#### 2.5.4.4 Pricing Issue

In a regular and fixed service, the question of pricing arises. Even if a validator were to be installed inside a shuttle, it would be difficult to automate the validation process at this stage of technological development. This leaves the possibility of carrying out control actions, in the same way as on the other vehicles of a public transport network. However, when considering the development of automated vehicles, we must consider the use cases that will minimise human intervention. In this context, the use of the on-demand transport service supported by a smartphone interface facilitates the act of payment of the service by the customer. It is easily imaginable that the customer can register their credit or debit card or their transport card in the application (depending on the fare choices of the PTO). In this case, only people who have a good knowledge of the current fare system will be able to make a reservation and use the service.

However, this is only part of the answer to the problem of fraud. It will also be necessary to investigate the control of access to the automated shuttles by people who take advantage of another person in a legal situation to get on board the automated shuttle.

The reasons given above are not intended to be exhaustive, but they do help to understand the interest in on-demand transport for automated vehicle technologies. It also allows us to observe that many questions arise with this type of service. This is why Keolis Lyon has committed to an evolution of the transport service initially proposed. The initial fixed line will have to increase its perimeter and offer a transport-on-demand service within a district that has been energised by the recent arrival of numerous centres generating daily passenger flows, with different motivations, and different travel times, not subject to the two peak full stops of the morning and evening.

## 2.6 Luxembourg

### *2.6.1 Pfaffenthal*

Pfaffenthal is a small, urban living area located in Luxembourg City, the capital of Luxembourg. This urban area with around 1300 inhabitants is based in a valley between the historical centre of Luxembourg City and Kirchberg, the business district that is home to the European Investment Bank and the Court of Justice of the European Union.

Pfaffenthal is connected to the city centre via a public elevator and to Kirchberg via a funicular railway. Several bus connections are available in the surroundings of the elevator entrance at the city centre level. The funicular is part of a multimodal station that has been newly implemented in Pfaffenthal. Besides the funicular station, this multimodal station consists of a train station, a stop for several bus lines, as well as a bike sharing station. A tram connection and additional bus stops are situated on the Kirchberg side (of the topography). Over the course of 1 day, a variety of individuals traverse the Pfaffenthal Valley via a range of transportation methods. During the morning and evening rush hours, the majority of individuals commuting through Pfaffenthal do so by tram (Fig. 2.22).

The specific characteristics of the Pfaffenthal Valley make it the ideal use case scenario for the demonstration of a first and last-mile mobility service. Prior to the beginning of the project, no interconnectivity between the various modes of transportation arriving in the distinct regions of Pfaffenthal existed. Furthermore, Pfaffenthal offers a very diverse traffic situation with all kinds of different road users. It is a showcase to see how an automated vehicle can be integrated into such a diverse environment.

**Fig. 2.22** Route and automated minibuses stops

### 2.6.1.1 Objectives

Until the beginning of the project, no transportation solution existed to overcome the distance between the residential area, the multimodal station, and the public elevator. The core objective is to fill this lack of transportation to connect the different means of transportation as well as the different areas of Luxembourg City with each other.

During the day, residents, employees, and a vast number of tourists are using the multimodal station in combination with the elevator to get to the different parts of Luxembourg City. The distance between the public elevator and the multimodal station is 500 m, and the distance between the residential area and the public elevator is 800 m. This corresponds to a 5–10-min walking distance. To connect the public elevator, the multimodal station, and the residential area with each other, the following route for the automated minibuses has been selected.

#### 2.6.1.2 Deployment

The automated minibus shares the street with all kinds of different traffic users from cyclists and pedestrians to trucks, buses, and individual cars. Even with the speed limit of 30 km/h, the automated minibus encounters numerous overtaking manoeuvres, which cause harsh breakings. Complex traffic situations around the automated minibuses cause rough driving behaviour. The shuttles need to identify other traffic and not only detect it. The automated minibuses speed of max. 18 km/h is slowing down traffic in Pfaffenthal, especially in the morning peak hours when the traffic is very dense, and the drivers seem to be very nervous and hectic. This is also the reason for the change of operational hours in 2019. It was decided to keep the automated minibuses out of the morning peak hours to prevent them from slowing down traffic. There is no need for an on-demand system because the shuttle only serves three stations. A safety operator has to be on board all the time by law (at all sites).

#### 2.6.1.3 Achievements and Key Success Factors

With more than 25,000 passengers and over 9000 km driven in around 18 months, Pfaffenthal is the most successful site regarding the deployment of automated minibuses. The presence of the tourist destination Pfaffenthal, along with the prime location between the two elevators, was the key contributor to the success. The shuttle service did not require any promotional campaigns as it enabled users to travel distances that would normally be traversed on foot, thus providing them with the opportunity to reduce their transit times and to experience new technology.

#### 2.6.1.4 Future Development

In the near future the city of Luxembourg is willing to restart the service. The restart on the existing route and with enough time to get all authorisations is a good way to get back to the service. Afterwards, the extension or implementation of other options and routes can be conducted, if required.

The vision for the trial in Pfaffenthal is to deploy more routes within Luxembourg City to establish a network and an on-demand service of automated minibuses that are linked to each other, to the different parts of Luxembourg City, and to the public transport of the city. SLA can implement the on-demand option because Esch has shown that this service can be offered in Luxemburg. During the time of the AVENUE project, there were plans to expand the route. The objective of this route extension is to provide a transport service to:

- Hospice de Pfaffenthal (limit in parking space).
- Youth Hostel (limit in parking space).
- Servior, a retirement home (limit in parking space).

These three future partners do not have enough parking facilities and no nearby bus stops for passengers taking the Panorama Lift from the upper town. An emerging factor is providing accessibility to different users where transportation is scarce. The city of Luxembourg wanted to get a few months of experience with the original route before giving the authorisations for on-demand and door-to-door service. Due to COVID and the full stop of the service until the end of the project, the extension has never been realised but is still a possible option in the future.

### *2.6.2 Contern*

Contern is a city located around 10 km southeast of Luxembourg City. An industrial zone with different companies has been implemented on its territory. A railway station and a stop for public buses are located on the border of the industrial zone. There are several other stops within the zone, but no direct connection exists from the train station to Campus Contern.

In this first phase, the automated minibus was operating between a real estate development company called "Campus Contern" with more than 300 employees and the train station.

#### 2.6.2.1 Objectives

The aim was to dispatch individuals arriving via public transit to various firms in the industrial section and to offer a transport solution within this area. The trial in the industrial zone of Contern was chosen for its different environment compared to Pfaffenthal. Whereas Pfaffenthal has a busy inner-city traffic situation with various types of road users, the traffic in the industrial area of Contern is primarily composed of industrial vehicles, trucks, and cars, with fewer cyclists and pedestrians. The morning and afternoon hours in Pfaffenthal are marked by a considerable rise in individual car traffic because of people going to and coming from work. This phenomenon is far less accentuated in Contern (Fig. 2.23).

The vast majority of the company's employees was using their private car for their work commute as well as for transfers inside the zone. The shuttle is a good alternative because it closes the last-mile gap from the train station during the rush

Fig. 2.23 Route Contern

hour. Frequent traffic with heavy vehicles and increased commuter activity during peak hours is observed in the industrial area.

#### 2.6.2.2 Deployment

The percentage of the automated mode with 50% is very low compared to other sites because it takes 30 min for the shuttle to get from Campus Contern to the train station and back. The train arrives every 30 min. To arrive in time, the safety operator switches into the manual mode and drives a little faster than in the automated mode. In Contern very few issues occurred during the operation that were caused by external factors. There is less traffic in the industrial zone of Contern compared to Pfaffenthal; thus the automated minibuses encountered less overtaking. However, there were more heavy vehicles, which made the overtaking more dangerous for the shuttle. The main issue encountered was vehicles parked illegally on the automated minibus's path. From the view of the other road users, the automated shuttle is driving comparatively slow. This can result in less acceptance of the automated minibuses as a new kind of road user and on the other hand to use the shuttle itself, as long as it is driving slower than their own cars. There is no need for an on-demand system because the shuttle only serves two stations.

#### 2.6.2.3 Achievements and Key Success Factors

There were over 850 passengers transported and 4000 km driven over 26 months. This is not as much as in Pfaffenthal, but there were a lot of breaks because of construction sites and COVID (including strict home office rules for a long time), and the people accepted the service and used it every time it was running. The length of the route and the slow speed of the shuttle prevented more people from using it.

As the automated minibus was operating exclusively for "Campus Contern", the total costs were paid by this company. On this extended route, the automated minibus is passing several other companies and thus could connect more of them to the railway station.

The continuation of the contract depends on other companies to join the service and to lower the costs for the participants. It is very important to notice that the operation of the automated minibuses is linked to very high costs—as long as a safety operator is on board and not supervising several shuttles from a central office. A lot of companies, cities, authorities, etc. are very interested in such an innovative mobility solution, but they are not able to bear the respective costs.

Other than in Pfaffenthal, one does not simply walk into the shuttle. The people have to know that there is a service because it is only two times an hour at the stops. Campus Contern engaged in extensive internal communication, and on average there was at least one person per ride in the shuttle. With more companies joining, the amount of passengers can rise and provide a good alternative for taking the car.

#### 2.6.2.4 Future Development

The industrial zone in Contern was seen as an appropriate use case for the establishment of an on-demand mobility system. The demand varies with the time of the day. During the peak hours in the morning and in the afternoon, the request for the automated minibuses is mainly done between the train station and the different companies. During non-peak hours, there is a shift in demand towards the interior of the industrial area between various businesses and locations of interest, such as restaurants.

It was hypothesised that the routes could be extended to the other side of the industrial area, as well as additional stops being added in front of companies that expressed a desire to use the service. Due to the stop of the service during COVID and the slow return of the employees from the home office, there were too few passengers to justify the (expensive) extension of the existing service. Nevertheless, other companies are still interested.

The vision for Contern is to establish a system of automated vehicles connected to the bus network and other shared transport modes like carsharing, establish the on-demand system, and implement the shuttles into the public transport information system. Like this, it is possible to synchronise the shuttle with the train in time, even in case of delays. Numerous shuttles could be available for the different parts of the industrial zone, which can be ordered on demand. This will facilitate the last mile of travel between public transportation stops and the workers' final destinations.

### *2.6.3 Esch-Sur-Alzette*

Esch-sur-Alzette with 35,000 inhabitants from more than 120 nationalities is the second most populated town in Luxembourg and is located on the border to France in the south of the country around 17 km away from the capital Luxembourg City. The introduction of a shuttle service to the longest shopping street in the country could facilitate the revitalisation of the area through improved mobility. The shuttle ends on one side near a bus stop, while at the other end, there is no public transport. In this way, the shuttle closes the gap between the public transport stop and the inner city.

#### 2.6.3.1 Objectives

In September 2021 the service started in the main pedestrian/shopping street. In this part of the city, no public transport can be provided, but the mobility of the people can be optimised, especially the elderly, wheelchairs, strollers, or people with heavy shopping bags by providing the shuttle service. On rainy days the shuttle service can get people dry from one spot of the shopping street to another.

In September 2022, the planned on-demand service was launched, extending service hours from 11:00 am to 9:00 pm. The shops are closed after 6:00 pm, but the cafés and restaurants are still open. The presence of a shuttle during the evening hours in the more quiet street can provide a sense of security to pedestrians due to its illumination and motion.

The long-term vision is an automated-based, door-to-door service in Esch-sur-Alzette. The objective is to start first trials of an automated minibus with dynamic

**Fig. 2.24** Route Esch

routing in a geographically defined area, without fixed bus lines or predefined timetables.

#### 2.6.3.2 Deployment

The shuttle is driving on a 1-km section of the street "Rue de l'Alzette" with several stops at the crossings. It is driving at around 5 km/h as it should not be a disturbing element in the street but merge with the pedestrians. Only pedestrians and bikes are allowed in the street. Before 11:00 am and after 6:00 pm, delivery traffic is allowed, too. The on-demand service started on 12 September 2022, extended the service hours to 9:00 pm, and added four virtual stops on the same route (Fig. 2.24).

In terms of the shuttles' operating environment, Esch is very different to Pfaffenthal and Contern. The very dense and narrow shopping street is only used by pedestrians and sometimes by bikes and scooters. Some people want to test the "reflexes" of the shuttles' systems, and others are not paying attention even if the automated minibus rings the bell. Occasionally, there were objects (e.g. chairs from restaurants or a fence of a long-term construction site) on the driving path of the automated minibus. Due to these circumstances, SLA was required to communicate the challenges for the service to restaurant owners or to the city of Esch. The testing and further development of the pedestrian mode resulted in less harsh braking.

Furthermore, the implementation of the on-demand service resulted in longer service hours and a demand-orientated service for the users.

#### 2.6.3.3 Achievements and Key Success Factors

With around 12,000 passengers in 12 months, the site in Esch is nearly as successful as Pfaffenthal.

In the summer of 2022, the mobility service had to be paused for 2 weeks due to technical difficulties. After the break, the passenger numbers were as high as before. This underlines that the people got used to the shuttle service, as they use it despite a longer operational break.

The street is wide enough for one vehicle, but with a lot of pedestrians, bicycles, maintenance cars, events, and construction sites, it can be difficult for the shuttle to drive all the time in automated mode. Intensive communication with the city of Esch is necessary to solve these problems, and it is working out very well.

The ODS started on 12 September 2022, extended the service hours to 9:00 pm, and added four virtual stops to the six existing ones. An app is required to book the shuttle for the desired time. The user cannot see the stops in the app but the service area. The user can choose the pickup and drop-off location freely within the area but is guided to one of the ten stops. These stops are very close to each other, so the user does not need to walk more than 1 min. Although it is a slight restriction, it allows the user to flexibly book the automated minibus.

Similar to the operation in Pfaffenthal, the people see the shuttle running and can use it to shorten their walking route. Since the service hours are the same as the shops' opening hours, no additional advertisement is required.

#### 2.6.3.4 Future Development

There was a long-term vision for an automated-based, door-to-door service in Esch-sur-Alzette, which should have covered the whole inner city. Due to COVID, there was a significant delay in the planned schedule. The normal shuttle service started one and a half years later than initially planned. Additionally, it was early obvious that for a useable service covering a whole city centre, much more shuttles are required. This requirement is going to be fulfilled within the ULTIMO project from 2023 on. Although this plan is desirable, the process of service implementation in Esch may have to be conducted with more steps.

The service should continue in 2023. It is also possible that the service hours will be extended. If people use the shuttle starting at 6:00 pm often and get used to it, it will show that the shuttle is part of their mobility and can maybe even drive until 10:00 pm—with or without the on-demand service. Another option is to add another street at the end of the existing route where a lot of restaurants are, enabling that the customers are closer to the shuttle service. In conclusion, the shuttle itself is not only an additional mobility option but is also a contributing factor in revitalising the city centre, making it safer, cleaner, and more liveable.

## 2.7 Lessons Learned

Automated vehicles for public transportation are seen as a promising solution to the growing urban mobility challenges. Despite significant investments in improving technology, the transition from a "technology offer" to a "service offer" remains unclear, and until today there has not been a real commercial deployment of automated vehicles in urban mobility. This is primarily due to the lack of experience on how to develop viable and sustainable business models, which, in turn, results from a lack of real large scale.

The several real-life deployments of the automated vehicles in different public transportation set-ups of the AVENUE project allowed us to identify the vast majority of issues and to imagine and implement solutions. In this chapter, we have described the AVENUE site deployments, and we have outlined the key elements that require further study and development. The key lessons learned are that, on one side, long-term planning needs to be established by public transportation operators (PTOs) and public transportation authorities (PTAs) over a decade, with the creation and implementation of new policies, infrastructures, and passenger services. On the other side, new open standards that support the interoperability of automated technologies must be developed and adopted. This will foster competition among technology players, including vehicle manufacturers, fleet management and orchestration providers, V2X technology developers, and more, thereby creating a sustainable ecosystem.

While the issues discussed in this chapter are significant, they are not the only challenges. Additional hurdles include retraining maintenance personnel from mechanical to IT and electronics expertise and addressing the transition of existing drivers to new roles, such as intervention teams or back-office operators. Furthermore, deploying automated vehicles with on-demand, door-to-door public transportation services may face opposition from other transportation actors, such as taxi services, who might view this innovation as a direct threat to their business. Additionally, the certification of automated drivers remains an open question since public transportation vehicles are currently certified for their mechanical capabilities, and human drivers are "certified" for their driving skills. Approaching these issues, along with any unforeseen challenges that arise during actual very large scale service deployment, will require a concentrated effort from PTOs, PTAs, legislators, and service users to develop workable solutions.

Ultimately, this is the objective of the European project ULTIMO, initiated in October 2022 based on the learnings of the AVENUE and other European projects. ULTIMO aims to lay the foundation for a large-scale deployment and develop the necessary standards, policies, and road maps to create a viable, user-oriented, and economically sustainable public transportation service using automated vehicles.

We expect that large-scale deployments in Europe and around the world will start appearing from 2026 and on, with the parallel introduction to the market of new types of highly performant SAE L4 vehicles.

**Acknowledgements** This project has received funding by the European Union's Horizon 2020 research and innovation program under grant agreement No. 769033.

## References

Alonso-González, M. J., Liu, T., Cats, O., van Oort, N., & Hoogendoorn, S. (2018). The potential of demand-responsive transport as a complement to public transport: An assessment framework and an empirical evaluation. *Transportation Research Record, 2672*(8), 879–889. https://doi.org/10.1177/0361198118790842

Antonialli, F., Boos, A., Fournier, G., Jaroudi, I., Mira-Bonnardel, S., & Thalhofer, M. (2022). D8.6 economic impact assessment (not approved). Available online at https://h2020-avenue.eu/public-delivrables/. Accessed 4 May 2023.

Coppe, A., & Gautier, A. (2004). Régulation et concurrence dans le transport collectif urbain. *Reflets et perspectives de la vie économique, 0*(4), 65–75. Available online at https://cerec.be/wp-content/uploads/2017/12/cerec2004_8.pdf

Davidsson, P., Hajinasab, B., Holmgren, J., Jevinger, Å., & Persson, J. (2016). The fourth wave of digitalization and public transport: Opportunities and challenges. *Sustainability, 8*(12), 1248. https://doi.org/10.3390/su8121248

European Commission. (2020, September 12). Communication from the Commission to the European Parliament, the Council, the European Economic and Social Committee and the Committee of the Regions. Sustainable and Smart Mobility Strategy—putting European transport on track for the future. Brussels. Available online at https://eur-lex.europa.eu/resource.html?uri=cellar:5e601657-3b06-11eb-b27b-01aa75ed71a1.0001.02/DOC_1&format=PDF

Konstantas, D. (2021). From demonstrator to public service: The AVENUE experience. In S. Mira-Bonnardel, F. Antonialli, & D. Attias (Eds.), *The robomobility revolution of urban public transport. A social sciences perspective* (1st ed., pp. 107–130). Springer International Publishing; Imprint Springer (Transportation Research, Economics and Policy).

Konstantas, D., & Fournier, G. (2023). Fully automated public transportation mobility: The hidden challenges for a commercial deployment. *Open Research Europe, 3*, 71. https://doi.org/10.12688/openreseurope.15751.1

Nieuwenhuijsen, M. J., & Khreis, H. (2016). Car free cities: Pathway to healthy urban living. *Environment International, 94*, 251–262. https://doi.org/10.1016/j.envint.2016.05.032

OECD/ITF. (2015). Urban mobility system upgrade. How shared self-driving cars could change city traffic. Corporate Partnership Board Report. With assistance of ITF, OECD. Available online at https://www.itf-oecd.org/sites/default/files/docs/15cpb_self-drivingcars.pdf. Accessed 23 Nov 2022.

OECD/ITF. (2016). Shared mobility. Innovation for liveable cities. Corporate Partnership Board Report. With assistance of ITF, OECD. Available online at https://www.itf-oecd.org/sites/default/files/docs/shared-mobility-liveable-cities.pdf. Accessed 23 Nov 2022.

OECD/ITF. (2017). Transition to shared mobility. How large cities can deliver inclusive transport services. Corporate Partnership Board Report. With assistance of ITF, OECD. Available online at https://www.itf-oecd.org/sites/default/files/docs/transition-shared-mobility.pdf. Accessed 23 Nov 2022.

OECD/ITF. (2018). The shared-use city. Managing the Curb. Corporate Partnership Board Report (56th ed.). With assistance of ITF, OECD. Edited by OECD Publishing. International Transport Forum Policy Papers.

SAE International. (2023, May 6). SAE J3016 automated-driving graphic. Available online at https://www.sae.org/news/2019/01/sae-updates-j3016-automated-driving-graphic. Updated on 6 May 2023, checked on 6 May 2023.
United Nations. (2018). World urbanization prospects: The 2018 revision. Available online at https://population.un.org/wup/Publications/Files/WUP2018-KeyFacts.pdf. Accessed 7 Mar 2019.
VDV. (2015, November). Scenarios for autonomous vehicles. Opportunities and Risks for Transport Companies. Position Paper.

# Chapter 3
# Automated Minibuses: State of the Art and Improvements Through AVENUE

**Pierre Chehwan**

**Abstract** This chapter looks at advances in automated vehicle (AV) technology, focusing on the central role of AVs in revolutionizing public transport as part of the AVENUE project. It traces the path of AV development, highlighting NAVYA's pioneering contributions and the significant technological advances that have been achieved. The chapter describes the evolution of NAVYA technology and the disruption of the transport ecosystem during the ambitious goals of the AVENUE project. The project's mission to integrate automated minibuses into the public transport ecosystem is critically analyzed, highlighting operational innovations, the integration of on-demand door-to-door services, and the comprehensive management of an AV ecosystem. Technology, legal, and regulatory challenges and continuous improvements in AV software and hardware are discussed, emphasizing NAVYA's influential position in the European AV landscape. The chapter concludes with a reflection on the implications of the AVENUE project for future technological directions, emphasizing the wider implications for automated mobility and the need to navigate the complex interplay of innovation, regulation, and public acceptance.

## 3.1 Introduction

The idea of AVENUE project was triggered in 2017 when the technology companies started to challenge the car original equipment manufacturers (OEMs) on their ground stating that the software will take control of the vehicles and change the way cars will be driven in the future. While car OEMs were working in a progressive V cycle way adding features one after the other to their cars, software companies

NAVYA assets were taken over on April 23, 2023, by Gama (Gaussin Macnica Mobility joint venture, 51% owned by Gaussin and 49% by Macnica)

P. Chehwan (✉)
NAVYA Group, Courbevoie, France

beti - Experts Mobilité Automatisée, Saint-Donat-sur-l'Herbasse, France



G. Fournier et al. (eds.), *Automated Vehicles as a Game Changer for Sustainable Mobility*, Contributions to Management Science,
https://doi.org/10.1007/978-3-031-61681-5_3

disrupted this process with an out of the box thinking, where they considered cars as connected computers on wheel driven by a software that is developed in fast delivery AGILE process with AI model.

NAVYA (founded in 2014 from a group, Induct, which had launched a driverless electric minibus called Navia for Navigation par Intelligence Artificielle) took part of this disruption, as a pioneer by introducing the first commercial automated shuttle in 2015. The Autonom® Shuttle was built without a steering wheel nor pedals to be SAE Level 4 by design. The strategy aimed to introduce shared and public automated shuttles in urban areas, thereby reducing reliance on cars for short trips, alleviating traffic congestion, freeing up parking spaces, and reducing greenhouse gas emissions. NAVYA became the reference in this automated public transport shuttle area showcasing its technology around the world, for public transport operators, advanced cities, universities and research around the world. At that time NAVYA was deploying the Autonom® Shuttle in geofenced areas with 1–2 vehicles per site with round trips of 1–2 km.

The AVENUE concept was to raise the challenge and to integrate the automated minibus in the public transport systems. By putting the services in the heart of the project, the technology should solve more operational situations, go into mixed traffic, connect to public transport infrastructure, bring new adapted and disruptive mobility services, and deliver smooth and seamless riding like any other public transport service. Among all, the on-demand (1) combined with a door-to-door (2) approach in public transport where a passenger can use a mobile application to order a ride and be hailed was the major disruption.

To raise this challenge, NAVYA developed advanced technologies for its automated driving system (ADS) and established open-standard connectivity from the shuttles to external IT systems, enhancing operational excellence for these new mobility services.

This chapter will detail these elements and what was done to make the AVENUE project one of the major automated vehicle (AV) projects in Europe for the European Commission. AVENUE paved the way for competition and changed the perspective on automated driving, not only in terms of technology but also in the realms of services and even in the legal and regulatory framework.

## 3.2 Automated Driving Context before Starting AVENUE

Before the AVENUE project in 2018, the hype on automated driving was predicting the arrival of driverless cars by 2022.

Technology leaders like Google, with a dedicated subsidiary WAYMO, planned huge investments and forecast that robotaxis will cover major cities around the world in 2022 with at least 3000 taxis per major city. Uber was stating that they will leverage their margin by reducing the cost of human drivers, and they started investing billions in the technology. Lyft, Apple, and many other tech giants jumped in the competition pushing car leaders to react and create dedicated departments for advanced driver assistance systems (ADAS) and autonomous driving system (ADS). All the ecosystem was looking to the future with the robotaxi point of view.

Although NAVYA was working on robotaxi vision following the trend, the core business was to focus on L4 automated shuttles for shared and public transport. The idea was to bring as quick as possible these shuttles on the market by managing the risk one after the other. This was also driven by the lack of investment. As you aim for higher performance, you will require a greater financial investment, potentially accelerating your go-to market timeline by several years. NAVYA's practical approach was to begin implementing automated driving in restricted geofenced areas, whether private or public, with limited features, low speeds, and minimal interactions with other road users. This strategy aimed to minimize safety risks, reduce costs, and address issues one by one. Once you begin, you can gradually scale up to higher performance, increase interactions, and add features incrementally. This incremental approach to the operational design domain (ODD) would enable AV shuttles to enter the market quickly, albeit with limited capabilities, while continuously building data and knowledge to help the technology learn, mature, and advance. This would allow NAVYA to start selling and engaging investments.

With the AVENUE project, the idea was to put in place several shuttles on open roads of several cities in Europe, connect them to the public transport network, and create new mobility services. One of the major concepts of AVENUE was to demonstrate a door-to-door on-demand service covering a large geofenced area with public and shared transport, rather than just having fixed bus stops and shuttles driving in a round trip. The vehicle can pick you up from where you are and drop you to where you want to go, and the same time it hails other passengers on the road as a normal public transport. It is a mix between taxi and public transport. The difference with a robotaxi is that a 15-passenger vehicle will pick up additional passengers along the route, even if their destinations differ. However, both pickups and drop-offs must be within the shuttle's range (ride pooling).

Such a challenge needs to manage the whole AV system and ecosystem and not only the automated driving technology of the vehicle. It needs technologies to drive the vehicle, technologies to assist and secure the passengers, and technologies to receive and manage missions to a fleet vehicle. All this should be done safely with respect to the legal framework in place.

From 2018 to 2022, within the 4 years of AVENUE project, high technological and operational investment was made to develop such technologies and to reach the target fixed by AVENUE. To better understand the technological and operational leap done, we need to put the project in the technological and economical context of 2018 to allow a clear vision. Then we will draw the technological advancements made during AVENUE project. We will finish by a projection action triggered during AVENUE and will build the technological long tale to show market needs.

### *3.2.1 Market Projection*

Before 2018, automated driving was at the top of the hype cycle for emerging technologies published by Panetta (2017). All major players and OEMs were projecting to have a prominent market in the coming years. The scale of the market projections was extreme, exceeding $60 billion by 2030 (Frost & Sullivan, 2018).

All actors wanted to be part of this and to take advantage. The AV technology, with the help of the advancement of artificial intelligence, built the ambition that it could be done fast and revenues can be available within a few years' time frame. This race to the tech market pushed actors to overcommunicate on their potentials, ambitions, and vision of the future, which lead to overinvestments and overexpectations. In its 2019 report, McKinsey and Company (2019) identified that "Since 2010, investors have poured $220 billion into more than 1,100 companies across ten technology clusters. Investors invested the first $100 billion of these funds by mid-2016 and the rest thereafter." Within a total of 120 billion investments in mobility, in the 24-month time frame (2016–2018), automated vehicle technologies were the most rapidly increasing technology sector. Today, mid-2023, the picture is much different. Expectations are even higher, but within longer time frames and with a more realistic approach, as clearly reported in the 2023 report of Deichmann et al. (2023). It is anticipated that automated driving will trigger even more revenues and will also benefit to ADAS market. Both ADS and ADAS are correlated and can benefit one to the other, and many technologies and features of ADS can be used for ADAS.

### 3.2.2 Automated Driving

Just as a reminder, when talking about driving autonomy, the standard SAE J3016 of the SAE International and ISO is the reference, with five levels of driving automation. Full automated driving System (ADS) is seen at Levels 4 and 5. Vehicles do not have driving wheels nor pedals, there is no driver, and the ADS system will fully drive the vehicle.

The advanced driver assistance system (ADAS) encompasses vehicles operating at Level 3 and below. In these levels, cars assist the drivers during the driving process. Even in Level 3, where the software can handle the vehicle in various scenarios, there will come a point when the system cannot handle the situation anymore and will request that the driver take control of the vehicle. At that moment, the system disengages technically passing full control to the driver. Level 3 automation introduces ambiguity by potentially misleading the driver about onboard responsibilities—whether the driver or the system is primarily responsible. This raises questions about legal liability in case of accidents during the transition from system control to driver control. A notable instance highlighting this issue is the well-known 2018 car accident involving an Uber self-driving car, which struck and killed a pedestrian (Said, 2018).

### 3.2.3 The Landscape of Automated Mobility

Significant technological players initiated an exceptional disruption within the traditional original equipment manufacturer (OEM) car market. While traditional OEMs were gradually incorporating ADAS features, advancing from level to level, tech giants such as Google/WAYMO, UBER, LIFT, BAIDU, and others recognized that technology was the true game changer. Instead of incrementally moving from L0 to L5, they took a more direct approach to working on SAE L5, even if, in practice, it resembled more of an L3 or L4. Their perspective envisions the future of automobiles as robustly connected computers on wheels, embodying the tech giants' vision, rather than mere vehicles with software components, as seen in the OEM's perspective.

The tech giants harnessed cutting-edge sensor technologies, including lidars, radars, and other advanced sensor systems. Additionally, they leveraged state-of-the-art AI technologies, whether in the realm of computer vision or AI prediction, to enhance the development of perception, trajectory planning, and driving capabilities within their advanced driver assistance systems (ADS). These companies were also well equipped with high-performance computing power, which they employed to a substantial extent. New emerging comers also followed this trend of top-down computer on wheels' approach.

In addition to those concentrating on vehicle or driving system development, a comprehensive ecosystem was also established, featuring numerous participants operating across all facets of the automated driving market. This includes players involved in hardware, electronic control units (ECUs), sensors, feature provisioning, AI solutions, fleet management, localization technologies, mapping, networked transport of RTCM via internet protocol (NTRIP) data, and more.

### 3.2.4 NAVYA before 2018

In 2015, a small French startup, NAVYA, took the pioneering step of introducing commercial automated shuttles for public transportation. At a time when most market competitors were engaged in the race for driverless personal cars, NAVYA shifted its focus to public transport. They recognized the numerous advantages that automated driving can offer to cities, rather than to individual drivers, such as reduced traffic congestion. Their goal was to develop automated shuttles capable of accommodating up to 15 passengers, offering a shared and public transport service that encourages people to leave their personal vehicles at home for short rides. By adopting this innovative approach, NAVYA emerged as one of the pioneers that significantly influenced the global landscape of automated vehicles.

#### 3.2.4.1 Hardware

As NAVYA investments were not at the same level as OEM or tech giants, NAVYA focused on solving the risk step by step, developing one feature after the other, solving ODD incrementally, and leveraging speed accordingly. No OEM, nor any industrial provider, was ready to build a vehicle platform able to be automative ready to receive the PCs and automated driving system. Thus, NAVYA built its own factory and its own mechanical platform called the Autonom® Shuttle (Fig. 3.1, NAVYA

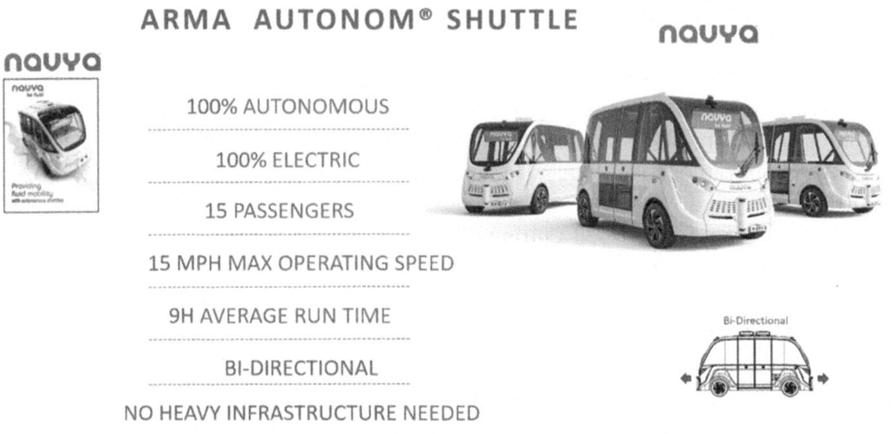


**Fig. 3.1** NAVYA ARMA© specifications

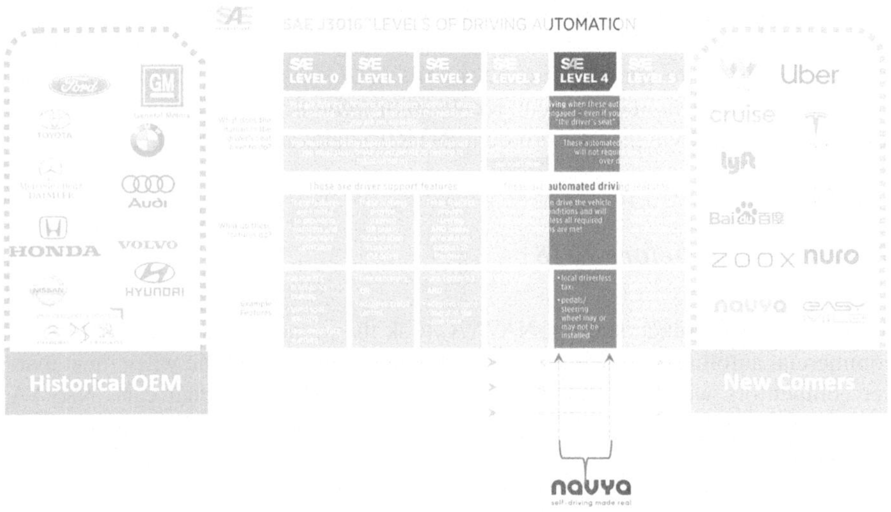


**Fig. 3.2** NAVYA automation positioning in the market, based on SAE J3016© levels of driving automation (Illustration by the author based on SAE J3016©, 2021)

ARMA specifications). The conception was targeted to be SAE L4 ready by design without steering wheel, without pedals, without car mirrors, and able to scale to full SAE L4 (Fig. 3.2, NAVYA ARMA sensor overview).

NAVYA successfully integrated lidars, sensors, and various components (Fig. 3.3, NAVYA ARMA sensor overview) that were readily available from the market, even though these parts had not yet reached mass production levels with the highest automotive industry standards, mainly because such standards were nonexistent in the market at that time. The majority of industry players utilized identical

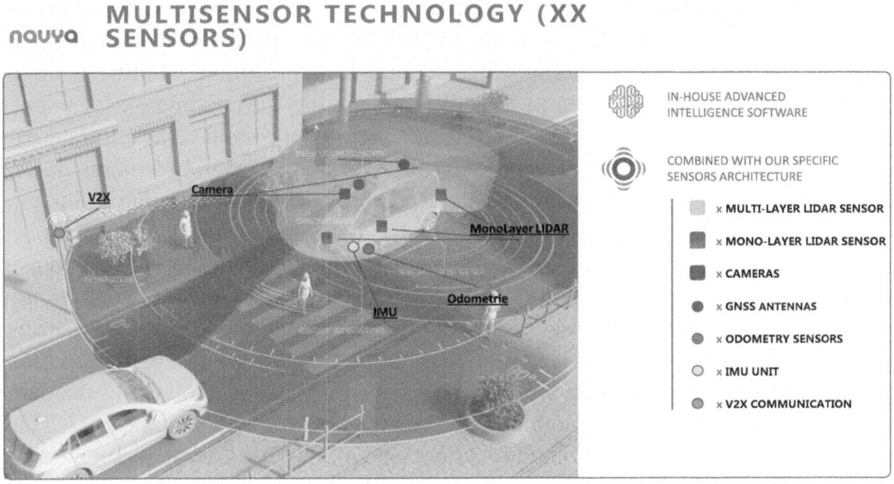


**Fig. 3.3** NAVYA ARMA© sensor overview

components, including the same lidars and sensor sets, for their prototypes and small-scale production runs (Fig. 3.3).

#### 3.2.4.2 Software

During that period, NAVYA demonstrated that their shuttle could be completely driven and operated by embedded software connected to a central supervision center. They employed localization technologies that relied on pre-mapped environments, utilizing global navigation satellite system real-time kinematic (GNSS RTK) technologies, among others. The maps used were basic 2D representations, and the commissioning and mapping technologies were relatively straightforward, typically involving simple round trips and linear routes. These systems effectively operated within a limited operational design domain (ODD).

The shuttle's primary routes consisted of straightforward journeys, such as round trips, covering short distances, typically about 1 or 2 km. There were minimal interactions, and if traffic lights were encountered, they were typically managed by the onboard operator.

#### 3.2.4.3 Services

To put shuttles in operation, a common work with the PTO is done to define the trajectory and the route. A radio GNSS base is installed on the site, open to the sky and covering the geofenced area. This base will get several satellite signals and will use GNSS RTK technologies to send correction to the shuttle giving precise 10-cm positioning. Subsequently, NAVYA implemented an in-house mapping hardware and software solution called mapping and localization measurement system (MMS). This system had the capability to be installed on any car, allowing the digitalization and creation of highly accurate route maps with precise positioning. Furthermore, it collected all the necessary data to digitally recreate the surrounding environment. The map was sent to the R&D for cleaning and taking out parked vehicles, pedestrians, dynamic elements, etc., to keep only the road and structural elements of the environment. A clean map was built by the R&D, and the commissioning teams can start working with the Autonom Shuttle® and a dedicated in-house tool on site to create and build the virtual rail, called the path. The path is what the shuttle will use to follow. It is built block by block, and each has its own parameters related to the real road (speed, intersection, curves, passenger crossing, etc.). Once the path is built, it is inserted in the vehicle system and put on the server. The shuttle can then use it to start the blank testing without passengers, driving and following the path and of course handling and reacting to all dynamic environments on the road. This will allow to fine-tune the path and to adjust also all the real driving parameters to have a smooth ride. NAVYA accomplished all these tasks using internally developed tools. Each of these steps was essential in establishing the mobility service they aimed to provide.

### 3.2.5 NAVYA Ecosystem

The success of NAVYA was based not only on the driving technology alone but is the whole system that makes it happen. NAVYA delivers to cities a full solution and service to put in place automated shuttles, to train the PTO, and to allow daily operation. The service value chain is public transport authority (PTA), to public transport operator (PTO), and to NAVYA AV system shuttle and supervision. NAVYA provides the technology to the PTO that will help him to operate the daily service for the PTA or the city.

From NAVYA shuttles, the value chain extends to the automotive industry, encompassing the suppliers of each hardware element of the shuttle. These hardware suppliers provide various elements, including fundamental components such as static or mechanical parts, batteries, battery management systems, sensors, and low-level command control systems, among others. Many of these suppliers lack prior experience in the field of automated vehicles, and the requirement for achieving a high level of safety in each component has not yet become common practice in their factories or production processes. Notably, basic elements like sensors remain relatively fragile, and their maturity is not yet at a stage where failure rates and false positives are optimal. Nonetheless, they are considered safe enough for deployment in vehicles under specific controlled conditions.

The entire ecosystem is collectively learning about its needs. Sensor suppliers require real data from manufacturers to enhance their products, while manufacturers seek greater reliability from the sensors. Public transport operators (PTOs) demand high-quality service and uptime from shuttle suppliers. These demands cascade down to the suppliers of various systems and subsystems. However, it is challenging to achieve a high level of safety as defined by Automotive Safety Integrity Level (ASIL) based on the regulatory standards outlined in automotive ISO 26262 and still make it a mass-market product. Everyone involved is on a learning curve as they work together to address these challenges and refine the ecosystem. Therefor even if the ADS of the shuttle would be the best of bread technology on the market, if the hardware subsystem is not reliable, the driving experience is not smooth. If sensors are providing false positive, stating there is an obstacle where there is none or just a leaf falling, the ADS will trigger harsh brakes to avoid accidents. That is why perfect correlation between hardware and software should always be harmonized. Moreover, to sell these shuttles on the market, it should be balanced in an accessible eco-financial equation.

### *3.2.6 Legal Boundaries*

In 2015, when NAVYA released its first commercial automated shuttles called The Autonom Shuttle® ARMA, no legal institution in any country had a good knowledge of such vehicles and technology. None was ready to deliver the authorization to put them on open road. Each vehicle on open road should have a registration plate, and the vehicle should be homologated by the authorities based on the related legal framework of its category (M1, M2, etc.). As the NAVYA shuttle does not have steering wheels nor pedals, it is seen as an unclassified vehicle. Authorities took into account that it is an innovative vehicle, an experimental vehicle, and they delivered partial homologation with a registration plate limited in the time of testing, e.g., 1 or 2 years, if the vehicle can prove that it is safe enough to operate on open road under certain conditions.

NAVYA had to operate on a country-by-country basis, collaborating closely with government authorities or entities representing the government. Their goal was to assist these authorities in comprehending the technology, the vehicle, and how to address the challenges posed by the absence of traditional elements such as a steering wheel, pedals, left and right rearview mirrors, bidirectional driving, and the associated lighting systems. This cooperative effort aimed to ensure the smooth integration of their automated shuttle technology into each region's regulatory framework and infrastructure.

In the absence of a standardized legal framework, the response to automated driving initiatives varied from one country to another. In nearly 20 countries, NAVYA took the pioneering step of being the first to introduce automated driving on open roads, where regulations and acceptance of this innovative technology were largely uncharted territory. In some, it took more than 1 year and half to get the registration plate, convincing the authorities that the AV shuttle is safe and secured for its innovation purpose on the dedicated road. Once the authorities gained the knowledge and established the procedures for partial homologation in collaboration with NAVYA, it paved the way for competitors to enter the country more swiftly. This scenario played out in countries like Denmark, as exemplified by the AVENUE project. Through the joint efforts of NAVYA, the local partner, AMobility (trademarked Holo), and AVENUE, we became the first to introduce an automated vehicle (AV) shuttle on open roads in Denmark, creating a precedent for others to follow.

## 3.3 Technology Improvements Through AVENUE

### *3.3.1 A Global View*

One specificity of the AVENUE project was that the AV operations started from day one. Four different cities in Europe put in operation a total of ten automated shuttles and connected them to their public transport system. AV shuttles were based on the technology available in 2018 and the operational design domain that NAVYA shuttles could manage at that time. The PTOs used these deployments to advance their process, build the IT, learn from operational deployment, and prepare to scale and reach the target built by AVENUE to have a smooth and seamless on-demand door-to-door service. Although some regions had prior experience with AV shuttle experimentation, embracing this technology for regular use represented a significant advancement and a substantial request for progress in terms of regulatory and operational readiness. The other project partners were able to get a deep dive into the technology, to understand the current limits and potential, and to ideate to support the delivery of all targeted services of the project.

To put in place automated shuttles services, a global view of the technology was needed to understand the enhancements at all levels. The AV shuttle operational capabilities were the starting point. If the ODD of the site is above these capabilities, the service is due to fail. For example, the AV shuttle cannot go faster than 25 km per hour due to safety issues. If in the route defined by the operator there is a part at 50 km/h, even a short one, the shuttle will not be able to follow, and it is better to decide not to build the service.

### *3.3.2 NAVYA Software*

Even though NAVYA is known as a manufacturer of automated shuttles, its core competencies are the ADS software and related technologies. NAVYA does conceive and build the vehicle and architect the whole set of sensors. The ADS, along with its related technologies, comprises the software suite that governs the automated shuttle, encompassing its full range of functionalities. It supports all aspects of the aforementioned system, from deploying to operating a mobility service using such automated shuttles. The software of NAVYA can be seen as three major parts: NavyaDrive®, NavyaOperate®, and public API (Fig. 3.4).

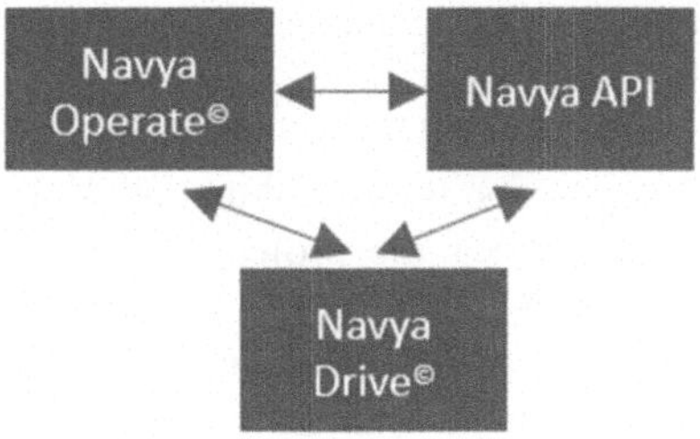


**Fig. 3.4** NAVYA ADS overview

**NavyaDrive®** is an embedded software installed on the computing system in each vehicle and allows to operate the vehicle. Internally, every new iteration of the *NavyaDrive®* software is referred to as a "Setup," which is linked to a specific version number.

**NavyaOperate®** is an off-board software installed on the cloud and allows to supervise and manage vehicles. NAVYA supervision center team uses this software to supervise sites, as well as to take actions on the vehicle like for maintenance, analysis, and other non-driving actions.

**Navya API** is a software installed on a cloud system that allows a third party to receive and send information to one or several vehicles. No direct connectivity to vehicles is allowed to third parties due to security and cybersecurity issues.

### *3.3.3 Automotive New Release Process*

The **NavyaDrive®** software is the main intelligence embedded in the shuttle. It is installed in the computing systems and connected to sensors from one side and to mechanical hardware from the other. The connection to mechanical hardware manages low-level command control system. It sends and receives all information to fully control vehicle dynamics. As AV shuttles are not yet full in industrial production with volumes, the providers made some evolution of existing hardware and mechanical parts to allow the use in AV. New software updates of these parts are made regularly, correcting bugs or adding some features. For this purpose, shuttles, even if they are all based on the same ARMA model, do have differences between each other.

For each new software version of NavyaDrive®, a comprehensive series of testing and validation procedures is essential to ensure compatibility and compliance with the various versions of the ARMA shuttles that have received authorization from regulatory authorities. This rigorous process is crucial to enable NavyaDrive® to fully utilize the capabilities of any deployed ARMA shuttle, offering a safe and reliable automated driving solution. With each new release of the NavyaDrive® software and setup, it is imperative to adhere to automotive standards during the testing and validation process. This ensures that no unintended regressions are introduced from previous versions, that all new features function as intended, and that the ARMA shuttles maintain the requisite safety standards to be deployed on the road. This meticulous process guarantees the reliability and safety of each software update. Adding a new feature or correcting even a small bug should go through this long process.

The AVENUE project played a crucial role in enhancing the standardization of this process. It ensured that all AVENUE shuttles adhered to the same service standards set for the project. Prior to this, NAVYA had been focused on promptly addressing client issues to maintain the continuity of their mobility services. In response to client-specific concerns, they often corrected bugs and released software updates after limited local testing, ensuring compatibility with the client's hardware version. However, this approach led to significant segmentation of the NavyaDrive software and introduced some unintended issues for other clients. Standardizing the process through projects like AVENUE helped streamline software development and ensure consistency in service across all deployments.

Due to the growing diversity in hardware among different clients, the approach of addressing client-specific issues had led to an accumulation of software discrepancies. With the AVENUE project involving ten vehicles and four public transport operators (PTOs), each facing unique scenarios and operational design domains (ODDs), the testing and validation process had to be meticulously fine-tuned and adapted to ensure that each software setup could run seamlessly on the various ARMA shuttles under different conditions.

While this approach improved overall software consistency, it occasionally resulted in frustration for the PTOs. The time required to go through this thorough process and reintroduce the service with new features could lead to operational challenges and delays. However, the aim was to strike a balance between software consistency and efficient service delivery. Each software setup iteration is a heavy specification, design, and validation process that takes several months to make sure that no safety issues are released to final customers. Fig. 3.5 (*NAVYA software update and validation process*) describes the software specification tasks and the test and validation process.

AVENUE operations commenced right from the project's inception, with both software and hardware versions already at an advanced stage. Over the course of the 4.5 years of the AVENUE project and due to the continuous evolution of hardware providers, several updates to both hardware and software became

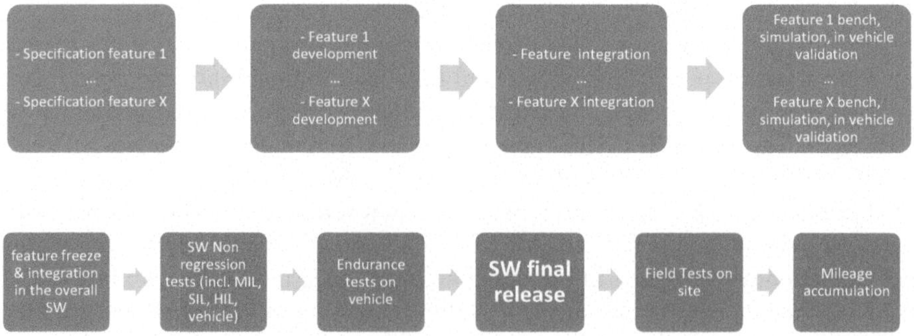


**Fig. 3.5** NAVYA software update and validation process

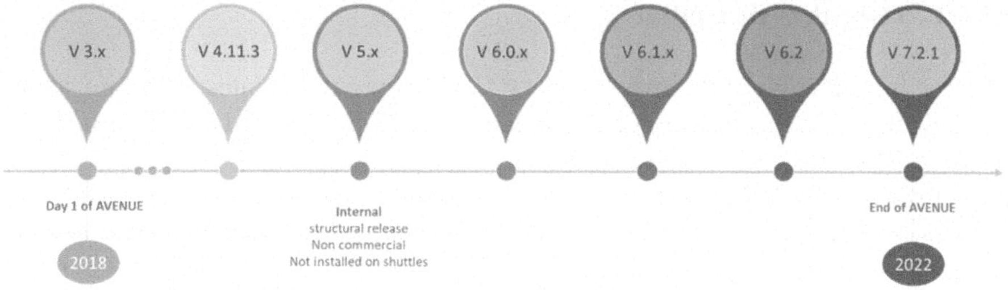


**Fig. 3.6** NAVYA software version setups

necessary. The AVENUE shuttles initially utilized version 3.X of the software and eventually progressed to the final release at version 7.X, reflecting the dynamic nature of the technology and the ongoing development efforts within NAVYA's research and development, driven by Agile development methodology. Above is a timeline of different major setups of NavyaDrive® during the AVENUE project (Fig. 3.6).

## *3.3.4 NavyaDrive® Evolutions*

### 3.3.4.1 The Operating System

In order to have a powerful and highly reactive driving system, NAVYA proceeded to refactoring the software architecture and migrate all from Windows to Linux operating system.

This migration was a major step to scale the NavyaDrive® system to higher level, bring more features, provide higher stability and performance, and structure the upcoming setups. A better homogeneity was also established with other connected hardware parts of shuttles like routers and other elements.

Enhancements at all levels were made to allow agility and resilience to cybersecurity attacks. The cybersecurity was not simple to manage when you are system dependent like on Windows. With Linux NAVYA did enhance the signature of PC images, the control of the access, the governance of the profiles management, and the cyphering and certificate management.

Several PTOs and authority assessors quickly noticed these changes by comparing not only the driving but also by comparing the cybersecurity audits done before and after this update.

This phase was difficult at several levels starting from the refactoring of the software till the installation on the ARMA shuttles. It must be done with the presence of a NAVYA engineer on site for each shuttle. It needs to travel to the site with the computing system, replace the previous ones with the new ones, update the software parameters depending on each site specificities, and run blank test to ensure non-regression, and only then can it go back to normal service.

#### 3.3.4.2 Over-the-Air Update

With the introduction of the over-the-air (OTA) feature in the latest software setup, the previously time-consuming and on-site installations for updates have become obsolete. This innovation enables software updates to be delivered through 3G or 4G networks, simplifying the process and enhancing safety compliance with EU regulations. Furthermore, OTA updates have accelerated the management capabilities of NAVYA's remote supervision center. It allows for remote diagnostics, investigation of events, retrieval of logs, assistance to onboard operators in various situations, vehicle maintenance diagnostics, and more, making the operation and maintenance of AV shuttles significantly more efficient and streamlined.

Now for each new setup, NAVYA can activate this feature remotely, transfer the new software, and check the integrity and health of software before putting back the AV shuttle in the hands of the PTO. It can additionally customize the display screens to include personalized elements such as sounds, animated images, icons, and more for PTOs.

#### 3.3.4.3 On-Demand Service

In AVENUE objectives, integrating on-demand/door-to-door system within the public transport service would be a game changer for urban mobility. This was a major technical challenge. A transformation of how the AV shuttle should conceive its route. The state of art of all AV shuttles at that time was to follow a virtual rail with a round trip. The shuttle starts in point A to finish at point D and then do the reverse from D to A. It should pass by bus station B and bus station C. This track is followed in a regular way according to a timetable, passing always by the same stops at the same time. Everything is done without any alterations or interaction with the passengers, except for stopping at the bus station.

In the AVENUE project, the on-demand system operates differently from traditional public transport services with fixed routes and timetables. Instead, any passenger can utilize a mobile application or an interactive bus stop to select any destination within a geofenced area. They can then wait until the AV shuttle arrives and pick them up. During the journey, the AV shuttle might make slight adjustments to its route to pick up another passenger heading in the same direction, even if they have a different final destination. This approach provides a more flexible and personalized transportation service.

This service requests several elements:

- An IT architecture where the IT system is connected to a fleet of vehicles covering the geofenced area
- An application, mobile, or interactive outdoor screen, connected to the system that a passenger can use request a travel
- A fleet management system to receive the travel orders, detect the nearest AV shuttle, and send the mission
- AV shuttles that can receive and operate the travel missions, sending back reel time information for tracking

NAVYA supported building the system architecture with technical partners and focused on developing the parts related to AV shuttle and external connectivity. Many steps were needed to build this flexible innovative system, described below.

**The driving route structure** was previously built with a fixed virtual rail (Fig. 3.7, fixed route/rail approach), called path, which was prepared by the commissioning team and with a dedicated map. The path is constructed with fixed stops. Once ready, the path is put inside the software of the AV shuttle. The driving system follows this path, guiding the shuttle accordingly. It executes the relevant driving instructions while also considering all new dynamic and static surrounding elements, such as pedestrians, traffic conditions, and other vehicles. If the shuttle gets out of its path, the embedded security system of the NavyaDrive® will stop the shuttle. So if there is any change to hail another passenger out of the path, it will immediately stop the shuttle. The idea of changing dynamically the defined path during the operation was at the time inconceivable (Fig. 3.8).

**The structure of the driving system** was refactored to take into account step by step these changes. The first step was to allow the onboard operator to define the pickup/destination points of the travel within a fixed path by using the onboard tactical screen, dashboard user interface (DUI).

**Fig. 3.7** Fixed route/rail approach (Navya©)

**Fig. 3.8** Fixed route as seen by the vehicle (Navya©)

**A mission management system** was built to prepare and control the way the driving system should treat pickup/destination requests. Tracking the mission is taken in account to follow, send information, and acknowledge the status of the missions. A mission can be modified, canceled, or replaced with a new one by the system. However, the system does not permit changes to the current driving if they introduce any safety risks. Additional features were added to allow the management of the events for the ride. **The event management system** is described later. It allows to receive missions with additional actions like opening the door and rolling the ramp once arrived. Such flexibility will enhance the service to be used easily by disabled passenger or elderly.

The next step in the development process involved creating a comprehensive **off-board system**, which was hosted on a cloud-based platform. This system is designed to receive mission instructions from external sources, verify them, and then transmit them to the shuttles. This ensures better control over external connections and prevents any compromise to driving safety. It will facilitate the creation of new functionalities for end users from external systems. Therefore, all security, cybersecurity features, access control, and stack management should be done to prevent hacking the driving system.

A secured API was created for the partners so they can connect from external IT system. The connectivity structure involves third parties interfacing with the NAVYA cloud, which then acts as an intermediary to communicate with the AV shuttles. This configuration ensures that direct connections from third parties to the shuttles are not permitted, enhancing security and control over the communication channels.

The software architecture described above enables partners to efficiently manage the fleet of shuttles on one side and oversee passenger applications on the other. This dual functionality streamlines the operation and coordination of the AV shuttle service, providing control over both the vehicles and the passenger experience. This step allowed the **first usage of on-demand service** in a round trip and provided a first connectivity between the passenger application and the shuttle connected to the fleet management system for the AVENUE sites.

To extend the coverage across the entire geofenced area, enabling the shuttle to deviate from round trips and pick up passengers anywhere within this zone, it is essential to have the capability to dynamically construct the driving path. This necessitates an adaptation in NavyaDrive®'s approach, shifting from predefined static paths to the construction of dynamic routes in real time. A representative site can serve as a testing ground to develop and refine this new path structure, allowing for greater flexibility in shuttle operations.

**The site of Belle Idée in Switzerland** (Fig. 3.9, the Belle-Idee site in Geneva) is the best use case to do so. It covers 38 acres, 9.2 km of roads, 24 buildings, 6 parking, and several points of interest, served by an extensive road mesh.

The ambition was that a passenger can request a pickup on his mobile app from any place to any destination within this area.

**The whole area was mapped**, and a clean and structural HD map was built for this purpose. NAVYA has always been innovative and active in developing its own

**Fig. 3.9** The Belle-Idee site in Geneva

technology of mapping and building a bench of advanced tools to do so. AVENUE helped enhancing such tools that allowed better knowledge and perception of the environment and brought a higher-level of localization.

Instead of building a path on a round trip, all the roads of the area were covered with a network of paths with different specificities as shown in Fig. 3.10 (*Belle-idee routes and virtual stops*).

**The NavyaDrive® navigation** system underwent a significant transformation. Instead of relying on a single static path to follow, it now possesses the capability to dynamically construct the path by assembling pieces of pre-mapped and prepared routes. This approach ensures that the navigation system can adapt and construct the optimal route for driving by piecing together predefined path segments as needed.

**For the door to-door service**, instead of fixed stops, virtual stops were used. Almost 73 virtual stops (Fig. 3.10) were decided with the PTO to allow safe boarding and dropping of the passengers in the most probable points of interest that are usually used. No physical bus stop was built. When a user selects a pickoff, there is always a virtual stop next to his/her position, and he/she was advised to go to this spot for safety issues.

The mission management system also was enhanced to allow:

- Add/cancel missions dynamically
- Dynamic rerouting
- Refuse any mission that could lead to dangerous maneuvers like changing at the last minute and then turning left and right when already engaged
- Status and event management

**A specific driving "Mode" in NavyaDrive®** system was built to allow AVENUE partners to connect and start engaging mission sending for the on-demand service.

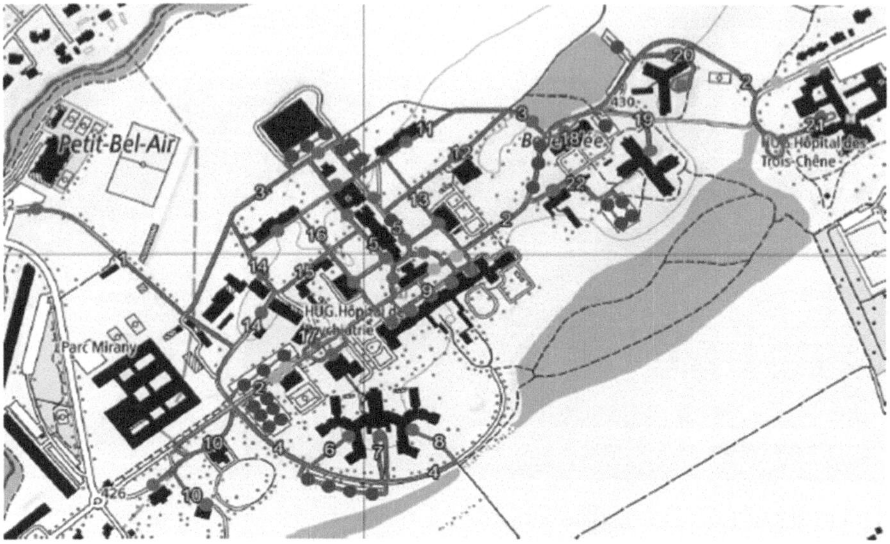


Fig. 3.10 Belle-Idee routes and virtual stops

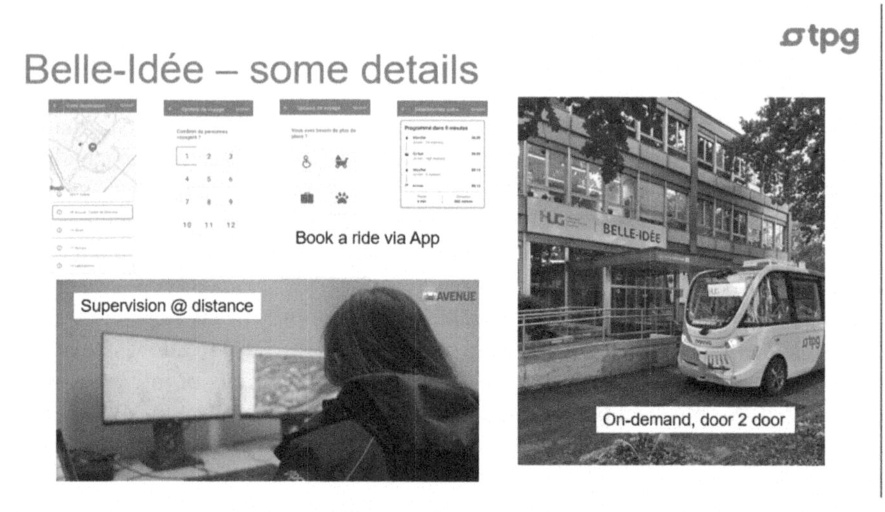


Fig. 3.11 The TPG control center, passenger app, and vehicle on the site

The system was first tested, trained, and fine-tuned by starting the operations without passengers on board. Connection to local supervision center was installed. Once validated, it was open to all passengers (Fig. 3.11).

The Belle-Idée site was at the forefront of implementing the on-demand system and incorporating all the latest features with the most recent setup, marking a significant milestone during the final year of the AVENUE project. Other sites activated the on-demand service and put it in operation, each differently, and with different application or fleet management provider. All were using the same innovation brought by NAVYA, and all demonstrated that such integration is a game changer for public transport. In Lyon, the deployment site was at Parc Olympic de Lyon (Parc OL) (Fig. 3.12, Lyon Parc OL deployment site) and in Copenhagen at the Slagelse site (Fig. 3.13, Copenhagen Slagelse site).

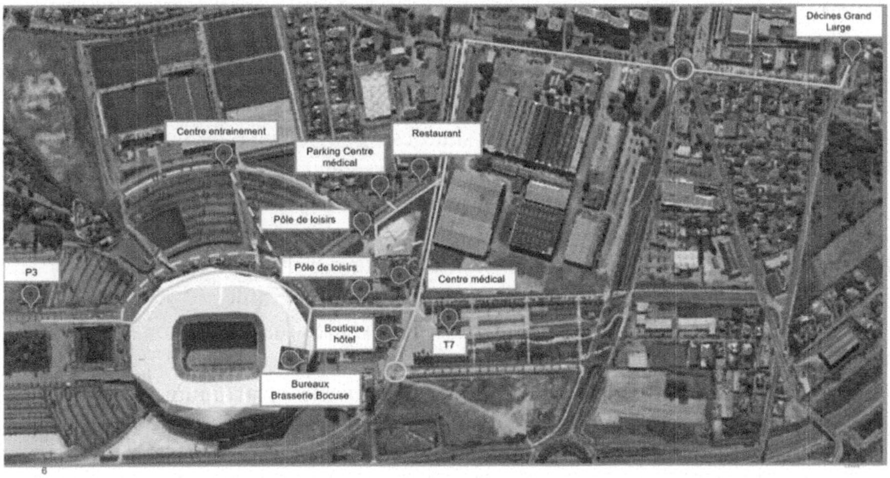


**Fig. 3.12** Lyon Parc OL deployment site (Keolis ©)

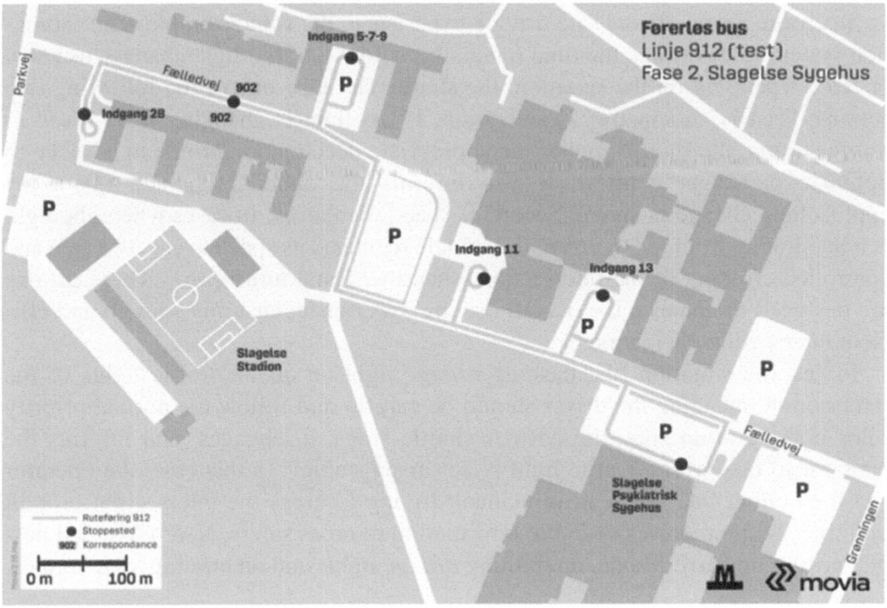


**Fig. 3.13** Copenhagen Slagelse site (Holo ©)

#### 3.3.4.4 V2X Traffic Light Management

The AVENUE project played a pivotal role in enhancing the driving experience by integrating vehicle-to-everything (V2X) technology. V2X technology was employed consistently, particularly at traffic light intersections, and occasionally in managing challenging turn situations. This technological integration improved overall traffic flow and maneuverability. As a reminder, in Europe traffic light goes from *green* to *orange* to *red* where only *green* allows to go on. In Europe the *orange* light appears just before the *red* light. Decisions are sometimes not easy to take when the vehicle is about to pass a *green* light and at the last minute the light changes to *orange*. Even normal drivers brakes aggressively when it happens, without stating those who accelerate instead. Previously NAVYA shuttles impulse harsh braking in such situations leading to uncomfortable situation for passenger and cars behind.

**NAVYA built a decision diagram** handling every phase of the traffic light, *green* to *orange* to *red* to *green*, taking into account the *time to change* to the next light, the traffic conditions, and the shuttle speed. All these elements together are considered to handle the right deceleration/acceleration profiles.

NAVYA's experience with V2X technology revealed that errors could occur, and at times, V2X might not transmit the correct signals. Although such instances were rare, they were a reality. To address the critical issue of avoiding running a red light, NAVYA developed a redundant solution that leverages camera-based recognition to assess the status of traffic lights in various lighting conditions. This approach is built on artificial intelligence models and was tested during the AVENUE project but subsequently integrated into production to enhance safety and reliability.

To ensure a smooth and safe driving experience, the vehicle employs an estimation system to determine the time remaining before a traffic light changes to *green* or *red*. Depending on the situation, the shuttle can take one of the following two actions: Frist is anticipatory deceleration. If the light is changing from *green* to *orange* to *red*, the shuttle applies the appropriate deceleration profile in advance to optimize passenger comfort. It also informs the onboard operator about the approaching change in signal. Second is speed regulation. In cases where the light is transitioning from *red* to *green*, the vehicle regulates its speed to ensure a safe and controlled crossing of the intersection when the signal turns *green*. These actions are designed to enhance the overall driving experience, ensuring smooth and efficient intersection navigation.

In specific situation like flashing *orange* light on all the traffic lights of the intersection, meaning the driver should be careful and follow the normal priority rules if there is no sign, the ARMA shuttle stops, disengages, and informs the operator to take over: "traffic light is not manageable." In this case, the operator can then either take the lead in manual driving to cross the intersection or wait until the shuttle receives readable information to cross safely. NAVYA added new diagnostics and errors code, on flashing *orange* lights and on missing V2X signals.

#### 3.3.4.5 V2X Solution for Complex Situations

The site in France, near Lyon, NAVYA and the PTO (Keolis) faced two major driving difficulties. First the shuttle must go through an important roundabout where more than 40,000 vehicles drive in a day. And second the shuttle must do an important turn left where traffic is difficult. The solution facing these two situations was to use an interactive V2X system with external partners.

For the roundabout, active traffic lights with V2X are installed on each of its entry. Before arriving to the roundabout, the ARMA shuttle sends information to the traffic light controller and requests the priority on the traffic. The first traffic light in front of the shuttles will take control and will allow the entry of the shuttle in the roundabout. The other traffic lights on the roundabout will turn *red* progressively one after the other on each entry of the roundabout and will block other vehicles from getting in the path of the shuttle. Once the shuttle passes through, the traffic lights behind get back to *green* to unleash the traffic allowing other vehicles to get in. In this way the shuttle will have a fluid traffic and will ensure a safe and smooth ride.

As for the left turn difficulty, the ARMA shuttle will send information to the traffic light controller when it is near the left turn. The shuttle will take the left lane, and the traffic light will turn to *green* smoothly and to *red* for the vehicles on the other side of the road. Therefore, the road will be clear, and the shuttle can engage safely the left turn without stopping. This V2X management was the world's first demonstration of how infrastructure and automated shuttles can actively interact to ensure secure traffic. Major developments were made on the NavyaDrive® software and the low level stack of V2X to ensure compatibility and non-regression.

#### 3.3.4.6 Driving Enhancement

Safe and smooth driving is the essence of driving activity, especially when it comes to automated driving. Many elements do interact and are associated to deliver such results including all hardware and software enhancements. As there are many done during AVENUE, it is difficult to go through all of them here, but we can get a peak on some.

**The acceleration and the deceleration profiles** are the major ones. Sudden and aggressive braking or acceleration can lead to accidents, cause discomfort for passengers due to jerky movements, and also lead to the disapproval from other road users and vehicles. As a result, the service can be stopped by the PTO if it does not deliver customer satisfaction at all levels, including comfort. The acceleration and deceleration profiles of the NavyaDrive® were enhanced from a setup to another to provide smooth driving. The purpose of this algorithm was to ensure the right trade-off between safety and comfort for both passengers inside and other road users outside. Acceleration was boosted upon departure, and intersections were efficiently navigated with increased acceleration. As for the deceleration and braking, they are softer, and the harsh brakes and jerks were minimized. The braking distances were modified, depending on the deceleration profile criteria.

"Adaptive cruise control" was also added to the NavyaDrive®. The ADS detects and evaluates the speed of the vehicle in front and adapts its acceleration and deceleration profiles accordingly. Adjustment of the speed of the shuttle on the speed of the vehicle in front was done automatically. This prevents the shuttle to always deploy different acceleration and deceleration profile or to do several braking. This is also valuable in traffic jams, a kind of traffic jam feature for ADAS.

**Obstacle tracking system** allows to anticipate the obstacle trajectory and to anticipate the entry of an obstacle in the path of the ARMA shuttle. This results in better behavior in several situations like on intersection and passenger crossing. For example, when the shuttle stops at a passenger crossing, without a traffic light, the shuttle tracks the movement of the passenger. It detects when the passenger is out of its path and when the passenger trajectory is getting farther away. So when the passenger passes the shuttle front lane, the shuttle gets back in movement carefully even if the passenger is always on the crossing. The same thing happens when the shuttle is driving and detects a moving obstacle on the left that is drifting and comes in the alert zone in front or around the shuttle.

**Overtaking** is a major difficulty for automated driving. Overtaking in all situations, expected or unexpected ones, and at any speed is not yet delivered by any ADS shuttle system by the time this document is written. Nevertheless, in some situations several OEMs have incorporated it in ADAS.

NAVYA during the AVENUE project made enhancement on this topic, and some overtaking was achieved. Overtaking of small objects (Fig. 3.14), objects parked half on the pavement and half on the road, and not well parked vehicles, were possible to avoid to stop or disengage. A maximal lateral distance of possible overtaking was set to keep safety first.

**Steering control** was enhanced, and a new software was implemented to improve responsiveness of the steering.

**Driving alignment (trajectory tracking)**: Maintaining precise trajectory and adhering to correct lines or curves are a fundamental aspect of manual driving. In the realm of automated driving, this aspect becomes even more critical and requires constant monitoring. It is important to recognize that deviations from the intended path can occur not only due to vehicle performance but also as a result of road conditions, such as potholes, uneven road surfaces, curves, road grip, and overall stability.

A trajectory tracker is of a high importance to keep safe driving, and NAVYA enhanced the trajectory tracker within the AVENUE project. This module improved accuracy of trajectory with higher control of the jerk. The jerk became a major measurement element of the comfort of the passenger and the smooth driving. Curve management is easy to build in software, but when hardware is involved, a 100% alignment is not simple. The hardware reactivity, curve angle, and speed while driving can lead to some misalignment. All drivers notice this curve's lateral deviation upon inertia while driving.

**Fig. 3.14** Overtaking a static obstacle

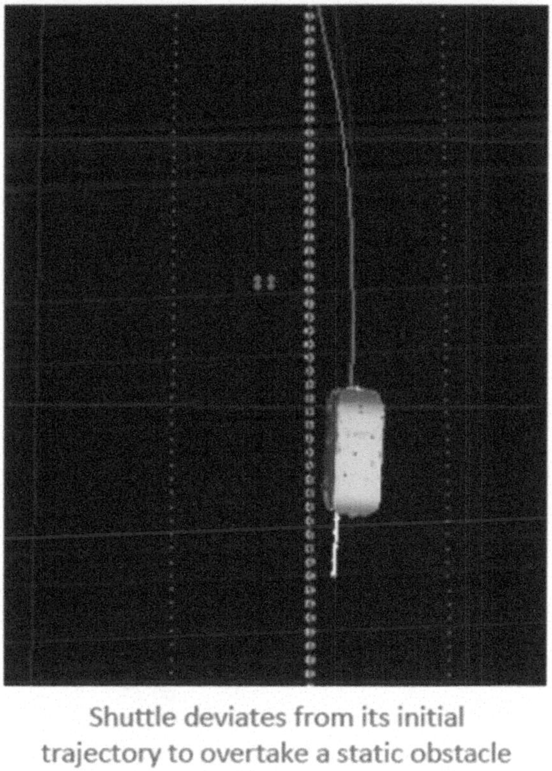


NAVYA developed a more precise trajectory tracking allowing a smooth driving on curves and better radius alignment on the road, reducing lateral deviation. This enhanced the safety in curves and prevented incidents due to out of path alerts (Fig. 3.15).

**Driving alignment (line/lane check):** Transitions from "MANUAL DRIVING" mode to "AUTO DRIVING" are sometimes difficult to manage depending on its position toward the road. While starting a journey, the shuttle should be correctly aligned within its path, with a centimeter-wise precision. A human-machine interface (HMI) has been developed to assist the operator in this peculiar situation (Fig. 3.16).

The perception viewer (Fig. 3.17) provides the operator with the exact position of the vehicle: lateral alignment and angular alignment error (offset angle and centimeter gap with the path).

**Vehicle behavior improvements:** As per the legal framework in effect during the AVENUE project, an onboard operator had to be present inside each AV shuttle when it was on open roads. The onboard operator must keep hands on the steering controls and remain standing to maintain an unobstructed view of the road ahead. This requirement was in accordance with the Vienna Convention, which mandates that every driving vehicle must have a driver on board. These regulations were implemented to

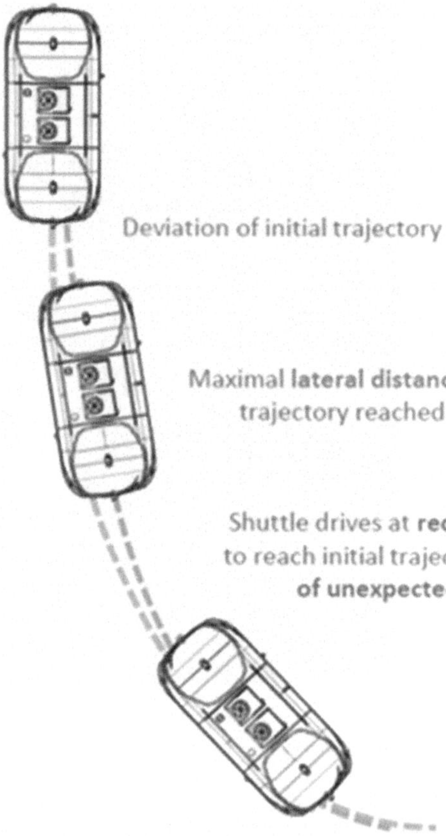


**Fig. 3.15** NAVYA trajectory tracker

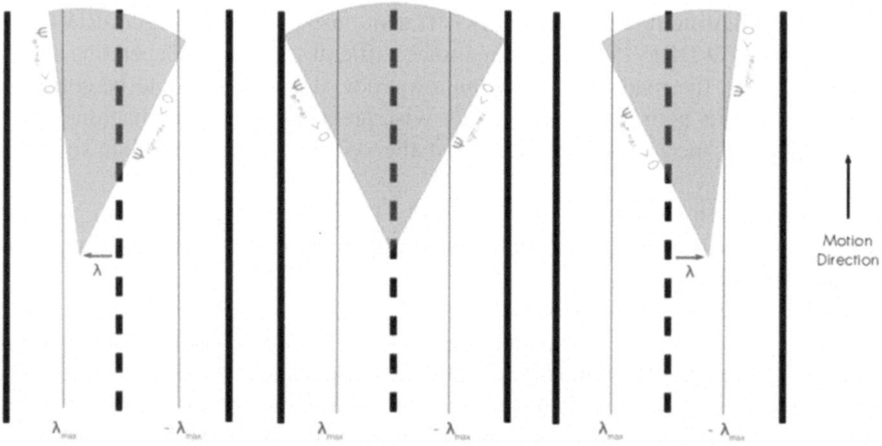


**Fig. 3.16** Median detection for confident departure feature

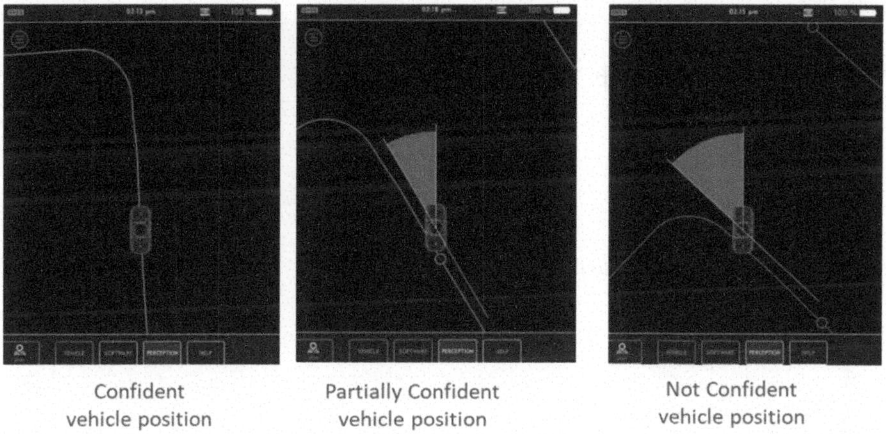


**Fig. 3.17** Perception viewer

ensure safety and align with established international driving standards. Until the change of this convention, which was done in 2022 as described in the chapter "Governance Impact Assessment" from Lionel Binz later in this book, in AV shuttle the driver is the onboard operator even if he does not do the driving operations.

The shuttle occasionally needs to disengage the ADS and allow the onboard safety operator to take over manual driving. This is necessary for situations where the ADS faces technological constraints, such as overtaking or reversing, or when the AV shuttle must exit the parking area and be positioned at the starting point of its automated route. Similarly, at the end of service, the shuttle may need to be manually driven back to the parking area or to a maintenance center. These maneuvers are to be done manually by the onboard operator with the steering controls. During manual driving, the ADS software and safety control system are deactivated. The onboard operator controls totally all the driving. Most of the time, incidents and minor accidents occur during such situations

Based on the requests of several PTOs, a new driving mode was introduced to allow the onboard operator to drive safely the AV shuttle. The assisted manual driving (AMD) feature assists the operator during all manual driving. In a later iteration activation became optional, but once activated, speed limitation, obstacle detection, and Autonomous Emergency Braking were enabled. With this mode also, a buzzer inside the vehicle alerts the onboard operator and the passengers on the presence of an obstacle (Fig. 3.18).

In manual driving mode, certain onboard operators provided feedback regarding the risk of passengers falling within the shuttle due to the detection of nearby obstacles, prompting the shuttle to initiate abrupt stops. As a result, NAVYA ceased to restrict the vehicle's speed in AMD mode, and braking procedures were adjusted accordingly. The decision to stop or continue driving remains with the onboard operator. Additionally, a new tone signal alerts the onboard operator to nearby obstacles, with frequency indicating proximity: the more frequent the tone, the closer the obstacle.

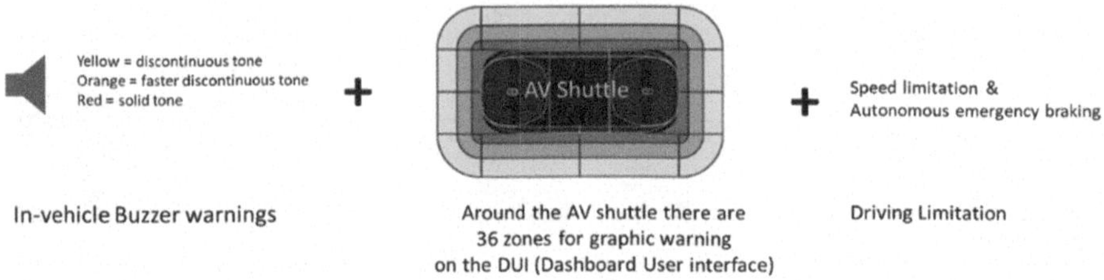


**Fig. 3.18** Information process to the onboard operator

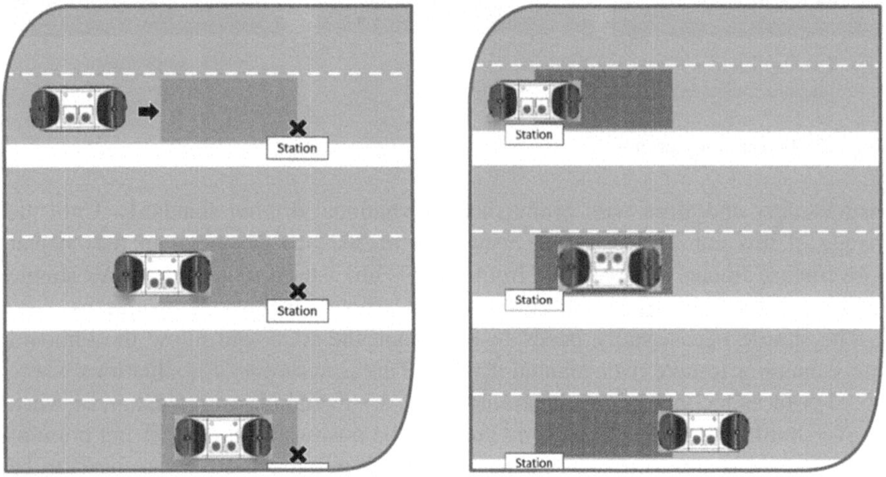


**Fig. 3.19** NAVYA blinking management

- Obstacle in the "close" zone = continuous tone signal
- Obstacle in the "middle" zone = very regular tone signal
- Obstacle in the "far" zone = irregular tone signal

**Blinker management:** A more operational blinker management was set up to allow better interaction with other vehicles and road users, especially during departure and arrival to stations. They stay activated on stations. Such feedback from the onboard operators and from AVENUE users allowed NAVYA to enhance such elements that cannot be easily seen in development or test and validation phases (Fig. 3.19).

**Blackbox improvements:** ARMA shuttles have been recording all their activities for a long time. Still enhancements were made within AVENUE by reducing significantly the recording time of the data and adding more data to allow a clearer view and to replay certain situations afterward. Reducing time of recording reduces the use of the CPU power and keeps the compute power to the NavyaDrive® system.

### 3.3.5 Supervision Improvements and NavyaOperate©

Automated public transport services, like automated metro, are connected to a control tower and a supervision center. All the vehicles are connected, facilitating the exchange of various types of information with the supervision center:

- Telematic information sent from the vehicles to the supervision center. This is the upstream where all details of the shuttles are viewed on a dashboard. It includes video stream inside and outside the vehicles, the position, speed, events, bus stops, unexpected stops, alerts, etc. All these information are needed to operate a daily service and are provided to the PTO either via the NavyaOperate® system or via a partner API. It allows access to the shuttle states and events, current and passed, with time stamps and associated status codes.
- Technical information exchange in both upstream and downstream. This level of exchange is under the control of NAVYA supervision and maintenance center. It allows to access the NavyaDrive® software of the shuttle, only under maintenance access and in stopped situation. The supervision center can get logs, diagnosis of the vehicle's vital information, and hardware details. It also allows the maintenance and the supervision teams to update over-the-air the software of the shuttles with a downstream channel.
- Mission management. This two-way stream is the one developed for the AVENUE on-demand system that allows a third-party connection to the shuttles and to send missions or event triggering. The upstream provides acknowledgment and tracking of the missions.

Numerous enhancements where done during the AVENUE project to allow a complete integration of the shuttles in the AVENUE technical architecture and to provide the complete on-demand service.

The AVENUE project emphasized the importance of having a richer dataset to facilitate more straightforward analysis. This dataset encompassed various components, including historical diagrams, diagnostic information, detailed event logs; highly technical data such as velocity, obstacle events, as well as comprehensive descriptions; and maintenance-related details like temperature and heat data. Additionally, it involved real-time streaming, complete mission information, and the ability to remotely trigger various actions such as honking, buzzing, controlling lights, stops, doors, and ramps, and monitoring interior vehicle information. The dataset also encompassed GNSS signals, hit ratios, and other signals, together with detailed site-specific information and synchronization with onboard data like bus stops, stop names, and more. They were made available, making easier the analysis. Some of this information was also put in *the record management system*.

There were three supervision modes developed:

- "PARTNERS" supervision mode: this mode is reserved only for PTOs that have access to the Navya API. PTO can send missions to the ARMA shuttle.

- "STAND-ALONE" supervision mode: through this feature, some other supervision solutions can send commands to the vehicle: STOP, GO, and OPEN AND CLOSE DOORS.
- "METRO" supervision mode: the supervision can now ask the vehicle to leave a station remotely in "METRO" mode.

### 3.3.6 Navya API

The connectivity between the shuttles and remote applications of third parties is due to the standardized public API put in place and enhanced during AVENUE. It is a major outcome of AVENUE. The architecture developed goes through the NAVYA cloud system before connecting to the shuttles as described earlier. This is not only for security reasons but also for functional and technical performance across all sites, all connected partners, and all vehicles in operation or not, avoiding to impede the main task of the shuttle, which is driving smoothly and safely. No partner can connect directly to a shuttle. All must pass through the NAVYA cloud system.

The API evolved a lot during the AVENUE project, and the connectivity protocols were simplified and standardized across the 4.5 years of the project. At the beginning only the partner Bestmile was able to connect, send, and receive information with a dedicated closed protocol. To add a new connection, it would take several months with this architecture. NAVYA changed the protocol to RESTful HTTP API with WebSockets, and with this, any new partner was able to connect its application within days. That is how three new AVENUE partners were able to quickly connect their fleet management system (FMS) and mobile application applications at all sites.

Also a training and support program was put in place to allow the new partners to onboard quickly. Before any connection to the real AV fleet, a simulation program with the digital twin site was developed and installed to train the FMS and to be sure that all sending and receiving are working. Once the FMS is working correctly with the simulator, a real testing phase with people on site is applied to be sure of real integration, and continuity is done between simulation and the real site.

With the AVENUE project, this public API got through different versions bringing many features like:

- Change of the exchange protocol to have a standard secured HTTP and a new architecture of WebSocket and associated low-level services
- Security and certificate management
- Access per vehicle per several vehicles per site
- Access mode management
- Enrich data that should be collected by the NavyaDrive® and then shared to the NAVYA cloud
- Video live streaming
- In-vehicle monitoring and information sharing
- Share event data

- Share control non-driving action on the shuttle (buzz, light, doors, ramp, sound, screens, etc.)
- Share diagnosis data (engine temperature, GNSS signals, lights and blinkers, batterie status, etc.)
- Mission management create, cancel, change, and track.
- Event triggering with mission management

Opening a new feature in the public API is more than just providing data through the connectivity and putting the API interface on the server. For each and every new feature or data delivered, development is needed on the NavyaDrive® software in the shuttle to collect, clean, secure, and send the data. And for each development on the NavyaDrive®, a new setup must be made, and the full test and validation process must be applied to be sure that there is no regression and no impact on the safety in operation.

The AVENUE project fed NAVYA with new features and ideas to enhance the API for fleet management systems, the supervision, and the shepherding features.

## 3.3.7 HMI and Experience Enhancement

### 3.3.7.1 Operator User Interface

Inside each shuttle, a screen is installed to replace the vehicle dashboard of a standard vehicle. This screen must respect legal regulations in place like for normal vehicle. and several mandatory elements are to be followed. This screen is called the dashboard user interface (DUI) (Fig. 3.20).

Several functions are to be support by the DUI:

- Provide and display of mandatory visual and control elements to the safety driver
- Provide tools to the safety driver to perceive, analyze, ad control the shuttle

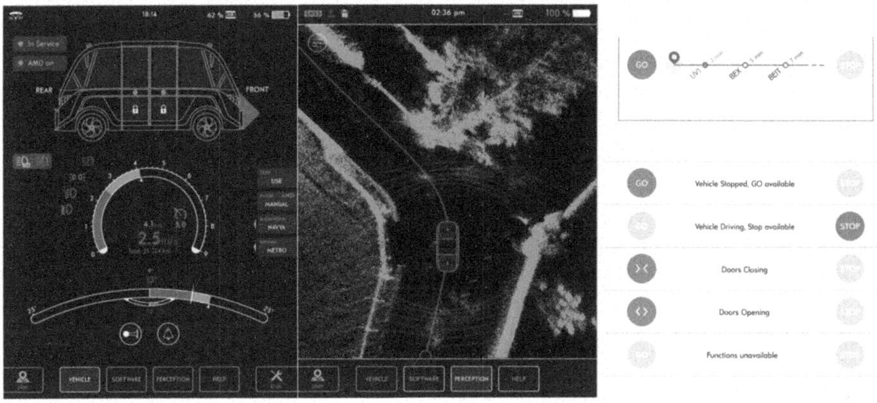


Fig. 3.20 NAVYA© DUI interface

- Provide passengers tools to define which bus stops are requested and to follow the status of the trip

As described earlier, in some cases the onboard operator needs to manually drive the shuttle. The DUI is also to help him in these maneuvers and to provide a clear vision. It helps to show the exact perception view as the NavyaDrive® is seeing it. Many advancements were done on the DUI based on the feedbacks provided by the PTOs of AVENUE during the operations. These AVENUE outcomes were important elements that could not be seen without feedback from the ground, from the people handling the real daily operation of the shuttles.

Below are some evolutions made on the DUI (example in Fig. 3.25):

- Authentication process.
- System status overview.
- Service mode control.
- Support-assisted manual driving mode.
- Diagnosis and analysis.
- Track and acknowledge any system event.
- Display help page.
- Alignment with regulation related to vehicle dashboard.
- A clear perception viewer indicating how NavyaDrive® is perceiving the environment.
- Several elements were added to the perception viewer like: positioning of the environment and all objects seen by the vehicles sensors, obstacles that are blocking the vehicle advancement clearly highlighted when needed, and dynamic point visibility.
- Visual color enhancement to clearly indicate elements like lidar hit ratio and % level.
- Better positioning of some buttons used by the onboard operator like "Navigator," "States," "GO," and "STOP" buttons.
- Door states displayed inside the GO button.

Asked by the PTO and for the comfort of the safety operators, a night mode of the DUI was added (Fig. 3.26). Three main possible choices are available: day mode, automatic mode (display of the night mode according to the time of the sunrise and sunset in the geographical area), and night mode.

Adding to the above, a new event management system was developed within NavyaDrive® during the AVENUE project. This allows to track events for the supervision center. For a better management between the onboard operator on site and the supervision center, an event viewer was added to the DUI allowing him/her to better diagnose the vehicle status (Fig. 3.21). An interactive interface helps to deep dive in the events, and solutions are directly proposed delivering more independency of the onboard operator.

Software and display changes in the "EVENTS" section (Fig. 3.22):

1. Icon indicating the severity of the event.
2. A short description of the event is automatically proposed.
3. A padlock indicates the impact of the event.

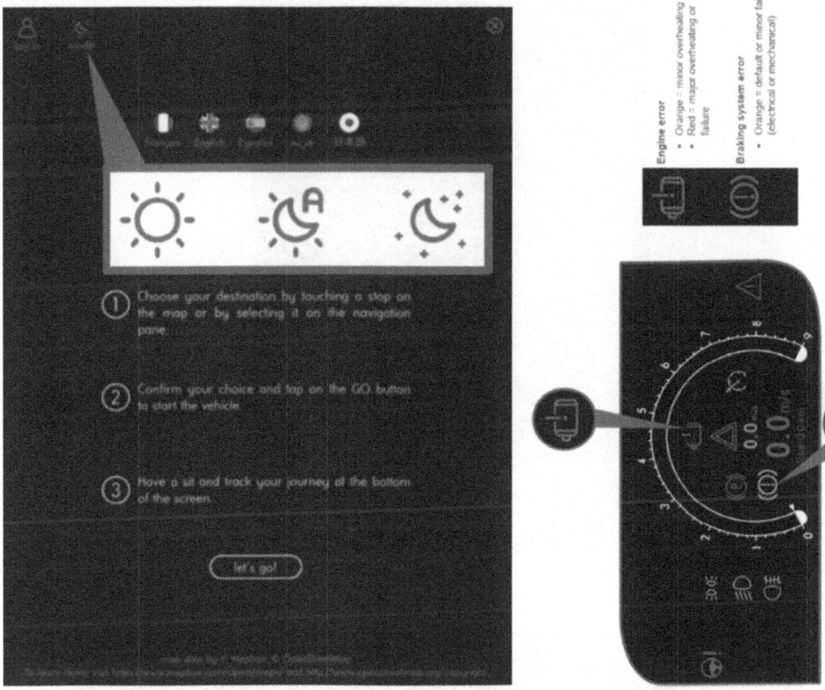

**Fig. 3.21** NAVTYA© DUI day mode selection and vehicle status

4. Detailed information by clicking on the event.
5. Procedure to be followed by the operator to solve the detected event.

#### 3.3.7.2 Event Triggering System

A real-time information with audio announcement was another outcome of AVENUE project. The AVENUE PTOs from different countries and with different languages are used to have inside their buses an information system for the passenger: visual indication of next stops, full itinerary display, audio announcements, etc.

NAVYA designed this software to have the capability to automatically initiate actions in response to event statuses. The event manager functions by alerting other systems when a particular event occurs, enabling those systems to execute corresponding actions. For instance, when the vehicle reaches a bus stop, the system can trigger the opening of doors and the deployment of the ramp. This real-time information display provides all the necessary passenger information, ensuring a smooth and well-informed experience.

Additionally, the system can also make adjustments to various supplementary elements within the vehicle and its interior. This includes modifications such as regulating ambient lighting, controlling low beams, making adjustments to the dashboard user interface (DUI), and deploying the ramp, among other features.

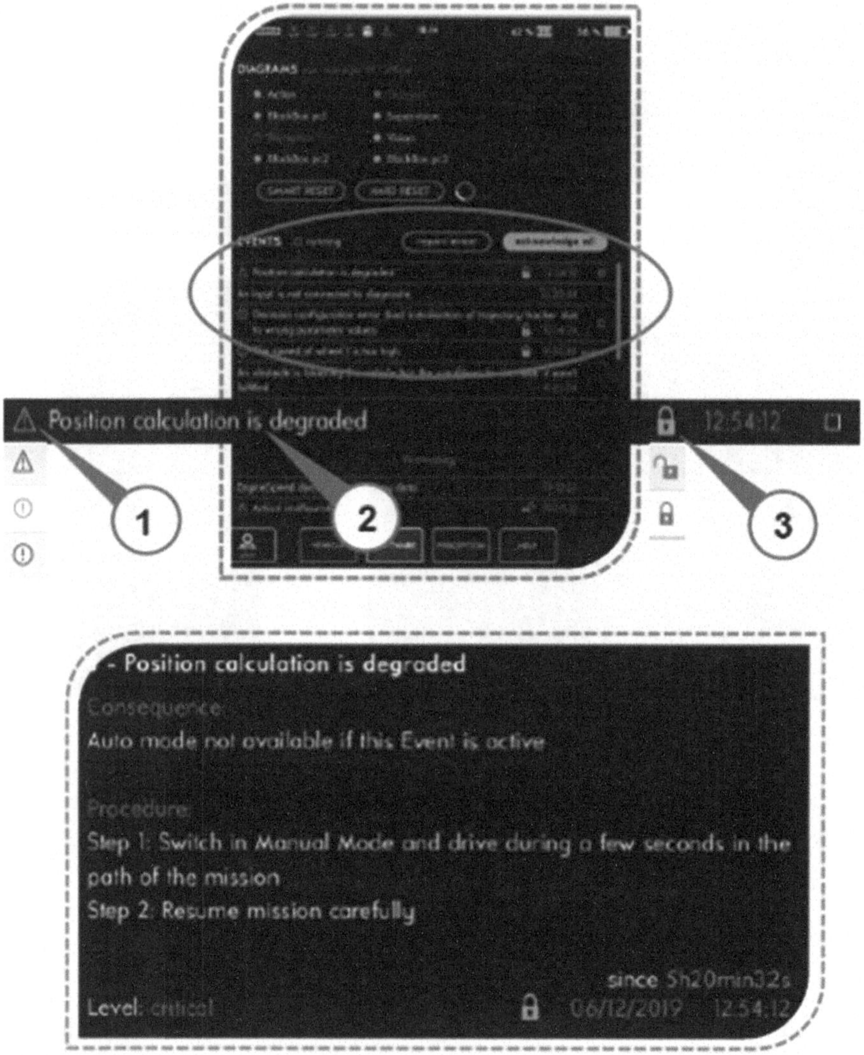


**Fig. 3.22** NAVYA© DUI event reporting

### 3.3.7.3 In-Vehicle Audio Announcements (UI)

The event triggering system is used also to automatically trigger audio announcement upon events, combining prerecorded messages as well as text-to-speech system. Audio announcements aim of course to assist disabled people. This can be done in multilanguage.

This system was first tested during the Dubai World Challenge for Self-Driving Transport for its first edition in 2019. Arabic and English languages were used to announce the bus stops and other information.

#### 3.3.7.4 Interactive Interface for Passengers (UI)

Based on the advancements made on the DUI stated previously for the onboard operator and the real-time information system, the internal screen was enhanced for the passengers. It was designed to assist users along their journey:

- Visualize the top-view map of the operation site
- Visualize the real-time position of the vehicle
- Visualize the list of available stations
- Visualize the current itinerary
- Visualize the estimated time of arrival
- Select a desired station when on-demand service is enabled
- Select the desired interface language
- Trigger the departure/stop of the vehicle
- Display help page

In AVENUE, all shuttles were ARMA shuttles, and all visual enhancements stated were done for DUI screen that is used for the onboard operator and the passengers. One outcome of AVENUE is to add to the next generation of NAVYA shuttles, called EVO, and a screen dedicated to passenger information. EVO will have a large lateral display facing the entry of the shuttle above the lateral window. This display will be used for all internal and real-time information. Such system was conceived, produced, and tested in a prototype during the Dubai World Challenge stated above. All visual and sound information were displayed and triggered in Arabic and in English.

#### 3.3.7.5 External Sound (UI)

Usually, NAVYA shuttles are operating in mixed traffic in cities and peri-urban areas where other vehicles are driving. Horns, sounds, and other external elements are of normal usage for these environments.

In AVENUE, the site of Belle-Idée is a hospital. Three ARMA shuttles are covering the whole area of the hospital. Cars, buses, and pedestrians are encountered, but all respect this quiet environment. NAVYA worked with the PTO to get all recommendations and adjustments needed to adapt the shuttles accordingly to this quiet environment. The features that can, in a way or another, disturb or make noise were deactivated, like busing when an obstacle is detected. These adjustments can be turned on and off easily on site.

#### 3.3.7.6 External Screen and Human-Machine Interface (HMI)

The interaction with external road users is necessary. The ARMA shuttle can include two dedicated screens. Each displays information through the front and back window. The event management described above allows to trigger displays on these screens upon the perception of the environment or to interact with the road users. For example, the system can display animations when pedestrians are crossing ahead, indicating to them that the vehicle has detected them and is stopping. Alternatively, it can show information to vehicles behind, informing them of various actions of the ARMA shuttle vehicle, such as boarding or disembarking passengers, turning left or right, displaying a stop or yield sign in front of the shuttle, indicating proximity to a traffic light, signaling manual driving, or alerting to the detection of an obstacle just ahead, etc. (Fig. 3.23).

### *3.3.8 Other Enhancements*

#### 3.3.8.1 Hardware Enhancement

NAVYA is enhancing the hardware constantly through commercial projects, upon customer feedback, and of course the needs of the AVENUE sites. AVENUE brought added value to this process as within AVENUE acceleration was requested in areas that were not explored before. The shuttle itself, mechanical parts, brakes and steering box, ECU and electronics, electrical network, sensor architecture, the safety concept, low-level control, etc., all were in constant enhancement, and AVENUE contributed to the assessment of needs. Several versions of ARMA shuttles went through this constant enhancement for AVENUE PTOs. It is difficult to highlight hardware enhancement here. Sometimes, small hardware details block the authorization to deploy a service on open road, and hardware enhancements must be made.

One specific issue was raised for the Copenhagen deployments. One mandatory point was requested and hardware changes in the structure of the vehicle so that the vehicles should handle the safety of wheelchairs according to the Danish law. In order to fix the wheelchairs, Q-straint has been installed in the shuttles. Q-straint works simply by having four mounted points in the floor, with seatbelts that can be hooked to the wheelchair. The seatbelts retract automatically and lock when necessary. When a wheelchair has to board the shuttle, the onboard operator mounts the four seatbelt heads on the floor. This means that the floor is empty when riding without a wheelchair. When carrying a wheelchair user, the three foldable seats cannot be used, and there is room for eight additional passengers (Fig. 3.24).

**Fig. 3.23** Images and animated figures can be personalized for the PTO

**Fig. 3.24** Attaching a wheelchair (Holo©)

#### 3.3.8.2 Mapping, Commissioning, and Tools

A whole set of tools, not directly related to driving the shuttles or supervising the operation, are needed and contribute a lot to deliver and AV service. The different levels of complexity of AVENUE sites contributed a lot in the enhancements of all these tools.

**The MMS tool**, related to the acquisition of the lidar cloud point and all environmental data of the road, was enhanced and got higher performance of lidars, sensors, and localization parts. A dedicated hardware and software tool was built to easily operate the system on site by the commissioning engineers on site.

**HD mapping** was developed to provide a higher level of precision in the perception of the environment and its knowledge. It brought higher localization precision while driving, and a notable enhancement was seen once the sites were mapped in HD mapping and updated. This enhancement is not only related to the acquisition of data. A full set of mapping tools and features were built by NAVYA engineers to manage this map, to rebuild this digital twin, to clean it, and to identify structural elements like driving zone, the road, etc.

Related to the commissioning, a dedicated **path construction tool** was built and enhanced to get the most of the HD map and to build the path with the easiest and the fastest way. Commissioning engineers can then quickly put a site in operation.

Also **diagnosis tools** accelerated the way the maintenance people can understand, correct, and put back shuttles in operation.

All these tools are fundamental to the AV ecosystem to deploy mobility service.

#### 3.3.8.3 Additional Tool Enhancements

A complete set of additional tools was required for other tasks, such as creating the path and shuttle maintenance. Having more than 13 shuttles in the project across 6 cities greatly contributed to the enhancement of these tools.

**Enhanced perception viewer** was developed for commission engineers and maintenance team to help them in their task. It provides a 360° view of the surroundings, the same view as the shuttle sees it. Such feature reduced the time to detect and solve issues from within the shuttle. Several information were delivered like the lidar map, perception, wheel angles, obstacles display in different angles, etc.

It also allowed to calibrate sensors with a specific process to follow and be sure that sensors are perfectly aligned (Fig. 3.25).

**Event registry,** as stated previously, is logged, aggregated, and managed through this software feature.

**Log management** is made simpler to allow the supervision center to isolate and retrieve information faster. The supervision center and R&D had also easier tools to analyze the different diagrams retrieved.

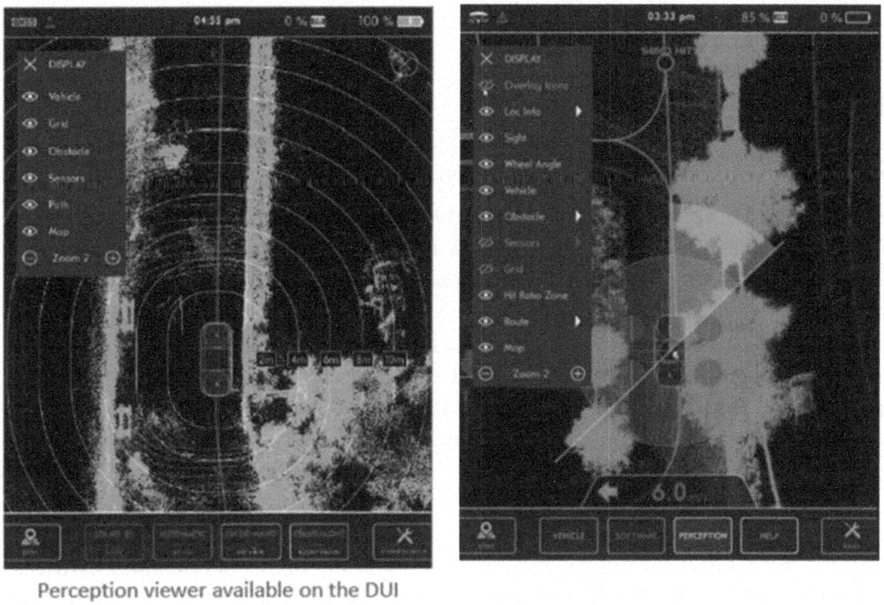


**Fig. 3.25** NAVYA© DUI 360 perception

**Fig. 3.26** NAVYA© booting status

**Deploy, sync, and store the configuration files** for faster commissioning and maintenance updates.

**Tool mode was created** for maintenance and deployment training purposes, with richer information (site name, vehicle technical details, hardware id, temperature of elements, etc.).

**NavyaDrive® boot** indication guides users through various startup stages, conducting autochecks and diagnostics, and displaying relevant information on the proper functioning of each step (Fig. 3.26).

## 3.4 Beyond Avenue

Developing profitable business models for automated shuttles requires substantial investment in research and development. Operational design domains (ODDs) vary, and with each step toward enhancing new features, there is a significant increase in the level of required resources and workload. Initially, many AVENUE public transport operators (PTOs) prioritized higher speeds, but as operational services were implemented, it became evident that speed was secondary to reliability and safety. Increasing speed affects all aspects of the software, sensors, and hardware components. It necessitates a more extensive capability to detect obstacles at greater distances, improve analysis and prediction, react more quickly, and ensure the reliability and safety of hardware at all levels.

Therefore, the hardware driving platform, an SAE L4 drive-by-wire system, is a crucial component. At present, no provider offers a fully homologated platform that closely aligns with automotive standards such as ISO 26262, and includes hardware redundancy features like brakes and steering. Additionally, this platform must incorporate a low-level security concept to ensure the vehicle can safely stop when the software system detects dangerous maneuvers or encounters issues that cause a loss of control.

Several universities, startups, and engineers demonstrated pieces of software that can control longitudinal and latitudinal maneuvers of a vehicle. However, all require a vehicle, and none can bring such software to market without having safe, homologated hardware. Producing such a vehicle cannot be done for very small series. It should be done in an industrial way and will require very high financial and resource investment that can be justified by high volumes (price/volume). All experts and OEM agree on this, and that is why, today, historical OEMs are working on incrementing features of the ADAS to go toward SAE L3 before getting to SAE L4.

Nevertheless, some are taking the challenge and are starting to invest in such drive-by-wire platform, and in CES 2022 some announced that they will have such vehicle by 2026. NAVYA has already started this process during the AVENUE project by building partnership with a bus OEM, Bluebus. This collaboration is made to build the first automated minbus SAE L4 ready and aligned with legal framework that is to be put in place. Since 2023 the hardware was out of Bluebus factory, and NAVYA connected the NavyaDrive® software in a few weeks and made first automated L4 tests without a driver on board. This L4 bus will officially hit the market in 2025 and will scale the automated vehicles for public transport to another level.

## 3.5 Conclusion

AVENUE was a game changer in the AV landscape in Europe and toward all the AV community. It changed the way the community and decision-makers are seeing the automated vehicles.

I remember when, in 2019, NAVYA launched the Autonom® Shuttle® in Brussels, right under the European Commission's headquarters, the Berlaymont. As the community experienced it, everyone saw the potential of this technology for public transport. It became clear that it is not just software, but an entire ecosystem capable of supporting the local socio-economic systems of EU countries.

This change paradigm from vehicle centric to ecosystem was needed, and the EU started to work on the legal framework. NAVYA was integrated in this legal work from the beginning, and the ADS act was released in 2022.

The work done in this area by the French government, where NAVYA was active since 2018, inspired and supported the ADS Act. The French government was the first to release in 2022 a full set of regulation on AV SAE L4, sowing the ecosystem dynamics. Just 1 month before, Germany released a part of its legal framework.

All are based on a double authorization as stated in other chapters: A type apporval of the automated vehicle to be put on the open road on one hand and an authorization of the AV service for a specific area with a specific ODD associated to the vehicle on the other. One delivered at a global level, the other at local level.

Before AVENUE all PTOs were waiting to receive their automated shuttles to put in place a service. But once they receive it, none was ready to put in place, and all stumbled on two elements: the legal authorization and how to manage the daily operations with their non-experienced team.

The role of public transport operators (PTOs) traditionally involves managing buses and drivers, overseeing driver competencies, handling client complaints, and maintaining schedules and frequencies. When traditional buses experience mechanical failures, their repair centers can manage these issues, as their personnel are trained for such maintenance. Similarly, transitioning from conventional fuel-powered buses to electric ones often requires minimal changes, as the drivers remain the same, with some necessary training.

However, the situation is markedly different when it comes to automated shuttles. Managing and maintaining automated shuttles present unique challenges that differ from traditional vehicles, necessitating specialized expertise and considerations.

Automated shuttles are digital objects, controlled by computers. At that time, they were experimental objects and were not industrialized with a high level of after-sales processes, local after-sales ecosystems in all countries, or trained repair centers. No driving competencies are needed to handle them but different expertise. Many public transport operators (PTOs) are not yet fully familiar with the digitalization of their profession and the related processes. Transitioning from traditional diesel-powered buses with fixed timetables to automated buses that don't require traditional driving skills poses a significant challenge. The move towards fully digitalized and connected buses, which operate without fixed timetables and instead offer an on-demand service, represents a disruptive shift.

In this context, when an automated shuttle encounters issues, the onboard operator from the PTO must, rather than calling his maintenance technician, engage in a digital process with the NAVYA supervision center. This process involves providing a detailed description and obtaining logs. Unlike conventional buses that rely on mechanical maintenance, solutions for automated shuttles are predominantly digital in nature. They may entail immediate digital troubleshooting or involve software updates that follow a dedicated process.

For instance, there were cases where certain onboard operators believed they could address incidents by attempting to manage or access the shuttle's computers. However, their efforts often resulted in exacerbating the issues rather than resolving them. In some instances, after attempting to independently rectify shuttle incidents, they resorted to manual driving to maneuver the shuttle. Yet, instead of moving forward, the bidirectional shuttle unexpectedly moved backward within the garage, ultimately colliding with a wall and causing damage to its lidars and sensors.

Also the authorities were not ready. They do not know what is this technological vehicle and in which legal "box" should they put it. They knew that it is game changer in the industry, but they wanted to take more time to understand it, to control all aspects of it to avoid any human or material damage, and to put all safety elements in a well-defined regulation.

The choice of the site, where the AV vehicle should drive, is an important part and role of the PTO. If the site's ODD is not fully aligned with the automated shuttles' capabilities, daily operations will struggle, and operational issues will start. It was seen in AVENUE as some of the sites that were decided on paper were not the most successful ones and brought many legal and daily complexity.

A shift in mindset was required among all partners, from public transport operators (PTO) to technical partners. AVENUE demonstrated that their "vehicle-centric" vision, solely focused on the level of driving autonomy of the shuttle, is not the optimal approach. Merely possessing an automated shuttle does not suffice to establish a safe and high-quality daily mobility service. The PTO must undertake several tasks: selecting suitable sites with the appropriate operational design domains (ODD) according to the shuttle capabilities, preparing and training staff, digitizing all systems, establishing connections with the supervision center, and developing operational processes. These processes, such as pre-operation checklists, procedures for starting and ending the service, incident management protocols, and passenger handling guidelines, are crucial components of a robust and sustainable AV service.

AVENUE allowed to put the focus on this ecosystem and all actions needed for that. A comprehensive vision was established, urging all partners to consider the "big picture" within urban planning and the public transport network, rather than focusing solely on AV technology. This encompassed questions on how AV shuttles should be integrated into Mobility as a Service (MaaS), identifying the external technologies required for integration, and evaluating in which use cases AV shuttles were more beneficial to citizens compared to traditional buses. AV shuttle should be, as any other transport vehicle, part of the whole mobility service of cities, bringing added value to all citizens.

Based on this knowledge and this "big picture" vision, the AVENUE team built a very ambitious follower project, ULTIMO project, to integrate AV in MaaS, in 3 major EU cities with at least 15 AV shuttles per site. This project of around 55 million Euros was the winning proposal of a major EU call and received funding from the European Commission in Horizon Europe program.

NAVYA significantly influenced all partners, the European community, cooperative connected and automated mobility (CCAM), competitors, researchers, and notably the European Commission. It demonstrated that even a smaller entity can make a difference and contribute to the establishment of an ecosystem that challenges industry giants. Many startups were created following NAVYA models. Big PTOs created dedicated departments, and today some actors of car giants created a strategic plan based on AV shuttles to produce in mass production around 2026.

We all hope that AV shuttles can become a regular part of our daily transportation, and every partner in AVENUE, along with NAVYA teams, hold strong confidence and take great pride in being part of this transformative disruption.

**Acknowledgements** This project has received funding by the European Union's Horizon 2020 research and innovation program under grant agreement No. 769033.

## References

Deichmann, J., Ebel, E., Heineke, K., Heuss, R., Kellner, M., & Stei-ner, F. (2023). Autonomous driving's future. Convenient and connected. McKinsey & Company. Available online at https://www.mckinsey.com/industries/automotive-and-assembly/our-insights/autonomous-drivings-future-convenient-and-connected. Accessed 30 Nov 2023.

Frost & Sullivan. (2018). Global Autonomous Driving Market Outlook, 2018. With assistance of Global Automotive & Transportation Research Team at Frost & Sullivan (K24A-18).

McKinsey & Company. (2019). The future of mobility is at our doorstep. Compendium 2019/2020. With assistance of McKinsey Center for Future Mobility.

Panetta, K. (2017). Top trends in the gartner hype cycle for emerging technologies 2017. Gartner. Available online at https://www.gartner.com/smarterwithgartner/top-trends-in-the-gartner-hype-cycle-for-emerging-technologies-2017. Accessed 30 Nov 2023.

SAE J3016. (2021). Taxonomy and definitions for terms related to driving automation systems for on-road motor vehicles. Available online at https://www.sae.org/blog/sae-j3016-update. Accessed 30 Nov 2023.

Said, C. (2018, March 20). Uber halts self-driving operations after car kills Arizona pedestrian. San Francisco Chronicle. Available online at https://www.sfchronicle.com/business/article/Uber-self-driving-car-strikes-kills-pedestrian-12764464.php. Accessed 30 Nov 2023.

# Chapter 4
# Safety, Security and Service Quality for Automated Minibuses: State of the Art, Technical Requirements and Data Privacy in Case of Incident

**Philippe Chrétien, Antonios Lalas, Konstantinos Votis, and Dimitrios Tzovaras**

**Abstract** The AVENUE project confirmed that the automated minibus technology is attractive to improve urban mobility services with contribution to the ecologic transition but at risk in terms of safety and service quality. Maintaining the public transportation level of injuries, based on heavy bus experience, is challenging and implies (i) an intermediate target for market introduction, gathering active safety and passive safety requirements, and (ii) an improvement process, based on anonymised and shared lesson learnt, under public power control. With resulting demands concerning reliability and maintainability, the safety improvement will contribute to the service quality and to the building of a sustainable business.

## 4.1 Introduction

Automated minibuses constitute an emerging technology that will transform public transportation in various ways. During the journey of this transformation from the conventional buses to the automated approach, a variety of safety, security and service quality needs are required to be addressed for achieving a successful transition. Sustainability and durability of the autonomous vision are also required for all stakeholders to engage acceptability. Traffic management, energy consumption, customer satisfaction are additional aspects have to be taken into account to accomplish sustainability. Mobility as a service (MaaS), which denotes mobility provided as a service by combining transportation services from several transportation

P. Chrétien (✉)
Centre Européen d'Etudes de Sécurité et d'Analyse des Risques, CEESAR, Nanterre, France
e-mail: philippe.chretien@ceesar.fr

A. Lalas · K. Votis · D. Tzovaras
Centre for Research and Technology Hellas, Information Technologies Institute, Thessaloniki, Greece

 
G. Fournier et al. (eds.), *Automated Vehicles as a Game Changer for Sustainable Mobility*, Contributions to Management Science,
https://doi.org/10.1007/978-3-031-61681-5_4

providers, is deemed critical to change the future of transportation by enabling more personalised transportation services. Data privacy and relevant regulations have to be taken into account to facilitate this procedure.

In this chapter, a shared sustainability and durability target for the society and the related companies is described, along with the conditions to achieve these objectives. Next, the critical path for market introduction of safe automated minibuses is presented, while the quality and safety state of the art for automated minibuses is discussed. A self-learning automated transport system at the European level is also revealed, whereas the data privacy of incident analysis and lesson learned sharing is explained. Finally, the automated minibus safety and service quality levers are summarised.

## 4.2 A Shared Sustainability and Durability Target for the Society and for Companies

With the different functionalities tested on the demonstration sites, the AVENUE project confirmed that automated minibus solution is attractive in terms of urban mobility services and now technically accessible, as presented in the previous chapter.

This is very important for the public power, as this technology can also contribute to the urgently required urban ecologic transition, if carefully integrated into the existing urban mobility systems, as a green and flexible complement to the conventional public transportation infrastructure. The best example of a conventional infrastructure is railway, which is adapted to massive transportation but not flexible: this introduction of automated minibuses is now a clear target for the European Union.

The following chapters will discuss the environmental and social impacts, which are positive when replacing private cars by such solutions, thanks to new attractive services: no doubt about that, but preliminary conditions have to be fulfilled to avoid any future regression.

This introduction of automated minibuses has to be organised carefully by the European Union. If not, the transition towards a more flexible and ecologic transportation system will not be sustainable in terms of societal acceptance, and the created business will not be durable for private actors, investing in this technology, which has to be financially amortised in the future.

## 4.3 The Conditions to Make it a Sustainable and Durable Solution

With these two goals in mind, the new automated minibus services will have to be not only attractive but also to confirm this good feeling with efficiency, reliability and safety, to establish year after year the customer satisfaction.

These new services will also have to get confidence from the public authorities, including other concerns like traffic management, energy consumption and health protection, with the associated mediatic and cost impacts.

### *4.3.1 Traffic Management and Energy Consumption*

On-demand and door-to-door services are crucial to get a flexible transportation service, which has to be inclusive for the elderly and persons with reduced mobility: these functionalities imply that automated minibuses will drive and park among other vehicles, including private cars, without any remarkable disturbance to the traffic of other vehicles, which does not mean any impact. If a conventional driver is autonomous in such situation, the vehicle has to be autonomous in the existing infrastructure and traffic: "a vehicle equipped with ADS aims to perform the entire dynamic driving task on a sustained basis within a defined operational design domain without driver involvement" (Standing General Order on Crash Reporting) (NHTSA, 2021, 2023).

In the future, the traffic performance will not be measured by the maximum speed available on a specific route. The average speed is more representative, but it will not be sufficient to measure the traffic performance: the target of traffic management will be to guarantee an acceptable travel time from A to B and so to limit the variability.

The responsibility associated to public transportation with automated driving will lead to a careful vehicle control, compared to private cars, with the hope of a quieter driving, without traffic jams, avoiding energy consumption and time wasting, with working loss, etc.

This quieter traffic could be slower, but the careful introduction of these automated minibuses has to avoid traffic disturbance and congestion, so the automated minibus will have to manage blocking situations, even when coming from driving rules' strict application.

A simple example is the necessity to cross the white line to pass a private car which is not correctly parked, which implies a human solution to be used: in such a case, the US strategy for Automated Driving (NHTSA, 2020) is a driver mimicry based on vehicle artificial intelligence, which will be sometimes necessary, maybe thanks to the quick arbitration of a distant operator.

### *4.3.2 "Customer" Durable Satisfaction, Including Safety*

The "customers" are service users but also other citizens, public transport authorities and more generally the different European states, with public concerns on emissions, energy economy, public health and cost.

Concerning the citizens who will pay in the future for this service (project demonstrators were free), they have explicit requirements in terms of comfort, punctuality, trustworthiness and resilience, which will have obviously to be satisfied.

In comparison, some other requirements are more difficult to manage as they are hidden, because they are implicit: it is generally quality and safety, with the associated conditions of technical reliability, cybersecurity, regulation compliance, insurance and juridical responsibility, which are behind quality and safety.

For these implicit requirements, the public power involvement is necessary: a roadmap has to be established from the realised demonstrations to the future durable customer-paid services.

To avoid future regressions, the roadmap has to anticipate the probable risks concerning traffic impact, service quality and road user safety, especially vulnerable users, and to organise the global improvement of these new services.

### *4.3.3 Safety Measurable Targets and Steps*

In addition to the fact that it is an implicit requirement, there is a second difficulty to manage safety, which is the difficulty to measure it: the fatality target has to be close to zero and not measurable, as it is clearly unacceptable that a city transportation vehicle kills a citizen!

The minimal requirement level is to maintain the global safety level of existing transport services, with coaches or buses (EU, 2023a). This fatality level is not easily measurable but can be estimated globally, with the help of massive data collected in terms of person travelled kilometres, which is the principal function of public transportation, but also of the traffic system, where comparison can be made between different transport modes, including individual vehicles.

In fact, this minimal requirement is already very ambitious, as buses and coaches are very heavy in comparison with the planned minibuses and with the existing vehicles, which is a mechanical advantage to protect passengers in case of an accident.

As a consequence, it can be discerned that the passive safety performance of buses and coaches is much higher than cars, even with lower passive safety technology.

The future automated minibuses will be smaller, as the absence of a paid driver allows us to reduce the number of passenger places, with benefit to transportation flexibility: the targeted weight for the urban automated minibuses is close to the weight of the coming electric SUVs, i.e. 2500 kg.

Based on that, two recommendations have been made to the European Community:

1. The safety measurement has to be done at a vehicle level and based on injuries compared to running hours, associated later to travelled kilometres. One condition is that public transport operators will be obliged to declare any accident and more generally any incident putting at risk safety or quality, based on a regulatory list to be defined (door blocked, vandalism, aggression, forgotten luggage, etc.).

2. A realistic target should be officialised for market introduction, which is $10^{-5}$ injury/hour, for example, less than 1 accident during 1000 running hours in terms of active safety performance and less than 1 injury among 100 accidents for passive safety. After the continuous improvement process, at the end of the roadmap, the target should be at the estimated level of buses and coaches: 1 passenger injury +1 vulnerable injury for one million hours: not easy!

Required in the USA as identical to conventional vehicles which is demanding, the passive safety level should be high, as citizens will consider they have to be protected from external vehicles when installed in the minibus, and as pedestrians will refuse that the minibus front face would be less protective, or more aggressive, than a conventional car.

Today, conventional cars are equipped with a front bonnet over the thermal engine, which is very useful in case of pedestrian crashes. Without this bonnet constraint, these front faces will be vertical, to limit the occupied space and maximise place number, with high risk of pedestrian projection and crossing, even at low speed.

At low speed, the projection is not far: the pedestrian can fall just in front of the running automated minibus, limited in terms of braking deceleration because of standing passengers inside the minibus, which is also called the "tramway or trolley dilemma" (Exeter, 2020).

Passive safety innovations will have to be financed and developed for door-to-door transportation in pedestrian streets, and new principles have been explored, as an inflatable structure integrated in the vertical front face, giving a better kinematic for pedestrian crash with energy absorption.

For years, researchers have been waiting for regulation demand and industry financing, to protect pedestrians but also cyclists and scooterists in such situations, against a front face, relatively high, which is already the case for a child in front of an SUV.

## 4.4 The Critical Path for Market Introduction of Safe Automated Minibuses

More than active safety systems, the passive safety requirements are structuring for the vehicle architecture and have to be introduced at the early beginning of development planning.

Two different usages will have to be dissociated, as one of them is compatible with a conventional architecture:

1. Downtown usage with derogative access to pedestrian streets requires vertical faces with a specific architecture to protect vulnerable street users and to facilitate vehicle stop and go with persons with reduced mobility (PRM) and standing passengers. To allow such vehicle developments (low production volume, limited investments), passive safety requirements have to be adapted: hopefully, in that case, performances can be limited (speed, acceleration, braking, etc.).

2. On the contrary, the liaison usage between suburbs and cities at higher speed (or the rural usages) must reuse industrial vehicle platforms, with conventional passive safety requirements. For technical protection in case of crash, the best battery position is under the cockpit, with impact to the platform height and to the quick access, especially for elderly or for parents of young children: this reuse will reduce the critical path, with applications available in the near future.

To develop correctly and safely the door-to-door services in the town centres, we will need a specific architecture, a low robotised platform adapted to elderly and PRM access, accepting standing passengers, represented by our H2020 AVENUE demonstrators, but they did not have to be homologated in terms of vehicle safety.

The critical path is longer for such application, which is the most important societal target: the first commercial application will probably use new electric robotised platforms, already under development for duty applications, but these vehicles will not be so optimised as H2020 AVENUE demonstrators in terms of access and vulnerable protection.

Their lesson learned will be useful for the later introduction of specific vehicles for downtown transportation, everywhere at low speed, respecting new passive safety requirements, to be urgently defined and integrated in the vehicle certification process.

## 4.5 Quality and Safety State of the Art for Automated Minibuses

The AVENUE experimentations were too short to measure anything in terms of quality and safety, and only attractivity measurement and characterisation were planned and done.

In any case, these experimentations showed that four preliminary conditions were not met, where the state of the art has to be improved before market introduction:

1. The technical reliability is not able to guarantee a quasi-nominal behaviour, which has to respect the ISO 26262 (ISO, 2018) automotive standard, which is available and manageable.
2. This level of reliability has to be obtained with an acceptable maintenance level, which is not the case today for sensor cleaning, as an example. Another risk to be taken into account is the consequence of light material accidents if sensors are not well protected.
3. The robustness towards meteorology and environment factor components is not sufficient, especially with the lack of reliability of the connected infrastructure in the urban and complex environment.

4. The protection against cybersecurity attacks of connected vehicles, which has to be built very early in the software and hardware architectures, is distributed among many different actors: standards have to be established and their application controlled, since a connected automated vehicle can be turned into a weapon, inside the city.

When these four basics will be mastered or better satisfied in service, the guaranteed nominal behaviour of the minibus will have to be confronted to risky scenarios, coming from real situations and human behaviours.

These risky scenarios have to be identified as they are leading to incidents and sometimes to accidents: so in addition to the ISO26262 (ISO, 2018) standard, the safety of the intended functionality (SOTIF, ISO21448) (ISO, 2022) will have to be checked in front of real driving situations and real human behaviours.

To guarantee a positive result at the end of the development, these risky scenarios will have to be specified at the beginning, using a scenario library as built by the MOSAR project in France (SystemX, 2023b) or automated driving scenarios (ADScene) (SystemX, 2023a), so they will be taken into account in the design, which implies numeric simulation of these scenarios.

## 4.6 A Self-Learning Automated Transport System at European Level

As explained concerning safety targets, a continuous improvement process will be necessary, based on systematic declarations of incidents and scientific analysis at transportation system level (including human management), with anonymised feedback to all actors, at the European level.

The proposed improvement process will need the implication of the public transport authorities to demand a systematic incident declaration when giving agreements, to organise the confidential analysis, to share the lessons learned, to get the corrective actions and to adapt norms and regulations. This field experience will help to establish future norms or to improve regulations, in cooperation with technical representatives, ensuring technical feasibility.

The establishment of this improvement process under the control of a public authority will allow us to accept an intermediate level of safety for market introduction. The market introduction with paid services will begin the first quality measurements: for sure, corrective actions will be necessary and will benefit from the same improvement process, sharing lessons learned and best practices, to the benefit of all actors.

## 4.7 Data Privacy of Incident Analysis and Lesson Learned Sharing

The absence of a human driver introduces new risks in terms of vehicle safety and passenger behaviour: to manage these new risks and to avoid bad incidents which could lead to a rejection of the solution, it is necessary to add a specific process when an incident occurs, as depicted in Fig. 4.1.

An incident will be defined as a situation that could compromise the safety or the service quality and lead later to a rejection and most probably to expensive corrective actions. It can be an accident but also a harsh braking, a difficulty to close the door, a vehicle blocked, vandalism facts and an aggression.

In case of an incident, the data collection includes personal data concerning passengers and other road users but also private information coming from technical products, the automated vehicle and its environment.

To establish robust lessons learned, we need all available data and relevant technical skills, which is common with the road accident analysis, for road safety improvement: the data privacy has to be protected, taking into account relevant regulations, such as General Data Protection Regulation (GDPR) (EU, 2023b; EUR-Lex, 2016).

Road safety methodologies can be reused to conciliate data availability with privacy, as presented in Fig. 4.2, but also to satisfy commonly legal treatment needs with data needs for scientific analysis:

- Beyond the system daily management, incidents will be collected, and data will be protected and crypted with restrained access.
- Data will be analysed by trusted experts in no-profit organisations: scenario coding, simulation and statistical analysis.
- The technical feedback will be shared, as anonymised, without any personal data and technical secret.

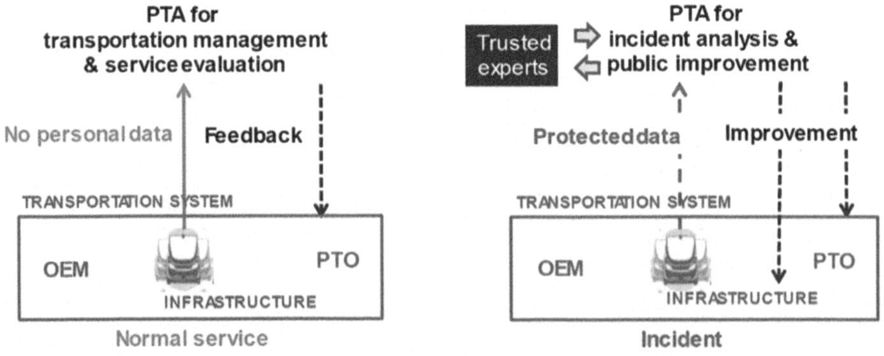


**Fig. 4.1** Data collection process in case of incident

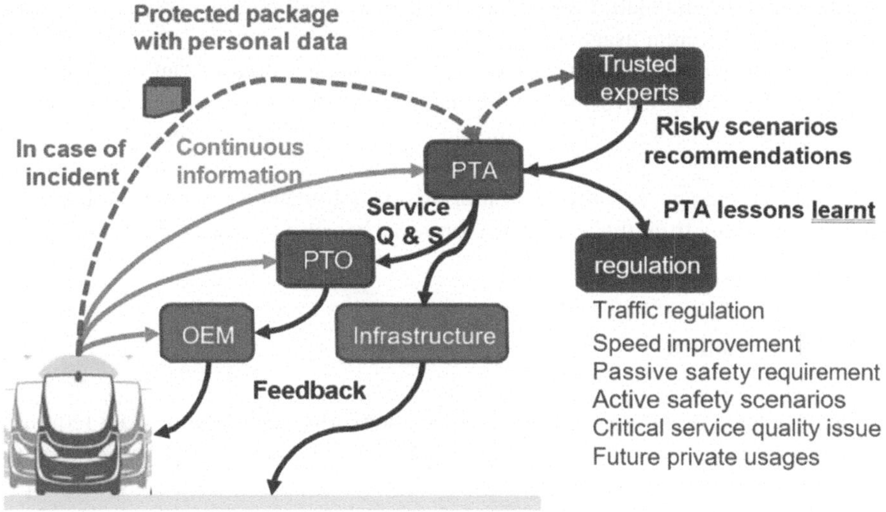


**Fig. 4.2** Lesson learnt process for anonymisation and sharing

Justified by safety and included in the AVENUE vision recommended to the European Community, this continuous improvement process will contribute to both safety and service quality improvement.

## 4.8 Automated Minibus Safety and Service Quality Levers

In addition to the societal benefits, the general target of the introduction of automated minibuses in the urban transportation system is to offer the same level of safety and service quality as conventional transportation, which is very ambitious for the explained technical reasons and dependent on many human factors not well known today.

This is very ambitious, and it cannot be reached without the anticipated implication of the public power to prepare our countries and towns to their future challenges.

Six levers will have to be used by the public transportation authority at the European level, but not only, as many agreements are delegated at the local level:

1. Continuous and general improvement process, based on known methodologies.
2. Systematic and limited incident data collection, feeding risky scenario database.
3. Functional safety and SOTIF ISO norms (26262 & 21448) (ISO, 2018, 2022), using this database.
4. Appropriate effort repartition between active and passive safety, with trade-offs.
5. Dissociation of urban needs from liaisons at higher speed and different vehicles.
6. Innovation for urban shuttle architecture, especially for vulnerable protection.

From now, the regulation work has to include active but also adequate passive safety for two different usage modes.

Secured by the public control on the applications, this coming "automated minibus step" is a valuable investment for general interest: it will not only contribute to the climatic and ecologic transition, but it will contribute to the future life in our towns.

It will also prepare the introduction of private automated cars in towns: it can be later with benefits from this experience or never because of unacceptable risks.

## 4.9 Conclusion

In the advancing of automated minibuses, sustainability and durability are required to enable wider adoption within Europe. Several safety, security and service quality needs are deemed also mandatory to be addressed in order to attain broad acceptability of the new services. In this chapter, the conditions to achieve these objectives are presented. The critical path for market introduction of safe automated minibuses is also discussed, along with quality and safety state of the art for automated minibuses. The data privacy of incident analysis and lesson learned sharing is introduced to enable further discussions about the evolution of the public transport system in Europe. Finally, the automated minibus safety and service quality levers are presented. This analysis envisions to adequately facilitate the adoption of automated minibuses by providing specific guidelines.

**Acknowledgements** This project has received funding by the European Union's Horizon 2020 research and innovation program under grant agreement No. 769033.

## References

EU, E. C. (2023a, December). Buses and coaches. Retrieved from Mobility & Transport - Road Safety: https://road-safety.transport.ec.europa.eu/eu-road-safety-policy/priorities/safe-vehicles/archive/buses-and-coaches_en

EU, E. C.. (2023b, December). GDPR European regulation. Retrieved from General Data Protection Regulation (GDPR)—Official Legal Text: https://gdpr-info.eu/

EUR-Lex. (2016, April 27). Regulation (EU) 2016/679 of the European Parliament and of the Council of 27 April 2016 on the protection of natural persons with regard to the processing of personal data and on the free movement of such data, and repealing Directive 95/46/EC (General Da. Retrieved from Document 32016R0679: https://eur-lex.europa.eu/eli/reg/2016/679/oj

Exeter, U. O. (2020, January 21). Trolley dilemma: When it's acceptable to sacrifice one person to save others is informed by culture. Retrieved from phys.org: https://phys.org/news/2020-01-trolley-dilemma-sacrifice-person-culture.html

ISO. (2018, December). Road vehicles—Functional safety. Retrieved from Standard Number: ISO 26262-1:2018: https://www.iso.org/standard/68383.html

ISO. (2022, June). Road vehicles—Safety of the intended functionality. Retrieved from Standard Number: ISO 21448:2022: https://www.iso.org/standard/77490.html
NHTSA, N. H. (2020, January). Automated driving systems. Retrieved from https://www.nhtsa.gov/vehicle-manufacturers/automated-driving-systems
NHTSA, N. H. (2021, June). Standing general order (National Highway Traffic Safety Administration (NHTSA)). Retrieved December 2023, from https://www.nhtsa.gov/laws-regulations/standing-general-order-crash-reporting
NHTSA, N. H. (2023, April). Standing general order 2021–01. Retrieved from https://www.nhtsa.gov/sites/nhtsa.gov/files/2023-04/Second-Amended-SGO-2021-01_2023-04-05_2.pdf
SystemX. (2023a, December 14). ADScene, a scenario database for the validation of automated driving and driving-assistance systems via digital simulation. Retrieved from https://www.irt-systemx.fr/en/adscene-a-scenario-database-for-the-validation-of-automated-driving-and-driving-assistance-systems-via-digital-simulation/
SystemX. (2023b). MOSAR platform (Methods and tools for operational safety assessment and robustness analysis of autonomous vehicles). Retrieved from Meeting the challenge of validating the autonomous vehicle: https://cahiers-transformation-numerique.irt-systemx.fr/accueil/designing-the-digital-world/meeting-the-challenge-of-validating-the-autonomous-vehicle-2/?lang=en

# Chapter 5
# In-Vehicle Services to Improve the User Experience and Security when Traveling with Automated Minibuses

**Dimitrios Tsiktsiris, Anastasios Vafeiadis, Antonios Lalas, Minas Dasygenis, Konstantinos Votis, Dimitrios Tzovaras, Christian Zinckernagel, and Kevin Salvi**

**Abstract** This chapter looks at the innovative development and deployment of in-vehicle services to improve both the user experience and safety in automated minibuses, a critical component of future mobility-as-a-service (MaaS) ecosystems. With the rise of driverless public transport, ensuring the safety and comfort of passengers becomes a primary concern. The chapter discusses the integration of cutting-edge artificial intelligence (AI) and deep learning technologies to address these challenges, focusing on services such as enhanced safety and confidence, automated passenger presence monitoring, and intelligent feedback systems. It outlines the AVENUE project's initiatives to create an AI-powered framework that not only promotes the widespread adoption of automated minibuses but also addresses social and personal safety concerns in the absence of a human driver. The services discussed aim to replace the driver's monitoring and interaction functions with technology-enabled solutions, thereby maintaining service quality and promoting

---

D. Tsiktsiris
Centre for Research and Technology Hellas, Information Technologies Institute, Thessaloniki, Greece

Department of Electrical and Computer Engineering, University of Western Macedonia, Kozani, Greece

A. Vafeiadis · A. Lalas (✉) · K. Votis · D. Tzovaras
Centre for Research and Technology Hellas, Information Technologies Institute, Thessaloniki, Greece
e-mail: lalas@iti.gr

M. Dasygenis
Department of Electrical and Computer Engineering, University of Western Macedonia, Kozani, Greece

C. Zinckernagel
Amobility, Copenhagen, Denmark

K. Salvi
Mobile Thinking, Geneva, Switzerland



G. Fournier et al. (eds.), *Automated Vehicles as a Game Changer for Sustainable Mobility*, Contributions to Management Science,
https://doi.org/10.1007/978-3-031-61681-5_5

passenger confidence. Using real-world deployments and pilot studies, the chapter illustrates how these technological interventions contribute to the overall efficiency, safety, and acceptance of automated vehicles in urban environments and provide valuable insights into the future of automated public transport systems.

## 5.1 Introduction

Automated vehicles offer significant assistance to drivers across various levels through multiple means. During the last decade, public transportation systems are progressively integrating different levels of automated vehicles into mobility as a service (MaaS) within their services. During the last few years, fully automated vehicles started appearing in the market, able to operate without a driver in complex road conditions. In this context, as there is no driver, no assistance can be provided to the passengers, and routes and destinations will be automatically determined by the associated fleet management system, based on the passenger trip requests. The ultimate goal of mobility as a service (MaaS) is to provide comprehensive, efficient, and seamless transportation solutions to users. MaaS aims to integrate various forms of transportation services into a single, accessible platform that allows users to plan, book, and pay for their journeys using different modes of transport, such as buses, trains, taxis, rideshares, bikes, and even micromobility options like scooters.

In our vision in the AVENUE project, one of the most promising mobility modes for MaaS services will be the use of fully automated vehicles (e.g., minibuses), providing the glue for all other transport modes of a successful MaaS in an urban environment.

However, the transition to fully automated public transportation vehicles is not seamless, and several obstacles arise in a real-world scenario.

Several concerns among end users are associated with the safety and reliability of fully automated vehicles (AVs), directly impacting the acceptance of this new technology. Two major concerns for potential passengers involve automated shuttle sharing with other passengers and trust in the technology itself. The absence of a driver in the bus raises various possible issues: for instance, no immediate personnel would be present to administer first aid in case of emergencies, and passengers might feel uneasy being alone, especially at night or in certain neighborhoods.

Additionally, the lack of an authoritative figure onboard could be a concern, especially when transporting groups such as schoolchildren, potentially leading to incidents like vandalism, theft, or altercations among passengers. With the removal of the driver and the introduction of external intervention teams, questions arise about how the formal and informal services provided by the driver could still be upheld while maintaining the current high-quality service standards.

Addressing these social and personal safety and security concerns within automated vehicles necessitates the implementation of specific services to replace the driver's presence. Within the AVENUE project, various IT-based solutions and services have been identified and developed. These solutions aim to provide passengers with similar services to those offered by a driver onboard. This includes the creation of a novel artificial intelligence (AI)-supported framework that facilitates the widespread adoption of these services. This approach shows great promise in significantly enhancing safety and security levels in automated public transportation and bolstering passenger trust.

The services that will be presented in this chapter are the following:

- Enhance the sense of security and trust
- Automated passenger presence
- Follow your kid/grandparents
- Shuttle environment assessment
- Smart feedback system

The reason for choosing these services is twofold: (a) They are all essential services necessary to remove the safety driver from the shuttle, as safety and efficient operation are critical factors. For example, ensuring safety for the travelers have been identified from end users as the single most important factor for choosing to ride with the automated shuttle. (b) They are all based on the same foundation of technology with cameras, sensors, and algorithms, meaning that testing multiple types of services is feasible, once the necessary equipment has been installed. The analysis is also similar, and therefore the results can be compiled together.

## 5.2 Service: Enhance the Sense of Security and Trust

The service "enhance the sense of security and trust" aims to address the new reality that is formed in automated shuttles' mobility infrastructures with the absence of a bus driver and the threat from criminal activities in European cities. Typically, drivers are trained to handle incidents of passengers' abnormal behavior, such as incidents of petty crimes, according to standard procedures adopted by the transport operator. Surveillance using sensors such as cameras (cameras of different technologies can be used so that passengers' privacy is protected) and microphones, as well as smart software in the bus, will maximize the feeling of security and the actual level of security. However, several concerns of the end users regarding the safety and robustness of the AVs, where a driver is not present, which are directly linked to the final user acceptance of the new technology, can be identified. Prospective passengers harbor apprehensions about potential scenarios in the absence of a driver in the bus, including:

- No one will be in the bus to perform first aid if required
- Feeling of discomfort being all alone in the bus at night, especially in certain neighborhoods
- No authority figure present to keep passengers calm (e.g., schoolkids)
- Vandalism, bag snatching, indoor fighting, and unaccompanied luggage

To address the aforementioned concerns on social and personal safety and security into the vehicle, certain measures need to be implemented. For example, the detection of unaccompanied luggage and of other personal belongings may raise a notification or an alert to the supervisor and/or the suitable authorities. This may be followed by appropriate notifications and/or instructions to the passengers, while the vehicle may also implement respective actions. Moreover, implementing a solution for enhancing the safety and security inside the automated buses will support safekeeping not only the users of the automated public bus but also the vehicle itself.

This section details the implementation of a video and audio analytics software module designed for an embedded security subsystem or for the cloud-based services within the system. Additionally, it covers the deployment and testing of this service at the pilot sites of the AVENUE project.

The service addresses the timely, accurate, robust, and automatic detection of various petty crime types or misdemeanors as well as the assistance of authorized end users toward the reidentification of any offenders. A misdemeanor is any "lesser" criminal act in some "common law" legal systems. Misdemeanors are generally punished less severely than felonies but theoretically more than administrative infractions (also known as minor, petty, or summary offences) and regulatory offences. Many misdemeanors are punished with monetary fines. The petty crimes that are targeted for identification by the sensors include petty theft like bag snatching and pickpocketing, vandalism, aggression, illegal consumption of cigarettes, public intoxication, simple assault, and disorderly conduct. These are explained in more detail:

- Petty theft: Theft is the taking of another person's property or services without that person's permission or consent with the intent to deprive the rightful owner of it.
- Vandalism: Vandalism is the action involving deliberate destruction of or damage to public or private property.
- Aggression: Aggression is overt or covert, often harmful, social interaction with the intention of inflicting damage or other unpleasantness upon another individual.
- Public intoxication: Public intoxication, also known as "drunk and disorderly" and drunk in public, is a summary offense in some countries rated to public cases or displays of drunkenness.
- Simple assault: An assault is the act of inflicting physical harm or unwanted physical contact upon a person or, in some specific legal definitions, a threat or attempt to commit such an action.
- Disorderly conduct: Disorderly conduct makes it a crime to be drunk in public, to "disturb the peace," or to loiter in certain areas.

In the "enhance the sense of security and trust" service, two distinct petty crime detection approaches are implemented: video analytics and audio analytics. The video analytics approach supports end-to-end detection of abnormal events; achieves real-time inference on modern hardware; offers flexibility with supervised, unsupervised, and semi-supervised learning to compensate the scarcity of data in the security domain; supports multiple camera types, positions, and angles; and is able to operate in embedded setup with limited power requirements. On the other hand, the audio approach uses information from the acoustic sensors of the shuttle for abnormal event detection by comparing different spectrogram representations and focusing on the effect of signal-to-noise ratio (SNR) to audio recognition and the potential of the generalization of a model in different SNR settings and datasets collected under different environments. More specifically, for the video analytics approach, a pose classification approach is developed that classifies the extracted skeleton key points from each dataset (Tsiktsiris et al., 2020). For training the proposed model, five distinct datasets were used, including data simulated in lab in CERTH/ITI facilities, data captured from Geneva Public Transport (TPG) shuttles in Geneva and HOLO shuttles in Copenhagen, data from the P-REACT project, the NTU-RGB-D (NTU-RGB-D is the name of the dataset from Nanyang Technological University (NTU) (https://www.ntu.edu.sg/rose) featuring RGB (Red-Green-Blue/color images) and D (Depth)) dataset by ROSE lab, and the UCSD Anomaly Detection Dataset. Three different models were tested for pose classification, namely, a stacked bidirectional long short-term memory (LSTM) network classifier, a spatiotemporal autoencoder, and a spatiotemporal LSTM classifier. The first approach consists of a stacked LSTM model as a classifier. An overview of the pipeline is depicted in Fig. 5.1. Overall, the classification is performed in four stages: (a) In the first stage, pose estimation techniques are applied to obtain skeleton key points. The generated pose proposals are refined by parametric pose non-maximum suppression to obtain the estimated human poses. (b) In the second stage, tracking is performed to match cross-frame poses and form pose flows. (c) In the third stage, features are generated from the detected and tracked human body key points and are forwarded into the network (d), which classifies the action into normal or abnormal.

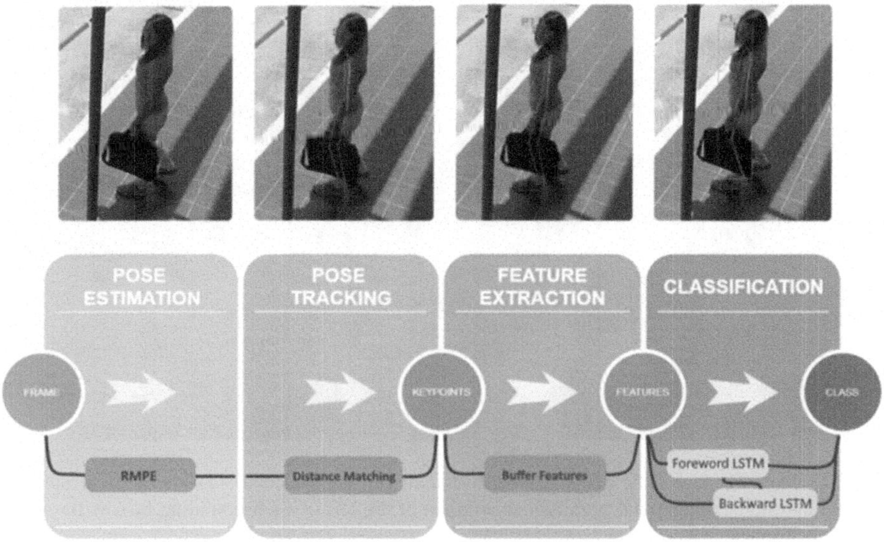


**Fig. 5.1** Pipeline of the pose classification (Tsiktsiris et al., 2020)

For the pose estimation stage, the regional multi-person pose estimation (RMPE) by Fang et al. (2017) was adopted with a pretrained VGG19 backend, a convolutional neural network model proposed by Simonyan and Zisserman (2014). In the tracking stage, a matching of cross-frame poses and form pose flows is performed, using a real-time algorithm that is based in a distance matrix. In addition, a pose flow non-maximum suppression is applied, in order to reduce unnecessary pose flows and relink temporal disjoint ones. This is an important step that associates poses indicating the same person across multiple frames. A skeleton tracking algorithm was implemented, in order to meet the performance requirements of a real-time service. The algorithm is sorting the skeletons based on the distance between neck and image center, from small to large. Certain heuristics are taken into consideration, such as the position of the joints, the average height of the person, and the height difference between frames. Height variation improves the ability of the algorithm to understand depth, since the analysis is based on two-dimensional input. Such parameters of the algorithms are optimized for in-shuttle space and fine-tuned to specified weights based on the camera calibration. A skeleton near the center will be processed first and be given a smaller human ID. Later on, each skeleton's features are matched based on its previous and current frame. The distance matrix (or cost) between the skeleton joints is the main criterion for the matching function. Skeletons with the less distance are paired between the frames and are given the same ID (Fig. 5.2).

**Fig. 5.2** Skeleton matching across two subsequent frames (blended) (Tsiktsiris et al., 2020). Notice that the passenger ID, highlighted in green at the left of each bounding box, is the same across the frames

For the spatiotemporal autoencoder, there are two stages to be formed: encoding and decoding. Autoencoders set the number of encoder input units to be less than the input; thus, they were first used to reduce dimensionality. Usually, unsupervised backpropagation is used for training, helping the reconstruction error of the decoding results from the original inputs to decrease. Generally, an autoencoder can extract more useful features when the activation function is nonlinear rather than some common linear transformation methods, such as principal component analysis (PCA).

In order to learn the regular events in training data, a spatiotemporal autoencoder was introduced. In particular, the spatial autoencoder consists of an encoder and a decoder that are composed of two convolutional and transpose convolutional layers, respectively, whereas the temporal encoder is comprised of three convolutional LSTM layers, as depicted in Fig. 5.3.

Finally, it was observed that even if the model is trained on thousands of data, some false positives will still be observed in certain occasions. As a result, it is possible to manually shift through the anomaly outputs and flag some of them as false positives, in order to let the previous autoencoder neural network model act as a high recaller. Semi-supervised learning is employed by decreasing the threshold, so that the majority of true anomalies (high recall) can be detected, as well as other false positives (low precision). To achieve the semi-supervised approach, a new model has been designed, which includes the previous encoder and an LSTM, which acts as a classifier as depicted in Fig. 5.4.

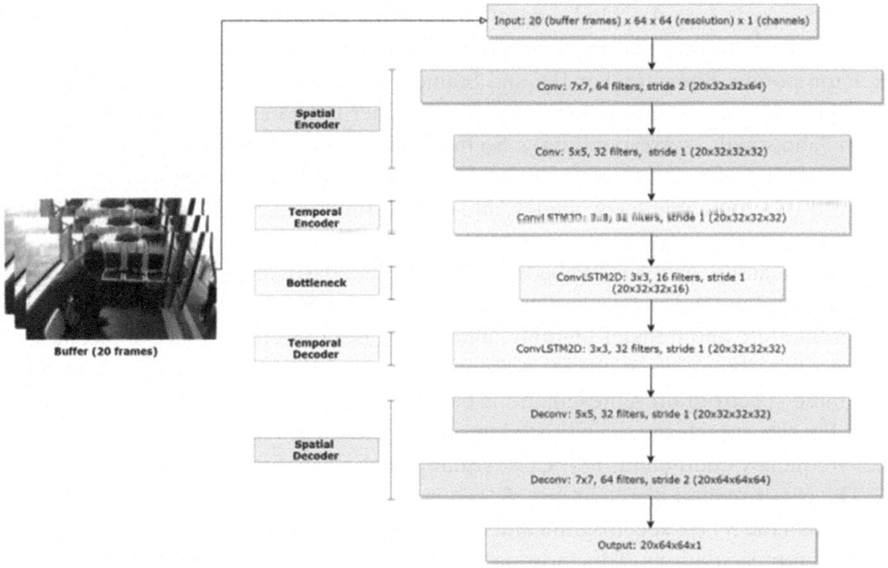


**Fig. 5.3** Model architecture of the autoencoder. The first two convolutional layers are spatial encoders, followed by temporal encoder and decoder. Between them, a ConvLSTM with reduced filters is used as a bottleneck to eliminate non-useful information. At the last two layers, spatial decoding is performed, reconstructing the input image to the same format (Tsiktsiris et al., 2020)

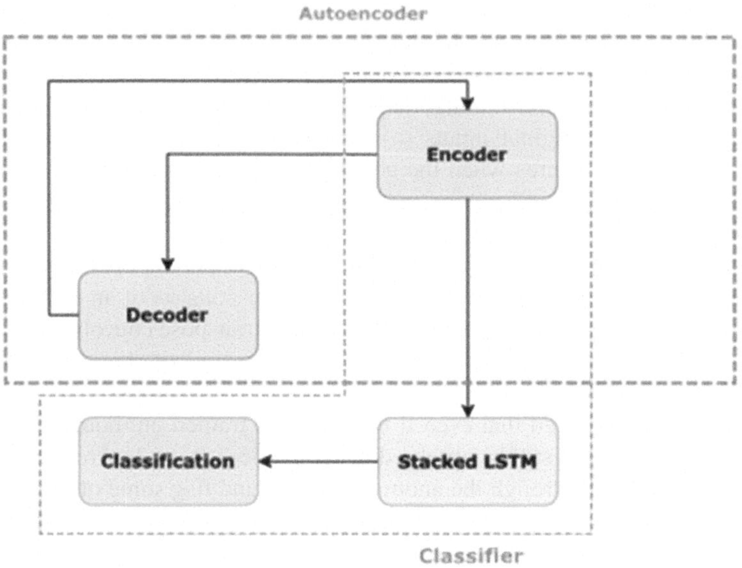


**Fig. 5.4** Model architecture of the hybrid model. The red container contains components of the previous autoencoder approach. The green components indicate the new hybrid model which acts as a classifier (Tsiktsiris et al., 2020)

The pose classification approach was tested on the NTU RGB + D dataset (Fig. 5.5a–c) and on the TPG dataset captured by CERTH inside the AV's shuttle (Fig. 5.5d, e). In the images, there are various debugging layers enabled, such as skeleton points, lines, tracker ID, and bounding boxes of each detection. The predicted result is marked as green when the classifier indicates it as "normal" and red when "abnormal," correspondingly. So far, NTU dataset samples were not included in the training set, so it is safe to assume that the derived model can generalize across different people, view angles, and events. Figures 5.6, 5.7, and 5.8 depict the aforementioned conditions and use cases.

For the audio analysis (Papadimitriou et al., 2020), the different procedures implemented include spectrograms, single-channel representation, multichannel representation, and transfer learning, and the dataset used for sound events classification (glass breaking, gunshot, screaming) is the MIVIA Audio Events dataset. The three spectrogram representations used were namely the single-channel short-time Fourier transform (STFT) (Salamon & Bello, 2017), the mel scale, and the mel-frequency coefficients (MFCCs) (Zhang et al., 2015).

The experimental results in Fig. 5.9 showed that the MFCC is able to generalize better than the STFT spectrogram and the mel spectrogram, when comparing single-channel representations. This most probably owes to the fact that this representation includes all the important information of the audio signal in the lowest MFCC features (e.g., first ten features) with regard to the concentrated energies and has minimum changes in the highest ones. Hence, it is suggested that it has its place in a

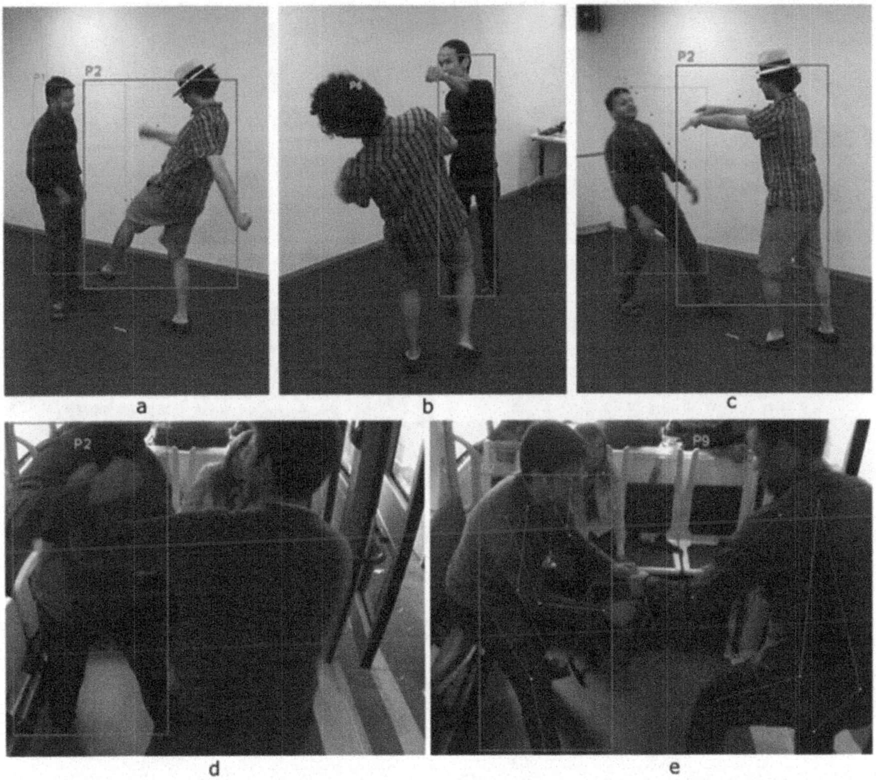

**Fig. 5.5** Evaluation on test data: (**a–c**). Abnormal event detection (violence/passengers are fighting) using different camera angles from the NTU-RGB dataset (**d**, **e**). Detection of fighting/bag snatching real-world scenarios inside the shuttle (Tsiktsiris et al., 2020)

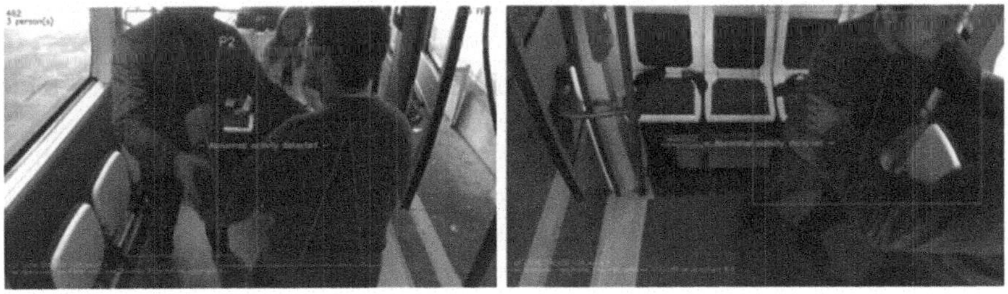

**Fig. 5.6** Evaluation on multiple camera angles, excessive occlusion, and partial presence (Tsiktsiris et al., 2020)

feature representation combination, and for that reason, it was indeed used in both methods of multichannel representation. With regard to the multichannel representation (Fig. 5.10), the stacked features method proved to be more generalizable compared to the concatenated features method, especially when training was

**Fig. 5.7** Evaluation across various scenarios (left to right): Bag snatching, fighting, vandalism, and unaccompanied luggage (Tsiktsiris et al., 2020)

**Fig. 5.8** Additional evaluation on the NTU-RGB dataset. Metrics at the top left depict the prediction scores for the P1 (Tsiktsiris et al., 2020)

carried out on higher signal-to-noise ratios (SNRs) and testing was carried out on lower ones. Neither the concatenated features method nor the separate single-channel spectrogram representations (STFT, mel, and MFCC) performed as well.

Finally, in Fig. 5.11, the generalization capabilities of the two multichannel methods are shown in terms of event-based recognition (GB, G, and S). As one moves along the sequence of the ten models, it is evident that the generalization capabilities of the stacked multichannel method are significantly better than the corresponding concatenated multichannel method. In both cases, the model that was trained in −5 dB and tested in 15 dB showed the best performance, with a recognition score of 91.51% for the concatenated method and 90.23% for the stacked method, with the lowest standard deviation, namely, 0.034 and 0.019, respectively.

Moving up in terms of SNR training (and model number), it became more difficult to generalize, especially in the case of zero and below SNRs. This is due to the fact that the lower SNR audio contains higher levels of noise and thus is more challenging, leading to more robust and generalizable classification.

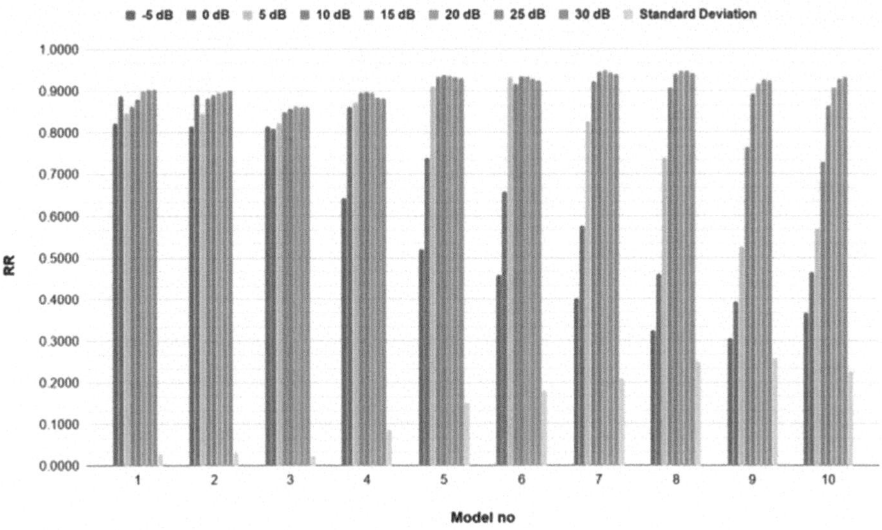


**Fig. 5.9** Frame-by-frame recognition rate (RR) for all models using concatenated features from STFT, mel, and MFCC spectrograms (single channel) validated for each signal-to-noise ratio (SNR): each column group refers to a specific model (Papadimitriou et al., 2020)

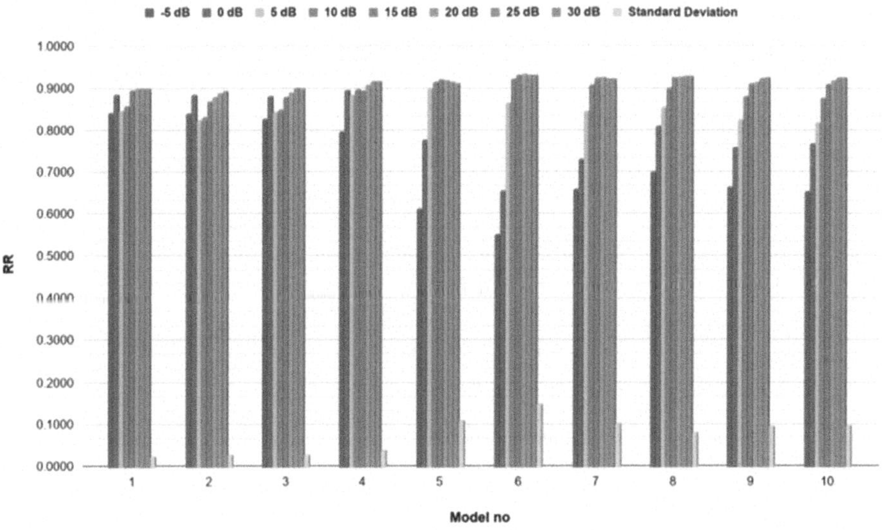


**Fig. 5.10** Frame-by-frame RR for all nodes using stacked features from STFT, mel, and MFCC spectrograms (multichannel) validated for each SNR: each column group refers to a specific model (Papadimitriou et al., 2020)

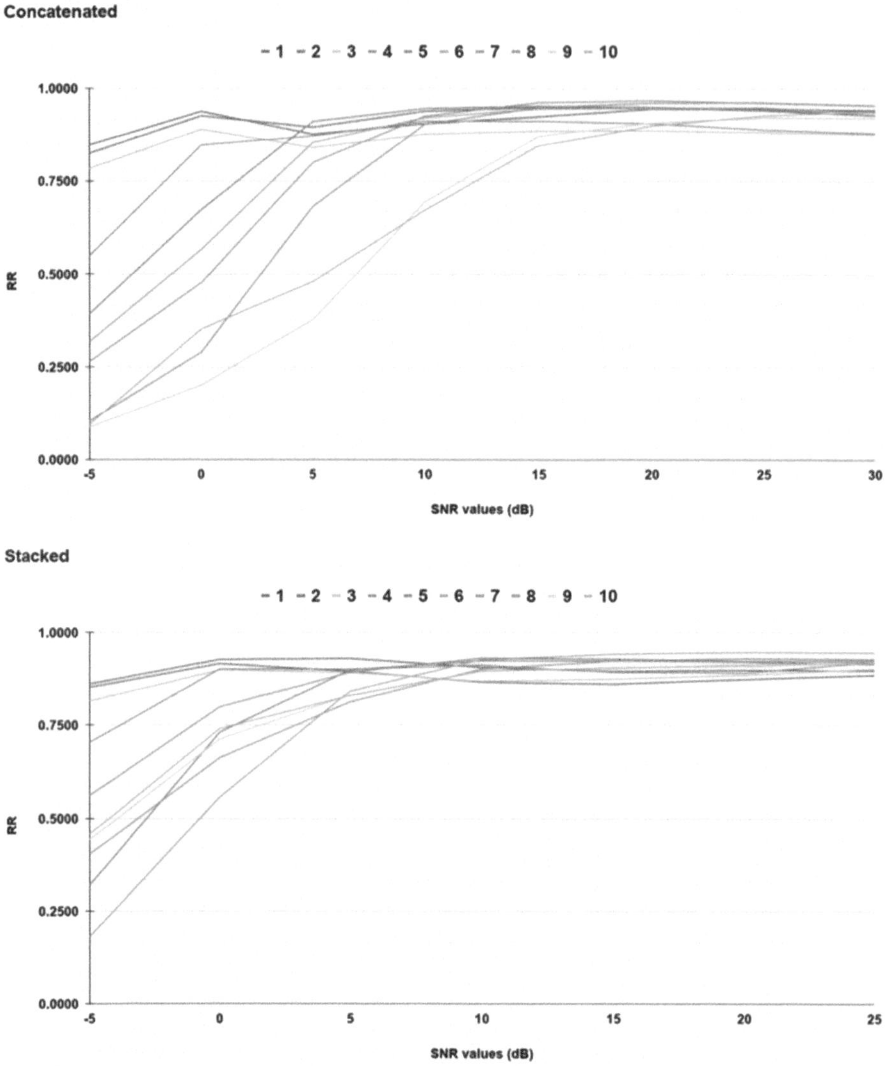


**Fig. 5.11** A comparison of the models trained with the concatenated (top) and stacked (bottom) features method with regard to event-based RR: the sequence of the models increases from 1 to 10 (Papadimitriou et al., 2020)

## 5.3 Service: Automated Passenger Presence

The service "automated passenger presence" aims to address a basic problem of operators' services, which is related to the occupation of their vehicles, as well as the awareness of the number of people onboard in order to schedule the routes. Furthermore, the passengers would like to know in advance if there is an available seat or enough space on a shuttle to plan their boarding. Traditionally, but also nowadays, passenger counting is conducted manually via passenger surveys or human ride checkers. Typically, the driver or inspectors are responsible for performing enumeration of the onboard passengers, something not feasible in an automated shuttle. Automatic passenger counting has been rapidly emerging in recent years to address similar needs. An automated system is introduced capable to detect passenger presence in real-time with high accuracy, count onboard passengers, and calculate vehicle occupancy. Surveillance using sensors such as cameras (cameras of different technologies can be used so that passengers' privacy is protected) and smart software in the bus will automate the detection of passenger presence.

Several concerns of the end users regarding the safety and robustness of the automated vehicles that are directly linked to the final user acceptance of the new technology can be identified. The prospective passengers may deal with several possible instances that could arise in case there is no staff inside the shuttle. Indicatively:

- No one will be in the bus to count the number of passengers with regard to the shuttle's capacity.
- There are continuous stops throughout the entire route, even in cases where the shuttle is fully occupied.
- No authority figure is present to alert passengers of their designated bus stop.

To address the aforementioned concerns on social and personal safety and quality of service into the vehicle, certain measures need to be implemented. For example, counting the number of passengers being inside the automated vehicle could help avoid overcrowding in the shuttle, as well as meaningless stops in cases where the bus is in full capacity. This may be followed by appropriate notifications and/or instructions to the passengers, while the vehicle may also implement respective actions.

The service provides a video analysis of the vehicle internals, using the onboard camera, in order to identify the vehicle occupation, vehicle free space, as well as counting people onboard. Automatic assessment of space occupation using the onboard cameras is enabled. Capacity is set as an absolute number of space units. For example, each space unit is associated with one standing passenger. Occupancy is set as an absolute number of space units currently in the shuttle. For the operation manager, occupancy is visible on the dashboard of the AVENUE platform, whereas occupancy is displayed as real-time information via the AVENUE mobile app, wherever the traveler is. Each passenger (normal, big size, wheelchair user, seated) can determine whether he/she can fit in or not. Assessment for different cases can be provided to assist the passengers on determining whether to request onboarding or not. Automatic counting of people using the onboard cameras is also provided. Moreover, occupancy marked with information for the different user cases is displayed as real-time information, wherever the traveler is; however it does not guarantee them a free spot by the time the shuttle reaches the station of their choice.

The implementation of a video analytics software module for an automated passenger presence counting subsystem or for cloud-based services of the system is described along with deployment and test of the service into the pilot sites of the AVENUE project, where the following use cases have been identified to be further examined and addressed:

- Passenger counting: The automated shuttle has a fixed capacity regarding the number of passengers it can carry. The video cameras installed in the automated shuttle acquire the color depth images, and the data are fed into the system's video analytics algorithms for further analysis. In case the algorithms identify that the total number of passengers is reached, the shuttle stops receiving any others, and appropriate notifications are sent to the AVENUE mobile app for the passengers that would like to board.
- Route optimization: Even though the shuttle is in full capacity, there may still be people waiting on a bus stop to go aboard. The bus only makes a stop when a passenger needs to get off, while the route is modified to save time and cost. The number of onboard passengers is always monitored, so that the new passengers could get on, in case of availability.
- Passenger awareness: Even though the automated vehicle has reached its terminal, there could still be passengers onboard. The shuttle counts the number of passengers to make sure there is no one left. If there are passengers, the bus alerts them to get off.

For the "automated passenger presence" service, a deep learning-based distance assessment service is proposed that uses an overhead perspective, which is able to function with high accuracy and low-power consumption in confined spaces, such as the inside of the automated shuttle (Tsiktsiris et al., 2022). For this purpose, a fisheye wide-angle camera with a top-down perspective is used. In order to timely and accurately detect human items, a pretrained RAPiD model was implemented that outputs bounding box coordinates used for computing their centroids.

More specifically, as already mentioned, the network architecture is inspired by RAPiD (Duan et al., 2020) and therefore consists of three stages: the backbone network, the Feature Pyramid Network (FPN), and the bounding box regression network. The backbone network works as a feature extractor that takes an image as input and outputs a list of features from different parts of the network. In the next stage, these features are passed into the FPN, in order to extract features related to object detection. Finally, at the last stage, a convolutional neural network (CNN) is applied to each feature extractor in order to produce a transformed version of the bounding box predictions, as depicted in Fig. 5.12.

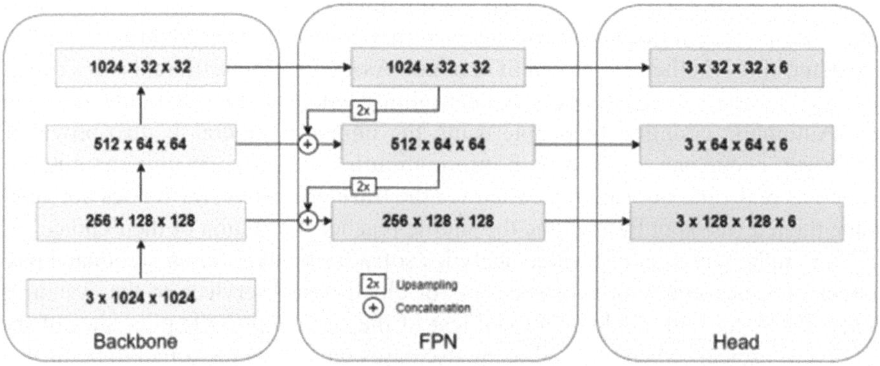


**Fig. 5.12** An illustration of multiple convolutional layers and multidimensional matrices, such as the feature maps with 1024 × 1024 input resolution (Tsiktsiris et al., 2022)

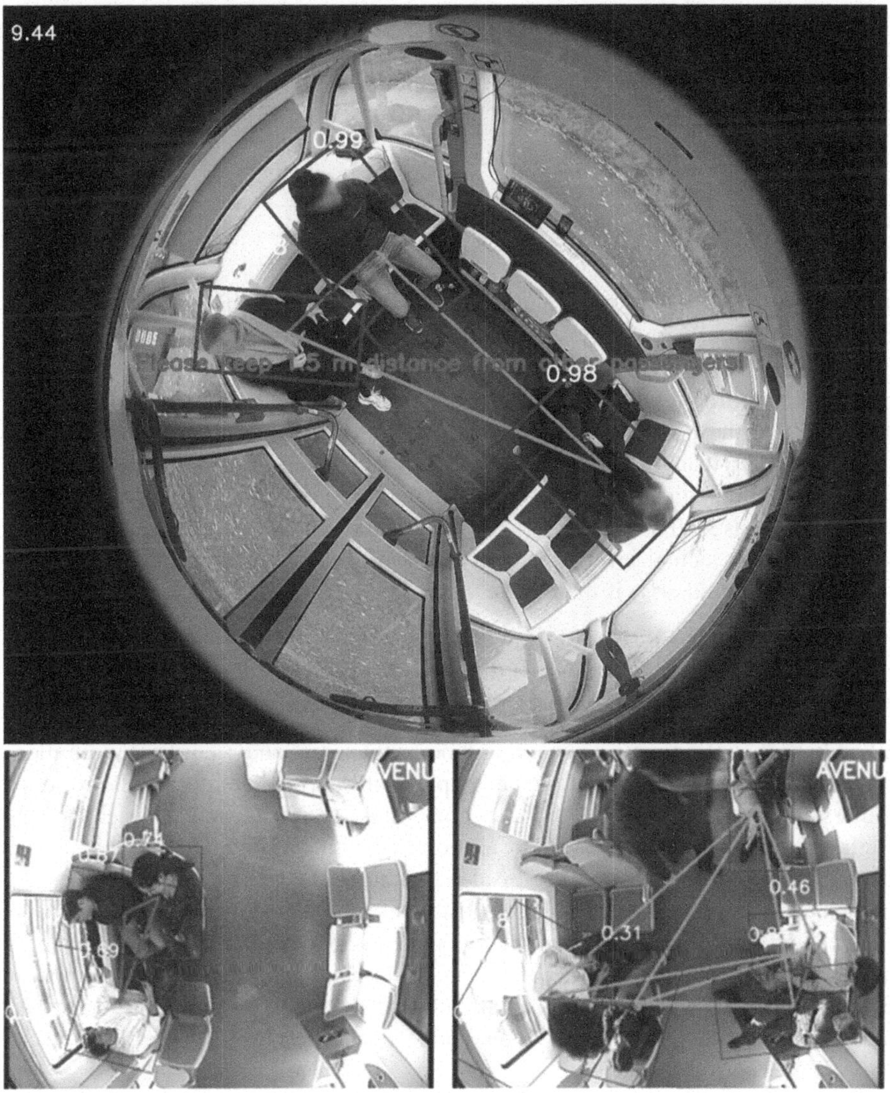


**Fig. 5.13** Results on unseen scenarios from a real pilot site and the BOSS dataset. Green lines represent a safe distance, while red lines an unsafe one (Tsiktsiris et al., 2022) (Color figure online)

Furthermore, the Euclidean distance formula is also implemented to compute the pairwise distances for each bounding box. This distance is later transformed into the real distance between passengers by multiplying it with a weight value that is defined via the camera calibration.

Experimental results indicated that the service efficiently identifies passengers with unsafe proximity, according to COVID-19 regulations (passengers must keep a distance of 1 m apart (Olivera-La Rosa et al., 2020)), as depicted in Fig. 5.13.

However, handling reflections is still a challenge on certain scenarios, as also happens in similar approaches. Passenger figures might appear to the windows of the shuttle as reflections, especially when the lighting is low. To mitigate this issue, a custom mask is applied on large reflective surfaces.

## 5.4 Service: Follow My Kid/Grandparents

The service "follow my kid/grandparents" is designed to increase autonomy of partly autonomous people (kids, grandparent(s), disabled people). It will allow carers or family members to be sure that their beloved family members are safe while moving around the city using public transport. On the other hand, it will increase confidence to the non-fully autonomous people to use public transports knowing that their family can "be with them." Surveillance using sensors such as cameras (different technologies can be used so that passengers' privacy is protected) and microphones, as well as smart software in the shuttle, will maximize the feeling of security and the actual level of security. The prospective passengers fear several instances that could arise if there is no driver in the bus:

- Passengers feeling discomfort travelling alone during nighttime
- Parents not being able to know if their kids have reached their destination safely
- Caregivers not being able to track passengers with dementia or other health issues

To address those concerns on social and personal safety and security, certain measures need to be implemented. For example, third parties monitoring the route of minors or passengers with health issues could make their route much easier and less frightening. This may be followed by appropriate notifications and/or instructions to the third party, while the vehicle may also implement respective actions. Moreover, implementing a solution for monitoring the routes of kids and patients will support safekeeping not only the users of the automated public shuttle but also the vehicle itself. The implementation of a video and audio analytics software module for an embedded security subsystem or for cloud-based services of the system is described along with appropriate the deployment and test of the service into the pilot sites of the AVENUE project.

The service and scenario propose a full-fledged solution that allows designated "guardians" to follow the automated public transport journeys of more vulnerable people, since the guardians can check the trip via a dashboard or mobile app, receive notifications, add people to their "guarded" list, and share trips/positions and estimated time of arrival (ETA) with others. In the context of AVENUE project, the following use cases have been identified to be further examined and addressed:

- Passenger monitoring: Travelling without a guardian during nighttime can be unsettling for a vulnerable person.
- Kids monitoring: Parents need to be able to track their kids.
- Patients monitoring: Caregivers need to track their patients, especially when they are not able to commute on their own.

The video cameras installed in the automated shuttle record the color images, and the data are fed into the system's video analytics algorithms for further analysis. When the automated bus's system identifies the passenger/kid/patient, the tracking begins, and the parents/caregivers can monitor their route.

For the "follow my kid/grandparents" service, an end-to-end service is developed based on deep learning models, for automated facial recognition inside the automated shuttle (Tsiktsiris et al., 2021). The techniques introduced in this service are based on attention to mitigate the occlusion issues introduced by face masks during the COVID-19 pandemic. More specifically, the first layer of the sensors connects to the hardware abstraction layer (HAL) that implements the IP and the USB cameras, respectively, but also requests raw data by the API end points to perform face recognition. The input data are then converted and transformed in a compatible format and passed into the analytics algorithms. The result is transferred via the API end points into the cloud. The user has access to the data and acts accordingly. A new passenger can be enrolled to the service using a single image of his/her face, which will be stored in a database, and by using it as a reference, the network will calculate the similarity of any new instances presented to it. An overview of the service is presented in Fig. 5.14.

As for the video analysis, facial recognition techniques identify human faces in images or videos by measuring specific facial characteristics. The extracted information is later combined to create a facial signature or a profile. When used for facial verification, a camera frame is compared to the recorded profile. More specifically, a multi-task cascaded convolutional network (MTCNN) (Zhang et al., 2016) receives the input frame to extract and align facial images. The facial images are then preprocessed and passed into a feature extractor (CNN backbone) linked with the explainable cosine (xCos) module that features an explainable cosine metric.

As current face verification models use fully connected layers, spatial information is lost along with the ability to understand the convolution features in a human sense. To address this obstacle, the plug-in xCos module is integrated as described below and depicted in Fig. 5.15, while experimental results are illustrated in Fig. 5.16.

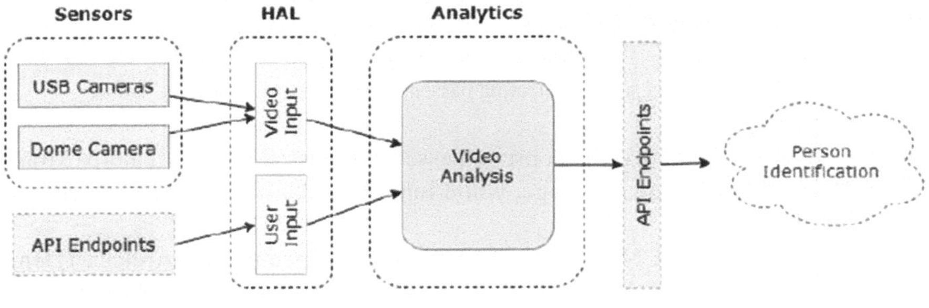


**Fig. 5.14** Overview of the "follow my kid" service (Tsiktsiris et al., 2021)

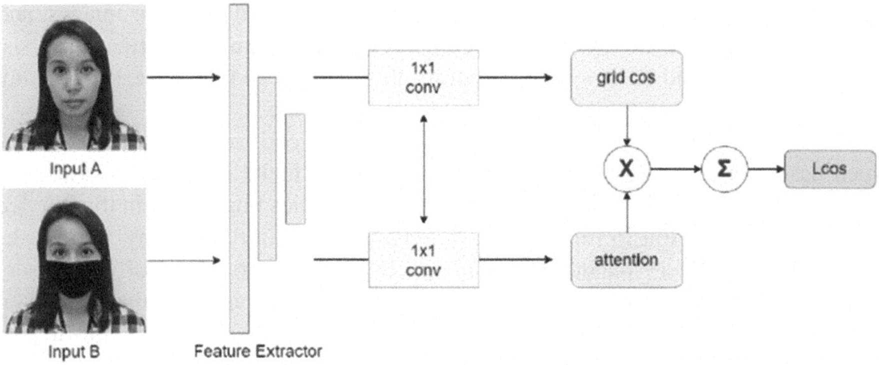


**Fig. 5.15** Network pipeline: the two input images are preprocessed and passed into the backbone CNN for feature extraction along with the plugin xCos module (Tsiktsiris et al., 2021)

**Fig. 5.16** The Follow-my-kid service detecting passengers in real time inside the shuttle

- Input: The two input images are preprocessed and passed into the feature extractor. Input A is the database image, while Input B is the image cropped from the video stream.
- Backbone: The same CNN feature extractor is implemented as in ArcFace (Deng et al., 2019). However, to employ the xCos module, the last fully connected layer and the previous flatten layer are replaced with a 1 × 1 convolutional layer.
- Lcos calculation (xCos): Patch-wise cosine similarity is multiplied by the attention maps and then summed to calculate the Lcos.

## 5.5 Service: Shuttle Environment Assessment

The service "shuttle environment assessment" aims to maintain the environmental conditions in the automated vehicle that may not be adequately controlled due to the absence of the shuttle driver at acceptable levels. Minimum acceptable conditions and comfort, such as good air quality, acceptable odors, and absence of smoke, are necessary for the safe transport of passengers, as well as the viability of the whole automated service, since lack of these conditions within the vehicle could significantly discourage potential users. After all, monitoring the environmental conditions could enable passengers' alert and warning services via notifications, thus enhancing the user experience and safety during their trips. Under these circumstances, there are several instances that must be considered in order for the prospective passengers to feel content and safe. While there would be no driver inside the vehicle, various problems might come up, such as the following:

- There will be no staff inside the bus to prevent someone from lighting a cigarette.
- In quite high or low temperatures, there will be no driver in order to regulate the air conditioning or heating system.
- In emergency situations, there will be no one in charge of informing the operators and the competent authorities.
- If the air concentration of $CO_2$ inside the bus is high, and someone might get dizzy or exhibit breathing difficulties, there will be no driver so as to either open the windows or stop the shuttle.

As far as it is considered, the buses should create a comfortable environment for all the passengers. This feeling could undoubtedly be strengthened by controlling the temperature inside the vehicle. Besides, heating, ventilation, and air conditioning control now belong to the standard equipment of city buses. As a result, it is crucial for temperature sensors to be positioned in specific locations inside the vehicle. Simultaneously, these sensors will be connected to the air conditioning system, and if the humidity/temperature exceeds a suitable limit, the air condition will be put into operation. In that way, the existing humidity/temperature will be automatically adjusted, to provide the appropriate indoor climate, neither too hot nor too cold for the passengers. In addition, detection of certain pollutants, such as $CO_2$, $NO_2$, or dust particles in the indoor environment, along with critical temperature variations, is critical for the condition of certain passengers, especially ill people, such as asthma patients. Smoke in the vehicle, i.e., from a person that lights a cigarette, will deteriorate the passenger experience but may also put the whole vehicle in danger (fire hazard). Detection of certain events (air quality deterioration, smoke) may raise a notification or an alert to the passengers along with instructions on how to handle this situation, while the vehicle may also implement respective actions. Moreover, smoke in the vehicle may also result in cancelling the automated transport service. Detection of certain events (air quality deterioration, smoke) may raise a notification or an alert to the supervisor and/or the suitable authorities (i.e., police, fire department). This may be followed by appropriate notifications and/or instructions to the passengers, while the vehicle may also implement respective actions.

The service is responsible for the timely, accurate, robust, and automatic detection of any change in the air quality and the presence of smoke or fire, inside the vehicle. In cases where there will be an alert on the system, notifications and instructions will be sent to the passengers, to the operators, and/or to the suitable authorities. Considering this, several possible situations could take place in counting from high-level $CO_2$ and $NO_2$ concentrations at the indoors air composition to presence of humidity, smoke, or even fire. Especially, exposure to carbon dioxide can produce a variety of health effects (Azuma, 2018). These may include headaches, dizziness, restlessness, a tingling or pin or needle prick feeling, difficulty breathing, sweating, tiredness, and increased heart rate. Furthermore, detection of certain air quality indexes and pollutants in the indoor environment, along with critical temperature variations, is necessary for providing a secure service to the passengers. A gas composition sensor could be used for checking the inside air quality. This sensor monitors the air intake of the heating, ventilation, and air conditioning system of the vehicle, while it detects undesirable gases and adjusts the system accordingly by shutting off the intake and recirculating the indoor air back to the outside. Despite that, another possible scenario, mostly observed during the rainy days of the year, is the fogging of the windows due to the increased humidity. This might have a negative impact on the passengers' attitude while reinforcing feeling of confinement. Consequently, including a fogging prevention sensor inside the vehicle might be an efficient solution. More specifically, fogging prevention sensors are used to prevent fogging of the windshield glass. These sensors consist of three sensing elements for sensing indoor temperature, windshield glass temperature, and cabin humidity. The fogging sensor feedback is used for adjusting the heating, ventilation, and air conditioning system to maintain the interior temperature higher than the windshield glass temperature. Hence, it prevents the windshield from fogging up.

To summarize, it is passengers' wish to travel in a clean and comfortable environment and be notified if the conditions deteriorate. Moreover, operators would like to be notified when environmental conditions are considered harmful for the passengers. Representative use cases are indicatively displayed as follow:

- Lighting a cigarette: Inside the shuttle, a passenger lights a cigarette. The smoke detection sensors detect the smoke coming out of the cigarette. The real-time sensor data is sent to a central PC, installed in the vehicle and in which the data processing takes place. With the real-time data processing, the PC decides that it is an emergency case and sends the message of smoke detection to the operators. The operator evaluates the criticality of the situation and decides how to intervene (with an announcement from the loudspeakers or by taking more drastically measures, i.e., stopping the bus).
- Exposure to carbon dioxide: While the bus is on its route, high levels of $CO_2$ are detected from the relevant sensor. The sensor data is sent in real time to the central PC and is processed. $CO_2$ in unusual levels of air concentrations might have an adverse effect on passengers' health. For instance, high levels of $CO_2$ are related to dizziness, restlessness or breathing difficulties, and increased heart rate. To prevent an event concerning these health issues, such as a fainting, the PC sends

**Fig. 5.17** Environmental assessment process

the vehicle's central system the command to open the windows, so that the air is refreshed and can come back to its normal composition. Also, the passengers are informed of the air composition through the mobile application in real time.

- High temperature on the automated vehicle: The sensor measures the temperature and sends the data to the central PC. When the temperature exceeds a predefined level, the PC sends the command to start the cooling system. At the same time, passengers can be informed for the temperature inside the vehicle through the mobile application.

For the "shuttle environment assessment" service, a set of sensors was used to determine the air quality inside the automated shuttle environment, as well as to detect any passengers smoking and to prevent any fogging, as depicted in Fig. 5.17. After the sensors were deployed, they measured several metrics, such as $CO_2$, $NO_2$, humidity, temperature, fog, dust, and smoke, for a sufficiently long period of time. The collected values are used to predict the indoor conditions regarding the air quality for the next couple of hours. These measurements, as well as the conclusions of the real-time assessment, are sent to both the passengers and the operators of the shuttle. In case of an alarm, such as possible fire, they are sent to the suitable authorities. Moreover, this may be followed by appropriate notifications and/or instructions to the passengers, while the vehicle may also implement respective actions.

## 5.6 Service: Smart Feedback System

The service "smart feedback system" aims to allow the travelers inside the shuttle to give easy and effortless feedback to the operators when the safety driver is no longer inside the shuttle. It is important for the operators to know if people are satisfied with the services and transportation. Currently, the safety driver talks to the travelers and with his/her presence also becomes the conversation channel between the operators and the travelers. When the safety driver is removed, knowing whether they are satisfied or disappointed can be even more important, as the safety driver is not there to support, hence assisting the travelers.

When removing the operator, automated services must perform the same level of service and interaction as he/she did while being in the shuttle. This service aims to allow travelers to give their feedback about the service experience as easily as possible. This will be done by instructing the travelers to give a hand gesture to one of the cameras inside the shuttle. This will allow the travelers to effortlessly say "I like" or "I don't like" the experience with a thumbs up or a thumbs down. The concept will be communicated to the travelers via stickers inside the shuttle. Camera technology is used to capture the thumbs up or thumbs down, but, if possible, sound sensors will also be tested to capture the experience/feedback from the travelers. The service will be communicated to the travelers as follows:

- Giving a thumbs up/down in light settings: Midday with sunlight. Good visibility for the cameras.
- Giving a thumbs up/down in dark settings: Early morning or night with no sunlight. Low visibility for the cameras.
- Giving a thumbs up/down in crowded settings: Many passengers inside the shuttle, both standing and seating. Low visibility for the cameras due to people standing close to the cameras.
- Giving thumbs up/down in empty settings (or few passengers): Little or no passengers inside the shuttle. Good visibility for the cameras, easy to see the hand gesture.

For the "smart feedback system" service, the model is trained end to end and regularized so that it distills the most compact profile of the normal patterns of training data and effectively detects the "thumbs up" and "thumbs down" gestures. The original images of the hand gestures are acquired through the USB camera that is inside the shuttle and then passed through a single-shot detector (SSD) for the detection of the bounding box of where the hand is and the corresponding cropped frame. The cropped frame of the hand is then passed to the CNN that predicts a class vector output of values between 0 and 1, as illustrated in Fig. 5.18. These values correspond to the probability of the frame to be one of the classes. Real-time results are depicted in Fig. 5.19.

The results were validated during a twofold live session. Staff from HOLO performed fighting, bag snatching, falling, and vandalism scenarios in vehicle P109

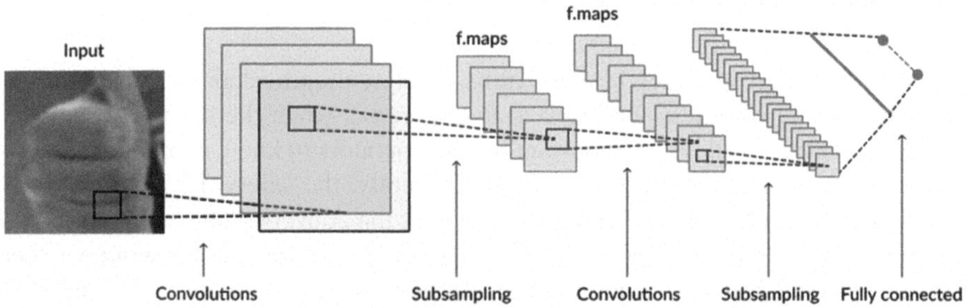


**Fig. 5.18** CNN architecture for thumb orientation detection

Fig. 5.19 The smart feedback service operating in real time inside the shuttle

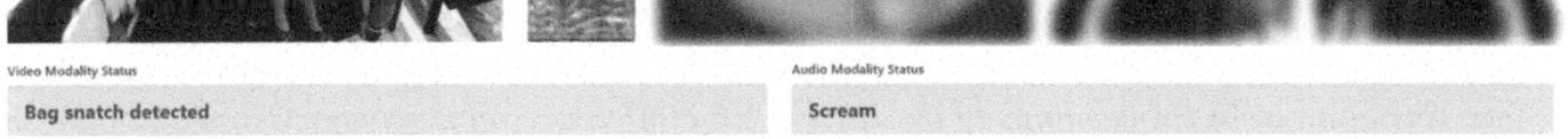


Fig. 5.20 Maintenance front-end interface regarding the service "enhance the sense of security and trust"

that serves the route Slagelse in Denmark. The algorithms were able to correctly identify the performed scenarios with 89% accuracy, and the appropriate notifications were also captured in the operator's dashboard. The two dashboards were correctly synchronized regarding real-time event detection. Figures 5.20 and 5.21 show the maintenance and operator dashboards regarding the results for the service "enhance the sense of security and trust." The validation of the automated passenger counting has been completed through manual comparison with data received from the operator phone's data stream. The automated count is considered precise in

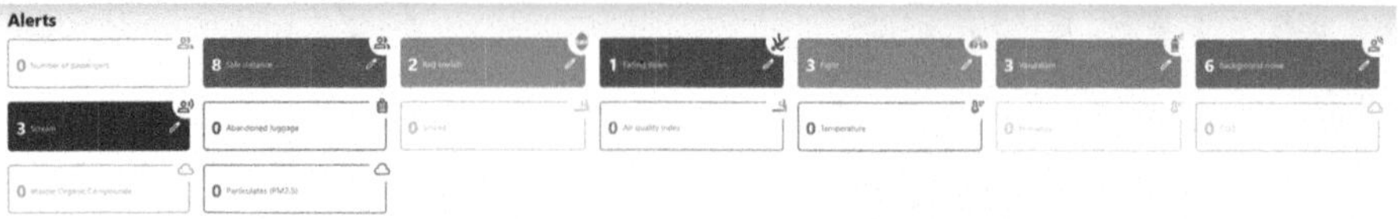

**Fig. 5.21** Operator dashboard regarding the service "enhance the sense of security and trust"

terms of timestamp and accurate in terms of the passenger count. Follow my kid, shuttle environment assessment, and smart feedback services were also validated successfully during the live session.

## 5.7 Conclusion

The transport service quality and passengers' safety and comfort are the major preoccupation of the public transportation operators. Thus, the automated AI-supported in-vehicle services will eventually substitute the role of the safety driver and the essential functions he/she currently provides inside the shuttle while enhancing the adherence of the passengers to the novel services and accelerating the adoption of automated mobility.

**Acknowledgements** This project has received funding by the European Union's Horizon 2020 research and innovation program under grant agreement No. 769033.

## References

Azuma, K. E. (2018). Effects of low-level inhalation exposure to carbon dioxide in indoor environments: A short review on human health and psychomotor performance. *Environment International, 121*, 51–56.

Deng, J., Guo, J., Xue, N., & Zafeiriou, S. (2019). Arcface: Additive angular margin loss for deep face recognition. In *Proceedings of the IEEE/CVF conference on computer vision and pattern recognition* (pp. 4690–4699).

Duan, Z., Tezcan, O., Nakamura, H., Ishwar, P., & Konrad, J. (2020). RAPiD: Rotation-aware people detection in overhead fisheye images. In *Proceedings of the IEEE/CVF conference on computer vision and pattern recognition workshops* (pp. 636–637).

Fang, H.-S., Xie, S., Tai, Y.-W., & Lu, C. (2017). Rmpe: Regional multi-person pose estimation. In *Proceedings of the IEEE international conference on computer vision* (pp. 2334–2343).

Olivera-La Rosa, A., Chuquichambi, E. G., & Ingram, G. P. (2020). Keep your (social) distance: Pathogen concerns and social perception in the time of COVID-19. *Personality and Individual Differences, 166*, 110200.

Papadimitriou, I., Vafeiadis, A., Lalas, A., Votis, K., & Tzovaras, D. (2020). Audio-based event detection at different SNR settings using two-dimensional spectrogram magnitude representations. *Electronics, 9*, 1593.

Salamon, J., & Bello, J. P. (2017). Deep convolutional neural networks and data augmentation for environmental sound classification. *IEEE Signal Processing Letters, 24*, 279–283.

Simonyan, K., & Zisserman, A. (2014). Very deep convolutional networks for large-scale image recognition. arXiv preprint arXiv:1409.1556.

Tsiktsiris, D., Dimitriou, N., Lalas, A., Dasygenis, M., Votis, K., & Tzovaras, D. (2020). Real-time abnormal event detection for enhanced security in autonomous shuttles mobility infrastructures. *Sensors, 20*, 4943.

Tsiktsiris, D., Lalas, A., Dasygenis, M., Votis, K., & Tzovaras, D. (2021). Enhanced security framework for enabling facial recognition in autonomous shuttles public transportation during COVID-19. In *IFIP international conference on artificial intelligence applications and innovations* (pp. 145–154).

Tsiktsiris, D., Lalas, A., Dasygenis, M., Votis, K., & Tzovaras, D. (2022). An efficient method for addressing COVID-19 proximity related issues in autonomous shuttles public transportation. In *IFIP international conference on artificial intelligence applications and innovations* (pp. 170–179).

Zhang, H., McLoughlin, I., & Song, Y. (2015). Robust sound event recognition using convolutional neural networks. In *2015 IEEE international conference on acoustics, speech and signal processing (ICASSP)* (pp. 559–563).

Zhang, K., Zhang, Z., Li, Z., & Qiao, Y. (2016). Joint face detection and alignment using multitask cascaded convolutional networks. *IEEE Signal Processing Letters, 23*, 1499–1503.

# Chapter 6
# Cybersecurity and Data Privacy: Stakeholders' Stand on Regulations and Standards

**Niels A. Nijdam, Meriem Benyahya, and Anastasija Collen**

**Abstract** AVENUE's technological interest lies in those state-of-the-art technologies and solutions that either are already commercial or close to the market and are expected to reach commercialisation within the following years. For the successful implementation of the project activities, a wide range of technological as well as non-technological elements will be employed, adapted, integrated, and, where required, partially developed. The present chapter focuses on those connected automated vehicle (CAV) technologies from the perspective of cybersecurity, delving into questions on in-vehicle, back-end, and infrastructure, including the communications between vehicle to vehicle (V2V), vehicle to infrastructure (V2I), vehicle to cloud (V2C), vehicle to everything (V2X), software safety, as well as security and privacy by design principles for the development of connected devices. Furthermore, non-technological issues cover stakeholder and user acceptance, regulatory and legislative requirements, a new standardisation progress, ethical considerations, and vehicle and technology certifications and licensing. The purpose of this chapter is to present the project context and relating it to the potential cyber assaults and data privacy threats. It further delineates the conducted assessment and the provided recommendations which were built based on the key standards and regulations wrapping together CAVs, cybersecurity, and personal data protection pursuits.

## 6.1 Introduction

More than ever, cybersecurity and data privacy are crucial to the introduction of new technologies, as everything tends to be *automated*, *autonomous*, and, most of all, *connected*. In this chapter, connected automated vehicle (CAV) technologies related

N. A. Nijdam (✉) · M. Benyahya · A. Collen
Centre Universitaire d'Informatique, Geneva School of Economics and Management, University of Geneva, Carouge, Switzerland
e-mail: niels.nijdam@unige.ch

 
G. Fournier et al. (eds.), *Automated Vehicles as a Game Changer for Sustainable Mobility*, Contributions to Management Science,
https://doi.org/10.1007/978-3-031-61681-5_6

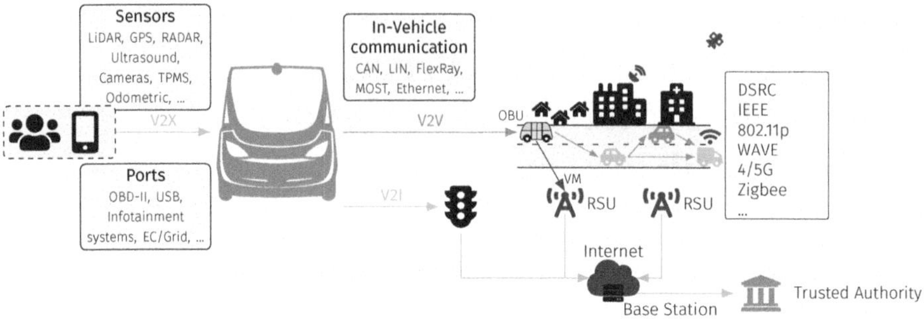


**Fig. 6.1** Attack vectors within the CAV ecosystem

to new digital services will be closer inspected to elaborate on a plurality of new threat vectors.

## *6.1.1 CAVs' Threats*

AVENUE involves a wide spectrum of technologies for a broad coverage and analysis of the existing advancements in a full ecosystem of the CAVs as depicted in Fig. 6.1:

- In-vehicle equipment: categorised in the present work as (1) sensors, representing the elementary devices through which the vehicle builds its perception and awareness model; (2) in-vehicle communication, the subsystems within the vehicle that serve for communication; and (3) ports, granting physical access to various parts of the vehicle systems.
- External communication technologies: from vehicle to vehicle (V2V) for fleet coordination to vehicle to infrastructure (V2I) and vehicle to everything (V2X) for infrastructure deployment in the cities and public transport operators (PTOs).

### **6.1.1.1 In-Vehicle Equipment**

To assure the CAV's highly automated navigation of levels 4 and 5 as defined by the Society of Automotive Engineering (SAE) (SAE, 2021), the vehicle intelligently compiles inputs collected from its internal equipment (including cameras, Global Positioning System (GPS), radio detection and ranging (RADAR), light detection and ranging (LiDAR), tire pressure monitoring system (TPMS), odometry, and ultrasound sensors). The processing and fusion of the collected data from such sensors support on providing accurate data on positioning, behaviour predictions, collision avoidance, pedestrian detection, and object recognition (Parkinson et al., 2017). Despite the sensors' crucial role to safe self-driving, they remain victims to attacks like spoofing, jamming, relay, and tampering, leading the vehicle to either hit non-perceived objects or consider non-existent obstacles (Benyahya, Collen, et al., 2022).

Furthermore, electronic control units (ECUs) are part of the essential in-vehicle components which control the CAV's systems by receiving and processing broadcast signals from sensors (Dibaei et al., 2020). On the same note, ECUs are connected to in-vehicle networks like controller area network (CAN), Local Interconnect Network (LIN), Media-Oriented System Transport (MOST), FlexRay, and Ethernet (Wu et al., 2020). Each protocol supports different communication within the in-vehicular network. However, they embed many security concerns and risks of potential attacks such as denial of service (DoS), packet injection, sniffing, eavesdropping, spoofing, relaying, and bus-off (El-Rewini et al., 2020).

Physical ports can broaden the points of entrance to an attack over the in-vehicle network. The on-board diagnostic (OBD) is a port that is normally used for the vehicle diagnosis or for ECU firmware upgrade (Elliott et al., 2019). Though if false messages are injected through it, the control over the automotive functions (like braking) can be granted to the attacker (Parkinson et al., 2017). Similarly, falsified data, malware, or a virus can be injected through Universal Serial Bus (USB) ports or the electric charging port (Bhusal et al., 2020).

#### 6.1.1.2 External Communication Technologies

External vehicular communications come to complement the in-vehicle equipment on the automated driving functions. CAVs communicate to their surroundings using V2V to broadcast and receive signals from other vehicles and V2I to exchange data with the infrastructure (Elliott et al., 2019). Besides, V2X embeds both V2V and V2I in addition to communication with any external devices like smartphones (El-Rewini et al., 2020). Such communication modes are taking place thanks to vehicular ad hoc networks (VANETs) where vehicles and infrastructure are referred to as nodes which can exchange traffic-related messages.

The VANET architecture is composed of on-board units (OBUs), which are located at the level of the CAV, roadside units (RSUs), which are placed within the infrastructure, and a trusted authority (TA), which is the base station connecting OBUs and RSUs to the core network and which distributes public/private keys and certificates among nodes (Noh et al., 2020). VANETs are supported by wireless technologies like IEEE 802.11p, which is the basis of dedicated short-range communication (DSRC) (Ali & Li, 2020). Maple et al. (2019) added that further Internet of Things (IoT) technologies such as IEEE 802.15.4 or ZigBee might be used to facilitate information exchange. However, if the VANETs' signals are eavesdropped by an attacker, the CAV can be compromised or maliciously tracked (Veitas & Delaere, 2018). Further attacks were identified and asserted to dangerously impact the traffic including sybil attack, distributed denial of service (DDOS), and man in the middle (MitM) (Dibaei et al., 2020).

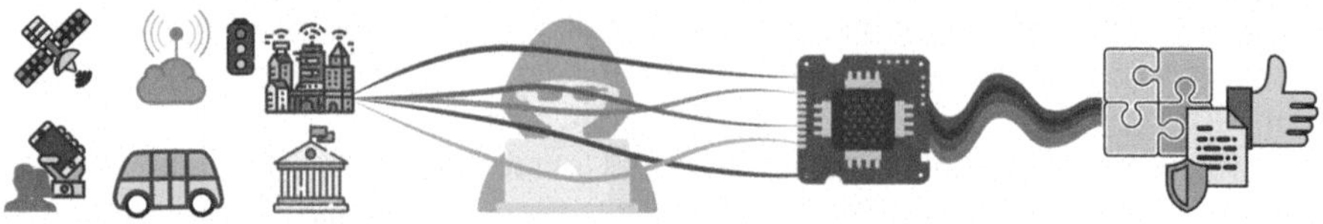

**Fig. 6.2** Investigation methodology

CAVs use a plethora of technologies to come to fruition of safe automated driving and high connectivity with their surroundings. Nevertheless, such minibuses end up inheriting security weaknesses and accumulating additional threat vectors that can be real showstoppers to the CAV's prevalence.

### *6.1.2 Motivation*

Bearing in mind the rampant CAV's threats, the different automation levels, and the multiple stakeholders within the AVENUE ecosystem, a cybersecurity and data privacy assessment and a guide of references regarding standards and regulations will be provided. Figure 6.2 depicts the adopted approach fusing thorough understanding of CAV's challenges, analysis of existing regulations and standards chasms, and an evaluation of the AVENUE landscape, resulting to what is presented as the standards coverage map (SCM).

This article addresses the following questions:

- Within the AVENUE ecosystem, how can cybersecurity and data privacy threats be efficiently mitigated?
- What are the key technical tools recommended by legal policies and standardisation bodies to countermeasure those threats?
- How are risk assessment, vulnerability analysis, and penetration testing considered by the partners to shield the AVENUE minibuses?
- By being compliant to the existing standards, how anchored would the CAV be from both cyber assaults and data leakages?

The remainder of this chapter is structured as follows. Section 6.2 sheds light on the most recent (up to 2021 fourth quarter) regulations and standards to be considered within the CAV's landscape. Moreover, it discusses global efforts which can infer some lessons learned. Section 6.3 outlines the cybersecurity assessment and data collection tools adopted within the AVENUE scope. Section 6.4 presents the means and input collection from the different PTOs and software providers. Section 6.5 depicts the key recommendations upon the identified shortcomings from the cybersecurity and data privacy perspectives. Finally, Sect. 6.6 provides concluding statements and future work orientations.

## 6.2 Regulations and Standards

Being crucial for the evolution, development, and deployment of CAVs, cybersecurity and data privacy challenges have attracted many stakeholders including automotive manufacturers, legal and regulatory bodies, information technology (IT) and telecommunication suppliers, operators of intelligent transport system (ITS), and mobility service providers to collaborate and come up with new laws, strategies, and guidelines. Figure 6.3 reflects an in-depth overview of the existing and forthcoming efforts from the key players in providing both mandatory and nice to have requirements.

The 4 years of the AVENUE project witnessed a major progress in regulating CAVs' deployment. In August 2022, the European Commission (EC) published the regulation (EU) 2022/1426 (Regulation (EU) 2022/1426, 2022) where technical specifications for the type approval of fully automated vehicles were defined, built upon the generic vehicular regulation 2019/2144 (Regulation (EU) 2019/2144, 2019). Furthermore, Network and Information Security (NIS) 1 and 2 (EU 2016/1148; EU 2020/1148) directives call the operators of IT service providers to take the appropriate measures to manage cyber risks posed to the security in a general scope.

While the EC regulations laid down on either generic type of approval or IT cybersecurity, the United Nations Economic Commission for Europe (UNECE) published acts joining cybersecurity measures to the CAVs' environment. The UNECE R155 (UNECE R155, 2020) and R156 (UNECE R155, 2020) came with the purpose to unify the automotive standards by requiring cybersecurity management system (CSMS) and software update management system (SUMS) certifications accordingly for the SAE level 3 onward. The two certifications cover the cybersecurity risk management, security by design, and security incident detection and mitigation and secure software updates over the CAV's life cycle including development, production, and post-production (Suh, 2020). The involvement of

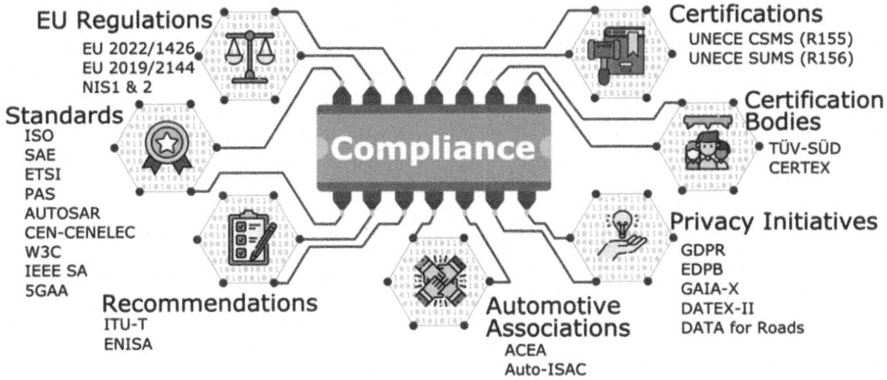


**Fig. 6.3** Taxonomy of regulation and standardisation of stakeholders

certification third parties, such as TÜV SÜD and CertX, can be compulsory to generate CSMS and SUMS either for the first time or for their renewal occurring every 3 years.

Under the auspices of standard development organisations (SDOs), efforts were made to shield the entire CAV's environment. The International Organisation for Standardisation (ISO) and SAE working groups claim to provide a complete cybersecurity management for the driverless landscape with a focus on the in-vehicle components (Schoitsch & Schmittner, 2020). The ISO/SAE 21434 (ISO/SAE 21434, 2021) and ISO/PAS 5112 (ISO/PAS 5112, 2021) represent the salient standards providing a high-level guidance on cybersecurity governance and auditing for the CAV's ecosystem. The European Telecommunication Standards Institute (ETSI), Publicly Available Specification (PAS), Automotive Open System Architecture (AUTOSAR), European Committee for Standardisation (CEN), World Wide Web Consortium (W3C), IEEE Standards Association (IEEE SA), and 5G Automotive Association (5GAA) institutions provide standards for securing vehicular communication (Kim & Shrestha, 2020).

As depicted in Fig. 6.3, further recommendations were provided by the International Telecommunication Union (ITU) outlining security threats definition, security guidelines for V2X, specification of secure software update procedure for ITS devices, and guidelines for intrusion and misbehaviour detection (ITU-T, 2020). Additionally, the European Union Agency for Cybersecurity (ENISA) published several reports spotlighting the CAVs' cybersecurity risks once deployed within smart cities. Besides, automotive associations such as the European Automobile Manufacturers Association (ACEA) (ACEA, 2019) and the Automotive Information Sharing and Analysis Centre (Auto-ISAC) (Auto-ISAC, n.d.) are orienting the original equipment manufacturers (OEMs) towards self-audit, testing, and deploying incident response plans.

### *6.2.1 CAVs Privacy Initiatives*

The General Data Protection Regulation (GDPR) is the fundamental privacy data law in the European Union (EU). It sets strict obligations related to personal data (PD) processing, rights for concerned individuals (data subjects), technical requirements to employ privacy preserving techniques (anonymisation and pseudonymisation), and Data Protection Impact Assessment (DPIA) as an assessment for any new technologies with privacy risks (Regulation (EU) 2016/679, 2016). Nevertheless, within the CAV's complex environment, GDPR implementation remains convoluted, as the stakeholder can accumulate multiple roles (e.g. being data processor and data controller simultaneously) (Benyahya, Kechagia, et al., 2022). Another limitation of the GDPR within the CAV context is considering anonymisation as a permanent solution. Though with minor reverse engineering efforts, the PD can be de-anonymised with no compliance violation to the GDPR (ENISA, 2022). Hence, a full compliance does not rhyme with an absolute PD protection.

To limit the GDPR pitfalls, the European Data Protection Board (EDPB) provided guidance on the processing of PD in the context of ITS, which highlighted privacy risks and provided recommendations on data protection by design and by default within the CAV's environment (European Data Protection Board, 2020). Moreover, the guidelines focused on consent as the legal basis for processing PD inside the vehicle and through V2X communications (European Data Protection Board, 2021).

Further European initiatives, such as GAIA-X, DATEX-II, and Data for Roads, aim to overcome the current privacy hurdles within the vehicular environment. GAIA-X provided proposals on data protection rules and architecture standards in many areas including location-based services (LBS) and mobility as a service (MaaS) in smart mobility (BMWi, 2020). DATEX II has been addressing traffic data sharing and transmission including transmitted data in cooperative and connected mobility (DATEX-II, n.d.). Data for Road Safety is another EU initiative who has been discussing connected vehicles of all automation levels and gathers partners from the European Commission, industries, and governments to reach cooperative, trustworthy, and free of charge vehicle data exchange with respect to the protection of PD (Data for Road Safety, 2021).

Within the AVENUE landscape, data should be processed with the consideration of the essential factors summarised in Fig. 6.4. It wraps up the GDPR and other aforementioned privacy initiatives recommendations. In other words, it reflects the data controllers' obligations regarding the data subjects' rights including the processing restrictions, privacy by design deployment, and consent implementation. From the privacy preserving perspectives, the figure depicts the anonymisation and pseudonymisation challenges. Finally, it calls back the high privacy risks linked to LBS and MaaS which are more likely to be deployed within the CAV's environment.

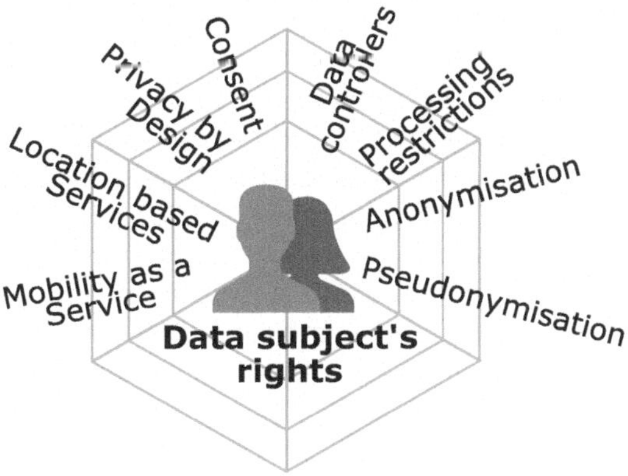


**Fig. 6.4** Data privacy challenges within the AVENUE environment

## 6.3 Methodology

An overall evaluation and recommendations are built based on the final in-vehicle and out-of-vehicle services implemented in AVENUE. A questionnaire was conducted to collect updated inputs from the different key stakeholders. Based on the collected data, the cybersecurity and data privacy mechanisms were assessed per deployment site and service provider. The questionnaire findings were evaluated further by matching the identified surface attacks to required standards from Sect. 6.2. Such mapping supported on providing the recommendations through an SCM defining the required standards to adopt per layer and sub-component of the minibuses' ecosystem.

To conduct the intended cybersecurity and data privacy assessment, it was crucial to identify the involved service providers and the final AVENUE services. Multiple data collection tools were deployed through several iterations, where questionnaires represented the most relevant source of extensive findings, summarised in Fig. 6.5. The orange entries were substituted by the respondent organisation name and related services. Questions related to hardware attributes were targeting the OEM, and questions on software patterns were designated to software service providers. The questionnaire was shared with several key stakeholders (henceforth providers):

(1) Three on-demand service providers, each providing services in one or more deployment sites (SP1, SP2, and SP3); (2) three software developers, developing in-vehicle and out-of-vehicle services (SD1, SD2, SD3); and (3) one OEM responsible for manufacturing the automated minibuses and vehicle operational software. Furthermore, the four public transport operators (TP1, TP2, TP3, TP4) for the deployment sites have been included as well.

The questionnaire consists of 19 questions that are split into 4 main parts. The first section spotlights the applications that are in the AVENUE scope and which are intended to be delivered by the end of the project. With the multiple types of respondents, the first section of the questionnaire aims to identify the site, the application that is or will be deployed, and by whom. Then multiple questions are raised regarding cybersecurity in the second part of the questionnaire with the purpose to highlight the cybersecurity tools, standards, and regulations adopted by the respondent to identify if the mandatory UNECE regulations (R155 and R156) and the key vehicular cybersecurity ISO standards are taken into consideration. The third part of the questionnaire collects information related to data privacy and to the GDPR compliance. Finally, the fourth part gathers any sketched architecture or design that the respondent aims to deploy within the AVENUE project.

Partner name

**Final Applications**

Q1 What are the applications that Partner is using and/or developing for AVENUE?

Q2 Please specify in which site your platform will be deployed within the AVENUE scope?

**Cybersecurity**

Q3 What are the security measures you have in place to prevent source code vulnerabilities?

Q4 Do you conduct internal pentests regularly or just if a risk is raised? Do you conduct hardware or software based pentests?

Q5 What is the minimum CVSS accepted?

Q6 How are Spoofing and Jamming attacks prevented by the GNSS receiver at the level of the vehicle GPS?

Q7 Do you have any vulnerability report? How often are security audits conducted over the implemented application? Would you like to share with us your latest audit report?

Q8 Do you coordinate with other partners' cybersecurity teams to assess risks on connecting your platform to the other servers/systems?

Q9 Do you provide training to vehicle operators/supervisors to face cybersecurity attacks?

Q10 Are you working on the CSMS certification required by the UN 155 regulation?

Q11 How is security handled while doing a software update? Is there any vulnerability scanning before?

Q12 Do you refer to any security standards/guidelines such as ISO9001, ISO27000, ISO/SAE21434...?

**Data Privacy**

Q13 Does your application use any personal data? Does it aggregate data from other service providers?

Q14 Is the location data, collected through your applications encrypted/anonymized before it is saved on the database? Do you process any IMEI/MEID ID? If yes, is it anonimysed/pseudonymised?

Q15 How do you secure the connection and the data exchange between your applications and data providers services (eg Navya/TPG,...) ?

Q16 How data processing and storage are handled with regard to the GDPR recommendations as well as the FDPIC acts? How long is personal data stored? and how often it is destroyed? Do you ask for the user's consent before storing the data?

Q17 Do you share any data with any LBS platforms?

Q18 Which procedure do you deploy in case of a data breach? Do you have a clearly elaborated procedure?

**Architecture**

Q19 Could you provide a view of your architecture as it is used within AVENUE?

**Fig. 6.5** Applications' service provider questionnaire

## 6.4 Findings

From the qualitative analysis perspective, it is noteworthy to highlight that the synthesised outputs from the questionnaire provided valuable findings regarding the final applications and the cybersecurity and privacy governance within AVENUE.

Table 6.1 gives an overview of the various application types and their usage by the operators. During the project's lifetime, the operators faced numerous challenges, which affected the selection and development of the services that were ultimately realised and utilised thereafter. The presented table abstracts this view and shows the mapping of the applications provided by the various service providers and software developers with the operators. It can be noted that only one service provider (SP1) was utilised by two PTOs. In most cases the service operators provided the client application (a mobile application) and also an application for the safety operator to receive new on-demand trip orders from the client. Meaning, the service provider may send an order through the OEM interface and act as the fleet manager or utilise a semimanual approach by informing the safety operator (from the PTO) to select the destination through the in-vehicle control service (using a touch screen).

In addition to the applications' mapping, the questionnaire inputs supported in assessing the cybersecurity governance generally and the code security measures specifically. Prevention, monitoring, and audit tools were gathered from the collected answers. Table 6.2 wraps up the key security measures adopted by every service provider. All respondents are highly concerned with the code vulnerabilities prevention. However, more efforts are still required to conduct continuous assessments and internal and external audits. From the standardisation perspective, valuable inspirations are retrieved from ISO and SAE key standards, while the mandatory UNECE R155 (UNECE R155, 2020) and R156 (UNECE R156, 2020) are still envisioned and planned before end of 2023. Though no accomplished certification or concrete compliance has been recorded yet.

**Table 6.1** Application type and usage by transport operators

| | | Operators | | | |
|---|---|---|---|---|---|
| Provider | Applications | TP1 | TP2 | TP3 | TP4 |
| SP1 | On-demand service (passenger and safety operator app), control centre | ✓ | ✓ | | |
| SP2 | On-demand service (passenger and safety operator app) | | | ✓ | |
| SP3 | On-demand service (passenger and safety operator app), control centre | | | | ✓ |
| SD1 | Data statistics dashboard | ✓ | ✓ | | ✓ |
| SD2 | In-vehicle dashboard (interface) | ✓ | | | ✓ |
| SD3 | In-vehicle dashboard (back-end) | ✓ | | | ✓ |
| OEM | Vehicle control, in-vehicle control service, fleet manager, data statistics (API) | ✓ | ✓ | ✓ | ✓ |

**Table 6.2** Code security measures indicated by providers

| | Security measures | | | | | |
|---|---|---|---|---|---|---|
| Provider | Prevention | Assessment | Penetration testing | Audit | Standards | Certification |
| SP1 | ✓ | SCA (Ferrara et al., 2021) | ✓ | ✓ | ISO 27001[a] | N/A |
| SP2 | ✓ | ✗ | ✓ | ✗ | ✗ | N/A |
| SP3 | ✓ | ✗ | ✗ | ✗ | ✗ | N/A |
| SD1 | ✓ | ✗ | ✗ | ✗ | ✗ | N/A |
| SD2 | ✓ | SAST (Synopsys, n.d.) | ✗ | ✗ | ✗ | N/A |
| SD3 | ✓ | CVSS V3.1 (NIST, n.d.) | ✓ | ✓ | ✗ | N/A |
| OEM | ✓ | MOABI (Moabi, n.d.)<br>Nessus (Tenable, n.d.) | ✓ | ✓ | SAE J3061<br>ISO/SAE 21,434<br>ISO 2700<br>IEC 62443 | R155 (UNECE R155, 2020)<br>R156 (UNECE R156, 2020) |

[a]Certified

**Table 6.3** Data privacy measures indicated by providers

| | Privacy measures | | | |
|---|---|---|---|---|
| Provider | Processed PD | Connectivity | GDPR | LBS sharing |
| SP1 | Location data | Encrypted HTTPS | User consent<br>Data storage<br>Data destruction | ✓ |
| SP2 | Name and phone number | Encrypted HTTPS | User consent<br>Data storage | ✗ |
| SP3 | Not collected | Tokenised | Not applicable | ✗ |
| SD1 | Not collected | Tokenised | Not applicable | ✗ |
| SD2 | Not collected | Encrypted HTTPS | User consent | ✗ |
| SD3 | Video and audio + facial biometrics | VPN + TLS certificates (encryption) | User consent<br>In-vehicle data storage | ✗ |
| OEM | Location data<br>Video streaming | Cyphered VPN | Data storage<br>Data destruction | ✗ [a] |

[a]Likely to occur

From the data privacy angle, Table 6.3 showcases how the providers are careful about the GDPR compliance in terms of respect to data processing principles. For key stakeholders who are collecting sensitive data such as personal and location data, their efforts are reflected through the user consent implementation, data storage limitation, and data destruction procedures. Additionally, worthy pseudonymisation techniques are deployed like encryption and tokenisation for authentication.

Moreover, at that stage of the mini-shuttle deployment, sharing data with LBS platforms remains limited without claiming any PD. Albeit laudable efforts are adopted by the service providers, data privacy assessments such as DPIA are omitted from the data protection procedures set in the AVENUE project.

## 6.5 Discussion and Recommendations

Per the conducted assessment, the cybersecurity and data privacy governance were evaluated including risk prevention, vulnerability management, events monitoring, and data leakage plans. From the aforementioned discussion, the governance strengths can be summarised as follows:

- The AVENUE partners are highly aware of the cybersecurity risks.
- Several "security by design" principles such as risk prevention are considered while implementing AVENUE services.
- The GDPR compliance is a high-priority obligation for all partners.
- The mandatory certifications (CSMS and SUMS) related to the UNECE R155 (UNECE R155, 2020) and R156 (UNECE R156, 2020) are envisioned and planned before end of 2023.
- The V2X immaturity represents an instinctive mitigation solution to the connectivity threats where the minibuses have limited V2I and V2V.

On the other hand, it is noteworthy to spotlight the following weaknesses:

- Albeit the high awareness about the cyber assaults, further thorough implementation of a cybersecurity culture is still required through all the system layers and among all the partners.
- For a flawless cybersecurity governance, the partners' efforts should not be limited to prevention tools, but it should be broadened to implement monitoring, continuous assessments, and risk mitigation strategies through the overall life cycle including the decommissioning stage.
- Despite the GDPR considerations within the project, an in-depth implementation of the law is still recommended as several obligations are still missing such data privacy assessments.
- There is a scarcity on deploying cybersecurity validation process as no penetration testing is conducted while testing the vehicles on the AVENUE sites.

### *6.5.1 Standards Coverage Map*

To overcome the asserted hurdles, the CAV's ecosystem SCM is suggested as a roadmap to be followed by the AVENUE partners upon their provided services and scope. The SCM is foreseen as a suitable approach on building a robust cybersecurity and data privacy governance based on the CAV's standards and regulation discussed in Sect. 6.2. The map combines both the technical and organisational audit avenues applied to the automated minibuses' landscape as depicted in Fig. 6.6. The map is classified into four layers, in vehicle, out of vehicle, applications, and organisation, where every layer groups the respective technical standards.

As a parent node of the four layers, ISO/SAE 21434(ISO/SAE 21434, 2021) is set as the core standard; however it is very generic to be followed per se. The combination of both generic and technical standards on the SCM is foreseen to overcome the broadening of the ISO/SAE 21434 leading to a more thorough assessment.

The in-vehicle layer addresses the potential attacks at the vehicle level (as discussed in Sect. 6.1.1) which were classified into six sub-layers. First, the sensors category defines the guidance on standardising the interfaces between the different sensors and the fusion unit leading to the automation navigation decisions. Second, the network buses category where standards propose guidelines on detecting intrusions and authentication measures within the in-vehicle communication networks. Third, the ECU standard aims to prevent non-authorised access to the vehicular software modules. Fourth, the software update outlines the directives on how to conduct secure software updates during the vehicle life cycle. Fifth, the artificial intelligence (AI) components standard provides guidance on secure usage of AI-based functions involved on the automation decision-making. Finally, the physical access specifies countermeasures against threats from plugged-in external devices.

Even with a limited V2X implementation within AVENUE, the communication-related recommendations for future development are anticipated. In Fig. 6.6, the out-of-vehicle layer relies on two main categories wrapping standards related to countermeasure CAV's Internet and V2X threats. To secure the vehicle's Internet access using DSRC, long-term evolution (LTE), and 5G, standards provided a set of secure channel models and through several use cases. Besides, the multiple V2X communications have been standardised by ISO, ETSI, and SAE. The security credential management standards, which sets V2X certificates security and privacy requirements, define the precise structure, format, and authentication schemes supporting the minibuses' communication to peer instances. It is noteworthy to mention that other V2X communications such as V2I and vehicle to grid (V2G) have been supported by dedicated standards, while others as per the vehicle to cloud (V2C) is still considered under the umbrella of broad standards like SAE J2735.

Moreover, the application layer consists of two sub-layers reflecting two types of applications: users and ITS. The minibuses' deployment is associated to the means of several services provided to the end user, such as the on-demand application, and to the smart city. The user application standards focus on data access and cryptography best practices to consider while building interfaces to the CAV's hardware or software. Likewise, the ITS application standards recommend mechanisms to determine permitted actions among the peer ITS applications to achieve security properties such as authorisation, integrity, and confidentially. Nevertheless, standards such as SAE J2735 and ISO/TS 21177 have a larger scope covering the V2X communication in general and, hence, other subcomponents from the second layer too.

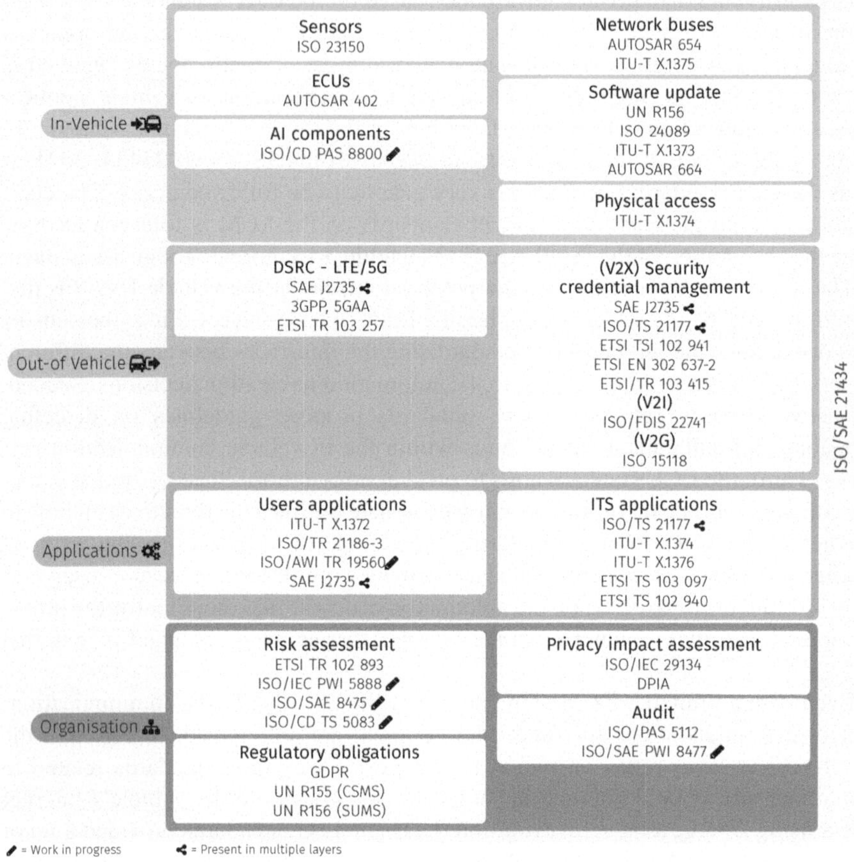


**Fig. 6.6** Standards coverage map

Finally, the organisation layer in Fig. 6.6 incorporates four procedural sub-layers supporting on mitigating the assessed weaknesses in the previous section. The risk assessment reflects evaluation procedures on quantifying the likelihood and impact of cybersecurity threats. The privacy impact assessment wraps standardised procedures, sample reports, and checklists to fulfil. The regulatory obligations sets the mandatory laws that the minibuses' environment has to comply with, which are summarised into the GDPR, the UNECE R155, and the UNECE R156. Finally, the associate's authorisation group points out the managerial methods that need to be conducted by the minibuses' trained associates while facing a cyber threat and processing any PD. The combination of both generic and technical standards on Fig. 6.6 is foreseen to provide a clear roadmap overcoming the cybersecurity and data privacy challenges discussed in the present work.

### 6.5.2 Further Recommendations

If a granular certification and standardisation, as recommended by the SCM, bring endeavours and extra efforts to AVENUE partners, we recommend to at least enforce the cybersecurity and data privacy culture through the following key actions:

- Accelerate the compliance process to the UNECE R155 (UNECE R155, 2020) and R156 (UNECE R156, 2020), the new mandatory regulations
- Deploy officially the intended mitigation strategies
- Plan cybersecurity assessments on the organisational and project levels
- Determine clear responsibilities and roles within the AVENUE partners and stakeholders by asserting who are data processors, sub-processors, data controllers, and joint controllers as defined by the GDPR (Regulation (EU) 2016/679, 2016)
- Unify the attack rating approaches among the partners, like threat analysis and risk assessment (TARA) and CVSS approaches
- Maintain permanent and up-to-date risk monitoring through risk matrices for the overall environment not only limited to the vehicle itself
- Conduct data privacy (such as DPIA) and event assessments
- Invest further resources on operators training on cyberattacks to deploy accurate countermeasures on real time while supervising the minibuses
- Identify clear procedures and plans in case of a data leakage

### 6.5.3 Assessment Limitations

With the CAVs' market evolution and rivalry, the automated minibuses are still foreseen as a black box for partners who are not OEMs or hardware providers. It is true that with less knowledge about the embedded technologies and the monitoring tools, the CAV can benefit from higher protection from cyberattacks. However, as security practices dictate, security by obscurity is not a viable solution in the long term. Furthermore, such shelter can represent a blocking wall for a granular cybersecurity and data privacy assessment.

Standards from various standardisation bodies at any stage, published or work in progress, change very often which requires a recurrent update of the SCM. As a matter of fact, the SCM should be updated frequently to cope with the evolving regulation and standardisation bodies publications.

## 6.6 Conclusion

As the public transport sector is dipping their toes into utilising the latest CAV technologies with the vision of offering on-demand, door-to-door, automated fleet of minibuses, this chapter brought forth the concerns regarding the cybersecurity and data privacy. As these vehicles are highly digitised and connected, they are not only prone to the same threats as traditional Internet connected devices but also affected by additional vulnerabilities due to their non-static behaviour and physical outdoor exposure. This was illustrated through two main attack vectors, namely, in-vehicle and external communications, and further briefly overviewed the related standards and regulations applicable in the domain.

As a use case the perceptiveness of the public transport operators, OEMs, and service providers (involved with or affiliated to the EU AVENUE project) was analysed through a questionnaire, with the outset to investigate if cybersecurity is taken thoroughly into account and integrated at an early stage. As per general observations, a great focus lies on getting the CAVs running and integrating the right services in place, where cybersecurity and privacy may easily become an afterthought. The outcome shows that cybersecurity and privacy are considered and several strategies are in place by the respective parties. However, instead of being pushed by regulations, more emphasis needs to be in place to naturally woven into the development process and life cycle of any related soft- and hardware development. For example, the OEM adheres the most for its implementation, but as the manufacturer of a CAV, they are under close observation of all the safety and security standards regulations, whereas any service developer and/or fleet managers who are providing higher level (software) functionalities are not.

From the assessment, a further set of recommendations is rationalised, and a standards coverage map is presented that organises known standards and legal policies into several layers (categories). It aims to provide a roadmap to ease the integration of cybersecurity and data privacy aspects. Finally, some limitations of the current work are pointed out attributed to the fast changing landscape of standards and regulations, as well as limited transparency due to highly competitive market and therefore reluctance of providing insights in ongoing/active developed products.

**Acknowledgements** This project has received funding by the European Union's Horizon 2020 research and innovation program under grant agreement No. 769033.

## References

Ali, I., & Li, F. (2020). An efficient conditional privacy-preserving authentication scheme for Vehicle-To-Infrastructure communication in VANETs. *Vehicular Communications, 22*, 100228.

Auto-ISAC. (n.d.). Best Practices. https://automotiveisac.com/best-practices/

Benyahya, M., Collen, A., Kechagia, S., & Nijdam, N. A. (2022). Automated city shuttles: Mapping the key challenges in cybersecurity, privacy and standards to future developments. *Computers & Security, 122*, 102904.

Benyahya, M., Kechagia, S., Collen, A., & Nijdam, N. A. (2022). The interface of privacy and data security in automated city shuttles: The GDPR analysis. *Applied Sciences, 12*(9), 4413.

Bhusal, N., Gautam, M., & Benidris, M. (2020). Cybersecurity of electric vehicle smart charging management systems. arXiv.

BMWi. (2020). *GAIA-X: The European project kicks off the next phase. Technical report.* Federal Ministry for Economic Affairs and Energy (BMWi). www.bmwi.de

Data for Road Safety. (2021, July). *Privacy statement-data for road safety. Technical report*. Data for Road Safety. https://www.dataforroadsafety.eu/

DATEX-II. (n.d.). Datex II specifications. https://datex2.eu/datex2/ specifications

Dibaei, M., Zheng, X., Jiang, K., Abbas, R., Liu, S., Zhang, Y., Xiang, Y., & Yu, S. (2020). Attacks and defences on intelligent connected vehicles: A survey. *Digital Communications and Networks, 6*(4), 399–421.

Directive (EU) 2016/1148. (2016, July). *Measures for a high common level of security of network and information systems. Technical report*. The European Parliament and of the Council.

Elliott, D., Keen, W., & Miao, L. (2019). Recent advances in connected and automated vehicles. *Journal of Traffic and Transportation Engineering (English Edition), 6*(2), 109–131.

El-Rewini, Z., Sadatsharan, K., Selvaraj, D. F., Plathottam, S. J., & Ranganathan, P. (2020). Cybersecurity challenges in vehicular communications. *Vehicular Communications, 23*, 100214.

European Automotive Manufacturers Association (ACEA). (2019). *Roadmap for the deployment of automated driving in the European Union. Technical report*. ACEA.

European Union Agency for Cybersecurity. (2022, January). *Data protection engineering. Technical report*. ENISA. www.enisa.europa.eu

Ferrara, P., Mandal, A. K., Cortesi, A., & Spoto, F. (2021). Static analysis for discovering IoT vulnerabilities. *International Journal on Software Tools for Technology Transfer, 23*(1), 71–88.

Guidelines 1/2020 on processing personal data in the context of connected vehicles and mobility related applications. Technical report. The European Data Protection Board (2020, January).

Guidelines 1/2020 on processing personal data in the context of connected vehicles and mobility related applications. Technical report. The European data Protection Board (2021, March).

ISO/PAS 5112. (2021). *Road vehicles - Guidelines for auditing cybersecurity engineering. Technical report*. ISO.

ISO/SAE 21434. (2021). *Road vehicles - Cybersecurity engineering. Technical report*. ISO.

ITU-T. (2020). *X. 1371 Security threats to connected vehicles. Technical report*. ITU-T.

Kim, S., & Shrestha, R. (2020). *Automotive cyber security*. Springer.

Maple, C., Bradbury, M., Le, A. T., & Ghirardello, K. (2019). A connected and autonomous vehicle reference architecture for attack surface analysis. *Applied Sciences (Switzerland), 9*(23), 5101.

Moabi. (n.d.). Connected vehicles. https://moabi.com/en/home/

NIST. (n.d.). NVD - CVSS v3 calculator. https://nvd.nist.gov/vuln-metrics/cvss/v3-calculator

Noh, J., Jeon, S., & Cho, S. (2020). Distributed blockchain-based message authentication scheme for connected vehicles. *Electronics (Switzerland), 9*(1), 74.

Parkinson, S., Ward, P., Wilson, K., & Miller, J. (2017). Cyber threats facing autonomous and connected vehicles: Future challenges. *IEEE Transactions on Intelligent Transportation Systems, 18*(11), 2898–2915.

Proposal of Directive (EU) 2016/1148. (2020, December). *Measures for a high common level of cybersecurity across the Union, repealing Directive (EU) 2016/1148. Technical report*. European Commission.

Regulation (EU) 2016/679. (2016, April). *On the protection of natural persons with regard to the processing of personal data and on the free movement of such data, and repealing Directive 95/46/EC (General Data Protection Regulation). Technical report*. The European Parliament and of the Council.

Regulation (EU) 2019/2144. (2019, November). *Type-approval requirements for motor vehicles and their trailers, and systems, components and separate technical units intended for such vehicles, as regards their general safety and the protection of vehicle occupants and vulnerable road users. Technical report*. The European Parliament and of the Council.

Regulation (EU) 2022/1426. (2022, August). *Uniform procedures and technical specifications for the type-approval of the automated driving system (ADS) of fully automated vehicles. Technical report*. The European Parliament and of the Council.

SAE. (2021, July). *Surface vehicle information report. Technical report*. SAE.

Schoitsch, E., & Schmittner, C. (2020). Ongoing cybersecurity and safety standardization activities related to highly automated/autonomous vehicles. In *AMAA 2020: Intelligent system solutions for auto mobility and beyond* (pp. 72–86). Springer.

Suh, S.-B. (2020). Understanding the UNECE WP.29 Cybersecurity regulation — PERSEUS. https://cyberperseus.com/understanding-the-unece-wp-29-cybersecurity-regulation/

Synopsys. (n.d.). What is SAST and how does static code analysis work? https://www.synopsys.com/glossary/what-is-sast.html

Tenable. (n.d.). Nessus vulnerability assessment. https://www.tenable.com/ products/nessus

UNECE: R155. (2020). *Technical report*. UNECE.

UNECE: R156. (2020, April). *Technical report*. UNECE.

Veitas, V.K., & Delaere, S. (2018, May). In-vehicle data recording, storage and access management in autonomous vehicles. arXiv.

Wu, W., Li, R., Xie, G., An, J., Bai, Y., Zhou, J., & Li, K. (2020). A survey of intrusion detection for in-vehicle networks. *IEEE Transactions on Intelligent Transportation Systems, 21*(3), 919–933.

# Chapter 7
# Technical Cybersecurity Implementation on Automated Minibuses with Security Information and Event Management (SIEM)

**Athanasios Papadakis, Antonios Lalas, Sofia Petridou, Konstantinos Votis, and Dimitrios Tzovaras**

**Abstract** This chapter examines the integration of security information and event management (SIEM) systems into the cybersecurity infrastructure of automated minibuses to emphasize their critical importance in today's urban mobility ecosystems. As automated minibuses become an integral part of smart city initiatives, ensuring their cybersecurity is critical in the context of increasingly sophisticated cyber threats. SIEM systems are becoming indispensable tools for detecting, responding to, and mitigating such threats through real-time monitoring, data collection, and analysis from multiple sources. This chapter examines various open-source SIEM solutions, their operational benefits, and the challenges associated with implementing these cybersecurity measures in automated mobility. It addresses the critical need for advanced security protocols, customization to meet specific operational requirements, and the balance between comprehensive cybersecurity coverage and resource constraints. Through an analysis of common SIEM platforms and their application to protect automated minibuses from cyber threats, the chapter highlights the critical role of SIEM in strengthening the cybersecurity defences of automated mobility ecosystems, ensuring passenger safety, and maintaining the integrity of smart urban transport networks.

A. Papadakis
Centre for Research and Technology Hellas, Information Technologies Institute, Thessaloniki, Greece

Department of Applied Informatics, University of Macedonia, Thessaloniki, Greece

A. Lalas (✉) · K. Votis · D. Tzovaras
Centre for Research and Technology Hellas, Information Technologies Institute, Thessaloniki, Greece
e-mail: lalas@iti.gr

S. Petridou
Department of Applied Informatics, University of Macedonia, Thessaloniki, Greece



G. Fournier et al. (eds.), *Automated Vehicles as a Game Changer for Sustainable Mobility*, Contributions to Management Science,
https://doi.org/10.1007/978-3-031-61681-5_7

## 7.1 Introduction

In a time marked by continuous technological advancement, the introduction of automated minibuses has revolutionized urban transportation. These vehicles rely on a sophisticated interplay of systems, sensors, and data streams to navigate and assist passengers. However, this technological breakthrough also brings forth a pressing concern: cybersecurity. The safety and security of both passengers and the broader transportation environment demand our utmost attention. Within this chapter, a security analysis is presented regarding the pivotal role that security information and event management (SIEM) solution plays in fortifying the defences of automated minibuses. SIEM solutions are suitable for real-time monitoring and response, gathering and analysing data from a multitude of sources within the vehicles' network. They provide a robust shield against threats such as data breaches, hacking attempts, and system vulnerabilities. Within the upcoming sections, an attempt is made to describe the indispensable necessity for SIEM in safeguarding an automated ecosystem which includes, among other, automated minibuses, enhancing their safety, and upholding the integrity of a continually evolving transportation domain.

## 7.2 Basics of a SIEM Software Solution

SIEM is a combination of two different security software: SIM and SEM. In 2005, Armit Williams described the SIEM term and presented the information gathered from network and security devices, the identity, and access management applications, from tools considering the vulnerability and policy of the monitored system, the operating system, the database, application logs, and, lastly, the external threat data (Williams & Nicolett, 2005). The mixture of the above assists the security administrators to utilize every possible source of information to detect and repel cyberattacks to their system. The importance of IT security grows exponentially every year, while the attacks and attackers are getting more sophisticated and less detectable. For this reason, security administrators should gather and analyse every possible source of information from each component of their system, as illustrated in Fig. 7.1.

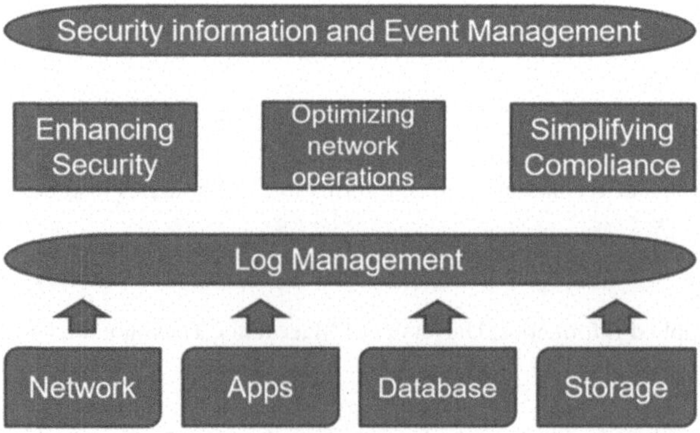


**Fig. 7.1** Main activities and resources for SIEM

## 7.3 Most Popular SIEM Open-Source Software

Open-source SIEM are software which have made available to the public their cybersecurity infrastructure and the architecture of their components. This allows the IT experts to adjust the available software to their needs and remove or lessen any limitations. This list takes into consideration the most popular implementations and by no means has a promoting or advertising nature. The candidates can be different depending on search criteria and occasionally are subject to the subjective point of view of each researcher. An overview of the most popular SIEM is presented in Table 7.1 and described next.

*Snort* is a popular network intrusion detection system (NIDS), which scans all traffic to mainly sniff, log, and perform real-time analysis on the network flows. There is a visualization of the real-time stream of packets and dump packets with the option to perform analysis (Roesch, 2011).

*MozDef* is an open-source SIEM created by Mozilla with the main objective to automate the security incident handling process and facilitate the real-time activities of incident handlers. It uses Elasticsearch (Gormley & Tong, 2015) for storing data and runs as a cluster, making it scalable and redundant.

*Wazuh* is a platform used for all three main objectives in cybersecurity, threat prevention, detection, and response. It includes endpoint security agents, deployed to the monitored systems and data visualization of the gathered information (Wazuh, 2023a).

*ELK Stack* fulfils the need for log management and analytics. Mainly the ELK stack is a combination of three different open-source products: Elasticsearch (Gormley & Tong, 2015), Logstash (Logstash, 2023), and Kibana (Kibana, 2023). These different components are most used for monitoring, troubleshooting, and securing IT environments (Elastic, 2023).

*AlienVault OSSIM* is also an open-source software which combines log storing functions and correlation capabilities. A basic attribute of OSSIM is the utilization of different other open-source projects to create a holistic SIEM solution (AlienVault, 2017).

*OSSEC* is a scalable, multi-platform, open-source host-based intrusion detection system (IDS) (OSSEC Project Team, 2019). It is commonly used for log analysis,

**Table 7.1** Different popular open source SIEM software solutions

| Name | Written In | Latest release | Works on |
|---|---|---|---|
| Snort | C++ | 3.0 | Cross platform |
| MozDef | JavaScript, Python | 3.1.2 | Cross platform |
| Wazuh | JavaScript, Python, C | 4.0.4 | Linux |
| ELK Stack | Shell, Go, Docker | 7.10 | Cross platform |
| OSSIM | Java, Python, C, PHP, Perl | 5.7.5 | Linux |
| OSSEC | C | 3.6.0 | Cross platform |
| Security Onion | Python | 2.3.21 | Linux |
| MISP | PHP/Python3 | 2.4.1.07 | Linux |

monitoring, and analysing firewalls, intrusion detection systems, web, servers, and authentication logs.

*Security Onion* uses a spectrum of open-source software with the main objective to ensure the proper intrusion detection, monitoring, and log management of the targeted system. The areas that Security Onion excels in are packet capture (PCAP) forensics, analyst virtual machine (VM), and as an external SIEM (S.O. Solutions, 2023).

*Malware Information Sharing Program (MISP)* collects, stores, and distributes security indicators and discovered threats. MISP is commonly used for fraud detection, information gathering, or threat hunting. Target users for this tool are security professionals (MISP, 2023).

## 7.4 SIEM Benefits for CAV Infrastructure

Not all the SIEM are created with the same design. As a result, there is not one that fits all needs. The main benefits of an average SIEM system can be sum up to log data management, compliance reporting, and threat intelligence.

Log collection not only enables real-time tracking of changes within the ecosystem but also allows for detailed monitoring of individual machines that may be compromised by attackers. This capability provides crucial insights into both the events that occurred and the subsequent analysis necessary to uncover vital evidence, such as the root cause of the exploitation. Normalization is an equally important tool for a security administrator which identifies and gathers complete information of the ecosystems' security status. The extraction of the essential data from the noise of the system is a major attribute of the SIEM. These data can be cross checked with known databases for malicious signatures, potential malware, and machine learning algorithms to identify patterns and propagation trends and map unknown categories. Notification and alerts are an inseparable part of the security concept, since without it the resource cost of the permanent SIEM supervision would outweigh its benefits.

The efficiency in incident response lies within the proper configuration and maintenance of the SIEM software. If this challenge is accurately addressed, then the incident handling activities will result in numerous benefits to the security administrators, such as saving resources and time. All the required information is going through one single interface making it easier to be examined and further analysed. In addition, some SIEM software also has the ability to manually include user and entity behavioural analytic, to potentially detect threats from both people and software before they cause any disturbance. Moreover, the intelligence produced from the analysis can be distributed to other platforms, which is a very useful characteristic considering CAV infrastructure. This can be easier to appreciate with the use of an example. Let's say that a CAV fleet has been targeted in a country from a group of attackers, with a zero-day attack. Regardless of the result from this kind of malicious action, all the gathered information and mechanisms will be distributed to the other cities which are also hosting CAVs' fleet. Automatically, the possibility of causing the same crisis with some similar kind of attack is close to zero.

Security incident detection is just the tip of the security iceberg. It generally identifies activities such as unauthorized access and use of the system, changes to system's firmware, and any other malicious actions. Threat response flow, which can be set and defined by the security administrators, is the key to properly deflect and minimize the hazard caused by an attack. A common observation is that companies, Internet of Things ecosystems, or even infrastructures like CAVs are better equipped with SIEM software than without it. Incident handling with the use of that kind of software is exponentially faster and with less human resources than having to examine all the logs to find the attacker's footsteps. The SIEM identifies and correlates those events, from a bird's eye view, while ensuring the efficiency of malicious tracking actions.

## 7.5 Limitations of SIEM

SIEM is a software solution that aggregates and analyses the activity of many different resources across the infrastructure under investigation. The automation of those activities conceals some complexities. One of them is the extensive configuration which must be done to the software to be tailored to the system's needs. Misconfiguration can cause several problems, depending how major is the accident. Nevertheless, this applies to a lot of software, and it is easily answered by having experienced staff executing such tasks. The above limitation leads to another disadvantage of the SIEM software. The perfect setting up of this kind of software is costly and time-consuming, something which will be reversed in the long run. However, when the first installation takes place, and for a short period after it, the need for time and modifications will be immense. A key part of the SIEM software is the definition and creation of rules, whereas the lack of accurate configuration is possible to lead to false positives. For that reason, constant monitoring and examining of those rules are needed from the security administrators to have an excellent result. The combination of the above reveals the need to have trained and around-the-clock staff for a smooth cybersecurity operation. The information which will be stored through the log management is usually enormous, and since it is about ecosystems such as CAVs, it will need sufficient staff who will not get distracted by the noise captured from the software.

## 7.6 Characteristics of the SIEM Platform

SIEM aggregates and analyses the activity of many different resources across our infrastructure (Kotenko & Chechulin, 2012). Security data are collected, and the software stores, normalizes, aggregates, and apply analytics to that data to

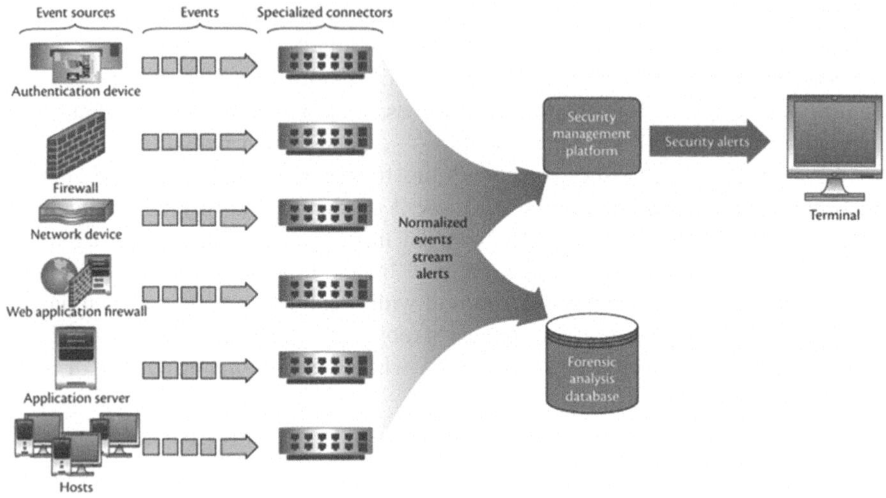


**Fig. 7.2** A typical SIEM architecture. The SIEM system accepts input from various security devices and sensors

discover trends, detect threats, and investigate alerts as it is shown in Fig. 7.2. This system could perform the analysis on the platform but also on the CAN bus. For instance, MISP is a sharing platform which stores and distributes security indicators and discovered threats (Wagner et al., 2016). This platform is useful for those involved with security incidents and malware research. The users benefit from having a well-tested platform to structure the vast number of data points available when it comes to security threats. The tooling allows interaction with other software, like IDS. MISP is commonly used for fraud detection, information gathering, or threat hunting. In MISP, a user can describe an event with multiple attributes while providing as much information as possible, or one can only put a minimum of information for an event. The pull mechanism allows a MISP instance to discover available events on a connected instance and download any new or modified events. It automatically goes through each of the event IDs that are eligible, converting them to MISP's JavaScript Object Notation (JSON) format and "POST request" them to the event creation API of the remote end. If the event already exists, it can be edited, while the remote side will match the event by universally unique identifier to a local event and return the URL that could be used to update the event. It shows an index, description, events, attributes, correlations found, proposals, active users, organizations, discussion threads, discussion posts, and number of instances to ease the usage of MISP. The Computer Incident Response Centre Luxembourg (CIRCL) provides a feed of events that can be easily shared, such as open-source intelligence (OSINT) events and attributes that are classified as unclassified information that can be distributed without any restriction.

## 7.7 Investigation on Diverse Implementations within AVENUE

Considering the advantages of the SIEM along with the necessity of advanced security mechanisms, it is highly recommended to have it embedded within the CAV ecosystem. The SIEM can be deployed in several positions within the CAV environment. It can be installed in the cloud, or it can be installed inside the CAV. Both placements have their use in the overall architecture of the system. Since the speed and the robustness of the security system are critical in any automated minibus, the proposed position is considered as an isolated computer unit inside the vehicle with the objective of monitoring the whole network of the CAV. It provides an additional layer of security in case the vehicle gets compromised since it will allow the software to be part of detection, prevention, and response mechanisms. As is commonly found, many of the components inside the vehicle are defined by a characteristic network behaviour, in addition to their unique log files performance when they are operating under normal circumstances. The in-vehicle network system is comprised of vehicle electronic control units (ECUs) interconnecting with each other to create an Internet of Things ecosystem. The security administrators that will monitor the SIEM software will be able to recognize and detect in time possible attacks. This solution will provide real-time analysis of security alerts generated by the applications and network hardware which are implemented inside the vehicle. In this context, two different SIEM solutions are presented. The first one is the Security Onion and the second the Wazuh.

**Security Onion** The installation of SecOnion for the simulated environment required approximately 15 GB of disk space, 2 GB of RAM, and 2 dedicated CPU cores of a 4.2-GHz processor. Each one of these is used for:

- Central processing unit (CPU): Used to parse incoming events, index incoming events, search metadata, capture PCAP, analyse packets, and run the front-end components. As data and event consumption increases, a greater amount of CPU will be required.
- Random-access memory (RAM): Used for Logstash, Elasticsearch, and disk cache for Lucene, Snort/Suricata, Zeek, Sguil, etc. The amount of available CPU will directly impact search speed and reliability, as well as ability to process and capture traffic.
- Disk: Used for storage of indexed metadata. A larger amount of storage allows for a longer retention period.

In this set-up, and as presented in Fig. 7.3, the Security Onion was able to run properly and without any operational restrictions. The system under monitoring was another VM from where the captured data were analysed and visualized in Squert (Visscher, 2014). Additionally, Sguil was used as an intuitive GUI that provided access to real-time events, session data, and raw packet captures (Visscher, 2014). Finally, the Security Onion is a suitable tool to facilitate the practice of network

**Fig. 7.3** Security Onion VM specifications

security monitoring and event-driven analysis. The information shown in Sguil window, as presented in Fig. 7.4, includes the alert's identification, date, and time the event took place, source and destination IP address, and the corresponding ports.

**Wazuh** An additional installation was performed to examine how user-friendly is the software, in addition to Security Onion where it was essential to have some experience in security domain. Wazuh provides a security solution capable of monitoring the infrastructure, detecting threats, intrusion attempts, system anomalies, poorly configured applications, and unauthorized user actions. It also provides a framework for incident response and regulatory compliance, as shown in Fig. 7.5.

In Fig. 7.6, the specifications for the Wazuh server are presented. In these settings Wazuh was able to run properly without any delays and restrictions. Wazuh server runs the manager and the API, where it collects and analyses data from the monitored agents. Additionally, there was an installation of two additional agents to

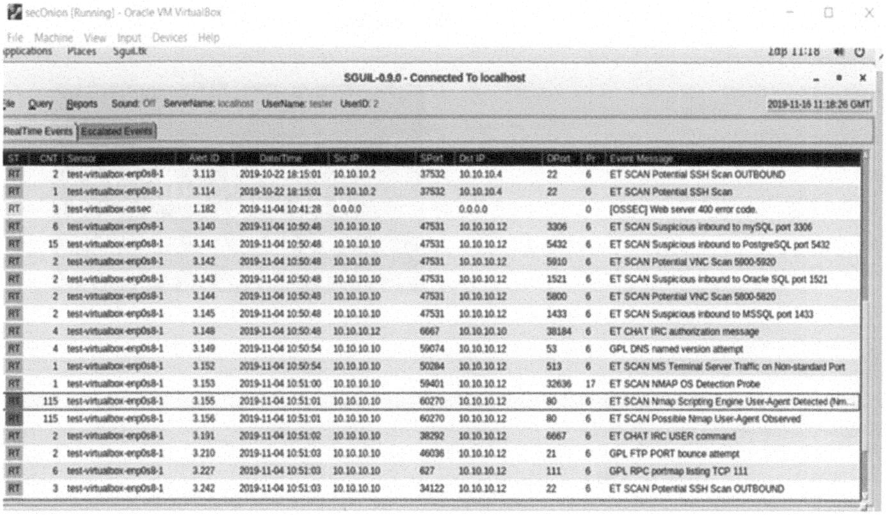


**Fig. 7.4** Sguil's window shows there are multiple events making available timestamps, destination and source, and event message information for analysis by the security administrators

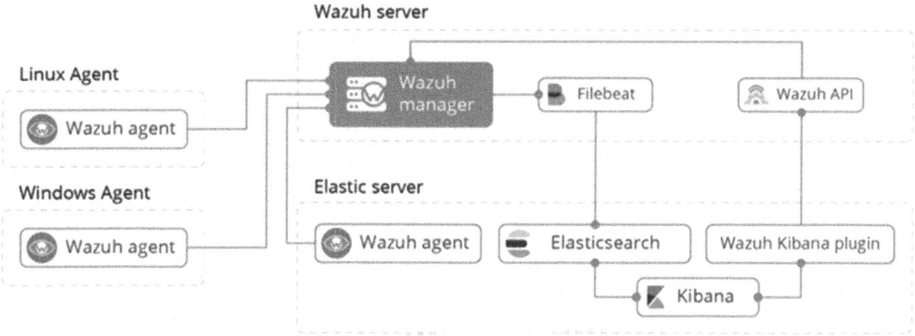


**Fig. 7.5** Wazuh server for monitoring, incident response, and regulatory compliance

examine the proper function of the SIEM software. In this case, the agents were two individual OS: Ubuntu 18.04, and Ubuntu 20.04. Wazuh agent runs on the monitored host, collecting system log and configuration data, allowing the detection of intrusions and anomalies. Communication with the server is mandatory, which forwards the collected data for further analysis. The compatibility between agent and manager/server is guaranteed when the manager version is greater or equal to the agent's version. Additional information for the installation and the requirements can be found at Wazuh's official site.

To test the Wazuh implementation in real scenario, the installation of Kali Linux was required. Kali can be installed using 128–512 MB of RAM and 2 GB of disk space. If 128 MB is available, it will be able to run only a basic SSH server with no desktop, and if it has 512 MB of RAM, it will run the default GNOME desktop with

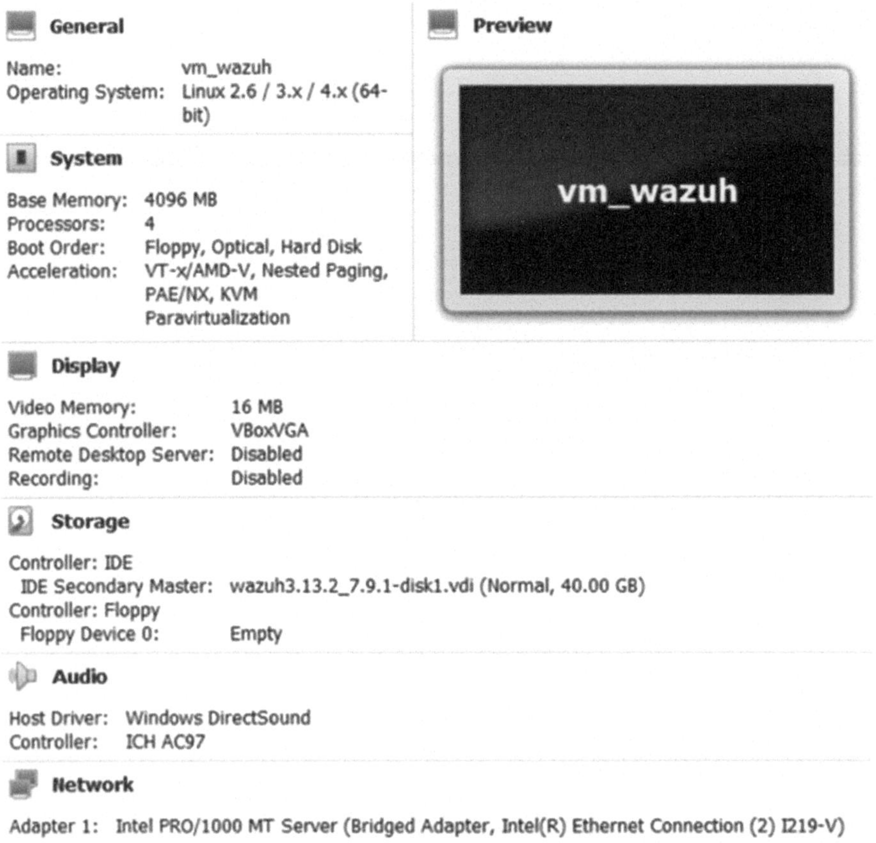


**Fig. 7.6** Wazuh server VM specifications

the full meta package. The recommended specifications are 2048 MB of RAM and 20 GB of disk space and at least one CPU supported by at least one of the amd64, i386, armel, armhf, or arm64 architectures.

The finalized test bench was implemented with four RAMs, two agents, one manager, and the attacker, as illustrated in Fig. 7.7. The attacker performed a brute force attack on the agents, which are the victims in this case, and the response of the system will be further analysed in the following figures.

The Wazuh databases store information related to agent keys and file integrity monitoring (FIM)/Root check event data. This information is highly optimized to be handled by the core. To provide well-structured data that can be accessed by the user or the Wazuh API, new SQLite-based databases have been introduced in the Wazuh manager (Wazuh, 2023b). The database synchronization module is a user-transparent component that collects the following information from the core:

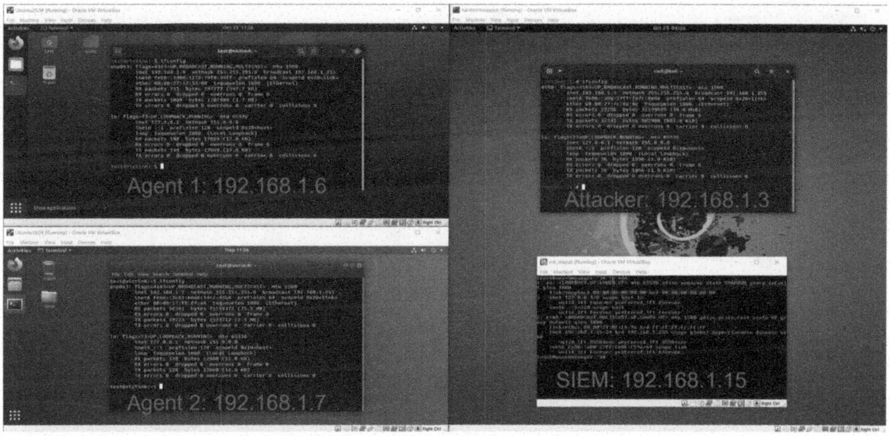


**Fig. 7.7** Testbench for Wazuh implementation

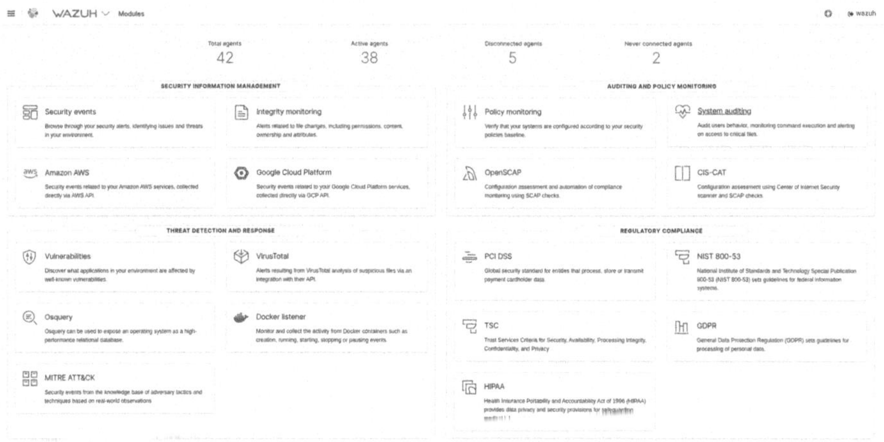


**Fig. 7.8** Wazuh user interface modules

- Agent information: name, address, encryption key, last connection time, operating system, version, and shared configuration hash
- FIM data: creation, modification, and deletion of regular files and Windows registry entries
- Root check-detected defects: issue message, first detection date, and last alert time
- Static core settings: maximum permitted agents or SSL being enabled for Authd

The above description is provided by the official Wazuh documentation as precise description of the dashboard, which is shown in Fig. 7.8.

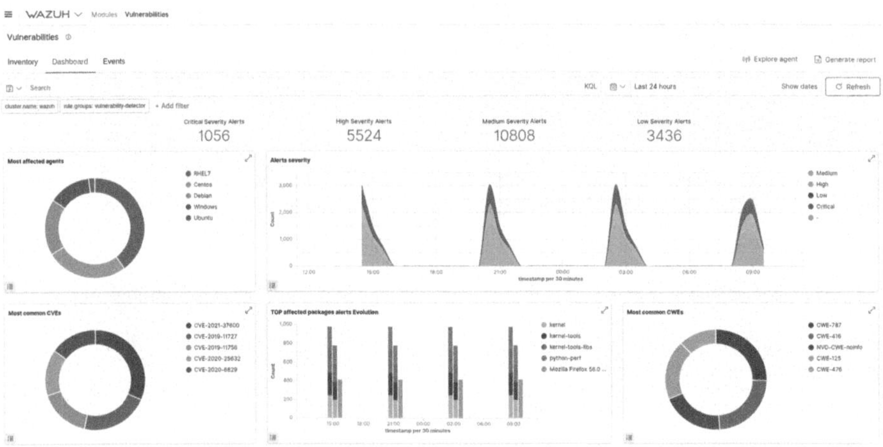


**Fig. 7.9** Security event dashboard

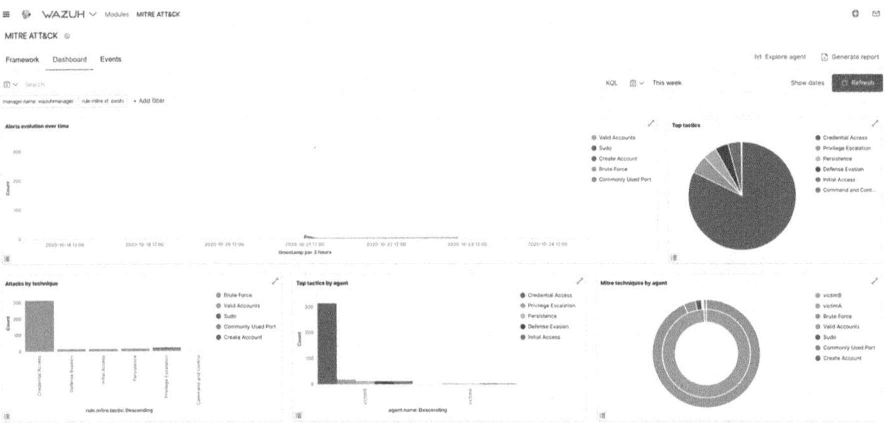


**Fig. 7.10** MITRE attack database analysis dashboard

Figure 7.9 presents a screenshot from the security event module. The information visualized here includes the count of the alert levels, the corresponding timestamps, and a pie chart with the most attacked agents, among other important information for the security administrator. Additionally, there is a second tab, called "Events," where a more detailed explanation is available.

MITRE attack (The MITRE Corporation, 2020) is a knowledge base of adversary tactics and techniques based on real-world cases of cybersecurity threats. It is globally accessible and documents the procedure of the attacks. Essentially is a database which answers to the question of "How" an adversary achieves a tactical objective by performing an attack. This feature allows the user to additionally customize and enhance the database, as there is an option to include specific information related to cyberattack techniques, as presented in Fig. 7.10.

## 7.8 Conclusion

In the advancing landscape of smart city infrastructure, the appearance of automated minibuses has revolutionized urban transportation, offering innovation and efficiency. However, since automated minibuses must be continuously connected to their surroundings, a critical matter has to be highlighted: cybersecurity. The discussed solution, SIEM, stands as a cornerstone of real-time monitoring and response, collecting and analysing data from a plethora of sources within the vehicles' system. Examining some of the most popular open-source SIEM platforms has revealed their diverse features and benefits. While SIEM offers numerous advantages, it is crucial to recognize its limitations, such as the need for extensive configuration, potential misconfigurations, and the resource-intensive nature of initial set-up. Considering the possible implementations makes it clear that SIEM should be a core part of the automated minibuses' infrastructure. An isolated deployment within each vehicle is suggested, as deployed in Copenhagen and Geneva pilot sites of the AVENUE project, since this solution allows comprehensive network monitoring while providing an additional layer of security and real-time analysis to detect and respond to potential malicious actions. In conclusion, the incorporation of SIEM as a defence mechanism is not just a necessity but a vital step towards ensuring the safety and security of passengers and the long-term sustainability of transportation systems.

**Acknowledgements** This project has received funding by the European Union's Horizon 2020 research and innovation program under grant agreement No. 769033.

## References

AlienVault. (2017). OSSIM: The Open Source SIEM | AlienVault.
Elastic. (2023). Recherche open source: les créateurs d'Elasticsearch, de la Suite ELK et de Kibana | Elastic.
Gormley, C., & Tong, Z. (2015). *Elasticsearch: The definitive guide: A distributed real-time search and analytics engine*. O'Reilly Media, Inc.
Kibana. (2023) Kibana: Exploration, visualisation et découverte des données | Elastic.
Kotenko, I., & Chechulin, A. (2012). Common framework for attack modeling and security evaluation in SIEM systems. In *Proceedings - 2012 IEEE Int. Conf. on Green Computing and Communications, GreenCom 2012, Conf. on Internet of Things, iThings 2012 and Conf. on Cyber, Physical and Social Computing, CPSCom 2012* (pp. 94–101).
Logstash. (2023). Logstash: collecte, transformation et analyse de logs | Elastic.
MISP. (2023). Open Source Threat Intelligence Platform & Open Standards For Threat Information Sharing (formerly known as Malware Information Sharing Platform).

OSSEC Project Team. (2019). OSSEC - World's most widely used host intrusion detection system - HIDS.
Roesch, M. (2011). Snort-network intrusion detection & prevention system.
S. O. Solutions. (2023). Security onion solutions.
The MITRE Corporation. (2020). Mitre Att&Ck.
Visscher, B. (2014). Sguil - Open source network security monitoring.
Wagner, C., Dulaunoy, A., Wagener, G., & Iklody, A. (2016). MISP: The design and implementation of a collaborative threat intelligence sharing platform. In *Workshop on information sharing and collaborative security (WISCS)* (pp. 49–56).
Wazuh. (2023a) Wazuh-modulesd - Daemons Wazuh 3.7 documentation.
Wazuh. (2023b) Wazuh: The Open Source Security Platform. Wazuh documentation overview, https://documentation.wazuh.com/current/
Williams, A., & Nicolett, M. (2005). Improve it security with vulnerability management. Gartner ID (G00127481).

# Chapter 8
# Persons with Reduced Mobility (PRM) Specific Requirements for Passenger Transportation Services

**Linda Mathé, Markus Dubielzig, and Matthias Lindemann**

**Abstract** The use of fully automated vehicles (AVs) in public transport has the potential to revolutionise the complete landscape of public transport (PT). Fully automated vehicles can provide the necessary frequency even in places with a low passenger volume. In addition, these vehicles can stop even in places which are not yet integrated into the public transport network, thus better meeting the needs of passengers as public transport becomes more and more individual. The absence of a driver, however, also implies that these vehicles must provide more services than conventional ones. This chapter presents the findings on passenger services for fully automated minibuses after 2 years of studies with passengers. These findings were gathered based on observations, workshops, questionnaires, and interviews with (disabled) passengers, safety drivers, bus drivers, PTOs, and associations of disabled persons.

## 8.1 Introduction

To enable all passengers, including those with special requirements, to benefit from the new public transportation services introduced by H2020 AVENUE, we need to fully understand their requirements and expectations, their needs, the issues and the problems, and also the personal tricks and strategies they have developed for themselves. Will they accept a driverless bus? What if there is no real person on board? What does the absence of a driver or conductor imply for older persons with disabilities and persons with reduced mobility (PRM) (European Commission, 2014)? What are passengers' expectations in their interaction with the vehicle?

L. Mathé (✉)
Siemens Technology, User Experience Design, Munich, Germany
e-mail: linda.mathe@siemens.com

M. Dubielzig · M. Lindemann
Siemens Technology, Accessibility Competence Center, Paderborn, Germany



G. Fournier et al. (eds.), *Automated Vehicles as a Game Changer for Sustainable Mobility*, Contributions to Management Science,
https://doi.org/10.1007/978-3-031-61681-5_8

## 8.2 Requirements of Passengers (Interview Results)

To gain a real understanding of the user, the AVENUE project used a variety of methods to acquire the needed information. Since people with disabilities have special requirements, they play an important role in gathering user requirements.

> Disability is part of the human condition. Almost everyone will be temporarily or permanently impaired at some point in life, and those who survive to old age will experience increasing difficulties in functioning. (World Health Organization, 2011)

Considering that every human will have some type of disability in his/her life or can be situationally induced disabled (e.g. wearing a headset), the requirements derived from people with disabilities will be useful for all (Fig. 8.1).

In a first step in 2018, interviews with public transport users were conducted in five European states. This survey had its major focus on public transport in general, as at this point in time hardly anyone has had real experiences with AVs.

In 2019 and 2020, when AVs had been in operation at several sites for a while, we were able to gather real passenger experiences with the AVs in six European cities. We conducted interviews with passengers as well as with safety operators in the busses, and we observed passengers in the busses and at the stops. Another information source were workshops with mobility experts (with and without special requirements) and public transport operators.

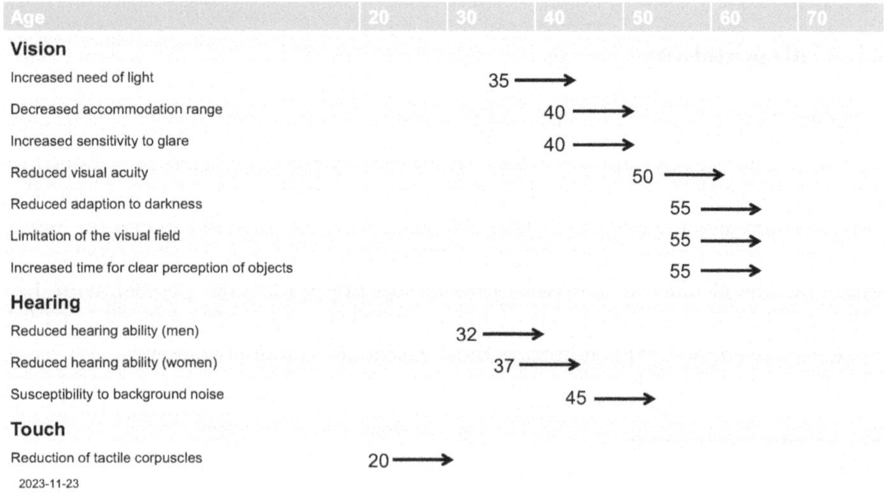


**Fig. 8.1** Deterioration of abilities in the process of ageing based on Saup (1993) (Translation by the author)

### 8.2.1 Phase 1 (July–September 2018)

In phase 1 (July–September 2018), together with the local AVENUE partners, we conducted interviews with public transport users in five countries: Denmark, France, Germany, Greece, and Luxembourg.

In this phase, we had our focus on public transport in general. Our goal was to fully understand the needs, the issues, and the problems of public transport users and the personal tricks and strategies they have developed for themselves.

The interview was divided into three parts:

- Requirements for public transport due to the well-known use of "classic" public transport.
- The aim of this part was to identify the status-quo and current issues.
- Evaluation of experience with fully automated vehicles (if existent), especially in comparison to conventional public transport.
- Wishes and expectations for future fully automated vehicles.
- In this part, all passengers were asked to imagine how future fully automated busses could change and improve public transport.

A total of 58 persons participated in the interviews, and 33 of them had some kind of disability, most of whom were visually impaired.

### 8.2.2 Public Transport in General

Most of the results of the surveys on public transport in general are as expected:

Passengers dislike delays, low frequency of service (and thus long waits), crowded vehicles/no seats, dirty and littered vehicles, lack of information, and unfriendly staff. They want on-time public transportation, affordable tickets, increased frequency and destinations, clean vehicles, and available and comfortable seating, including guaranteed seating for people with special needs. Accurate and accessible information, e.g. audibly understandable announcements, accurate (not delayed) announcements of next stops, estimated time of arrival, and current location of the bus are always requested.

It would be nice if the stops were more flexible: For some participants, it would be a real advantage to be picked up from home and dropped off at their actual destination, rather than at a nearby (or not so nearby) bus stop. This way, they would no longer have to walk all the way to a bus stop and could live a more mobile and independent life.

### 8.2.3 *Attitude Towards Fully Automated Public Transport (Unexperienced Pax)*

Most interviewees in 2018 have never used or even seen an AV. An evaluation of their experience with AVs was thus not possible. Nevertheless, the interviewees were asked about their attitude towards fully automated vehicles. Many interview partners are quite sceptical and are not convinced that the technology is already mature enough to be trusted. They consider traffic situations too complex to be handled by technology.

Some even say that they would never use a fully automated bus (7 persons out of 58).

A majority needs to gain more trust in this technology before they are ready to get in. Only four interviewees are ready to hop on a fully automated bus at the day of the interview.

But it turns out that fully automated public transport is accepted more easily with increasing distance to the next stop (Fig. 8.2) and with decreasing frequency of public transport.

Unexpectedly age seems to play no role (Fig. 8.3).

Safety is an important topic for the interviewees. Most of them have heard of accidents by fully automated vehicles, and they stress that this technology needs extensive testing before it can be put into operation. They would use fully automated vehicles only if they are convinced that they are safe.

While they think that technology like sensors, etc. provides advantages, many users nevertheless fear accidents. Some believe that other drivers will crash into the bus as they will not be able to anticipate its behaviour. Others believe that the technology is not fail-safe and/or advanced enough to handle complex traffic situations, and the fully automated bus will have accidents without a driver. Only a minority of interviewees is of the opinion that safety will be increased.

Many interview partners are afraid that the use of fully automated busses in the field will lead to more delays and failures due to unstable technology and because traffic situations are too complex to be handled by technology in general.

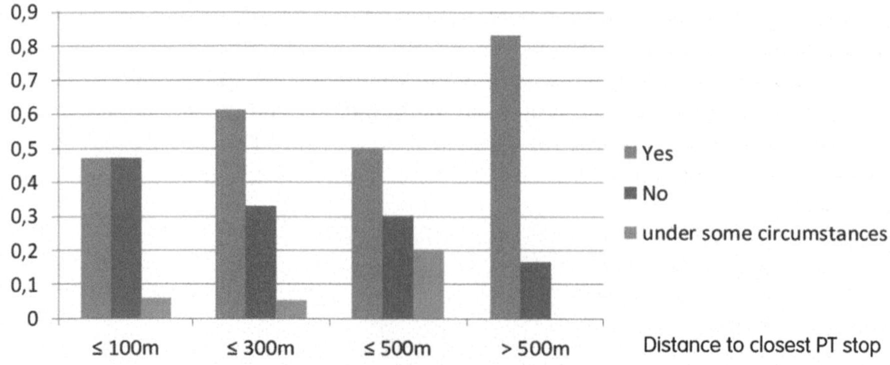


**Fig. 8.2** "Would you take a driverless bus?" in relation to the distance to the next bus stop

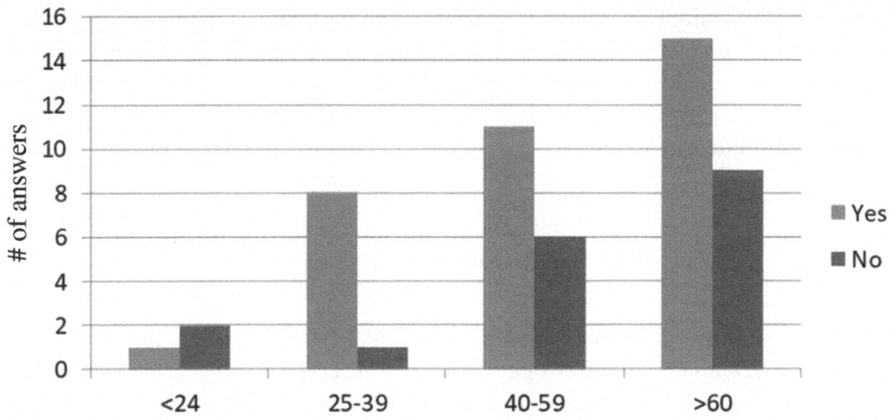


**Fig. 8.3** "Would you take a driverless bus?" in relation to age group

Some are even worried about cyberattacks against a fully automated vehicle: if there is no driver to interfere, hackers could make the bus go faster or drive off a bridge or into oncoming traffic.

The overall attitude/scepticism towards fully automated vehicles makes clear to us that trust has to be gained first, and even little incidents or accidents are likely to destroy that trust.

There is a great empathy for drivers who might lose their jobs. Most respondents cannot imagine fully automated vehicles without having this in mind. This is especially reflected in the fact that when talking about positive experiences in public transport, respondents include other people, especially the bus driver (receiving help, amusing passengers, a funny bus driver, etc.) and in the fact that 60% of participants mention talking to the driver (half of them for a conversation).

For most interviewees, a "safety operator" who can interfere or take over and acts as an authority figure is essential. Passengers also want someone in the bus who can answer questions, provide information, and help them getting on or off the bus when necessary. They are afraid that vandalism or even robberies or assaults could be a problem if there is no supervisor in the bus.

Besides all the scepticism, the interview partners also have some expectations that reflect the above-mentioned wishes for public transport: They assume that there will be connections where there are none today, because they are not profitable. Many also expect a smoother ride by the machine than by a real driver and cheaper tickets and that busses will not have to be cancelled due to lack of available drivers.

### *8.2.4 Phase 2 (June 2019–February 2020)*

While our focus in 2018 was public transport in general and anticipated experiences with AVs (due to lack of real experiences), 1 year later our aim was to go beyond anticipated experiences and to figure out the issues, the problems, and the missing services in fully automated vehicles. Therefore, all participants in this second series of studies have already gathered some experiences with AVs. To get a wide range of information, we decided to follow several different approaches to gather information: interviews with safety operators, interviews with (disabled) passengers, and observation studies.

### *8.2.5 Interviews with Safety Operators*

Currently, every fully automated vehicle in the AVENUE test sites (and the visited non-AVENUE sites) has a safety operator on board. These operators are responsible for the safety onboard and can control the bus manually if necessary. Furthermore, they are contact persons for the passengers. The safety operators are a valuable source of information as they have direct contact to passengers and experience their reactions, open questions, requirements, and problems every day. There is probably no other person who has had more user contact and knows the questions and problems of the users better.

In order to obtain the most information from this valuable source, we have decided to give questionnaires to the safety operators of all participating PT providers and additionally conducted some interviews with safety operators in non-AVENUE sites like Berlin, Bad Birnbach, and Vienna.

A total of 27 safety operators (5 from non-AVENUE sides) have been interviewed and asked about their impression and experiences made with passengers onboard of the vehicle. All of them mention a positive acceptance of the vehicles by nearly all passengers. Most users (especially the anxious ones) talk with the safety operators about the technology. It is often the first topic of conversation. Questions like how the bus knows where it is and how it works are dominating here. The better understanding of the technology helps them to relax and enjoy the ride. One operator mentions that even those passengers with a negative view change their mind while driving with the vehicle.

Nearly all passengers like the vehicle, and also the arrangement of the seats is welcomed because it encourages communication. However, all are complaining about the hard and unforeseeable braking. So smoother braking is one of the most wanted features, besides more speed.

Typical situations in which the drivers help the passengers are getting in and out (including opening and closing the doors) and buckling up. The high boarding height of the minibus is a real disadvantage and frequently requires the support of the safety operator. Some drivers report that they have to calm passengers after one of the above-mentioned hard brakes.

The question about the operators' needs and wishes results in interesting answers and ideas, and the most popular answers and suggestions are:

- Colour coding of the seat belts (which are currently not marked clearly)
- A smoother braking behaviour (has been solved already)
- A stand or a seat with seatbelt for the safety operator (back pain is already an issue)

### *8.2.6 Interviews with Experienced Passengers: Attitude Towards Fully Automated Public Transport*

In our interviews in 2019, we only asked persons that have already gathered some experiences with AVs, i.e. they have made the decision to use an AV. So it is not surprising that the answers are much more positive and less sceptical.

We interviewed 18 passengers in Berlin and Vienna, 7 of them with reduced mobility.

The interviewed passengers are very interested in the shuttle and its technology. They want to know things like What can the shuttle "see"? Why and when does the operator drive?

Many passengers complain about the low speed and that the vehicle comes to a stop too often. In order to be a useful means of transport, it needs to go faster. Everybody wants a smoother braking behaviour—as mentioned above, this issue seems to be solved by now.

A huge topic is information. Our interview partners miss a real passenger information system (PIS). The monitor in the shuttle is too small, its position not ideal (cannot be seen from every seat), and combined with the used font size and contrast, it is not considered usable.

Some passengers complain about the missing announcement of the next stop.

As there have been many complaints about missing information, we asked users more precisely what kind of information they would expect. They clearly state that they want at least the classical information one gets in standard PT as well as some information specifically required in this new type of transport:

- Announcement and display of stops (next stop/current stop)
- Transfer options in combination with time to transfer
- Information on how to request a stop
- A warning that the vehicle is about to brake
- Info on shuttle's battery: battery status and required charging time

The request for information on the vehicle's battery might seem unexpected at first. One reason for this could be that some passengers have experienced that the service had to be stopped due to an empty battery caused by the heavy usage of the air condition. Another reason could be that e-mobility is becoming more common, and people are therefore simply curious about such details.

Our interview partners also ask for information at the bus stop or from outside the bus:

- Information (at the bus stop) when next shuttle will arrive
- Number of free seats in the bus

The request for this information might be due to the fact that the vehicles are rather small and therefore the number of seats are limited.

The passengers appreciate the presence of a safety operator on board the AV. They think it is good to have someone in the shuttle who can intervene in an emergency and who can help to get on the bus with a stroller, etc.

Overall, passengers enjoy the experience, they trust the technology, and they would use the shuttle again at any time and recommend it to friends.

Most interviewed passengers say that they would use the shuttle.

The PRMs are even more willing to do so as they rely on public transport and are willing to take advantage of everything that makes life easier. All passengers agree that they would consider using the fully automated shuttle on a regular basis if

- The shuttle went faster
- The routes were longer
- There was a regular service with frequent departure times
- The shuttle could deal with everyday traffic situations
- There was an on-demand service with flexible routes (including bus stop in the immediate vicinity of the place of residence)
- Information was available at bus stop when next shuttle will arrive
- Infrastructure was improved: coordination with other means of transport like subway
- There was more space for strollers, etc. (currently only space for one stroller, buggy, etc.)

Many users like the idea of a shuttle service that does not depend on fixed bus stops. However, they raise the topic that the shuttle may stop only at secure places. It is not clear how the passenger could communicate these places to the shuttle and how the shuttle could confirm this. Other interviewees are concerned that the bus could not keep to the schedule without fixed stops. Some people could make the bus stop all the time just for fun.

Of the 18 interview partners, 11 instantly say they would use the fully automated shuttle even if there was not a safety operator on board. They would like to have a hotline available in the shuttle so that passengers can call if they feel the need to do so.

Some interviewees hesitate and would drive with the shuttle without safety operator only if the shuttle was more reliable and did not constantly stop whenever there is an unexpected situation like an illegally parked car.

Four interviewees are firmly convinced that the shuttle without an operator is not possible at the present time as this was too dangerous in road traffic. Some think that a human attendant will always be required in the shuttle. What if the bus suddenly stops and does not drive on? What if someone gets stuck in the door? What if there is an accident?

## 8.3 A Blind Users' Perspective on Automated Vehicles

To illustrate the challenges for PRMs, we accompany our persona Klaus-Dieter on his first trip alone with an automated e-minibus.

Klaus-Dieter, a 59-year-old male person, has good experiences in terms of mobility. In his hometown, a medium-sized city in Germany, he daily uses public transport completely on his own, from home to work and vice versa.

### *8.3.1 Bus Stops on the Course*

The automated bus, which is tested by Klaus-Dieter, runs in a kind of circular course and provides ten different bus stops where the vehicle automatically stops to allow the passengers to get in or out.

Regarding Klaus-Dieter, the test environment is completely unknown for him, and therefore he is not able to find his first bus stop on his own without an assistance, even though the address of the corresponding bus stop was given to him in the lead-up to his test.

All bus stops are represented by normal traffic signs with additionally installed timetables, directly mounted at eye level on the pillars of the traffic signs. The bus stops are neither equipped with audio signals (e.g. attention signals on traffic lights) nor with tactile markers on the walkway, which could help blind users to find and quickly locate specific places on their own.

The next difficulty occurs while Klaus-Dieter tries to get the arrival and departure times from the timetable. The table is not provided in braille, and an audio-based passenger info system, e.g. a standard loud speaker, is not available. Nevertheless, online-provided timetables as well as the usage of personal mobile phones together with appropriate OCR apps can solve these kinds of problems in the future.

The automated bus is not equipped with AVAS; therefore the arrival of the bus is difficult to detect for Klaus-Dieter, who is not able to perceive road noise from the bus due to environmental sounds.

### *8.3.2 Boarding Process*

The bus directly stops at the bus stop sign but without any attention signal. The door does not open automatically, and Klaus-Dieter has to locate the corresponding sensor button that is mounted on one site of the door of the bus but without any kind of tactile markers. Moreover, there are two different sensor buttons available, the second one is for wheelchair users who can request a special ramp for boarding.

First challenge, which of the buttons is the right one? There is neither braille nor tactile markers available to answer this question in an accessible way. Audio signals which could help blind users to quickly locate the door as well as to find the appropriate sensor button are also absent, and Klaus-Dieter has to cope with this issue for himself by trial and error.

There is no audio-based passenger information system installed at the bus stop which could enable Klaus-Dieter to identify the arriving bus as his correct line.

### 8.3.3 Interior Situation and Bus Ride

All passengers of the bus are assigned to take their seats during the bus ride to reduce any potential transport risks. Standing passengers are not allowed. Klaus-Dieter assesses the number of grips, handles, and handrails to be too low to ensure a safe movement or orientation within the interior of the bus. He also has problems to quickly find his seat without assistance. He was disappointed, as there is no additional storage space for hand luggage, shopping bags, or guide dogs. The riding process is comfortable, there are no abrupt braking actions at all, and all speed-up or slow-down actions lead Klaus-Dieter to a good driving experience in terms of security requirements.

The automated bus is configured to operate in metro mode; this means that the bus automatically stops at all available bus stops at the circular course. There is no audio-based passenger information system installed which could tell blind users the next bus stop in an accessible manner. Therefore, Klaus-Dieter has to know the stops at the circular course and moreover to count the already visited bus stops to get out at the right one.

The security driver explains Klaus-Dieter that there are no complex traffic situations within the bus course, e.g. complex set of traffic lights or difficult crossings. Potential stops of the automated bus are only caused by two situations: the vehicle reaches the bus stop or the vehicle detects obstacles on the course. In both cases the bus indicates the stop by a clear and good perceivable audio signal but without any audio information of the corresponding reason. Arrival points are not announced by speech. Moreover, it is the same signal for both situations, and Klaus-Dieter is not able to differentiate between the above-mentioned triggering events. As a result, he tries several times to get out of the bus on the course but without success.

### 8.3.4 Getting Out of the Bus

In metro mode, the bus stops at every stop, but the doors have to be opened manually, in terms of accessibility blind users are faced with several challenges.

Despite the already mentioned missing acoustical information, the blind user needs to figure out how to open the door. Klaus-Dieter remarks that the placement of the different sensor buttons, each one equipped with a specific function, will lead blind persons to operating errors. These difficulties are caused by the sensitivity of the sensor buttons that instantaneously trigger in case of any finger contact, but finger contact is required for this user group to read braille as well as to identify other tactile markings.

Due to these problems, Klaus-Dieter is unable to distinguish between the three halt functions, provided by the different sensor buttons: SOS, halt request for wheelchair users (which automatically activates a ramp), and the default exit button that opens the door. In particular, the exit button raises additional problems because the door must be released by the automated bus security system before passengers are able to successfully trigger the exit button to leave the bus. And exactly this release feedback is not perceivable by Klaus-Dieter who is only able to get out of the bus after pressing several times the different buttons to open the door.

### 8.3.5 *Klaus-Dieter's Summary*

Some of the above-mentioned accessibility issues can be resolved by using the capabilities of a modern mobile phone, e.g. GPS localisation, and online timetables could help users to successfully deal with these specific problems. But regarding the described topics, there are a lot of additional improvements required at the busses to answer the needs of this user group and to ensure a comfortable public transport with automated vehicles.

## 8.4 Situation-Based Impairments of Different Passenger Groups

Additional mobile and audio information is required for the following user groups on the go, at the bus stop as well as in the bus:

- International tourists or other passengers who are not familiar with the local language.
- People with dyslexia.
- Persons with misted-up glasses, particularly in winter times or with an FFP2 mask.
- Younger children who better cope with audio than with visible feedback; audio feedback grants a higher real-time information awareness for this specific passenger group.
- All persons who use their mobile phones to read articles (or use the Internet) during the bus ride; appropriate audio info raise their attention to get the next bus stop just in time.
- Smaller groups of passengers who intensively discuss with each other often get lost within their conversations and therefore miss the exit at their final destinations; appropriate audio feedback can avoid such situations.
- Persons with hearing impairments need an appropriate kind of suitable feedback in all above-mentioned situations, e.g. vibration or other haptic feedback, to raise their awareness in corresponding situations.

## 8.5 Proposed Implementation of User Requirements

The project co-created a mock-up for an accessible app with disabled users for fully automated public transport. Furthermore, a proposal for an in-vehicle information display to meet the user requirements in a fully automated vehicle was created.

### 8.5.1 *Mock-Up for an Accessible App for Fully Automated Public Transport*

Based on the challenges Klaus-Dieter has experienced, the project co-created with disabled users a mock-up for an accessible app that supports all users to use fully automated public transport (Fig. 8.4).

The app covers the complete passenger workflow:

- The way to the bus stop
- Waiting at the bus stop
- Identifying the correct bus
- Entering the bus
- Riding on the bus
- Leaving the bus at the desired destination

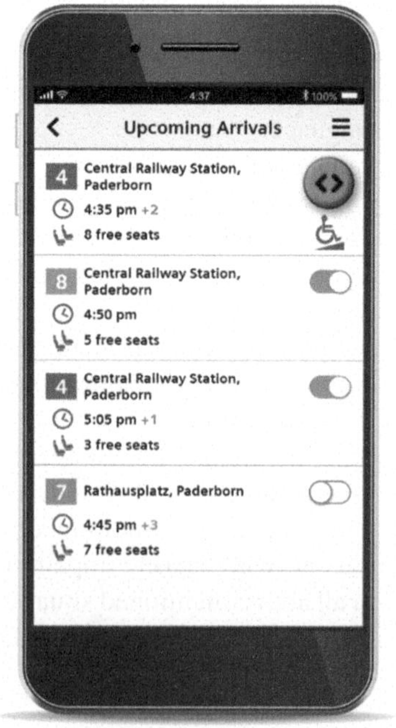


**Fig. 8.4** Mock-up for an accessible app

### 8.5.2 Information Display in the Vehicle

As most interviewed passengers ask for a passenger information system (PIS) that goes beyond the classical information we know from public transport today, we developed a concept for displays in the fully automated e-minibus (Fig. 8.5).

It consists of three major components:

1. Classical PIS
2. Sensor view
3. Map view

The passenger information system provides a list of upcoming stops including information on actual (not planned) arrival times and delays. If there are any incidents or restrictions on the route, this is also displayed here. The number and destination of the bus are shown as well.

The sensor view allows the passenger to "see through the eyes of the system". Stewards report that many passengers ask questions about the capabilities of the vehicle, especially regarding the sensors, so by indicating the sensor range and detected objects, we can satisfy their curiosity and at the same time build trust in the technology.

The map view takes up the most monitor space. It answers the questions "Where am I right now?", "Where am I going?", and "What is nearby?". This especially benefits passengers who are not that familiar with the area. For them, the name of the bus stop is often not meaningful. The map with marked landmarks can support their orientation in the area and help them find the best way to their destination and the stop where they should leave the bus.

In addition, sights, restaurants, and stores along the route can also be displayed on the map so passengers can see what they can reach from the bus (Fig. 8.6).

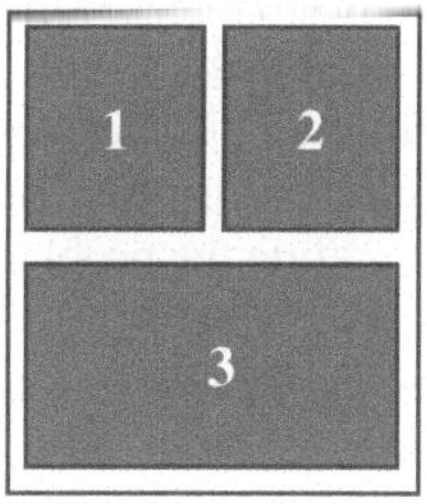


Fig. 8.5 Wireframe of information display

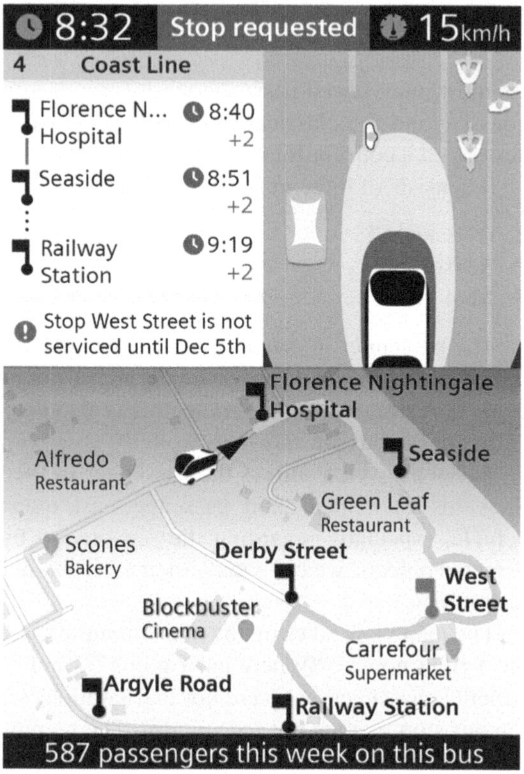


**Fig. 8.6** A suggested information display in the vehicle

## 8.6 Conclusions

The user acceptance increases to the same extent as the fully automated vehicles are optimised and adapted to fit the different user requirements.

Passengers with and without disabilities differ in their needs, but both user groups expect a significant added value in the use of automated minibuses for public transport, especially in areas with poor public transport coverage.

While the problems of people with (temporary) physical or sensory impairments are more related to the practical use of the technology, e.g. finding the right bus, finding and activating the door used, and missing PIS, other passengers tend to see difficulties in buying their tickets and in connecting to other public transport systems.

In order to make automated public transport a solution for *all*, all the services provided by the driver must be provided in some other way according to the dual-channel principle.

This also includes the perceived safety and security.

In addition, low-level access is essential, especially for passengers with walkers or buggies.

However, the prerequisite for use of automated vehicles is a general trust in the technology.

**Acknowledgements** This project has received funding by the European Union's Horizon 2020 research and innovation program under grant agreement No. 769033.

## Appendix: Mobile Apps for Blind and Low-Vision Public Transport Travellers

**Matthias Lindemann**
**matthias.lindemann@siemens.com**

The following list represents an exemplary set of public transport iOS apps for common tasks, e.g. localisation of bus stops, obstacle detection, orientation, and others. All apps can be found in the Apple App Store, and functionally similar apps are available in the Google Play Store.

Important Note: None of the apps can substitute required assistive Aids like, e.g. Guide Dogs or White Canes. All apps should only be used as additional tools to ease travel activities.

### *List of Mobile Applications*

#### GoodMaps Outdoors

GoodMaps Outdoors is a turn-by-turn GPS navigation app with a special focus on visually impaired travellers. The app provides audio instructions (directional or clockwise) as well as haptic feedback for directional guidance. The app supports an easy, collaborative points of interest (POI) handling, e.g. users are able to share safe or critical waypoints like accessible crossings and bus stops or interceptions with their friends and family. GoodMaps Outdoor is mostly used to overcome the distance between start/destination points and their corresponding bus stops.

#### BlindSquare

Blind Square The BlindSquare app allows visually impaired users to get important surrounding information for a preselected distance range, e.g. 200 m. These information includes streets, interceptions, restaurants, physicians, groceries, pharmacies, and a lot of other points of interest. For all POIs, the cardinal directions (compass style or clockwise infos) and their air distances are provided to allow a kind of spatial perception as well as a map-like orientation for this target group. All available

POIs can be used within their preferred GPS navigation apps. BlindSquare automatically detects whether one is travelling as a pedestrian or by vehicle and adjusts the spoken audio cues accordingly. With BlindSquare, visually impaired passengers are able to follow the route as if they were looking out of the window on the bus.

#### myfinder

myfinder uses the camera of the mobile phone and acts as an obstacle detection tool. Using special AI algorithms, the app scans the environment, categorises all objects found, and offers either a generic scene description or informs on surrounding objects. Continuously calculating directional and distance infos, myfinder supports a quick navigation to nearby objects. Audio instructions and haptic guidance directly lead the user to an object or allow alleged obstacles to be avoided.

#### Seeing AI

Seeing AI offers a set of different tools, e.g. the currency recognition feature, when dealing with ticket vending machines. A more important function presents the short text recognition function that enables users to scan important travel information, e.g. door signs, arrival and departure timetables, or the content of passenger information system displays. Achieving good results requires sufficient lightning conditions and a direct view of the objects to be scanned. The short text recognition starts automatically and is very responsive to what eases the comfortable handling of the app while boarding.

Note: The above download URLs refer only to the Apple App Store in Germany, but all apps are also available in other App Stores.

Despite their versatility, the use of the above apps does neither guarantee comfortable nor trouble-free public transport. To ensure this, the vehicles must be equipped with additional assistive technologies. Some examples: providing an easy localisation of entries for safe boarding and a reliable identification of destination stops to prevent the change over into a wrong line.

Regarding the deployment of fully automated vehicle fleets, only the provision of sophisticated assistance systems will ensure reliable public transport for all potential target groups.

## References

GoodMaps Outdoors. App Icon & Download URL: https://apps.apple.com/de/app/goodmaps-outdoors/id945756779

BlindSquare. App Icon & Download URL: https://apps.apple.com/de/app/blind-square/id500557255

myfinder. App Icon & Download URL myfinder: https://apps.apple.com/de/app/myfinder/id1457760748

Seeing AI. App Icon & Download URL: https://apps.apple.com/de/app/myfinder/id1457760748

Apple AppStore Germany URL: https://www.apple.com/de/app-store/

## References

European Commission. (2014). COMMISSION REGULATION (EU) No 1300/2014 on the technical specifications for interoperability relating to accessibility of the Union's rail system for persons with disabilities and persons with reduced mobility L 356/110. Available online at https://eur-lex.europa.eu/eli/reg/2014/1300/oj

Saup, W. (1993). *Alter und Umwelt. Eine Einführung in die ökologische Gerontologie*. W. Kohlhammer.

World Health Organization. (2011). *World report on disability*. WHO; The World Bank. Available online at https://www.who.int/teams/noncommunicable-diseases/sensory-functions-disability-and-rehabilitation/world-report-on-disability. Accessed 12 Jan 2021

# Chapter 9
# Stakeholder Analysis and AVENUE Strategies

**Dorien Duffner-Korbee, Eliane Horschutz Nemoto, Ines Jaroudi, Nicole van den Boom, Tobias Viere, Gabriele Naderer, and Guy Fournier**

**Abstract** This chapter contains the final stakeholder analysis. It showcases the key actors involved in testing and deploying automated minibuses for public transport in European cities. This assessment helps construct a strategic overview of the expectations, needs, and impacts of the stakeholders and the connections between them. This stakeholder analysis relied on qualitative methods and involved four steps. As a first step, an initial stakeholder scan was conducted. Through a literature review, including both academic and grey literature, and stakeholder mapping techniques, the main stakeholder groups were defined. The second step involved semi-structured interviews to gain a thorough understanding of the objections, perceptions, and information-seeking behaviour of these stakeholder groups. In the third step, a content analysis of the interviews was conducted, which led to the depiction of six key themes. The initial three steps are centred on conducting an analysis at the EU level.

D. Duffner-Korbee (✉)
Fraunhofer ISI, Karlsruhe, Germany

Pforzheim University, Pforzheim, Germany
e-mail: dorien.duffner-korbee@isi.fraunhofer.de

E. H. Nemoto
Pforzheim University, Pforzheim, Germany

CentraleSupélec, LGI, Gif-sur-Yvette, France

ICLEI Europe, Freiburg im Breisgau, Germany

I. Jaroudi
Fraunhofer ISI, Karlsruhe, Germany

Pforzheim University, Pforzheim, Germany

CentraleSupélec, LGI, Gif-sur-Yvette, France

N. van den Boom · G. Fournier
Pforzheim University, Pforzheim, Germany

CentraleSupélec, LGI, Gif-sur-Yvette, France

T. Viere · G. Naderer
Pforzheim University, Pforzheim, Germany



G. Fournier et al. (eds.), *Automated Vehicles as a Game Changer for Sustainable Mobility*, Contributions to Management Science,
https://doi.org/10.1007/978-3-031-61681-5_9

The fourth step took a project-level perspective. A literature review and expert interviews were conducted to determine the main relations and themes for the AVENUE ecosystem, based on which the AVENUE stakeholder and mobility services map was developed. The analysis delivers valuable findings:

- The AVENUE-centred map shows that automated minibuses are expected to be integrated into a multi- and intermodal mobility system and offered as Mobility as a Service (MaaS). There is a need for collective action to settle one legal framework and specific guidelines for automated vehicles (AV) within the EU. Finally, the integration of AV into the mobility system must be promoted in synergy with the goals of Sustainable Urban Mobility Planning (SUMP).
- The stakeholder map at EU level demonstrates that not all actors possess equal influence on the advancement of automated minibuses in urban public transportation systems. Primary actors comprise city governments, public transport operators (PTO), manufacturers, software developers, the European Union, and citizens/end-users. The relationship between the city government and the PTO is central.
- The strategy presents key recommendations for stakeholders, such as:
  - More emphasis should be given to the crucial role of the government and the legal framework in the deployment of automated minibuses.
  - PTOs need more support from the local government, and they should reach out to other key stakeholders such as NGOs and policymakers.
  - A MaaS platform should be operated in collaboration with private and public partners.
  - There should be active involvement from citizens and civil society; this is done through strategic tools for citizens' participation and inclusion in the debate about AV in general.
  - For AVENUE and similar projects, it is important to involve stakeholders such as civil society organizations, e.g. as driver unions and environmental NGOs. Also, they need to cooperate more with city governments and focus on user-centric services and not only technological achievements.

## 9.1 Introduction

### *9.1.1 Research Aim*

Deploying and integrating automated minibuses in public transport following the vision of AVENUE require a social-technical transition. Technically, the automated minibuses and the digital infrastructure should operate flawlessly. Socially, a multitude of organizations, networks, and institutions must accept the use of automated minibuses and adapt their current system. To support this transition, this study aimed to identify the stakeholder environment in which the automated minibuses services will operate. A stakeholder analysis is a tool to gain insights into a multi-actor issue. It is important for the identification of public interests and concerns and becomes even more important due to the increasing interconnectedness of today's world (Bryson, 2004). A stakeholder can be defined as 'any group or individual who can affect or is affected by the achievement of the organization's objectives' (Bryson, 2004; Freeman, 1984). There is no fixed strategy for conducting a stakeholder analysis; the applied method depends on the goal of the analysis.

### 9.1.2 *Research Approach*

The research methodology for this stakeholder analysis relies on two key components: an empirical approach (the EU-level stakeholder map) and a conceptual approach (the AVENUE stakeholder and mobility services map) as shown in Fig. 9.1. Based on the outcomes of these two approaches, a thorough analysis of stakeholder synergies and comparisons was conducted, thereby enabling the identification of stakeholder synergies.

The stakeholder map at the EU level was developed using an empirical approach, consisting of three consecutive steps: an initial stakeholder scan, semi-structured interviews, and content analysis. The initial stakeholder scan identified all potential stakeholders, through a brainstorming session and literature review. Hereafter, a focused literature review was conducted to gain information on the identified stakeholder groups, and afterwards, stakeholder mapping techniques were applied (e.g.

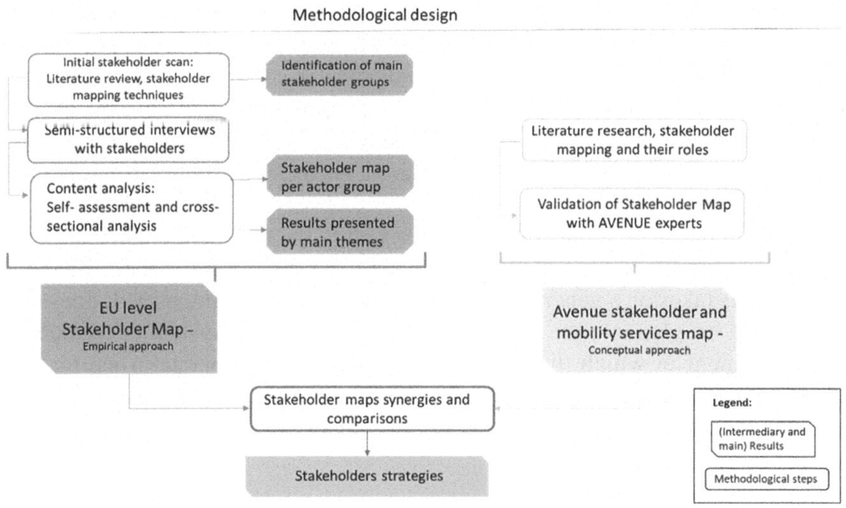


**Fig. 9.1** Methodological approach

**Table 9.1** Sample structure empirical stakeholder survey

| Stakeholder group | Number of interviewees | Associated IDs |
|---|---|---|
| Public transport operators | 4 | ID 20; ID21; ID22; ID23 |
| Manufacturers (OEM) | 4 | ID1; ID2; ID3; ID14; |
| Software developers | 2 | ID4; ID5; |
| Driver unions | 3 | ID6; ID7; ID11 |
| Environmental NGOs | 3 | ID8; ID9; ID10 |
| Policymakers/city government | 3 | ID12; ID15; ID19 |
| Consumer/citizen organization | 4 | ID13; ID16; ID17; ID18 |
| Safety operators on the automated minibuses | 7 | ID24, ID25, ID26, ID27, ID28, ID29, ID30 |
| Total interviews | 30 | |

power-interest grid). Through these analyses, the primary stakeholder groups that were to be examined through semi-structured interviews were identified.

In the subsequent step, the interviews were conducted with representatives of the eight groups identified as primary stakeholders in the initial stakeholder scan. Where appropriate, AVENUE partners were included as interviewees. The interviews were semi-structured and guided by a topic list consisting of five central themes (Appendix A: Topic list). A total of 30 interviews were conducted; see Table 9.1. As anonymity was promised during the interviews, each of the interviewees was given an ID to identify the quotes in the remainder of this deliverable. The interviews were recorded, and transcripts were made.

The transcripts of the interviews were analysed using qualitative content analysis. First, an analysis was made for each of the identified stakeholder groups. Therefore, the transcripts were grouped to gain insight in the role and responsibilities for each of the studied stakeholder groups. The results of this first analysis step were maps that depict a bottom-up self-reflection from the perspective of the stakeholder groups. Hereafter, a cross-sectional analysis was conducted. The data from the interviews was compressed and compared using bottom-up coding. This analysis aimed to gain insights into the overall stakeholder setting. As a result of this analysis, seven main topics were defined. Based on these insights, an integrated stakeholder map was developed.

As visualized in Fig. 9.1, the empirical approach to the stakeholder analysis was complemented by a conceptual AVENUE stakeholder and mobility services analysis. This analysis takes on a general perspective on automated minibuses. It does not directly provide insights into the structure of the AVENUE ecosystem or the stakeholder environment in which the automated minibuses will operate. Therefore, a second study was conducted, focusing on the AVENUE stakeholder and services. The associated map is presented in Sect. 3 of this chapter.

## 9.2 Empirical Stakeholder Analysis

This chapter presents the results of the EU-level stakeholder analysis. Section 2.1 presents the results of the initial stakeholder scan and thereby main stakeholder groups that were part of this analysis. Hereafter, the outcomes of the qualitative interviews with the main stakeholder groups are presented. Section 2.2 presents the insights and self-perception per stakeholder groups, and Sect. 2.3 presents the results of the cross-sectional analysis.

### *9.2.1 Results of the Initial Stakeholder Scan*

The following stakeholder groups were identified and included in the analysis:

- Manufacturers of automated minibuses
- Original equipment manufacturers (OEM)
- Software providers who offer platforms that enable the intelligent operation and optimization of automated mobility services, managing fixed-route and on-demand services
- Public transport operators who are responsible for the public transport system function
- The European Union (European Commission) as they can set regulations that can hamper or support the implementation of the technology
- National governments that develop regulations, incentives, and rules
- Local-level governments that set local mobility policies and are responsible for road infrastructure, etc.
- Citizen associations that represent the needs and demands of (potential) users
- Insurance companies that will financially cover costs in case of an accident
- Electricity charging infrastructure providers that will provide the charging infrastructure for the electric minibuses
- Energy providers that can influence the energy mix for the electric minibuses
- Environmental non-governmental organizations (ENGO) that can influence acceptance and use of the automated minibuses and promote the sustainability aspect
- Industry lobbies (such as the Society of Automotive Engineers (SAE))

An important stakeholder group that is not included in the analysis is the potential users of the system (Kyriakidis et al., 2015; Litman, 2019; Nordhoff et al., 2018; Wicki & Bernauer, 2018). Potential users are not included in this analysis, as they were the prime target group for the social impact assessment of the AVENUE project (see Chap. 14); (Korbee et al., 2019; Korbee et al., 2024).

#### 9.2.1.1 Power-Interest and Impact-Attribute Grid

In a power-interest grid, the power and interests of stakeholders are used to classify different actors (Hermans & Cunningham, 2018). It is dividing the stakeholders into four quadrants. Actors in the quadrant in the upper right are key players and should be taken along in the analysis. Actors in the quadrant in the bottom right are so-called context-setters and can—depending on the boundaries of the analysis—be taken along. The actors on the left side of the grid can be left out (Hermans & Cunningham, 2018).

The analysis indicates that key players who are supportive of AV include governmental actors, both at the national and EU levels, as well as public transport operators, software developers, and manufacturers (OEM). Municipalities show an equally high interest but are less powerful to influence the AV development compared to states, the EU, and OEM. Furthermore, new competitors, as well as legislators and assessment agencies, possess high power, but they do not have a high interest in the development of AV (see Fig. 9.2).

The impact-attitude grid on the other hand places the actors according to their opposing, neutral, or supportive attitudes towards a project and the high or low impact of the integration of AVs on mobility. This analysis shows that the majority of the included stakeholder groups are supportive of AV. The driver unions and the environmental NGOs are two stakeholder groups that are opposing further AV development (Fig. 9.3).

The national governments and the EU have the high institutional power to influence and set policies, regulations, and incentives to support the implementation of automated minibuses. Software providers and manufacturers are strategic for technical feasibility and daily improvements. The PTOs are key operational actors and the bridge between the new mobility technology and society. New competitors are proposing new services and products for mobility. Legislators are responsible for setting the laws and specific conditions for the implementation of automated

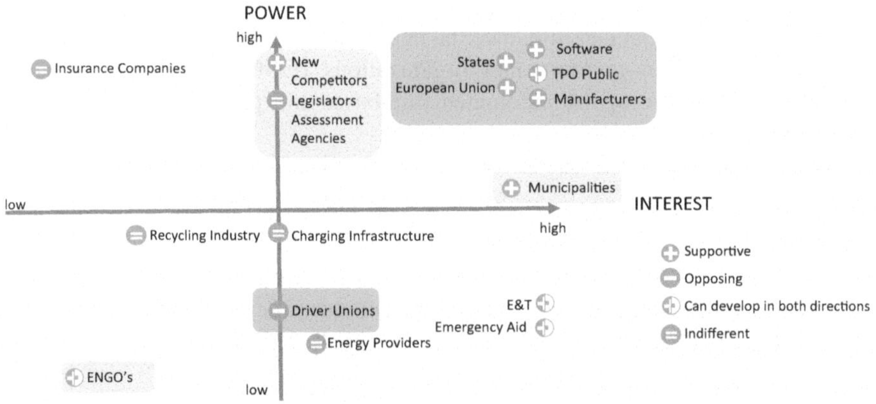


**Fig. 9.2** Power-interest grid towards the implementation of AVs in the public transport system

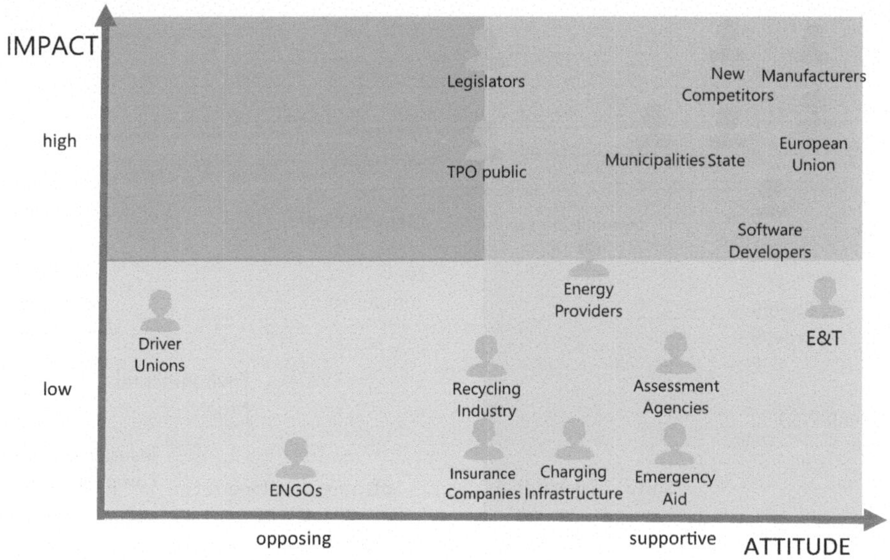


**Fig. 9.3** Impact-attitude grid regarding the implementation of AVs in public transport systems

vehicles. Municipalities see AVs as a source of innovation, attractiveness, sustainability, and improvements for the transport system. Driver unions and environmental NGOs as the only actors presenting an opposing attitude must be taken on board in discussions and decision-making to mitigate potential negative impacts on society and the environment. No stakeholder with a high impact and opposing attitude was identified; therefore no one is threatening the project's success.

### 9.2.1.2 Onion Diagram

The onion diagram in Fig. 9.4 shows different stakeholder levels. At the centre, the primary level shows stakeholders with significant influence on the project and strong control over essential resources regarding AVs and public transport (Bodnar et al., 2015; Czischke, 2018). At the secondary level, stakeholders with relevant importance and medium control over resources are placed. The wider environment is represented at the tertiary level, including stakeholders with weak control over resources that affect the project on a low scale (Bodnar et al., 2015; Czischke, 2018).

Stakeholders at the primary level are market actors and government actors. Environmental NGOs can be classified in the second level, as they are active in the dialogue with multi-stakeholders, in social awareness, and in influencing policymakers. The tertiary-level actors do not have a direct impact or a strong influence on AV decision-making and implementation. Some of these actors, such as the insurance companies, are still awaiting more results before setting their strategies.

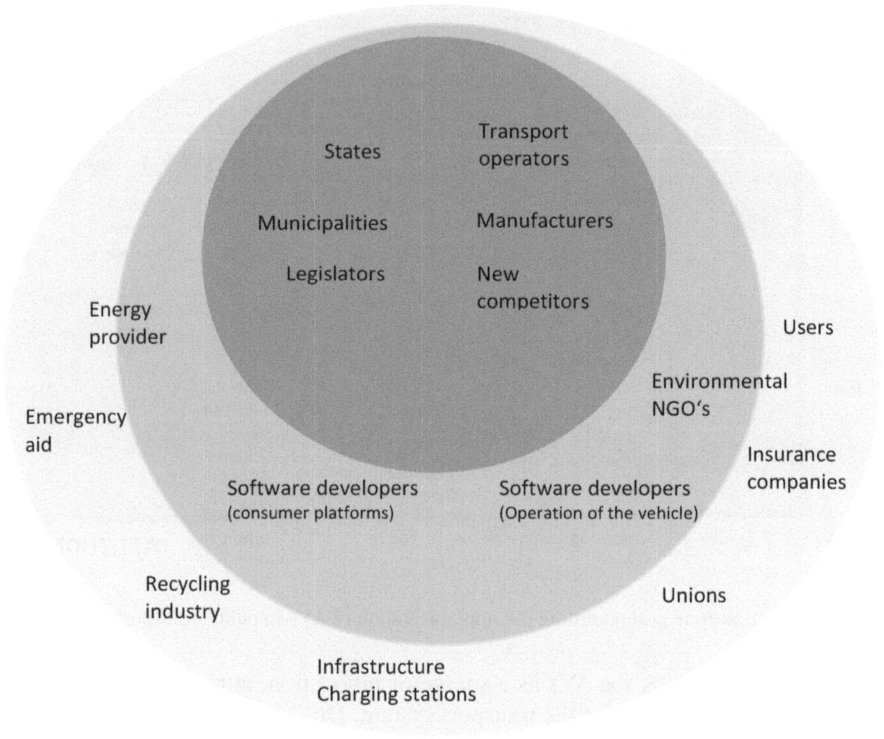


**Fig. 9.4** Onion diagram concerning the implementation of AVs in the public transport system

#### 9.2.1.3 Selection of Stakeholder Groups

Based on the three analyses, key stakeholder groups can be defined. They are introduced in the following paragraphs:

- *Manufacturers* of automated minibuses are important stakeholders. The primary goal of NAVYA, the manufacturer of the minibuses in the AVENUE project, is to offer new mobility solutions, the establishment of a good market position, and consumer confidence/acceptance. To do so, they focus on implementing their products in public systems as soon as possible. Well-drafted standards can increase the rate of development and reduce overall system cost per vehicle.
- *Software providers* offer platforms that enable the intelligent operation and optimization of automated mobility services, managing fixed-route, and on-demand services. They are crucial stakeholders, as an automated system cannot function without a proper software platform.
- The prime objective for *public transport operators (PTOs)* is to seek a high market share and good market positioning. To stay competitive, they must innovate and reduce costs. In their perspective, automated minibuses could be a solution to pursue both goals. However, public transport operators are also holding back,

as recent studies show that the development of automated vehicles could result in public transport losing its attractiveness due to innovative services, such as on-demand taxi services and private car-pooling. They are important stakeholders but could be overtaken by new competitors that share similar objectives. All involved competitors are focusing on developing the most innovative, secure, and trustworthy vehicle for the market (Boersma et al., 2018).
- A current barrier to the establishment of fully functional automated minibus systems is regulations that require stewards on board (Ainsalu et al., 2018). The *European Union (EU)* is an important stakeholder in this respect. Promoting automated driving will challenge the EU to create incentives and regulations and remove possible barriers that can stop the development of this technology. The EU can develop general guidelines and policies, but it is up to the *government* of the member states (from now on referred to as countries) to implement them. There is great differentiation in the level of application between different *countries and cities*. On a city level, local incentive structures—mobility policy, willingness to adapt road infrastructure, etc.—are crucial for a successful system of automated minibuses.
- A strong opposing position is taken by *unions of transport operators*, as bus drivers fear losing their jobs with automated minibuses. The unions would favour an automatic unemployment assurance, provide education and retraining for the transport operators to find comparable or even better jobs, and expand support for displaced workers to start and sustain their own businesses.
- Other stakeholders were identified but will not be further detailed in this part, as their influence is, not yet, crucial for the project's success.

## 9.2.2 Self-Assessment Stakeholder Groups

Following a series of 30 interviews with representatives of the primary stakeholder groups, self-assessments were conducted for these groups. In these self-assessments, the main roles and strategic objectives of these groups are distilled. The resulting maps portray the stakeholder environment from the perspective of the central stakeholder group. Therefore, the stakeholder maps differ for each stakeholder group. The purpose of these maps is not to create one overall map but to gain insights into how each of the stakeholder groups is perceiving its relation to other stakeholders.

### 9.2.2.1 Public Transport Operators

The PTOs are responsible for providing public transport services in cities. Interviewees stressed the strong need to be competitive in the future. They expect automated public transportation to contribute to societal benefits such as better quality of life and improved health and environmental conditions due to reduced pollution. PTOs perceive automated vehicles as a 'key topic for the future' and as a decisive element of competition (Fig. 9.5).

#### 9.2.2.2 Manufacturers

Manufacturers of automated minibuses define the following goals: (1) offering new mobility solutions, (2) establishing a good market position, and (3) creating consumer confidence and acceptance. To accomplish these goals, they focus on implementing their products in public systems as soon as possible. Well-drafted standards can increase the rate of development and reduce overall system cost per vehicle. The manufacturers all have similar ideas of what future mobility should look like and pursue a similar strategy.

The automated minibuses are currently implemented in pilot projects all over the world to learn more about the requirements of the environment and their use in various practical scenarios. There is still a need to further develop and optimize the products and to extend them to other areas of application. The manufacturers are convinced that with their vision of automated driving in public transport and on short distances, they have developed a concept for the future that will transform traffic in cities in the future with the properties of automated, shared, connected, and electrically driven vehicles.

With their vehicles, manufacturers want to meet the needs of their partners, improve transportation services, play an active role in shaping the future of mobility, and drive this new technology. Figure 9.6 presents how the manufacturers see their interaction with other stakeholders within the mobility ecosystem.

#### 9.2.2.3 Software Providers

Software providers offer platforms that enable intelligent operation and optimization of automated mobility services, managing fixed-route and on-demand services. The platforms should function as the interface between vehicles, travellers, and mobility providers. Software developers aim to offer their services to the greatest number of vehicles possible. Therefore, partnerships with governments or big fleet managers are crucial. Another common strategic objective of software developers is to change the mindset about mobility systems. The time that is now spent on transportation can be used for activities that add value to personal life.

The main objectives and responsibilities of software providers are clear: assure safety, efficiency, and punctuality for end-users. Like the previous parts, the map presented in Fig. 9.7 presents the software developers' perspective on their relation to other key stakeholders.

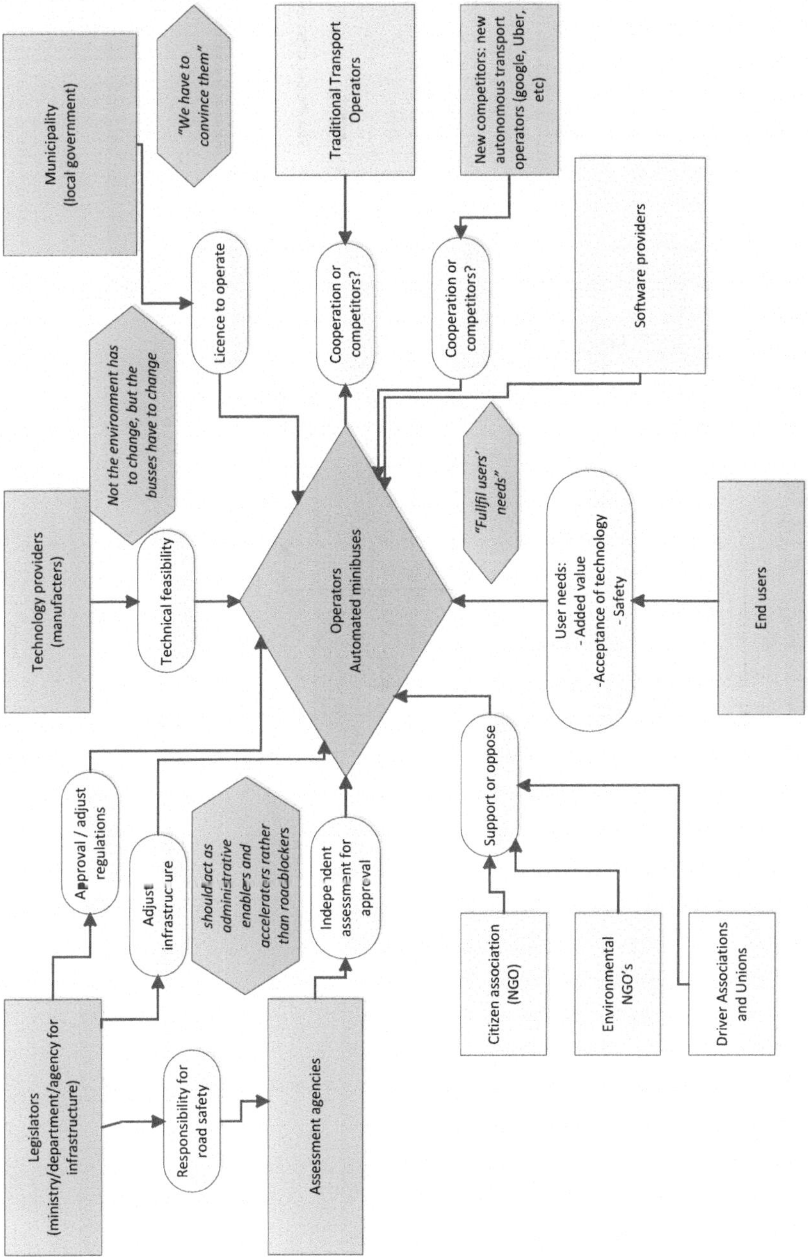


**Fig. 9.5** Stakeholder map from transport operators' perspective

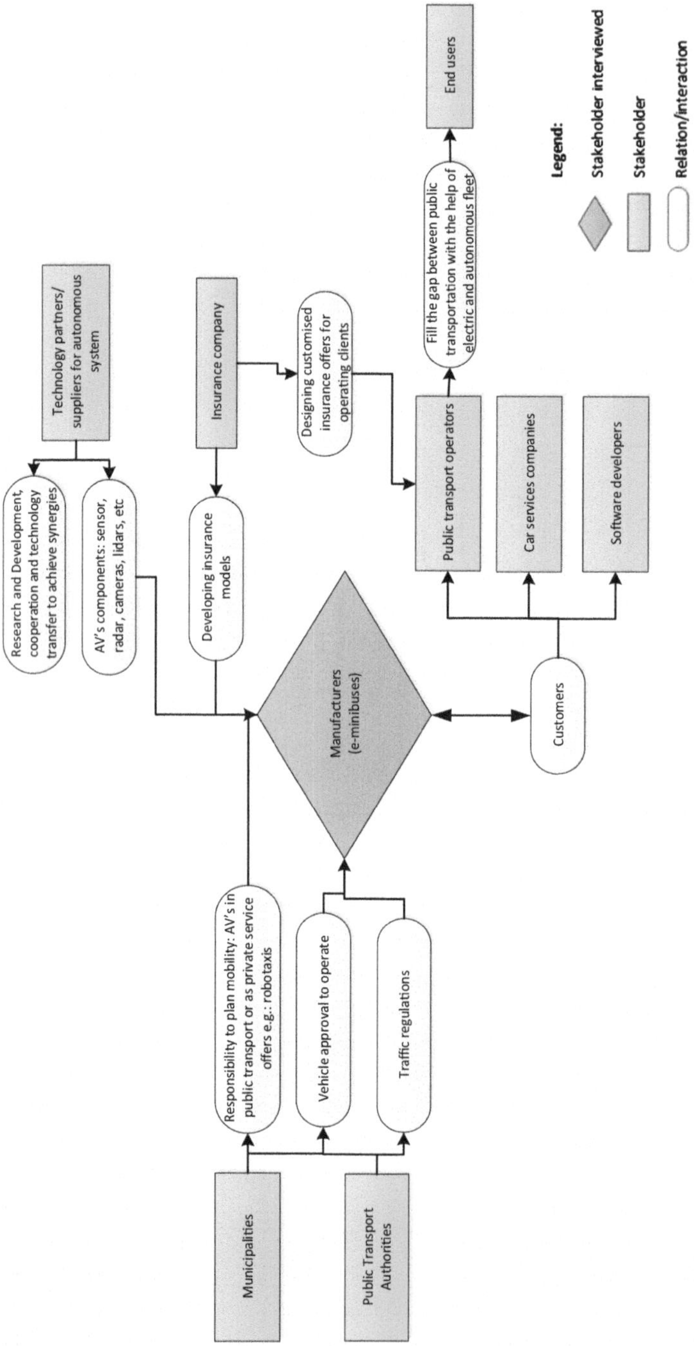


**Fig. 9.6** Stakeholder map from the manufacturers' perspective

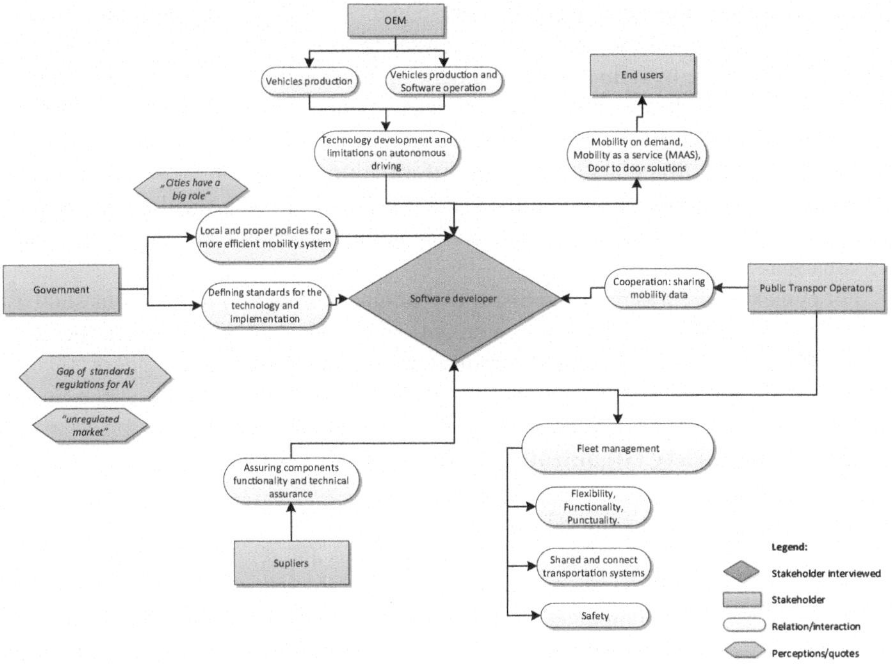


**Fig. 9.7** Stakeholder map from the software developers' perspective

#### 9.2.2.4 Driver Unions

Driver unions address topics like the drivers' re-education and formation for work, better road conditions, safety assurance, and agreements with employers' organizations and governments. The interviewed driver unions are well-consolidated organizations with many years of experience, a significant number of members, and local, national, and international networks. One of their primary actions is to be part of the dialogues in the transport sector and to negotiate collective agreements for their members. Priority actions address improving the drivers' working conditions. The interviewed driver unions highlighted the strategic importance and their focus on professional education, informing and training.

#### 9.2.2.5 Policymakers

The interviews with the policymakers are those conducted with the government representatives of two European cities and a national road directory. Policymakers have a centric role, as they have interactions with most key stakeholders (PTOs, users, manufacturers, NGOs). The policymakers are primarily concerned with urban planning, regulating automated vehicles, and the effects of mobility on future cities. They recognized that they should have a more active role in shaping the city of the future rather than only reacting to the consequences of deploying automated vehicles (ID12). The minibuses have the potential to affect and change the design of urban infrastructure by reducing needed parking spots and urban roads (ID12; ID15; ID19). The change leads to more urban spaces being dedicated to green and blue infrastructure (ID15; ID19). But it will also require incorporating specific structures for AVs such as fibreglass cables or information signs for crossings (ID12). Moreover, the interviewees explained the difficulty of drafting regulations for AVs due to concerns about liability, safety, and the ability to react to complex mixed traffic situations (ID15).

Policymakers are aware of the disruptive nature of transport innovations such as automated minibuses and their potential role in shaping cities' and users' perceptions. Thus, they rely on insights from pilot projects and other stakeholders (ID15).

#### 9.2.2.6 Civil Society Organizations/Citizen Organizations

The citizen organizations interviewed are working at city, national, and EU levels. They address issues concerning transport and citizens and represent end-users. They interact mostly with other transport associations, policymakers and government representatives, transport companies, and citizens. The civil society focuses on promoting public transportation. In the interviews, they expressed the urgency of developing and innovating public transport (ID17; ID16).

They want to reduce individual and public transport and especially active modes of transportation to become the first choice (ID16; ID17). The organizations are working on improving services for the users with customized public transport services such as MaaS and first-and-last-mile solutions. As the automated minibuses are the focus of the interviewed organizations, they also displayed concerns about how these minibuses will be deployed. Mostly, there is a fear that they will be used as an individual mode of transport and compete with active mobility and public transport (ID 16; ID 17). In general, there was a consensus on using automated driving as an opportunity to make the transport ecosystem more robust and sustainable and to focus on the users' needs (ID13; D16; ID17; ID18). Finally, infrastructure was also an important topic in the interviews; the organizations see that cities should focus less on long-term megaprojects and focus on solution-targeted mobility.

### *9.2.3 Results from the Cross-Sectional Analysis*

In this section, the results of the cross-sectional analysis are presented. These results are based on compressed data using bottom-up coding. As a result of this analysis, five main topics were defined.

#### 9.2.3.1 The Crucial Role of City Government

The introduction of new mobility innovations into the transportation network showcases the importance of the government's role in dealing with transportation challenges. The cities are responsible for regulating the usage of automated minibuses in the built environment, setting requirements for the public transport system, and issuing concessions for public transport operators. Therefore, based on the interviews, the importance of the government's position in deploying future mobility and the relationship between the PTOs and local government were important issues.

The city governments are responsible for the public transport system within their city. They define the requirements of public transport based on their transport vision, financial resources, and requirements from the national government. Based on these requirements, concessions to operate public transport within a city are issued (ID21). This includes the percentage of electric vehicles that should be deployed, the frequency of operation, as well as price (ID21; ID22). Different contracts between the government and PTOs lead to more opportunities for testing and achieving robust results concerning the deployment of different forms of automated mobility.

Testing new mobility innovations requires the financial support of the government. The PTOs cannot afford to procure the needed equipment for such operations. It does not offer substantial revenue because the service is free and requires dedicated resources to manage the pilot site. The PTOs are dependent on governmental funds. Thus, cities should seek to diversify mobility innovation testing in collaboration with the PTO (Nuttall et al., 2018).

The collaboration between the city and the PTOs hinges on the city's understanding of the importance of rolling out innovations within the transportation ecosystem. Mobility trends can provide solutions to urban planning challenges. However, governments are reluctant to approve the testing due to safety and regulation concerns. Whereas AVENUE aims to deploy the automated minibuses as an on-demand, door-to-door service, this is not accepted by all governments, as this service would be classified as taxis and not public transport (ID22). There is a need for participatory actions to involve all stakeholders to reach a common understanding of the best practices for deploying automated driving (ID11).

The city is also responsible for the infrastructure and urban planning needed for deploying these forms of transportation (ID5). Urban planners are focused on reducing urban space dedicated to private cars. This presents the motivation for cities to work with the PTOs to reimagine future cities. The use of automated minibuses could reduce parking and road space. The shift in urban planning is happening from long-term, big-budget infrastructure projects to short-term plans that account for the potential disruption of mobility trends.

Currently, the trials of deploying automated minibuses require adaptation of the infrastructure to accommodate the needs of automated vehicles, such as road sensors and special signalling devices, additional digital infrastructure, and charging posts (ID13; ID20; ID22; ID 11; ID10; ID12). The lack of infrastructure and the focus on developing infrastructure that does not account for future mobility trends present obstacles to the deployment of automated minibuses.

#### 9.2.3.2 Technology Development and Legal Regulations

Rolling out new mobility technologies is intertwined with legal regulations. The legal system should ensure that these innovations are not harmful and unruly. However, the regulatory framework could act as an obstacle to innovations in the public sphere. Interviewees state that, because of the innovativeness of the technology of automated minibuses, regulators do not have the experience nor the knowledge of the technical specifications. Hence, it is difficult to explain the important technological considerations to lawmakers and to fit them within existing regulations. Issues related to reliability, insurance, and interoperability for the connected systems are still not clearly defined. Moreover, the effects on other road users also pose a struggle for lawmakers. It gets more complicated to consider the technical regulations within the overall transportation system regulations such as the no-car zones and road pricing. Hence, legal regulations are seen as a barrier by most of our interviewees.

The government's primary concern remains the safety of the passengers and the citizens in the circulation areas. The need for a safety operator on board is mandatory in all the AVENUE trials, especially for vehicles operating in mixed traffic. Thus, policymakers and operators should work together to prioritize safety while not restricting the advancement of the technology of automated minibuses (ID15, ID18, ID1).

Our interviewees stress that the technology of automated minibuses should be considered a solution and not a barrier (ID10, ID15). In the testing phase of the technology, it is advisable to bring the technology into the public sphere, to better understand the opportunities and barriers of the technology (ID4). The regulatory framework should anticipate the consequences of the deployment of automated minibuses rather than just react to them.

The regulations also vary between countries. This has an impact on the advancement, testing, and deployment of automated minibuses. The strategy for automated minibuses depends on the national goals and trends for mobility. The deployment of AVs in the EU is managed on a national level. Some countries have appointed councils and committees on these subjects; others are just trying to fit the technology within the existing regulation for transportation (ID4). A unified EU legal framework could facilitate the development of the technology and reduce confusion for the manufacturers of the vehicles and software developers. It could also improve the competitiveness of European countries in the automotive sector (ID10, ID6).

The lack of standards and regulations is also a bottleneck commonly quoted (ID12; ID15). In addition to the fact that it is difficult to regulate the use of new mobility modes, specifically AVs, more research and results are required. This includes not only regulations for AVs but should also take on a holistic approach to mobility as a whole (ID10).

#### 9.2.3.3 Restructuring the Mobility Industry

Traditionally, innovation in the mobility industry was the realm of the vehicle manufacturers (referred to as OEMs). The processes of digitalization and automation are changing the industry. Companies specialized in software development set requirements for innovation and production of vehicles. Our interviewees state that most OEMs still focus on being vehicle suppliers but are starting to develop their own software as well. Vehicle manufacturers are dependent on the software providers, but the software providers are less dependent on the OEMs as they do not have a specific vehicle. Nevertheless, they can't build the vehicles entirely by themselves and require input from OEMs. The renewal of the mobility industry has not yet settled; new types of cooperation, competition, and the rise of innovative companies require constant renewal. The development of automated vehicles is pushing for this restructuring of the mobility industry and more cooperation. However, Mobility as a Service (MaaS) is pushing this development a step further. The key innovation and requirement for this system is a functioning software that allows end-users to mix several public and private mobility systems.

#### 9.2.3.4 Social Acceptance and Environmental Aspects

This section is based on the opinions of the interviewed stakeholders and not the (potential) users.

Automated minibuses are an innovation that most of the end-users have no experience with, other than hearing about it in the news. To enhance social acceptance, users must gain familiarity with the technology. This could be with as many pilot projects in public spaces as possible for the development of the technology and to address technical barriers (ID14). Furthermore, it is claimed to be important how the minibus is communicated; 'self-driving vehicle' and 'automated assisted system' seem more promising than other terms (ID1). Other interviewees stated that the technology presented to end-users should be incident- and accident-free (ID2). Users expect a normal bus without drivers (ID21) but find that the service is not as advanced as expected which leads to disappointment (ID21). Based on past and current pilot projects, it is not expected that social acceptance will be problematic (ID16). Most users (97%) of a pilot in France reported being happy and confident. Other pilots show that although users are content, a problem lies in convincing the potential users in the service area of the use and advantages of the service (ID21). In addition, the respondents addressed diverse current societal issues that reflect the concerns associated with automated driving, with data protection and data privacy being the most cited (ID8; ID10; ID7). Other concerns relate to cybersecurity and hacking risks (ID14; ID6) and ethical questions that are still open and need to be addressed (ID5; ID8).

Automated minibuses could be used as a conventional bus, but there are more possible applications. They could provide an on-demand, door-to-door service and could service additional lines. Our interviewees state that, for the end-users of the automated minibuses, the service provided is important, and not so much whether the service is automated or conventional (ID18). These insights are confirmed by the outcomes of the social impact studies, conducted for WP8. These studies show that (potential) users expect a higher level of flexibility from automated minibuses. This flexibility is defining the willingness to use them, not the fact that these buses are driverless (Naderer & Korbee, 2021). Public transport is on a decline in Europe, although not consistent in all countries. Hence, there is a quest for improving the quality of public transport (ID18). This could be done via the integration of automated minibuses (ID18). Automated minibus services could resolve mobility gaps by linking unserved areas, therefore adding lines instead of replacing them (ID10, ID21). They could enforce the development of MaaS systems and seamless traffic flow. Nevertheless, it seems to be unclear what the need and acceptance of such a system are and how to overcome technical barriers (ID16). Part of the respondents also claim the need for more citizen involvement to understand if citizens agree with and would want the use of those new technologies (ID12; ID16; ID8; ID10).

Respondents believe that the deployment of automated vehicles will occur in the long term, which would give time for drivers to be re-educated and learn new skills (ID11). Three respondents explained that automated driving would create new and better jobs (ID11; ID16; ID7). Those higher qualifications entail higher wages. Therefore, personnel costs could increase due to increasing qualification needs and the number of automated vehicles. In addition, one respondent highlighted that it is important for driver unions to be part of the discussions before disruptive changes take place in the sector (ID11). Some respondents point to the fact that the experience that a bus driver gets over the years is very difficult to compensate for with automated driving (ID6). A respondent drew attention to the fact that the elderly and people with reduced mobility (PRM) may rely on the safety drivers' help onboard the minibus, which cannot be replaced by technology (ID7). A solution could be to book rides with safety drivers through the app (ID20). Due to this kind of consideration, interviewees report that the role of bus drivers will become (even) more important (ID11).

Electric automated vehicles are often promoted as a sustainable solution to urban transportation problems, such as congestion and pollution. Our interviewees have different opinions on that. Some interviewees stated that additional research is necessary (ID8) and others that this innovation does indeed entail possibilities to alleviate environmental pressure from the urban transportation system when fulfilling certain requirements. Automated minibuses should, for instance, replace the use of cars, rather than substituting walking or cycling. A potential disadvantage could be that automated vehicles will not be operated in a shared manner but will replace individual cars and be privately owned although this will not result in elevating environmental pressures.

#### 9.2.3.5 Future Scenarios

The vision and perspective of many respondents converge on a transition of mobility services towards a more connected, electric, automated, and shared mobility (ID1; ID4; ID10). AVs can be seen as an opportunity to reduce emissions, to make mobility more flexible, attractive, and shared, and as an opportunity to design our cities as we want them (ID10). AVs can also redefine the traffic flow in cities through shared mobility (ID3). In addition, AVs are expected to transform urban mobility, providing more reliable and scalable services, which meet individual needs (ID5). However, one of the respondents stated that AVs are perceived to come in a very far future. They also pointed out that the vehicles operate at a very low speed and are not made for changing road traffic, therefore, requiring even more street space (ID16). The respondent highlights that public transport must be the first choice and that investments in the public transport sector cannot be neglected (ID16). Most of the stakeholders agree that shared AVs are the best solution, more beneficial and affordable, in contrast to individual AVs (ID1; ID3; ID4; ID5; ID8; ID14). In this regard, one respondent commented that OEMs predict automated driving to be the future, but at the same time, they want individual mobility to be part of a future mobility system (ID14). When asked about the role that they foresee regarding the automated minibuses, the respondents presented two different perspectives:

- Deployment of the automated minibuses on a local and small scale would aim to fill the mobility gaps, with a focus on first and last miles (ID1; ID12; ID15; ID16).
- Deployment of automated minibuses to increase driven distances and drive on more complex roads (ID2, ID3).

### *9.2.4 Stakeholder Map*

In the previous sections of this chapter, the results of the stakeholder analysis on the field of automated minibuses in European urban public transport systems were presented. Based on the presented analysis, a ‘stakeholder map’ was defined that shows how the field is *currently* seen by the main EU stakeholders (see Fig. 9.8).

#### 9.2.4.1 Structure of the Stakeholder Map

Through the analysis, it became clear that not all actors have an equal influence on the development of automated minibuses in urban public transportation systems. Therefore, three types of actors were defined: primary, secondary, and tertiary actors:

- Primary actors can influence this development and deployment directly and have a large stake in the direction of the development. These include the city governments, public transport operators, manufacturers, software developers, the European Union, and citizens/end-users. There are strong relations between

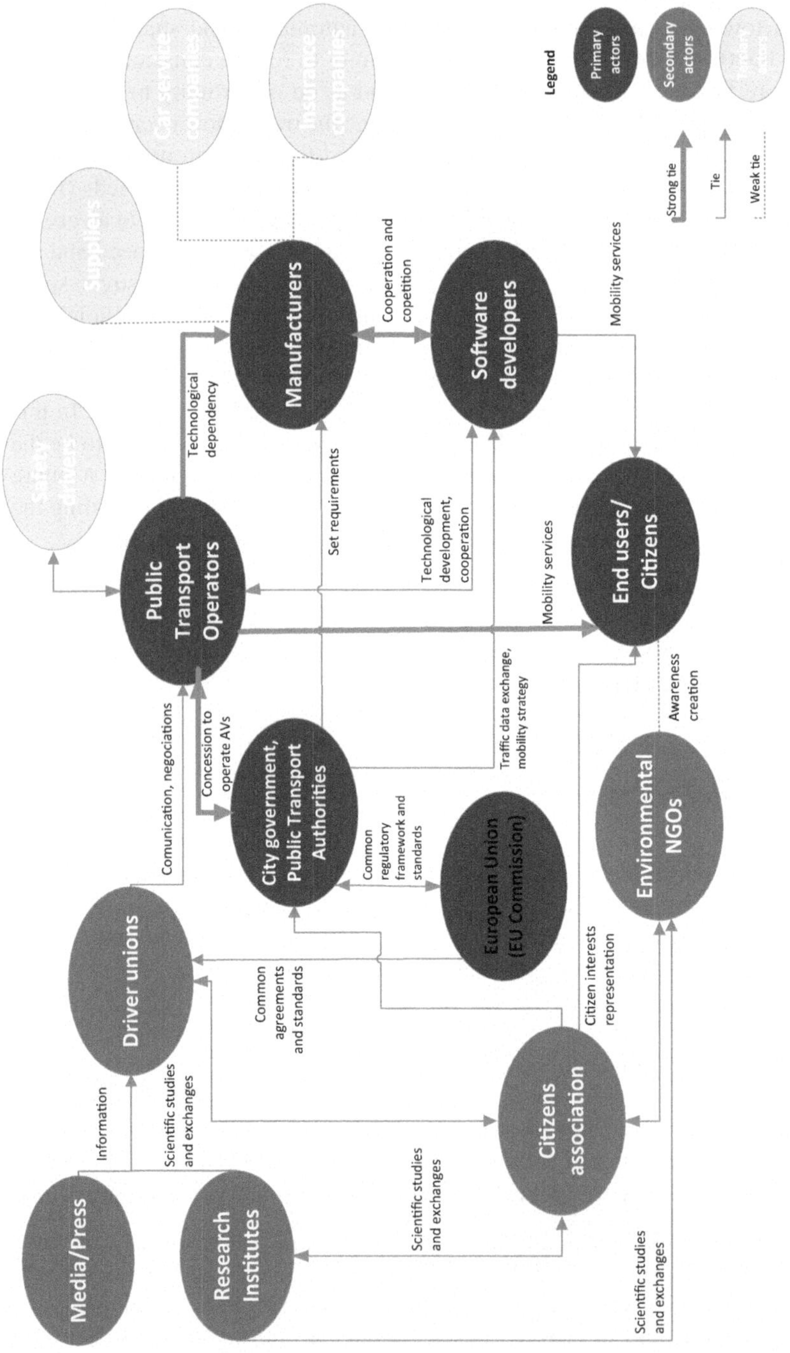


**Fig. 9.8** EU-level stakeholder map

these primary actors, with the city government and the PTO as a central axe. The testing and future deployment of the automated minibuses hinge on the cooperation between these stakeholders, specifically the PTO and the government.

- The secondary actors influence the development but, more indirectly, through influencing the primary actors. They are civil society organizations (driver unions, environmental NGOs, citizen associations, research institutes, and the media). This actor group is well connected but primarily to the groups of primary actors via the PTOs, the city government, and the citizens/end-users. They play an important role in supporting or opposing the key players.
- The tertiary actors are important for the technical development of the system (suppliers), as well as for the operation of the system (safety drivers and insurance companies). They provide constraining and enabling conditions but are not actively involved in shaping the system.

The division of strengths in ties between the actors results from the type of connection between them. Furthermore, the number of times that the connection was mentioned by the stakeholders and the weight attached to it by them was considered.

#### 9.2.4.2 Insights from the Stakeholder Map

The stakeholder map presents insights into the relations between the involved stakeholders in the field of automated minibus deployment. A group of six stakeholders are key to the development of a public transportation system, in which automated minibuses are integrated as on-demand services.

The group of *manufacturers and software developers* strive to develop the necessary technologies and push for their implementation. To do so, they are dependent on the other four stakeholder groups. Close collaboration with the PTOs is necessary. These organizations will have to operate the services, so their commitment to this new service and the transition to a new system are necessary. The *PTOs* provide transportation services on behalf of the (city) government. They define the requirements of public transportation, for allowing automated minibuses on public roads, but also more practical issues such as the service hours of public transport. The *city governments* and public transport authorities are dependent on the decisions of the *European Commission* and *national governments* to define the room to manoeuvre for them to enhance and stimulate the development of a revolutionized public transport setting. The final group of stakeholders are the *end-users* of the system. A technical innovation, such as automated driving, can only be successfully implemented once accepted by the (potential) users. Based on the result of this study and confirmed by the AVENUE social impact assessment, there is a high level of goodwill towards automated minibuses. To assure that end-users will also change their mobility behaviour, the system should provide a high level of flexibility with innovative software solutions and driverless automated shuttles. These interdependencies show that a renewal of the public transportation system towards a flexible system that focuses on providing tailor-made mobility solutions requires cooperation between and commitment of the main stakeholder groups. Another requirement is the creation of a common discourse, a common wording and an understanding of the desired system. This means that the development of innovative techniques, the software and data-sharing platforms, the required regulatory alterations, and the social and behavioural changes should be aligned. It should be accepted that the deployment of automated minibuses is not only an extension of the existing public transport system but is intended to revolutionize the system. This new system is not structured by fixed routes from the supply side but is guided by the travel request from the demand site. Only through the acceptance of this common frame necessary changes can be made.

The distinction between primary, secondary, and tertiary stakeholders does not mean that some groups are not of importance. Each of the stakeholder groups can alter and change the perception and position of the primary stakeholders. Pilot projects, such as the AVENUE operations, are not only important to develop and test technical innovations but also to create goodwill and science-based facts that civil society organizations use to inform citizens. The stakeholder map shows that the implementation of automated minibuses in the public transport sector is only partly dependent on the technical system. This technical system is embedded in a larger system that includes not only governmental actors, end-users, and civil society organizations but also competitors that envision another restructuring of the (public) transport system. To gain better insight into these two systems and how the interactions between the two systems influence the work and vision of the AVENUE project, Sect. 3 focuses on an AVENUE stakeholder and mobility services map.

## 9.3 Conceptual AVENUE Stakeholder and Mobility Services Analysis

The AVENUE stakeholder and mobility services map were developed to gain insights into the consequences of deploying automated minibuses in the public transport sector. It aims to show the complex ecosystem that this deployment will become part of. It aims to identify main mobility trends, strategic actors, and their interactions in the process of implementation and integration of automated vehicles in the transport systems of European cities. The map is based on a comprehensive literature review, which is validated by experts and AVENUE project partners.

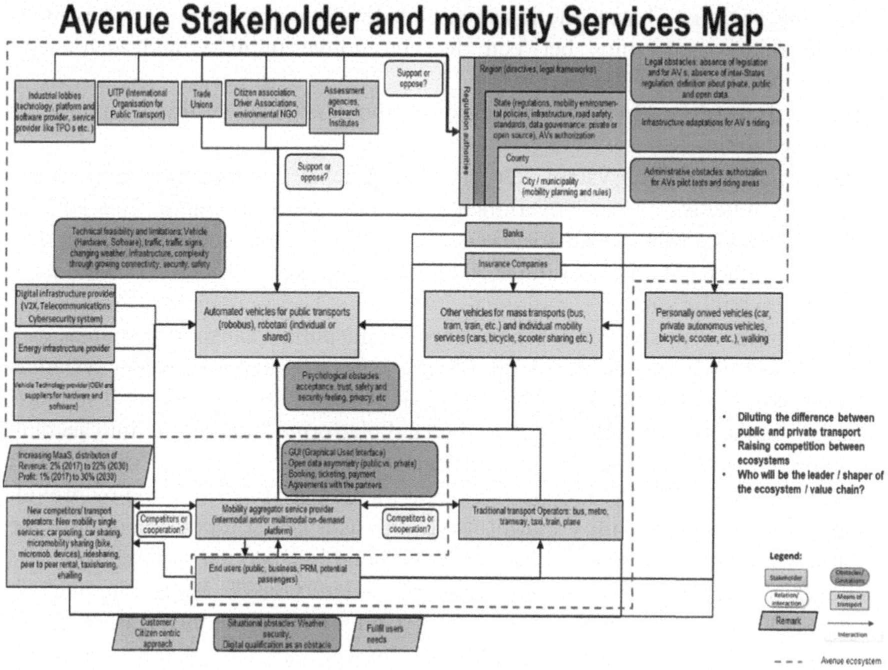


**Fig. 9.9** AVENUE stakeholder and mobility services map

## *9.3.1 AVENUE Stakeholders and Mobility Services*

The starting point of this map is the vision of AVENUE that automated minibuses could be integrated into a multi- and intermodal mobility system and be offered as Mobility as a Service (MaaS). AVs and MaaS could be game changers in mobility as AVs combined with other means of transport could individualize public transport. This approach follows the people-centric vision of mobility of the EU (EU and EFTA Ministers of Transport, 2020; UITP, 2020) and the SUMP approach.

At the centre of the map presented in Fig. 9.9 are three means of transport (in light orange): automated vehicles for public transport, vehicles for mass transport and individual mobility services, and privately owned automated vehicles. The automated vehicles for public transport are connected to four subsystems: infrastructure (left side), civil society (top left), government (top right), and the transport sector (bottom). These subsystems each have their own stakeholders and goals and present obstacles and limitations to the integration of automated minibuses in public transport:

- Stakeholders in the infrastructure subsystem include digital infrastructure providers that set the requirements and comprehend telecommunications, data centres, cybersecurity systems, vehicle intersections (V2V, V2I, V2X), and energy infrastructure providers that set requirements for infrastructure for EVs and charging stations. Hence, the research and development for those innovations entail more partnerships between the automotive industry and technology providers.

- The civil society subsystem includes stakeholders such as trade unions and industry lobbies.
- The governmental subsystem consists of multilevel governmental institutions, city governments, country governments, and the European Union. This governmental subsystem is crucial for the deployment of automated vehicles, as they can constitute legal obstacles, must adapt infrastructure, and authorize automated driving on public streets.
- The mobility subsystem includes operators of the public transport system as well as those privately owned. A crucial role in the development of a MaaS system is foreseen in the mobility aggregator, a service provider that creates and supports an intermodal and/or multimodal on-demand platform. Thus, taking into account a scenario with multimodal mobility and connected vehicles (Attias, 2017; Fielden et al., 2017), the mobility aggregators play an important role by providing the Integrated Mobility Platforms (IMPs) as a key solution to simplify the journey planning and payment and providing highly customer-tailored solutions (Baron et al., 2018). The current main players are Google Maps, Citymapper, Omio, Qixxit, and Moovel, among others. Connecting these means of transport also means connecting both the traditional transport operators and new competitors. The end-users are an integral part of this subsystem, as it is developed to be a 'customer-centric approach to mobility services' (Fournier, 2017). In establishing this system, the end-users have several options regarding the means of transport for short and long journeys in European cities.

Banks and insurance companies are stakeholders included in the map that do not directly relate to one of the subsystems. The insurance companies consider that initially, automated vehicles will increase insurance rates because they will become more complex and more expensive to fix (Noble, 2018). Later on, it is expected that the insurance prices for automated vehicles decrease, considering that the frequency of claims might reduce and the percentage of automated vehicles on the roads might increase (Noble, 2018). In addition, when addressing automated vehicle technology, potential new insurance market fields emerge, such as cyber risk, software and hardware, and infrastructure (Karp et al., 2017).

From the banks' perspective, automated vehicle technology can trigger significant changes in financial services. Therefore, banks should embrace those changes and emerging new technologies (e.g. artificial intelligence, blockchain, distributed ledger technology) and re-think financial services according to them (e.g. IoT, sensors, connected cars) (Pinto, 2018). Important transformations mentioned by Pinto (2018) include the sharing economy and how banks model their financial services, considering the shift from ownership to sharing-based models, open banking, and how users will do their payments, loans, credits, risks, security concerning privacy and hacking cars, and customer trust. Banks will also play a key role when AVs are combined with other means of transport within a MaaS system for ticketing.

### 9.3.2 MaaS Strategies

In AVENUE, the deployment of automated minibuses is part of a technical ecosystem, aiming to demonstrate that automated minibuses can operate as an integral part of the public transport system. The stakeholders in this ecosystem should cooperate and work towards achieving this goal. However, there is also a bigger ecosystem, that of the entire urban mobility, including private and shared mobility as well.

The AVENUE stakeholder and mobility map, therefore, defined three means of transport:

- Automated vehicles for public transport
- Vehicles for mass transport and individual mobility services
- Privately owned automated vehicles

As is shown in Fig. 9.9, the development of privately owned automated vehicles is excluded from the AVENUE scenario, as these automated vehicles will only replace conventional vehicles. An important aspect is whether the means of transport are part of a public or private transportation system. A taxi, for instance, is not part of the public transport system but is shared between users, whereas a bus is both shared between users and part of the public transport system. The same division can be made for automated vehicles. Defining two systems (a private and a public transport system) in which automated vehicles can be deployed on-demand and door-to-door in an intermodal mobility concept invites for a discussion on how these possibilities relate to each other and to the goals and vision of the AVENUE project. The central question therefore is what are the possibilities and consequences for the technical innovations of automated vehicles to revolutionize transportation?

We see three main considerations that are important for future scenarios. First, private automated vehicles could be introduced and developed. This scenario is not part of our AVENUE focus and is not very likely as the price is expected to be rather high in the next 10 years (Fournier et al., 2020). Cities would furthermore like to avoid additional traffic and continue to have a policy that tries to keep individual vehicles out of cities (Buehler et al., 2017; Marsden, 2006).

Secondly, private, unimodal robotaxis (such as Waymo/Google, ATG/Uber, Zoox/Amazon) could provide individual mobility services. These services could later integrate preferred private partners and public means of transport through open data (Directive 2019/1024 of the European Parliament and of the Council of June 2019) and offer customer-centric mobility services. These companies could then create a private MaaS system. This can be called a 'laissez-faire' scenario where authorities do not try or are not fast enough to anticipate the digital and automation transformation in mobility. This scenario would allow few private companies to become dominant (through the so-called 'winner-takes-it-all' effect), capture the created values, and displace public transport (see, e.g. Clewlow & Mishra, 2017). Huge rebound effects could be the result of more traffic and fewer benefits for the environment and the citizens. These scenarios have been further developed in WP8 and should be deepened in future research.

In a third scenario, the linkage between the private and public platforms is established to create a system where on-demand door-to-door automated minibuses are combined with other means of transport through MaaS. In this scenario, public transport is made individual and is providing a real alternative to privately owned vehicles. For that purpose, different combinations of complementary means of transport of all public transport are bundled into one trip and thus most effective and sustainable for stakeholders:

- Citizens could have more choices and a better on-demand and door-to-door service (customer-centric approach).
- PTO could use the existing network capacities better (open seats in private and public vehicles, on-demand service delivery) and offer their passengers more information.
- The city could provide a more efficient transport system through:
  - Better understanding of the behaviour of passengers
  - More complementarity between means of transport
  - More competition and innovation
  - Cost savings with better use of transport resources (rolling stock, roads, etc.) and lowering of negative externalities if certain conditions are satisfied

These scenarios are presented in Chap. 14 of this book. The representative survey as presented in Chap. 15 provides insights into the expectations of citizens. A consequence of the third scenario, the division between public and private transport systems could be blurred. The consequences, hereof, should be deepened in future research.

Admitting the wider ecosystem of urban transport in which the automated minibuses will be deployed influences the role, position, and support of involved stakeholders. Automated minibuses can be deployed to strengthen public transport, but the same technology can be used by private companies to optimize their own ecosystem, which will result in additional competition for the public transport system. To gain support for the deployment and development of automated vehicles, it is therefore crucial to invest in creating intermodal transport systems that rely on open data and open API (application programming interfaces).

### *9.3.3 Open Data and Open API for Mobility*

A crucial topic in the integration of automated minibuses into multi- and intermodal mobility that requires new connections between different stakeholders is open data and open API. API brings data to the potential users to inform them about the location of an automated minibus and when it could arrive. When the user books a mobility demand, an API transforms this demand into a mission for the automated minibus to pick up the passenger. Data availability is a key issue for users to easily plan and pay for their journey using one single platform. In the past, end-users had to juggle among different mobility platforms to plan multimodal journeys (Steinmann, 2019), but now, mobility aggregators have developed Integrated Mobility Platforms (IMPs) aiming to win customers by fulfilling this gap and simplifying route planning and the travelling experience (Baron et al., 2018). However, the encouragement of the emergence of platforms that would centralize all mobility services offered in one region involves strategic decisions and regulations regarding mobility open data (Steinmann, 2019).

From the state and local mobility authorities' perspective, enlarging the mobility open data would contribute to a better overview of transport flows especially private mobility flows in cities, allowing them to adapt public provision according to the customer needs (Hassani, 2018). Nevertheless, the discussions around the mandatory opening of mobility data present discontents and uncertainties for both sides: mobility companies and public transport operators. Mobility companies agree to cooperate with information for local authorities to improve their transport supply, but they also express the concern that opening access to anyone poses a competitive problem (Hassani, 2018). Additionally, open data legislation can also be critical for public transport operators once they fear that it would facilitate the hegemony of non-European digital giants (e.g. Google, Uber, Apple). Operators consider the risk of being deprived of contact with the customer, including ticket selling, in favour of the 'integrators' and their Integrated Mobility Platforms (Julien, 2019). Hence, the mobility open-data regulation is a current strategic topic for private and public stakeholders to plan their next steps.

## 9.4 Stakeholder Strategies

In this section, strategies and recommendations for the main AVENUE stakeholder groups are defined. The strategies are based on the two stakeholder maps developed in this deliverable: the EU-level stakeholder map (Fig. 9.8) and the AVENUE stakeholder and mobility services map (Fig. 9.9). Due to different goals and starting points, the maps differ, in the level of detail, in scope, and in how they can be used. Nevertheless, the maps complement each other.

### *9.4.1 City Governments*

The current pilots of AVENUE provided important insights into how to proceed in the future to support the testing and the deployment of automated vehicles for public transport. As experienced in Copenhagen, the cities could work in collaboration with key stakeholders to update the homologation guidelines for automated vehicle technology. The guidelines must be adapted to the novelty of innovation while keeping the citizens' safety in mind. Moreover, running the automated minibuses in public areas must go beyond the testing phase to the actual roll-out to serve the city in the long run (Eden et al., 2017) and fill real social mobility needs. The long-term planning ensures constant investment and development of the technology and the services that could provide funding support for the PTOs. To achieve that, the city government should commit to involving key stakeholders and especially the citizens.

Furthermore, future deployment and their impacts must be present in the mobility master plan and future transportation agenda. The SUMP guidelines could provide reliable support on how to manage automated vehicles to benefit all stakeholders. The emphasis should also be put on promoting sustainable active modes of transport such as walking and biking. Moreover, the responsibility lies in the hands of city planners to ensure that the infrastructure is adequate for the circulation of the automated minibuses but that it remains in line with sustainable urban planning (Sime, 1986; Vleugel & Bal, 2017).

Finally, collaboration with other cities lays the foundation for a network of support. Membership in cooperative networks could accelerate adaptation planning for AV roll-out. The benefits of networking between cities were explained in research on climate change adaptation for networks (Heikkinen et al., 2020). A similar consortium to AVENUE could be beneficial. An alternative could be to integrate AVENUE projects into the focus of C40 cities to support their sustainable mobility policies.

### 9.4.2 Transport Operators

The PTOs in AVENUE prepared for the pilots by researching the existing public transport offer in the area, the city's agendas and goals, and AV's potential technological trends. All of that built a solid foundation for the testing of the automated minibus. However, a more detailed assessment of the regulatory framework could have further reinforced the project. This showcases the importance of cooperating with key stakeholders such as the local policymakers and government. The regulations' analysis could have provided further insights into the city's mobility strategy, infrastructure restrictions, and transportation demand management, e.g. if there are pull and push policies: no car zones and road pricing (TUMI, 2011).

Furthermore, the PTOs should work more with the citizens. The use of surveys or workshops could help them understand more the needs of the users before defining a test area or in the testing areas. Thus, they will provide the service to meet demand rather than only test the services of the automated minibus.

Such collaborations with other stakeholders provide a holistic vision of needed services. It also promotes private-public partnerships. The automated minibuses should support public transport but not compete with active modes of transport. The NGO's recommendations could support smooth deployment operations, increase the acceptability of the automated minibuses, and promote the services. Even more, the PTOs could focus on the integration of automated minibuses within the transportation ecosystem. In the future, they could collaborate more with software and MaaS operators to better implement one platform where passengers could select the on-demand automated minibuses for public transport and book the entire trip with one ticket.

### 9.4.3 MaaS Stakeholders

The MaaS provider role could be occupied by private, public, or both transport operators. The recommendation for them is to focus on providing integrated customer-centric data to the passengers. Public and private transport operators should work together as one integrator for MaaS services to define the standards for sharing data between all key stakeholders. The access of the PTOs to transport regulations and transport authorities makes it easier to offer a range of public transport services on a MaaS platform quicker, yet the business competitiveness and profit-driven model of private operators could help the MaaS market to grow faster (Kamargianni & Matyas, 2017). Thus, the collaboration between both sides could help develop the MaaS services.

To better incorporate the automated minibuses in the transportation ecosystem, data sharing should be open and fair and API a prerequisite to enable interoperability. The results of the AVENUE social impact assessment show that the acceptance to use data is high among (potential) users and fears regarding data misuse are low (Korbee et al., 2021). Further, the information, ticketing, and billing to ride the automated minibus must be integrated into the public transport system of cities and MaaS platforms. Consequently, the services' quality will be improved, and the impacts of automated minibuses for public transport will be beneficial to all (Pickford & Chung, 2019). The collaboration between private and public transport operators under the care of public authorities leads to a unified platform for MaaS.

### 9.4.4 Citizens and Civil Society

The deployment of automated minibuses in public transport entails a promise to revolutionize our mobility and enable people-centric mobility. Not only does it aim to enhance access to mobility, but it also intends to redefine how we use different mobility systems. As this socio-technological transition has a lot of implications for citizens, they must be actively involved in it. Equally important is the creation of spaces for dialogue, public discussion, and living labs for interactions of citizens and technologies and innovations. This active involvement is not only required for citizens; an important role is reserved for civil society organizations. Environmental NGOs have currently not taken a clear position in this debate but stress the need for better studies to be able to assess the environmental implications.

## 9.5 Conclusions

The AVENUE project aimed to demonstrate the use of automated minibuses in public transport. The stakeholder analysis presented in this chapter aimed to provide insight into the social setting in which the automated minibuses will operate. The social-environmental assessment consists of a multitude of organizations, networks, and institutions. Each of these stakeholders has its own set of expectations and its own specific role. Based on the conducted stakeholder analysis, a conclusion can be drawn on three levels. The first set of conclusions can be drawn concerning the main actors and the level playing field of the deployment of automated minibuses. The second set of conclusions can be drawn on how stakeholders currently assess the development of deploying automated minibuses in public transport systems. The third set of conclusions focuses on how AVENUE partners can increase the success of the deployment of automated minibuses in public transport through a specific stakeholder approach.

### *9.5.1 Automated Minibuses in Public Transport*

Deploying automated minibuses in the public transport system takes place in a complex stakeholder environment. The stakeholder maps show that a plurality of stakeholders involved do not share similar goals and expectations. Furthermore, developing a system in which the automated minibuses provide an on-demand, door-to-door service creates an entanglement with the mobility system in general, as it is a service that competes with private taxi services and personal use of cars. This also results in an altered level playing field; the division between public and private transport is diluting.

### *9.5.2 Stakeholder Positions*

The interviewed stakeholders picture the outlook for automated vehicles in very different ways, meaning that each stakeholder is focusing on different issues to be addressed with the implementation of automated vehicles. In addition to different topics that stakeholders have in mind, they also have different attitudes towards the same topic. A good illustration of that is the future role of bus drivers in a system where automated vehicles will be applied. While the bus drivers themselves picture their current responsibilities as job enrichment (they expect to take over more ambiguous tasks), others perceive the job of bus drivers as not required for the operation of automated minibuses.

The development of fully automated vehicles has also resulted in a renewal of the mobility industry, which is still ongoing. Stakeholders like manufacturers have always played a leading role in market competition, and as they would not like to give up this position easily, they try to steer in one direction. Classical OEMs used to be market leaders in the automotive industry, but with the increasing extent of automation, new competitors dropped into this market. Another point to consider here is the investment linked with fully automated driving. To support investments, it is desirable that policies and regulations in support of this new technology are in place. That would give manufacturers security in their investment. Introducing a legal framework supporting this innovative technology does present a dilemma for governments as they should keep the needs and wishes of society and citizens in mind. In addition, there is a discrepancy in place between technology development (fast and dynamic) and legal framework development (slow and long term).

Before scaling up the deployment of fully automated vehicles, some barriers need to be solved. These barriers are not only of a technological nature but do especially address legal and social issues. Finding solutions for barriers is not in the response of manufacturers and software developers alone, as they will not possess sufficient resources. Especially, social acceptance can become a great threat to putting automated minibuses in place. Another fact to be considered when analysing stakeholders is to look at the resource equipment each stakeholder group has. Each stakeholder group is rich in similar resources, differing in quantity. Additionally, many interviewees raised the missing legislation and regulations for implementing automated driving and claimed that governments partly fail to put appropriate regulations in place.

### *9.5.3 AVENUE Stakeholder Strategies*

To stimulate a successful integration of automated minibuses in the public transport system, partners of the AVENUE project should include new stakeholders in their network. Stakeholders that are currently under-represented are civil society organizations, such as driver unions, environmental NGOs, and citizen associations. These actors are crucial in a user-centred urban mobility system. Another group that is not represented in the AVENUE consortium are the city governments. These governmental bodies are crucial stakeholders in public transportation, as they define the requirements of the public transport network in their cities. An example that shows the importance of the city government is the refusal of the city of Luxembourg to allow an on-demand, door-to-door service to be offered as a public transport service. The third group of actors that are recommended to be included in the AVENUE consortium are the partners deploying existing modes of transport that the automated minibuses will be connected to. Therefore, the AVENUE consortium should focus on:

- Creating stronger cooperation between city government, PTOs, and software developers to develop common strategies to deploy automated minibuses that serve a general interest. The deployment should become an integral part of the city's goals for mobility, as well as for the concession process for public transport services. Cooperation and non-synchrony are key factors to success.

- Developing tools for citizens' participation and inclusion in the debate beyond a user vision.
- Stimulating more interaction and exchange between scientific research institutes, policymakers, and decision-makers.

Finally, a user-centric system (as is envisioned in the AVENUE vision) requires a user-centric approach, more than technology development. This means, for instance, that the selection of services that will be deployed depends on real social mobility needs rather than technological possibilities. Bottom-up strategies, such as living labs and citizen forums, enable more participatory approaches, debates, and dynamic interaction between society and innovations. Moreover, automated vehicles combined with a MaaS open new opportunities to fulfil the best mobility needs and help to develop people-centric transport policies that serve the general interest. This could raise the acceptance of citizens for innovation.

**Acknowledgements** This project has received funding by the European Union's Horizon 2020 research and innovation program under grant agreement No. 769033.

## References

Ainsalu, J., Arffman, V., Bellone, M., Ellner, M., Haapamäki, T., Haavisto, N., Josefson, E., Ismailogullari, A., Pilli-Sihvola, E., Madland, O., Madzulaitis, R., Muur, J., Mäkinen, S., Nousiainen, V., Rutanen, E., Sahala, S., Schønfeldt, B., Smolnicki, P., Soe, R.-M., Sääski, J., Szymańska, M., Vaskinn, I., & Åman, M. (2018). State of the art of automated buses. *Sustainability, 10*(9), 3118. https://doi.org/10.20944/preprints201807.0218.v2

Attias, D. (2017). *The automobile world in a state of change: From the automobile to the concept of auto-mobility*. Springer International Publishing.

Baron, R., Zintel, M., Schemken, N., & Uferer, C. (2018). Integrated mobility platforms: How to win the customer in tomorrow's world of smart mobility. https://www.adlittle.de/en/insights/viewpoints/integrated-mobility-platforms. Accessed 16 Mar 2021.

Bodnar, P., Brown, J., & Nakhooda, S. (2015). *What counts: Tools to help define and understand progress towards the $100 billion climate finance commitment*. WRI, CPI, ODI. https://www.wri.org/publication/what-counts-tools-help-define-and-understand-progress-towards-100-billion-climate. Accessed 24 Aug 2020

Boersma, R., van Arem, B., & Rieck, F. (2018). Application of driverless electric automated shuttles for public transport in villages: The case of Appelscha. *World Electric Vehicle Journal, 9*, 15.

Bryson, J. M. (2004). What to do when stakeholders matter. *Public Management Review, 6*, 21–53.

Buehler, R., Pucher, J., Gerike, R., & Götschi, T. (2017). Reducing car dependence in the heart of Europe: Lessons from Germany, Austria, and Switzerland. *Transport Reviews, 37*, 4–28.

Clewlow, R. R., & Mishra, G. S. (2017). *Disruptive transportation: The adoption, utilization, and impacts of ride-hailing in the United States*. Institute of Transportation Studies, University of California. Research Report UCD-ITS-RR-17-07. https://escholarship.org/uc/item/82w2z91j

Czischke, D. (2018). Collaborative housing and housing providers: Towards an analytical framework of multi-stakeholder collaboration in housing co-production. *International Journal of Housing Policy, 18*, 55–81.

Eden, G., Nanchen, B., Ramseyer, R., Evéquoz, F. (Eds.), 2017. On the road with an autonomous passenger shuttle.

EU and EFTA Ministers of Transport. (2020). Smart deal for mobility: Shaping the mobility of the future with digitalisation—sustainable, safe, secure and efficient. https://www.google.com/url?sa=t&rct=j&q=&esrc=s&source=web&cd=&ved=2ahUKEwjk59zIsoDvAhXIhqQKHeQEApwQFjAAegQIAxAD&url=https%3A%2F%2Fwww.bmvi.de%2FSharedDocs%2FEN%2FDocuments%2FK%2Fpassau-declaration.pdf%3F__blob%3DpublicationFile&usg=AOvVaw1KVKs33ex7GsPniCUCLi52

Fielden, P., Davidson, F., & Arcadis. (2017). Sustainable cities mobility index 2017 (28 pp.). https://www.arcadis.com/en/global/our-perspectives/sustainable-cities-mobility-index-2017/. Accessed 4 Apr 2019.

Fournier, G. (2017). *The new mobility paradigm. Transformation of value chain and value proposition through innovations*. Springer.

Fournier, G., Boos, A., Wörner, R., Jaroudi, I., Morozova, I., & Nemoto, E. H. (2020). Substituting individual mobility by mobility on demand using autonomous vehicles: A sustainable assessment simulation of Berlin and Stuttgart. *International Journal of Automotive Technology and Management, 20*, 369–407.

Freeman, R. E. (1984). *Strategic management: A stakeholder approach*. Pitman.

Hassani, J. E. (2018). Les entreprises de mobilités bientôt dépossédées de leurs données ? *Journal du Net*. https://www.journaldunet.com/economie/transport/1417359-loi-mobilites-ouverture-donnees-transport-open-data/

Heikkinen, M., Karimo, A., Klein, J., Juhola, S., & Ylä-Anttila, T. (2020). Transnational municipal networks and climate change adaptation: A study of 377 cities. *Journal of Cleaner Production, 257*, 120474.

Hermans, L. M., & Cunningham, S. W. (2018). *Actor and strategy models: Practical applications and step-wise approaches*. John Wiley & Sons.

Julien, B. (2019). La LOM calme les appétits des GAFAM. Autoactu. http://www.autoactu.com/la-lom-calme-les-appetits-des-gafam.shtml.

Kamargianni, M., & Matyas, M. (Eds.). (2017). *The business ecosystem of mobility as a service*. Transportation Research Board.

Karp, L., Kim, R., & Liu, C. (2017). *Insuring autonomous vehicles. An 81$ billion opportunity between now and 2025*. Accenture, Stevens Institute of Technology. https://www.google.com/url?sa=t&rct=j&q=&esrc=s&source=web&cd=&ved=2ahUKEwijkNKTw7TvAhWJ8BQKHfL7A3wQFjAAegQIBBAD&url=https%3A%2F%2Fwww.accenture.com%2F_acnmedia%2Fpdf-60%2Faccenture-insurance-autonomous-vehicles-pov.pdf&usg=AOvVaw2LtBhi0wpDJ5zmUU8WOQn6

Korbee, D., Naderer, G., Nemoto, E.H., 2019. Deliverable 8.8 first report on social impact assessment.

Korbee, D., Naderer, G., Nemoto, E.H., Frank, L., Fournier, G., (2024). Who will accept travelling with automated shuttle buses? Defining target groups for potential users. (submitted).

Kyriakidis, M., Happee, R., & Winter, J. d. (2015). Public opinion on automated driving: Results of an international questionnaire among 5000 respondents. *Transportation Research Part F: Traffic Psychology and Behaviour, 32*, 127–140.

Litman, T. (2019). *Autonomous vehicle implementation predictions implications for transport planning*. Victoria Transport Policy Institute.

Marsden, G. (2006). The evidence base for parking policies—A review. *Transport Policy, 13*, 447–457.

Naderer, G., & Korbee, D. (2021). Deliverable 8.7 Second iteration social impact assessment. *AVENUE.*. https://h2020-avenue.eu/wp-content/uploads/2023/03/D8.7-Second-iteration-social-impact.pdf

Noble, B. (2018). *Self-driving cars expected to shake up insurance industry*. The Detroit News. https://eu.detroitnews.com/story/business/autos/2018/10/29/self-driving-cars-insurance-industry/1617277002/. Accessed 16 Mar 2021

Nordhoff, S., Winter, J. d., Madigan, R., Merat, N., van Arem, B., & Happee, R. (2018). User acceptance of automated shuttles in Berlin-Schöneberg: A questionnaire study. *Transportation Research Part F: Traffic Psychology and Behaviour, 58*, 843–854.

Nuttall, K., Arbuckle, T., & Haworth, L. (2018). *Harnessing the future of mobility: How governments can enable a better transportation experience for all citizens*. Deloitte. https://www2.deloitte.com/us/en/insights/focus/future-of-mobility/government-and-the-future-of-mobility.html

Pickford, A., & Chung, E. (2019). The shape of MaaS: The potential for MaaS lite. *IATSS Research, 43*, 219–225.

Pinto, H. (2018). How autonomous vehicles will change the banking industry.

Sime, J. D. (1986). Creating places or designing spaces? *Journal of Environmental Psychology, 6*, 49–63.

Steinmann, L. (2019). Le projet du gouvernement qui inquiète le monde du transport. https://www.lesechos.fr/industrie-services/tourisme-transport/le-projet-du-gouvernement-qui-inquiete-le-monde-du-transport-1023525#xtor=CS1-3046

TUMI. (2011). *Transport demand management*. TUMI. https://www.transformative-mobility.org/assets/publications/Transport-Demand-Management-Push-and-Pull.pdf

UITP. (2020). Sustainable and smart mobility: UITP input to the European strategy. https://www.google.com/url?sa=t&rct=j&q=&esrc=s&source=web&cd=&ved=2ahUKEwih8vLhtYDvAhVKP-wKHSGtARsQFjAAegQIBRAD&url=https%3A%2F%2Fcms.uitp.org%2Fwp%2Fwp-content%2Fuploads%2F2020%2F09%2F20200924_UITP_contribution_EUSSMS_final.pdf&usg=AOvVaw3jAz2pZfPLwxIytH7_EV9S

Vleugel, J. M., & Bal, F. (2017). More space and improved living conditions in cities with autonomous vehicles. *International Journal of Design & Nature and Ecodynamics, 12*, 505–515.

Wicki, M., Bernauer, T., 2018. Public opinion on route 12 interim report on the first survey on the pilot experiment of an automated bus service in Neuhausen am Rheinfall.

# Part II
# Impact Assessment of AVENUE

# Chapter 10
# Research Approach: Introduction to SUMP and AVENUE Methodology

**Eliane Horschutz Nemoto, Nicole van den Boom, Michael Thalhofer, and Guy Fournier**

**Abstract** This chapter introduces the AVENUE sustainability assessment methodology, which is rooted in the Sustainable Urban Mobility Plan (SUMP) framework to formulate a comprehensive approach to an urban mobility assessment. The chapter explores how the AVENUE project is adapting to SUMP by improving public transport and incorporating innovative mobility solutions like automated vehicles (AVs) and Mobility as a Service (MaaS). Through an interdisciplinary lens, this chapter details how the different work packages of the AVENUE project contribute to an integrated and sustainable urban mobility ecosystem assessment. It emphasises the crucial role of stakeholder engagement in shaping mobility solutions that are not only technologically advanced, but also environmentally sustainable, economically viable and socially inclusive. The chapter demonstrates the practical application of SUMP principles in different urban environments, providing valuable insights for policymakers, urban planners and the mobility sector in general. This research underlines the importance of a holistic assessment framework that includes environmental, economic and social impact assessments and also technical and governance issues. It provides a possible interdisciplinary methodology to assess all the facets of future sustainable urban mobility developments.

---

E. H. Nemoto (✉)
CentraleSupélec, LGI, Gif-sur-Yvette, France

Pforzheim University, Pforzheim, Germany

ICLEI Europe, Freiburg im Breisgau, Germany
e-mail: eliane.nemoto@iclei.org

N. van den Boom · G. Fournier
CentraleSupélec, LGI, Gif-sur-Yvette, France

Pforzheim University, Pforzheim, Germany

M. Thalhofer
Software-driven Ecosystems and Market Places, Siemens Technology, Munich, Germany



G. Fournier et al. (eds.), *Automated Vehicles as a Game Changer for Sustainable Mobility*, Contributions to Management Science,
https://doi.org/10.1007/978-3-031-61681-5_10

## 10.1 Introduction

This chapter sets forth an explication of the AVENUE sustainability assessment methodology, starting with an examination of the theoretical underpinnings of Sustainable Urban Mobility Plan (SUMP) concepts and the methodological framework applied for rigorous assessment.

Subsequently, Sect. 2 presents a scholarly synthesis highlighting the principal outcomes emerging from the endeavours within Work Package 8. These results encapsulate the insights from the environmental impact assessment, the economic impact assessment and the social impact assessment, thus contributing to the scholarly discourse on sustainable urban mobility. This approach has been completed by technical and governance issues as they have indeed economic, environmental and social impact which should not be neglected. In contrast to conventional assessments, this evaluation adopts thus a comprehensive, holistic but also challenging approach, thereby striving to contribute a novel methodology that blends diverse facets of sustainability into a unified framework.

## 10.2 SUMP as a Framework

The concept of the Sustainable Urban Mobility Plan (SUMP) aims at a 'new planning paradigm' in mobility, which comprehends a shift from planning for motorised roads and infrastructure to planning for people (Arsenio et al. 2016). SUMP's approach has been widely recognised, targeting sustainable and integrative planning processes to deal with the complexity and dynamicity of urban mobility (Eltis 2021). Hence, it embraces new modes of transport, e.g. micromobility, automated and connected vehicles and new concepts such as Mobility as a Service (MaaS), shared mobility and so on.

The concept of SUMP comprehends the integration of all modes of transport, public and private, motorised and nonmotorised and a long-term planning vision. It targets to improve mobility accessibility, sustainability and citizens' well-being (European Commission 2013).

Furthermore, SUMP provides general guidelines for planning and implementation. It is composed of four main phases:

1. Preparation and context analysis.
2. Strategy development.
3. Measure planning.
4. Implementation and monitoring.

SUMP has been implemented in a number of cities and countries and in diverse settings. For instance, in the city of Koprivnica, Croatia, the municipality carried out a status analysis of its mobility situation; for this, an extensive consultation process engaged a range of stakeholders and a public survey (Mobility Plans, n.d.). In Cambridgeshire, UK, the Local Transport Plan (LTP) 2011–2026 defined indicators and targets to monitor progress towards the plan's objectives, which were aligned with the long-term strategy for transport (ibid).

Russo and Rindone (2023) explore how European cities are becoming smarter by focusing on urban mobility, energy and Information and Communication Technologies. It highlights the importance of SUMPs in achieving this goal. By implementing MaaS plans, cities aim to better understand mobility patterns and their energy implications. The paper offers insights for researchers, planners, and decision-makers into European efforts towards sustainable urban development and improved mobility systems.

Mück et al. (2019) describe the living labs as an innovative approach to foster sustainable mobility planning in Munich. Such living labs aim to demonstrate innovative solutions on mobility, to provide user experiences and to reduce potential gaps between long-term urban planning and the current development of mobility in the city (ibid).

Sampaio et al. (2020) carried out an economic and environmental analysis of measures from a SUMP in a small-sized city. The study compared the transport emissions and external costs of the baseline scenario with the status after the SUMP measures were implemented. The measures consisted of (M1) promoting cycling, (M2) modernisation of the local fleet and (M3) trucks logistic optimisation. According to the study, all measures presented a potential to reduce emissions, in particular the modernisation of the local fleet, with a potential reduction of $CO_2$ emissions by 9% and the reduction of external costs by 11%. The study from Arsenio et al. (2016) reviewed a sample of 40 case studies of SUMPs in Portugal, focusing on climate change goals and equity issues on accessibility. The main findings point that SUMP guidelines remain very broad and general, and there is an absence of specific guidance. For instance, there are gaps of guidance on methods to account for GHG emissions and monitoring indicators to measure the progress in different issues.

Such examples illustrate the SUMPS adoption and implementation in different phases: decision and planning, developing vision and strategies with stakeholders, setting targets and indicators and assessing the impacts of measures.

After this detailed explanation of SUMP, the next chapter shows how it was adapted and applied in AVENUE.

## 10.3 SUMP Concept and the AVENUE Project

The AVENUE project aims at deploying automated minibuses as an innovative and safe mobility solution to strengthen the public transport system of European cities. The automated minibus is electric and shared, and it is expected to improve accessibility, attractiveness and environmental performance of public transport (flexible on-demand, door-to-door services) to fill gaps in mobility and foster intermodal seamless individual mobility. The scope of the project also aims to critically assess the impacts of the introduction of these new technologies in the urban mobility system. The assessments investigate technical obstacles, potential environmental and climate emissions impacts, social acceptance of users and potential users, business models and economic impacts, safety and security issues and the development of regulations, standards and policies for AVs.

AVENUE project and the SUMP concept are aligned by embracing new and alternative modes of transport and new concepts such as Mobility as a Service (MaaS), integrated and shared mobility and intermodal mobility. Such innovations could support the future shift from private car and individual trips to on-demand public transport and shared rides. It could thus address the mobility gaps and needs existing today which makes it necessary to use privately owned cars.

Furthermore, the AVENUE social, environmental and economic impact assessments will provide key findings to guide the integration and implementation of AV in the urban mobility system while endorsing the sustainable planning, strategies and goals of cities. The assessment studies are important to support a long-term vision, design and planning of mobility. Although the pilot projects are deployed on a small scale and with a technological focus, aspects of being strengthened are the citizens' participation (e.g. citizen forums and discussions), as well as the active participation and partnership with the local municipality.

Moreover, the integration of automated minibuses in public transport has to be done accordingly to the specificities of each territory and the different mobility needs, aiming to cover real gaps in mobility (in particular in rural and suburban areas) to a real contribution to better accessibility, affordability and environment-friendly mobility.

Finally, by aiming for a transition towards greener and sustainable transport, it is crucial that AVs deployment be consistent with the Sustainable Development Goals (SDGs), namely, SDG 9 targeting to build resilient infrastructure and foster innovation, SDG 11 on sustainable cities and communities and SDG 13 Climate Change (United Nations, 2015).

## 10.4 AVENUE Impact Assessment Framework

The AVENUE impact assessment offers a thorough review of the AVENUE trials, encompassing evaluations in technical, environmental, economic, societal and governance dimensions. Utilising an interdisciplinary methodology, it delves into multiple analyses to understand the complexities of introducing novel urban mobility and its integration within the transportation framework. For example, the economic and societal evaluations yield insights crucial for forecasting scenarios related to automated vehicles and assessing both direct and indirect costs. The Life Cycle Assessment (LCA) provides environmental metrics essential for calculating environmental externalities. Moreover, findings from the environmental, societal and economic evaluations are integrated into sustainability assessment metrics. To better understand the relationships between these components, the AVENUE assessment framework is illustrated in Fig. 10.1.

The framework outlines three principal axes: firstly, the comprehensive spectrum encompassing data input, methodologies and analyses; secondly, the multifaceted

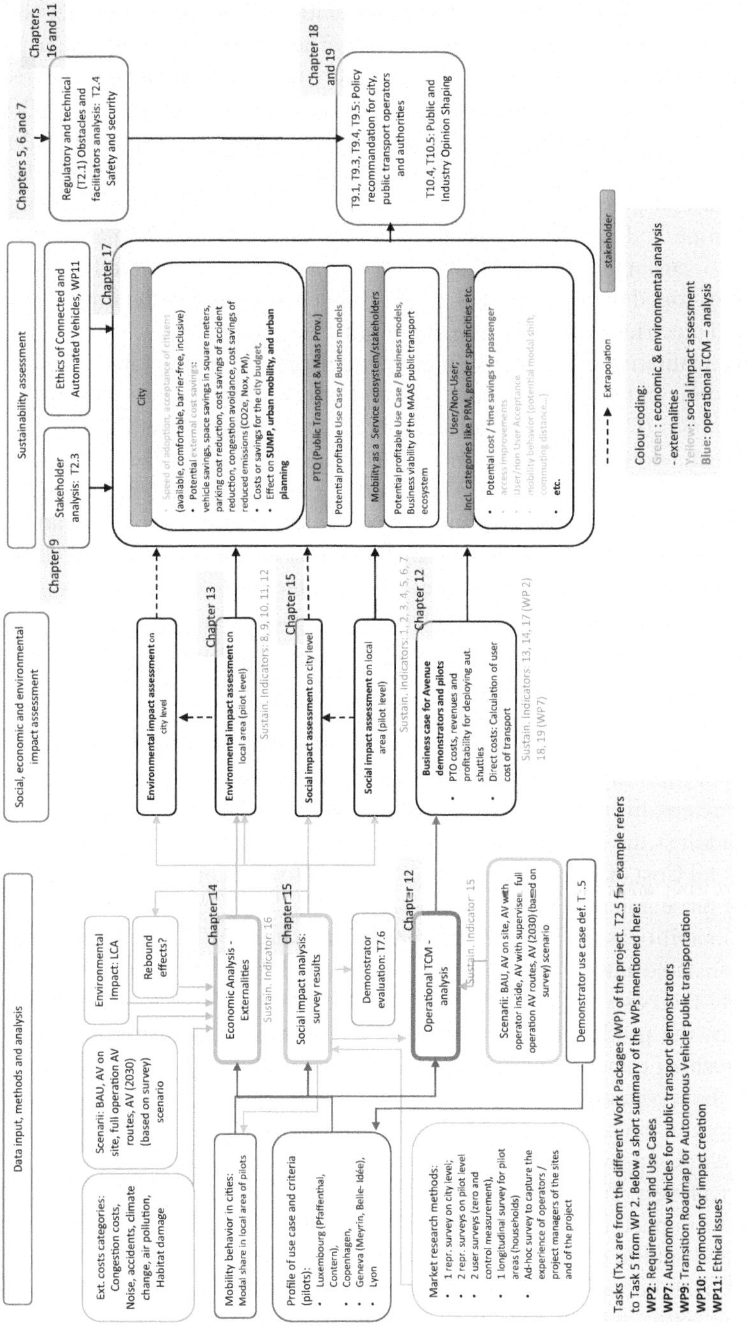


**Fig. 10.1** Framework of the AVENUE impact assessments

assessments covering social, economic, environmental and sustainability dimensions; and thirdly, the complex interconnections with tasks across other work packages.

The array of arrows signifies the intricate interrelationships and mutual dependencies among various segments of this assessment. The economic analysis (Chap. 14), particularly focusing on externalities, is intricately influenced by diverse previously formulated scenarios, including Business as Usual (BAU) and an AV 2030 Vision scenario derived from a survey. Moreover, a Life Cycle Assessment (LCA) of automated minibuses (Chap. 13) underpinned this analysis by supporting the calculation of various external cost categories such as those related to climate change, air pollution and habitat degradation, among others. The outcome of this analysis fed into sustainability Indicator 16, a framework articulated during AVENUE and developed by Nemoto et al. (2021), aimed at assessing the impacts of shared automated electric vehicles on urban mobility (see Chap. 17).

Additionally, qualitative discussions on rebound effects were integrated into the analysis of economic externalities. These rebound effects were gathered from the social impact analysis (Chap. 15), notably a survey on mobility attitudes and possible changes in behaviour in cities that shed light on modal shares in the local areas of the pilots and potential modal shifts. These surveys encompassed representative surveys at the city and pilot levels, user surveys, a longitudinal survey of pilot areas and an ad hoc survey of passengers and operators on site across all four pilot sites. Representative surveys are in particular relevant to understand the acceptance and possible changes in behaviour for user but also for non-user as car drivers, children, persons with reduced mobility, etc. The results were then utilised to calculate the modal shift and further the environmental and economic impact like externalities or possible rebound effects depending on the used scenarios.

Having addressed the environmental and social dimensions, a business operational Total Cost of Mobility (TCM) approach, grounded in the same pre-established scenarios as the externality analysis, was introduced (Chap. 12). The TCM approach is necessary to address the seamless journey. It has to be seen as complementary to the traditional Total Cost of Ownership approach which focuses on the ownership of vehicles and not on seamless service-based journeys. This assessment also contributed to the sustainability assessment framework as Indicator 15. The TCM analysis subsequently translated into a business case for the AVENUE demonstrators and pilots, explaining the costs for Public Transport Operators (PTOs), their revenues and, significantly, the profitability of deploying automated shuttles. Furthermore, the direct user costs of transport were factored into this analysis.

Conversely, the economic externalities analysis informed the environmental impact assessment at both city and local area levels (pilot level). Similarly, the social impact analysis provided insights for both the social impact assessment at city and local area levels. Together, these three assessments at varying scales constitute the AVENUE impact assessment framework, aiming to illustrate the repercussions of automated shuttle deployment on diverse stakeholders.

Among the stakeholders identified in the stakeholder analysis of Chap. 9, cities emerge as crucial actors, with factors such as adoption speed and citizen acceptance of the new service playing pivotal roles in social impacts (see colour yellow in the sustainability assessment box of Fig. 10.1). Noteworthy potential cost savings include space optimisation, accident reduction and emissions reduction. Economic impacts on cities entail identified cost savings in municipal budgets, alongside implications for Sustainable Urban Mobility Plans (SUMP), urban mobility and planning.

Public Transport Operators (PTOs) represent another significant stakeholder poised to benefit from the business models developed in the economic assessment. The MaaS ecosystem, with its intricate network of stakeholders, constitutes yet another vital stakeholder in this sustainability assessment, thriving on a profitable business model and the demonstrated viability of the AV public transport ecosystem. Lastly, the user/non-user category stands out, with potential benefits including cost and time savings and improved accessibility, while their acceptance and mobility behaviour significantly influence the impact of automated mobility. The holistic analysis together with the stakeholder analysis allows then for the formulation of policy recommendations for the city and the public transportation operators and authorities (Chaps. 18 and 19). These recommendations additionally include regulatory and technical obstacles and facilitator analysis of Work Package 2 (Chaps. 11 and 16) which have been conducted based on input from Chaps. 5, 6 and 7.

This comprehensive overview serves as a guide for delving into the nuances encapsulated within Fig. 10.1. of the framework.

## 10.5 Conclusion

In conclusion, this chapter has provided a detailed explication of the AVENUE sustainability assessment methodology, rooted in the theoretical foundations of Sustainable Urban Mobility Plans (SUMPs) and a rigorous methodological framework. By synthesising scholarly insights from environmental, economic and social impact assessments, this chapter contributes significantly to the discourse on sustainable urban mobility. The innovative approach of AVENUE's evaluation, which integrates diverse facets of sustainability into a unified framework, marks a step forward in assessing the impacts of urban mobility solutions on a holistic way.

The discussion elucidates the concept of SUMPs as a framework for sustainable and integrative planning in urban mobility, emphasising the importance of embracing new modes of transport and innovative concepts like MaaS and shared mobility. Case studies presented underscore the diverse implementations of SUMPs across different cities and settings, highlighting the necessity for more specific guidance and methods in future iterations.

Furthermore, the alignment of the AVENUE project with the SUMP concept is evident in its aim to deploy automated minibuses as a sustainable mobility solution, while critically assessing their impacts on the environment, society and economy. The integration of AVENUE's impact assessment framework, encompassing environmental, economic and societal dimensions but also technical ad governance issues. It offers crucial insights for guiding the integration and implementation of automated vehicles in urban mobility systems.

Through a comprehensive analysis, the chapter delineates the complex interrelationships among various components of the AVENUE assessment framework, elucidating the implications for diverse stakeholders such as cities, Public Transport Operators and users. Ultimately, this chapter lays the groundwork for a deeper understanding of the multifaceted impacts of automated mobility solutions and underscores the importance of aligning urban transport innovations with the goals of sustainable development. The result will be integrated in the FAME initiative[1] which consolidates and shares in the European Union all the methodologies used for evaluation of implementing automated vehicles in large-scale demonstration sites.

**Acknowledgements** This project has received funding by the European Union's Horizon 2020 research and innovation program under grant agreement No. 769033.

## References

Arsenio, E., Martens, K., & Di Ciommo, F. (2016). Sustainable urban mobility plans: Bridging climate change and equity targets? *Research in Transportation Economics, 55*, 30–39. https://doi.org/10.1016/j.retrec.2016.04.008

Connected Automated Driving. (2022). FAME - Connected automated driving. Available online at https://www.connectedautomateddriving.eu/about/fame/. Updated on 4/28/2023, checked on 2/26/2024.

Eltis. (2021). The SUMP concept. Available online at https://www.eltis.org/mobility-plans/sump-concept

European Commission. (2013). Annex 1: A concept for sustainable urban mobility plans. Available online at https://ec.europa.eu/transport/themes/urban/urban-mobility/urban-mobility-package_en

Mobility Plans. (n.d.). Sustainable urban mobility plans—Planning for people. Available online at https://www.google.com/url?sa=t&rct=j&q=&esrc=s&source=web&cd=&ved=2ahUKEwim8sX85_rsAhVi8uAKHdkcDkEQFjACegQIAxAC&url=http%3A%2F%2Fwww.rupprecht-consult.eu%2Fuploads%2Ftx_rupprecht%2FSUMP_Brochure_final.pdf&usg=AOvVaw0_usuHRNAt4WzXZqG5xebM

Mück, M., Helf, C., & Lindenau, M. (Eds.). (2019). *Urban living labs fostering sustainable mobility planning in Munich* (Vol. 41).

Nemoto, E. H., Issaoui, R., Korbee, D., Jaroudi, I., & Fournier, G. (2021). How to measure the impacts of shared automated electric vehicles on urban mobility. *Transportation Research Part D: Transport and Environment, 93*, 102766. https://doi.org/10.1016/j.trd.2021.102766

Russo, F., & Rindone, C. (2023). Smart city for sustainable development: Applied processes from SUMP to MaaS at European level. *Applied Sciences-Basel, 13*(3), 1773. https://doi.org/10.3390/app13031773

Sampaio, C., Macedo, E., Coelho, M. C., & Bandeira, J. M. (Eds.). (2020). *Economic and environmental analysis of measures from a sustainability urban mobility plan—Application to a small sized city* (Vol. 48).

United Nations. (2015). *The 17 sustainable development goals*. United Nations. Available online at https://sdgs.un.org/goals

[1] See Connected Automated Driving (2022).

# Chapter 11
# Technical Impact Assessment: Obstacles and Developments of Automated Minibuses for Public Transport

**Charly Beye, Christian Zinckernagel, and Guy Fournier**

**Abstract** In recent years, automated minibuses have emerged as an innovative addition to public transport systems worldwide, generating interest through various pilot projects. While these initiatives have shown promising technical advancements, the full-scale implementation of such vehicles faces significant technical hurdles. This study builds upon prior research by delving into the technical developments and challenges encountered within the AVENUE project, drawing insights from discussions with manufacturers, operators, and mobility platform providers operating automated minibuses in European metropolitan areas. Specifically, the examination highlights the complexities associated with vehicle sensors, data transmission for trajectory management, and external environmental factors that can influence vehicle operation. The progress within the AVENUE project has shown that a critical evaluation underscores the ongoing need for comprehensive solutions to overcome technical obstacles and ensure the seamless integration of automated minibuses into public transportation networks. Furthermore, the findings highlight the necessity of continually enhancing the safety, efficiency, and reliability of automated minibuses.

## 11.1 Introduction

There are several technologies applied for automated vehicles. Within the scope of automated driving, different scenarios and trends are currently considered within this technological framework (Wang et al., 2020; Zhang et al., 2018). Areas such as the localization and positioning of vehicles in certain traffic environments are

C. Beye (✉) · G. Fournier
Pforzheim University, Pforzheim, Germany

CentraleSupélec, Université Paris-Saclay, Gif-sur-Yvette, France
e-mail: charly.beye@hs-pforzheim.de

C. Zinckernagel
Holo A/S, Copenhagen, Denmark



G. Fournier et al. (eds.), *Automated Vehicles as a Game Changer for Sustainable Mobility*, Contributions to Management Science,
https://doi.org/10.1007/978-3-031-61681-5_11

associated with GPS, GNSS and RTK,[1] fusion of sensor technologies, and transmission of data required to ensure the implementation of automated driving (Englund et al., 2018).With the transmission of measurement data from sensors, estimates of the vehicle's relative position can be calculated. Although this approach has a certain error sensitivity, the precision of the localization can be incrementally improved by integrating additional sensor types (Toledo et al., 2018). For increasing the localization accuracy from the meter range in lower centimetre ranges (Englund et al., 2018), several approaches are available. Above all, the use of artificial intelligence and Internet of Things (Iot) holds a high significance, with which ideally complex systems are to be dynamically controlled in public transport (Kappel et al., 2019; Kuo et al., 2023). There is likewise the potential to improve the ability of an automated vehicle to perceive objects in its environment (Gupta et al., 2021; Netter, 2017).

Despite the application of different technologies, there are still technical limitations impacting the trajectory and the operation of automated vehicles (Benmimoun, 2017; Muzahid et al., 2023; Tengilimoglu et al., 2023). Obstacles can among others be identified in automated shuttle capabilities, environment, and human involvement. Achieving level 4 autonomy to eliminate the need for an on-board driver carries, e.g. significant implications for AV service costs and is closely connected to regulatory requirements regarding safety demonstrations of AV operation.

The research question guiding this study is as follows: What technical obstacles impact the implementation of automated minibuses for public transport, and what ongoing advancements aim to address these challenges in the field of automated driving? In order to examine the technical obstacles and their relevance within the AVENUE project, several experts from the respective consortium were interviewed. Furthermore, the technical developments in the project were investigated. Based on the project progress, several evaluation criteria were identified to frame this study.

## 11.2 Assessment Taxonomy

The taxonomy for the technical assessment comprises three major areas of consideration. The first one is "automated shuttle capabilities", which underlines the technical capabilities of the applied automated vehicle. The second area is "automated shuttle environment". This aspect includes diverse criteria concerning the external limits that influence the operation and trajectory of the vehicle. The superordinated criteria "human involvement" refers to the need of an on-board supervisor. Besides these three groups of criteria, additional technical findings were identified, which are addressed in Subchapter 4.2. "Remaining Technical Obstacles". The applied assessment taxonomy based on (Bürkle, 2019) is shown in Table 11.1.

[1] RTK (real-time kinematic) positioning is a method using data from an RTK base station that is transmitted over the Internet in particular through 3G, 4G, or 5G (Kim & Kim, 2022). RTK was used in Geneva to assure a more precise positioning in an environment with high buildings.

**Table 11.1** Taxonomy for assessing automated minibuses

| Criteria group | Included subordinated criteria |
|---|---|
| Automated shuttle capabilities | General aspects<br>Hardware-related limits<br>Software-related limits |
| Automated shuttle environment | General aspects<br>Weather-related limits<br>City infrastructural limits<br>Digital infrastructural limits |
| Human involvement | Necessity of on-board attendant |
| Additional technical findings | Processing power<br>Speed of the shuttle<br>Cybersecurity<br>5G networks<br>Number of shuttles on the same route<br>Safety demonstration<br>Disabled access |

## 11.3 Methodology

### *11.3.1 Assessment of Technical Obstacles*

The initial situation is based on previously examined technical obstacles to the implementation and use of automated shuttles from various pilot projects (Bürkle, 2019). For re-evaluation purposes, a summary of the technical obstacles that refer to the first examination and reporting was distributed to selected experts from the public transport operators within the AVENUE consortium. The aim was to receive feedback on the current technical obstacles. Existing changes and newly identified obstacles are listed and described in a summary. With the provided summary, the experts were asked to assess the relevance of the described technical obstacles for the implementation of AVENUE. The applied rating scale is equal to the previous report conducted to identify and describe potential changes coherently. By asking the experts to rate the relevance of the technical barriers in addition to their description, two steps of the method were covered. The experts were asked to rate the technical obstacles (existing and new ones) from 1 (not relevant) to 5 (very relevant). The average ratings were calculated based on the experts' contributions. The assessment results formed the basis for further expert interviews (Table 11.2).

### *11.3.2 Expert Interviews*

As mentioned in the previous section, the results from the first step of the method formed the basis for expert interviews. The subsequent phase encompasses interviews conducted with diverse experts from the AVENUE consortium, in which more specific attention is devoted to the obstacles and technical advancements

**Table 11.2** Rating scale for the evaluation of technical obstacles

| Rating scale | Description |
|---|---|
| 1: Not relevant (not) | The obstacle is solved or has no influence on the implementation and operation of the automated minibuses |
| 2: Less relevant (less) | The obstacle exists but does not really influence the implementation and operation of the automated minibuses |
| 3: Medium relevant (medium) | The obstacle exists and influences the implementation and operation of the minibuses from time to time |
| 4: Very relevant (very) | The obstacle influences the implementation and operation of the minibuses. Once in a while, the operation of the minibuses must be stopped due to the obstacle |
| 5: Most relevant (most) | The obstacle influences the implementation and operation of the minibuses significantly. Due to the obstacle, the operation cannot be guaranteed |

achieved through the AVENUE project. With the gathered results, the questions for the expert interviews were prepared accordingly. Additionally, as the subject of technical obstacles and advancements constitutes a considerable complexity, the semi-structured interview approach was applied (Galletta, 2012). This approach was chosen in order to have a reliable basis to address this complexity and to benefit from the experts' knowledge for the interpretation of the results (Galletta, 2012). The expert interviews were conducted in two separate parts. The first part relates to obtaining new information on technical obstacles as well as on the technical developments within AVENUE. Regarding the implementation of an on-demand service, the second part includes the challenges and obstacles that occur with the deployment of the application programming interface. Based on this separation, the public transport operators (PTOs) were considered in the first part of the interviews. In the second part, representatives of the vehicle manufacturer and provider of the fleet orchestration platform were interviewed. For preparation reasons, a guideline including questions and consent was provided to each expert. Two guidelines with different questions were created due to the separation of the interviews.

### 11.3.3 Expert Selection

For the first part of the interviews, public transport operators were taken into consideration as they had in-depth knowledge of the progress of the AVENUE project. This aspect justified the selection of the experts in the sense that they were also interviewed in the first examination or later started participating in AVENUE. Hence, they can show technical advancements as well as existing and potential new obstacles for the deployment of automated minibuses in public transport. The experts were from Keolis Lyon, Transport Public Genevois (TPG), Sales-Lentz, and Autonomous Mobility (AM) respectively Holo and Bestmile.

### *11.3.4 Data Analysis*

The consolidation and analysis of the data were based on expert evaluations and interviews (Part One) and expert interviews regarding API-specific barriers (Part Two).

**Analysis of Part One: Obstacles Related to Vehicle Operation** As mentioned in, providing a summary of the technical barriers from the last examination provided a basis for obtaining an update on technical obstacles. With the PTO's evaluation of already known and new technical obstacles, a profile emerged regarding their relevance. Through subsequent expert interviews, this topic was dealt with in more detail. The results of the expert interviews included technical obstacles and technical advances achieved at each demonstration site.

**Analysis of Part Two: Obstacles Related to API** With regard to API-specific obstacles, the results from the associated expert interviews in part two are outlined. The elaborations of the experts were linked in such a manner that the current obstacles to enabling an overall on-demand driving system are concisely described. Overall, the analysis comprised two parts, each addressing the results of the expert interviews. For better understanding, the methodology is presented in Fig. 11.1.

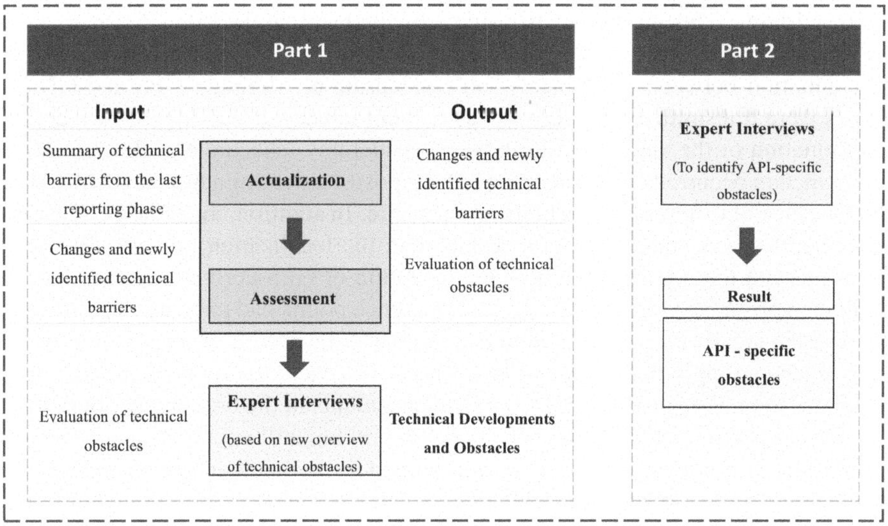


**Fig. 11.1** Methodological procedure

## 11.4 Technical Assessment of Automated Vehicles in AVENUE

The first subchapter (4.1) describes the initial capabilities of automated vehicles in AVENUE, which were derived from an earlier project phase. In order to present the latest findings of the examination, the remaining technical obstacles (4.2) and the remaining specific obstacles and developments related to the API (4.3) are then described.

### *11.4.1 Initial Capabilities of Automated Vehicles in AVENUE*

The initial capabilities of the applied automated vehicles are based on a previous examination from an earlier project phase (Bürkle, 2019). Several technical circumstances influenced the operation of automated minibuses during the initial stages of the AVENUE project. In particular, manoeuvres involving left turns and driving in difficult and blind intersections could only be performed using predefined stops implemented in virtual maps of the minibuses. Furthermore, an investigation of technical obstacles revealed that the vehicles had no capability to overtake low-velocity or static objects. Referring to the perception abilities of the automated vehicles, it was identified that the distinction between diverse participants on the road constituted a great limitation and challenge to overcome, as the minibuses were not able to differentiate between multiple road participants. Evident from the earlier investigation was that the automated vehicle did not possess the ability to drive under certain weather conditions. In the case of wind, rain, or snowfall, vehicles stopped driving. This problem was justified by the fact that the LiDAR sensors—at the time of investigation—were not equipped with filtering systems enabling the differentiation between real objects and condensation. Although GPS and GNSS signals were transmitted via audio waves without latency, it was identified that the determination of the vehicle position was considerably inaccurate as deviations of 2–3 m meters occurred. The inaccuracy of the position determination is justified by the existence of different layers in the atmosphere. In addition, the number of available satellites was stated as a crucial factor in the localization process. The first report showed that the transmission and reception of GPS correction data poses a significant obstacle. Occasionally, virtual reference points had to be used in order to improve the localization of the automated minibus. The driving strategy of the vehicle was also strongly dependent on how the sensors were interplayed, and the disseminated data were processed. Despite enhancements in the braking, acceleration, and driving performance of the vehicle, harsh braking still occurred under certain circumstances. Another decisive issue was human involvement in vehicle operation. In an earlier project phase, the vehicles could not perform all driving tasks. Thus, their operations had to be completed by an attendant on board. In particular, the observance of blind spots was conducted by an on-board attendant, as there were still technical limitations regarding the sensors. Because the vehicle could not independently monitor dead angles, the intervention of the operator onboard was necessary for safety reasons.

### 11.4.2 Remaining Technical Obstacles

In Fig. 11.2, the remaining technical obstacles and their evaluated relevance are consolidated.

The assessment showed that the implementation of automated minibuses in public transport is dependent on numerous technical obstacles and success factors that need to be addressed and overcome. The deployment of automated driving in various settings, both urban and rural, requires that the technology is capable of handling all safety-related scenarios and that it is effective, intelligent, connected, and fully integrated into the digital traffic infrastructure.

In terms of automated shuttle capabilities and environment, LiDAR and camera technologies in particular have been classified as key drivers for the implementation of automated vehicles. The importance of these technologies lies mainly in the fact that they detect and avoid obstacles and are ultimately decisive for the trajectory of the vehicle. The use of machine learning algorithms and AI technologies will also play an important role in enabling vehicle positioning in diverse traffic patterns and infrastructures. The perception and determination capabilities of automated vehicles will also be considered here as object differentiation in traffic becomes increasingly important, according to experts. The combination of the various technologies and a high processing performance in terms of sensor data is the determining factor in this context in order to achieve safety-relevant functions (e.g. braking) and thus make the vehicle more reliable and safer in its driving behaviour. All sensors should be functional under all conditions. Currently, heavy rain and snow cause disturbances to the sensors and lead to many impairments of the operation. GNNS and the ability to locate the vehicle with precise measurements also need continuous improvement. However, through the use of an Ntrip real-time kinematic (RTK) service, with which differential GNSS data is disseminated via the Internet, the localization and the positioning of the vehicle could be enhanced and therefore supported the smooth operation of the vehicle. The urban infrastructure is another key factor for the implementation of automated minibuses and should be considered in local land use planning. Moreover, cybersecurity measures are imperative to safeguard automated vehicles against potential threats, necessitating the evolution of laws and standards to address emerging challenges that come with increased processing power and complex sensory signal streams.

Additionally, human involvement remains crucial in the operation of automated minibuses, as safety operator had to intervene in various situations, including blind spots or when overtaking parked vehicles. However, a safety operator is generally required, due to legal considerations. For the future the transition from an on-board safety driver to remotely working supervisors has to be developed and tested accordingly.

| Criteria | Keolis | TPG | SLA | AM | Ø | How relevant |
|---|---|---|---|---|---|---|
| **1. Automated Shuttle Capabilities** | | | | | | |
| Construction Quality | 5 | 4 | 2 | 4 | **3,8** | Very |
| Traffic Regulation & Choice of Roadway | 3 | 3 | | 3 | **3** | Medium |
| Perception & Ability to determine | 5 | 4 | | 3 | **4** | Very |
| Driving Strategy | 3 | 2 | 1 | 3 | **2,3** | Less |
| Interoperability | | 2 | | | **2** | Less |
| 1.1. Hardware-related Factors | | | | | | |
| Sensor-Position (LIDAR 360°) | 3 | 4 | 5 | 4 | **4** | Very |
| Sensor-Resolution (VLP16 LIDAR) | 3 | 4 | 2 | 3 | **3** | Medium |
| UMTS & 3G-Modem | 3 | 3 | 1 | 2 | **2,3** | Less |
| Charging Time | 3 | 2 | 1 | | **2** | Less |
| Switch | 3 | 2 | 1 | | **2** | Less |
| 1.2. Software-related Factors | | | | | | |
| Hit-Ratio | 3 | 3 | | 1 | **2,3** | Less |
| Gateway to Fleet Management System | 5 | 4 | 1 | 3 | **3,3** | Medium |
| **2. Automated Shuttle Environment** | | | | | | |
| Mapping & Modification of Routes | 5 | 4 | 5 | 4 | **4,5** | Most |
| 2.1. Weather-related Limits | | | | | | |
| Sensor-related | 5 | 4 | 5 | 3 | **4,3** | Very |
| Battery-related | 3 | 3 | 5 | 1 | **3** | Medium |
| Power-related | 3 | 3 | 5 | 1 | **3** | Medium |
| 2.2. City Infrastructural Limits | | | | | | |
| Reference Points | 3 | 4 | 5 | 4 | **4** | Very |
| Road Markings | 3 | 3 | 1 | | **2,3** | Less |
| Surface of Roadway | 4 | 3 | 1 | 3 | **2,8** | Medium |
| 2.3. Digital Infrastructural Limits | | | | | | |
| GPS & GNSS Signal | 3 | 4 | | 4 | **3,7** | Very |
| 3G/4G Signal | 3 | 4 | 5 | 2 | **3,5** | Very |
| Data Transfer & Updates | 5 | 3 | 1 | | **3** | Medium |
| **3. Human Involvement** | | | | | | |
| Necessity of on-board attendant | 5 | 5 | | 5 | **5** | Most |
| **4. Additional Technical Findings** | | | | | | |
| Processing Power | 5 | 5 | | 3 | **4,3** | Very |
| Speed of the Shuttles | 3 | 3 | | 4 | **3,3** | Medium |
| Cyber Security | 5 | 5 | | | **5** | Most |
| 5G Networks | 3 | 3 | | 2 | **2,7** | Medium |
| Number of Shuttles on the same Route | 3 | 1 | | 4 | **2,7** | Medium |
| Safety demonstration | 5 | | | | **5** | Most |
| Disabled access | | 5 | | | **5** | Most |

**Fig. 11.2** Evaluation of technical obstacles

### *11.4.3 Remaining Specific Obstacle-Related API*

**The Manufacturer's API**

Based on the elaboration of the experts, the so-called Navya API is an interface that enables the communication between Navya vehicles and external systems. This API makes the link between the Navya drive software and the items that are necessary to share with the fleet management system. With this linkage, the fleet management system is able to monitor and command shuttles and missions. Regarding the on-demand features, the expert stated that Navya proposes a specific mode that allows an external partner to send punctual on-demand missions to the vehicles. Those missions are executed among other missions that are automatically created by manufacturer's system itself. Nevertheless, there is now a new mode (called partner mode) available within the latest software version. According to the experts, this new mode enables the fleet management system to not consider predefined missions but only those sent by the operator. Hence there are two options with which a partner can send missions to the automated minibus:

1. The certain partner can be the only mission provider and have control over the fleet.
2. The partner can be a provider of punctual missions that complete a predefined service with on-demand missions.

These missions are then stacked in a queue. The operator can modify this queue (e.g. by removing them). Nevertheless, it was stated that the overriding of an on-going mission is not possible once a mission is assigned to a vehicle. Within the AVENUE project, the assignment of missions through the Navya API is possible. A conducted update of API to the vehicles in 2021 was released, enabling some basic on-demand functionalities. However, the maturity of the technologies is still a considerable aspect to stabilize and make on-demand a fully functional product for all PTO's applying Navya vehicles. Since now, multiple operators (such as AM and TPG) within AVENUE have been trying to use the automatic assigning missions features and received important learnings to prepare and organize for this future way of orchestrating vehicle start and end destinations.

**On-Demand Challenges**

With the reference to on-demand services, several challenges are existing for its full deployment. Hereby it was outlined that overwriting of ongoing missions is not possible. This limitation encompasses the cancelling of ongoing missions and dynamic mission changes which are currently not implementable. According to the experts, this aspect is the key limitation for enabling shared rides as it has significant operational impact and is the most critical feature for the purpose of on-demand mobility services. Especially when a mission is already executed, no other mission can be inserted to modify the vehicle's plan regarding its current route. Besides these aspects, there are additional challenges restricting the implementation of on-demand transportation services. Currently it is not possible for the fleet management platform to control the routing of the vehicle. Additionally, it is conductible to detect that the person entering the vehicle actually booked his or her place. The start of the missions is also currently restricted as the vehicle cannot automatically recognize when a passenger enters. Consequently, the automated minibus was not able to detect when it has to close the doors and start the underlying mission. Furthermore, recognizing that the right persons exit the vehicle—before starting a new mission (or going back to its garage)—is also a functionality that is currently not enabled. However, it was stated that there will be a new software set-up enabling to override an ongoing mission. Regarding this set-up, it is being worked on providing more accurate information of a mission, as a prerequisite to override an ongoing mission.

## 11.5 Technical Improvements Through AVENUE

Proven by this study, technical advances have been made in the use of automated minibuses within the scope of the AVENUE project. These advances are concisely summarized in Table 11.3.

The AVENUE project demonstrates notable advancements in the capabilities of automated minibuses, enhancing their functionality and reliability within the public transport domain. Among these advancements is the integration of on-demand features, facilitating a more flexible assignment of missions within the project framework. This development allowed for greater adaptability to changing passenger demand and traffic conditions, ultimately improving the efficiency of the minibus service. Moreover, significant progress has been made in object recognition technology, enabling minibuses to detect overtaking vehicles on the road with greater accuracy and precision. This enhancement enhances passenger safety and contributes to a smoother and more efficient flow of traffic. In addition to improved object recognition, software enhancements have enabled the differentiation of objects based on their speed vectors. This capability

**Table 11.3** Technical improvements through AVENUE

| Aspect | Description of improvement |
|---|---|
| On-demand automated Minibus service | There are some on-demand features working within the project (e.g. assignment of missions) |
| Recognition of overtaking vehicles | Not as earlier, the minibus can now detect overtaking vehicles on the road |
| Object differentiation based on speed vectors | Some software features now allow the speed vector-based differentiation of objects (and road participants) |
| V2I<br>Communication at traffic light section | A bidirectional communication between the automated minibus and a traffic light section is possible |
| System stability | Higher operational stability and a lower system downtime rate |
| Smoothness of driving | The vehicle's driving behaviour is perceived to be more human-alike |
| Ntrip RTK correction | Better navigation and localization of the automated minibuses with the transmission of differential RTK correction data via the internet |

enhances the minibuses' ability to navigate complex traffic scenarios by providing more nuanced and context-sensitive responses to their surroundings. Another noteworthy development in the AVENUE project is the establishment of bidirectional communication between minibuses and traffic light sections. This communication enables real-time coordination between minibuses and traffic signals, optimizing traffic flow and reducing congestion on roadways. Furthermore, advancements in system stability have been achieved, resulting in higher operational stability and a reduced system downtime rate. These improvements enhance the overall reliability of the minibus service, ensuring a more consistent transportation experience for passengers.

## 11.6 Conclusion and Outlook

The Vienna Convention on Road Traffic composed by the United Nations in 1968 is an international treaty aimed at increasing safety in international road traffic, primarily through uniform regulations (United Nations, 1968). As of 2021, this road transport convention has so far been ratified by 84 countries (United Nations, 2021). Since 2014, it is permitted to use automated and assistance systems (UN Economic and Social Council, 2014). This change was enacted in 2016 (United Nations, 2015). Even if it is allowed to use such automated systems, which influence the driving of vehicles, it must still be able to be switched off and controlled by the driver (UN Economic and Social Council, 2014). In terms of this convention and the related need for monitoring and possible control of these systems by the driver, the application of automated vehicles is still limited to a level 3 autonomy. This underlying research has indicated that, above all, the demonstration of safe driving in road traffic is a success-critical factor for automated driving. All experts rated this aspect as significant for convincing the authorities on the safety of automated vehicles in public transport and thus fully validating their deployment. The safety of automated driving nevertheless depends strongly on the maturity of the technologies applied to enable overall reliable navigation of the vehicle. Accordingly, it was also pointed out that public transport operators act as customers throughout the ecosystem, specifying requirements for their use cases to the automated minibuses manufacturers. Due to the limitations of the technologies used, their optimized use by the automated driving software was thus recommended. With an increasing reinforcement of the software-hardware interaction, cybersecurity issues may be of higher importance, even if no hacking incidents are known within the AVENUE project yet. It was demonstrated that there have been technical developments that have advanced AVENUE. Nevertheless, technical obstacles to the use of the automated minibuses remain: in certain situations the intervention of the safety operator is still required. Demonstrating the safety of the vehicles is highly important not only from a regulatory perspective but also potentially for public perception, which may influence the acceptance of automated minibuses. For the course of possible further similar projects, it thus remains to continuously review what technical developments are made and what obstacles exist to the full implementation of automated minibuses in public transport.

The study showed that the demonstration of safe vehicle operation is important in terms of local authority regulations. The principle of safe automated driving is closely related to the extent to which a safety driver must intervene. Due to technical limitations (e.g. sensors), there are still interruptions in operation that require the vehicle to be controlled manually. Another recommendation in relation to this challenge is the integration of further sensors, although this task would depend on the manufacturer. Despite positive developments in the localization and navigation of the automated minibuses with the Ntrip RTK correction data service, operational interruptions and system failures still occur. It is important to document and analyse these cases in detail. This may facilitate the identification of potential sources of error and the development of solutions. However, there is a risk that some solutions may be limited to the technical capabilities of the hardware. Another important aspect, which was also outlined by the experts, is the access to the automated minibuses for elderly people or people with a restricted mobility. Associated with the vehicle, it is a significant task for the PTOs and also for the manufacturer to ensure that these people are capable of accessing the transportation service without any inconveniences.

There are additional limitations to the full implementation of an on-demand transport service, mainly related to the API and fleet orchestration platform. It is important to focus on the extension of on-demand functionalities because dynamic cancellation and overwriting of started vehicle missions were not possible. In order to provide a flexible minibus service to potential passengers, the routing must be made more flexible. Here, too, the exchange of information between the vehicle manufacturer and the mobility platform provider is of enormous importance. The exact description of the API problem, starting from the booking of the vehicle by a customer and the provision of the minibus is therefore necessary. As the objective is to deploy and integrate automated minibuses in public transport, the respective service has to be provided in a way that is flexible for potential clients and that it is connectable to other transport means in the multi- and intermodal mobility environment. Due to this aspect, it is suggested focus on enabling the flexible route planning and dynamic control of the vehicle based on passenger bookings. This may include new software configurations to provide the full compatibility to the fleet orchestration platform. Ultimately, the availability and effective utilization of data will be essential in overcoming these challenges and realizing the full potential of automated minibus services by integrating them into extended systems such as mobility-as-a-service, which incorporates different travel modes in one application (Arias-Molinares & García-Palomares, 2020; Narayanan & Antoniou, 2023). The integration of automated minibuses in an intelligent transport system (ITS) would enable further advantages for the stakeholders like self-learning systems but also raise new technical challenges which will have to be addressed (see Chap. 18).

**Acknowledgements** We would like to express our sincere gratitude to Jeroen Beukers from Transports Publics Genevois, Quentin Zuttre from Keolis Lyon, Johan Larsson from Holo Copenhagen, and Daniel Kaeding from Luxembourg who contributed to the successful completion of this study. Their dedication, expertise, and commitment were instrumental in the realization of our research objectives.

This project has received funding by the European Union's Horizon 2020 research and innovation program under grant agreement No. 769033.

## References

Arias-Molinares, D., & García-Palomares, J. C. (2020). The Ws of MaaS: Understanding mobility as a service fromaliterature review. *IATSS Research, 44*(3), 253–263. https://doi.org/10.1016/j.iatssr.2020.02.001

Benmimoun, M. (2017). Effective evaluation of automated driving systems. In *SAE technical paper series, WCX™ 17: SAE world congress experience, APR. 04, 2017*. SAE International.

Bürkle, T. (2019). Autonomous shuttles. Technical obstacles for the diffusion, implementation and the deployment.

Englund, C., Estrada, J., Jaaskelainen, J., Misener, J., Satyavolu, S., Serna, F., & Sundararajan, S. (2018). Enabling technologies for road vehicle automation. In G. Meyer & S. Beiker (Eds.), *Road vehicle automation 4* (pp. 177–185). Springer.

Galletta, A. (2012). *Mastering the semi-structured interview and beyond: From research design to analysis and publication (qualitative studies in psychology)*. New York University Press.

Gupta, A., Anpalagan, A., Guan, L., & Khwaja, A. S. (2021). Deep learning for object detection and scene perception in self-driving cars: Survey, challenges, and open issues. *Array, 10*, 100057. https://doi.org/10.1016/j.array.2021.100057

Kappel, M., Krune, E., Waldburger, M., & Wilsch, B. (2019). Die Rolle der KI beim automatisierten Fahren. In *Künstliche Intelligenz. Technologie, Anwendung, Gesellschaft* (pp. 176–193). Springer Vieweg.

Kim, E., & Kim, S.-K. (2022). Global navigation satellite system real-time kinematic positioning framework for precise operation of a swarm of moving vehicles. *Sensors (Basel, Switzerland), 22*(20), 7939. https://doi.org/10.3390/s22207939

Kuo, Y.-H., Leung, J. M. Y., & Yan, Y. (2023). Public transport for smart cities: Recent innovations and future challenges. *European Journal of Operational Research, 306*(3), 1001–1026. https://doi.org/10.1016/j.ejor.2022.06.057

Muzahid, A. J. M., Kamarulzaman, S. F., Rahman, M. A., Murad, S. A., Kamal, M. A. S., & Alenezi, A. H. (2023). Multiple vehicle cooperation and collision avoidance in automated vehicles: Survey and an AI-enabled conceptual framework. *Scientific Reports, 13*(1), 603. https://doi.org/10.1038/s41598-022-27026-9

Narayanan, S., & Antoniou, C. (2023). Shared mobility services towards mobility as a service (MaaS): What, who and when? *Transportation Research Part A: Policy and Practice, 168*, 103581. https://doi.org/10.1016/j.tra.2023.103581

Netter, F. (2017). Künstliche Intelligenz im Auto — Applikationen, Technologien und Herausforderungen. *ATZelektronik, 12*(S1), 20–25. https://doi.org/10.1007/s35658-017-0064-8

Tengilimoglu, O., Carsten, O., & Wadud, Z. (2023). Infrastructure requirements for the safe operation of automated vehicles: Opinions from experts and stakeholders. *Transport Policy, 133*, 209–222. https://doi.org/10.1016/j.tranpol.2023.02.001

Toledo, J., Piñeiro, J. D., Arnay, R., Acosta, D., & Acosta, L. (2018). Improving odometric accuracy for an autonomous electric cart. *Sensors (Basel, Switzerland), 18*(1), 200. https://doi.org/10.3390/s18010200

UN Economic and Social Council. (2014). Report of the sixty-eighth session of the Working Party on Road Traffic Safety. Available online at https://unece.org/fileadmin/DAM/trans/doc/2014/wp1/ECE-TRANS-WP1-145e.pdf. Accessed 19 July 2021.

United Nations. (1968). 19. Convention on Road Traffic. Vienna, 8 November 1968. Available online at https://treaties.un.org/doc/Treaties/1977/05/19770524%2000-13%20AM/Ch_XI_B_19.pdf. Accessed 19 July 2021.

United Nations. (2015). CN.529.2015.Reissued.06102015-Eng. Available online at https://treaties.un.org/doc/Publication/CN/2015/CN.529.2015.Reissued.06102015-Eng.pdf. Accessed 19 July 2021.

United Nations. (2021). United Nations Treaty Collection. Available online at https://treaties.un.org/pages/ViewDetailsIII.aspx?src=TREATY&mtdsg_no=XI-B-19&chapter=11&Temp=mtdsg3&lang=en. Accessed 19 July 2021.

Wang, Z., Wu, Y., & Niu, Q. (2020). Multi-sensor fusion in automated driving: A survey. *IEEE Access, 8*, 2847–2868. https://doi.org/10.1109/access.2019.2962554

Zhang, S., Chen, J., Lyu, F., Cheng, N., Shi, W., & Shen, X. (2018). Vehicular communication networks in the automated driving era. *IEEE Communications Magazine, 56*(9), 26–32. https://doi.org/10.1109/MCOM.2018.1701171

# Chapter 12
# Economic Impact Assessment: Local Service Costs of Automated Vehicles for Public Transport

**Fabio Antonialli, Sylvie Mira-Bonnardel, and Julie Bulteau**

**Abstract** While most of the current research on automated public transport focuses on improving operational and technical aspects, as well as tackling policy and user behavioural factors, the integration of these vehicles into public networks is mainly dependent on costs and breakeven points. Research quantifying costs and return on investment specifically in academic settings are sparse. This chapter aims to introduce an economic assessment tool, designed as a decision-making tool to support public policies on the decision of implementing innovative mobility services. The simulation contributes to the Total Cost of Mobility (TCM), a vision that accounts all alternative modes of travel and is therefore more adapted to today's fleets which are evolving towards automated electro mobility, pool management, etc. The presented assessment tool was tested with real data from pilot sites within the AVENUE project, and the results prove it to be a relevant tool to aid decision-makers on whether to adopt services with such vehicles.

## 12.1 Introduction

Technology and social acceptance are no longer the major obstacles to the deployment of automated vehicles for collective transports (Mira-Bonnardel, 2021). However, the question of their economic impact remains. For that reason, one of the AVENUE project's ambitions was to propose a methodology to assess the economic feasibility of an automated-based public transport service.

F. Antonialli (✉)
PPA Business School, Paris, France

École Centrale Lyon, Écully, France
e-mail: fantonialli1@reseau-ges.fr

S. Mira-Bonnardel
École Centrale Lyon, Écully, France

J. Bulteau
Université de Versailles Saint-Quentin-en-Yvelines, Versailles, France



G. Fournier et al. (eds.), *Automated Vehicles as a Game Changer for Sustainable Mobility*, Contributions to Management Science,
https://doi.org/10.1007/978-3-031-61681-5_12

Based on the studies from Bösch et al. (2018), Henderson et al. (2017), and Kalakuntla (2017) as well as by applying the Total Cost of Ownership (TCO) approach as done by Ongel et al. (2019), a simulation tool for assessing the economic impact of services with Automated Mobility (AM) was developed.

The tool, named EASI-AV proposes an *E*conomic *A*ssessment of *S*ervices with *I*ntelligent *A*utomated *V*ehicles by providing the fleet dimensioning for the service, calculating the total service costs (accounting both investments (CAPEX) and operating costs (OPEX) and comparing those with a given baseline vehicle, as well as calculating the local external costs for the communities where the vehicles are to be deployed (also in a comparative manner with a baseline vehicle).

EASI-AV was designed with the objective of helping policymakers in cities and regions, Public Transport Operators and Authorities (PTOs and PTAs), and even other interested stakeholders that may wish to implement services with Automated Mobility. The tool aims to assess the economic impact of different implementation scenarios like supply-pushed or demand-pulled strategies, fixed-routes, or geofenced on-demand services. Thus, a comparison between the automated service and other transport modes can be drawn up.

The overall scope, methodology, and previous results obtained with EASI-AV are elaborated in detail in Antonialli et al. (2021). The simulation tool was developed and validated by the PTOs from the demonstrator cities in the AVENUE project and is freely available online on the project's website.[1] In this chapter, we focus on analysing the results of the local economic simulation.

As stated by Nemoto et al. (2021), for the economic impact analysis, the Total Cost of Ownership (TCO), Total Cost of Mobility (TCM), costs and revenues analysis, and calculation of mobility externalities are methods that can support estimating the costs and economic attractiveness for users, public transport operators, transport services providers, and municipalities.

As stated by Nemoto et al. (2021), and PwC (2019), the concept of Total Cost of Mobility brings a more recent and holistic understanding of mobility, which has not been well defined until now, going beyond the fact of owning and operating a fleet of vehicles; it is a response to the changes towards current and future service-oriented mobility and presents a dedicated emphasis on the demand (passenger/citizen and trip) perspective. This change of perspective is closer to the reality of citizen mobility use case. It allows evaluating customer-/citizen-centric business strategies.

The benefits and costs of integrating Automated Mobility in a transport system can be found in Jaroudi (2021) and on part 5 of the AVENUE project deliverable D8.6.[2]

The chapter is structured as follows. Section 12.2 provides a literature review by presenting a brief overview of AVCT's economic impact assessment for the applicative domain. The tool, EASI-AV, is presented in Sect. 12.3. Section 12.4 presents the

[1] EASI-AV web application: https://h2020-avenue.eu/avenue-economic-calculator/. Copyright by École CentraleSupélec (Université Paris-Saclay)

[2] AVENUE D8.6 Economic Impact, available at: https://h2020-avenue.eu/wp-content/uploads/2023/03/20220512_D8.6-Economic-Impact_WP8_final-version_DK-not-approved.pdf.

economic assessment for the four AVENUE demonstrator sites, and different scenarios are discussed, and in Sect. 12.5, a conclusion is derived.

## 12.2 Theoretical Framework: Automated Mobility Impact Assessment

With automated vehicles expected to be an accepted technology by 2030 (Litman, 2018), their market penetration rate is dependent on investment costs as well as operating costs. By not requiring a driver and with expected lower energy consumption due to smoother driving, AVCTs may have lower operating costs than their human-driven counterparts (Fagnant & Kockelman, 2015; Fournier et al., 2020). However, their current embedded autonomy package constitutes the major cost components—with LIDARs, sensors, cameras, processing unit, and V2X equipment ranging from around 25,000 to 30,000 dollars, not to mention that automated vehicles are generally equipped with an electric battery and powertrain, which also increases the costs and reduces the overall lifecycle of the product to currently around 5 years (Ongel et al., 2019).

On the other hand, it is expected that the prices for the automation package as well as battery prices will decrease over time and that, if necessary, batteries and other fast-moving parts will become replaceable, so that the buses could have a much longer service life. Hence automated vehicles may become cost-effective compared to conventional vehicles in the long term (Bansal & Kockelman, 2017; KPMG, 2015). Consequently, cities and PTOs/PTAs should consider the costs and benefits of implementing a public transport service using automated vehicles over traditional services, and several recent studies have sought to provide answers to these demands.

In order to investigate possible changes in the urban mobility behaviour in the cities of Berlin and Stuttgart (Germany), Fournier et al. (2020) proposed an analytical model that simulates the impacts of a shared automated electric vehicles fleet (AEV) versus private vehicles with an internal combustion engine. Their results showed that a shared AEV fleet system could reduce externalities (accident avoidance, traffic jams, free spaces, parking costs, and lifetime losses) in cities and generate cost benefits for customers.

Kalakuntla (2017) carried out a prospective comparative study of costs and benefits of Automated Mobility fleets versus traditional regular diesel buses for the city of Austin (Texas, USA) with the aim of guiding PTOs/PTAs on the feasibility of Automated Vehicles (AV). The author concluded that AVs could save PTOs/PTAs from capital and operational costs, reduce the environmental effects, and increase the quality of life of the people.

The study carried out by Henderson et al. (2017) aimed at finding useful and efficient ways to use AVs in the campus of the Ohio State University (USA); the authors conducted an analysis to compare the current fleet of traditional vehicles used on campus to the costs of purchasing and maintaining a fleet of AVs (in their case the shuttle Olli from Local Motors). It was concluded that the automated shuttle exceeded the fleet of traditional vehicles in several categories. The costs and the carbon emissions per mile (0.91 lbs) as well as the annual maintenance costs ($600/yr) were comparatively lower. However, the automated shuttle was currently not cost-effective due to its high initial price in contrast to traditional shuttles.

Bösch et al. (2018) carried out a substantial cost-based analysis comprising of a bottom-up calculation of the cost structures (including besides the fixed costs, the overhead costs of shared services) for different types of AVs in various operation models, such as dynamic ride-sharing, taxi, shared vehicles fleets and AVs. The authors stated that their methodology allows the determination of different cost components' importance and differentiation of vehicle automation effects on individual cost components. Their results showed that more than half of AVs' fleet operating costs will be service and management costs. Furthermore, they have concluded that automated driving technology will allow taxi services and buses to be operated at substantially lower costs.

At last, the study from Ongel et al. (2019) aimed at determining the Total Cost of Ownership (TCO) of AVs and comparing them to regular internal combustion engine buses and minibuses. Their TCO analysis included three major cost components: acquisition costs, operating costs, and end-of-life costs. Their simulations have shown that, although the acquisition costs of AVs are higher than those of conventional buses, they can reduce the TCO per passenger-km up to 75% and 60% compared to conventional minibuses and regular buses, respectively.

Although bringing several promising and interesting results regarding the economic feasibility of services with AVs, none of the studies proposed a holistic methodology for dimensioning and assessing the economic impact of AVs services, which could be easily applied by decision makers in the economic evaluation of AVs.

Therefore, the simulation tool EASI-AV was designed that helps to assess the economic impact of AVs integration into public transport networks and to simulate different scenarios by allowing the users to adjust cost variables and revenue variables. The next section explains how EASI-AV is structured and how its different parts work.

## 12.3 The EASI-AV Simulation Tool

### *12.3.1 Design Methodology*

The *E*conomic *A*ssessment of *S*ervices with *I*ntelligent *A*utomated *V*ehicles (EASI-AV) tool was developed as a support tool to assist decision-makers in cities, as well as transport operators and other organizations to estimate the economic assessment of implementing a service with AVs. EASI-AV has been developed within the European project AVENUE and in collaboration with the transport operators in charge of the collective transport networks which were responsible for the demonstrators in the four cities of the project (Copenhagen, Geneva, Lyon, and Luxembourg). Their data on the automated service as well as on traditional human-driven services were collected in order to test the tool and the reliability of its algorithms.

## *12.3.2 EASY-AV Structure and Implementation in the AVENUE Program*

The EASI-AV tool provides different types of assessments in a comparative manner (between the shuttle and different baseline transport modes), such as the total service costs (based on the TCO analysis)—including investment costs and operational costs, the local impact of externalities, as well as the global impact assessment. EASI-AV is composed of five different parts that may be carried out sequentially or independently according to the needs of the user.

### 12.3.2.1 Part 1: Service Contextualization

This part focuses on qualitatively defining the local context envisioned for the new services with AVs. Contextualizing the service helps to build more accurate scenarios and allows decision-makers to have a holistic view of the service context to be implemented. The tool helps to properly frame the territorial typology (urban, peri-urban, rural), the zoning (residential, commercial, industrial, or mixed areas), and define the public transport supply (if there is already existing public transport in the area) and the area's population density as well as surface area and extension of roads. Figure 12.1 illustrates the EASI-AV web application data entry for one of the AVENUE project's testing sites (Antonialli et al., 2021).

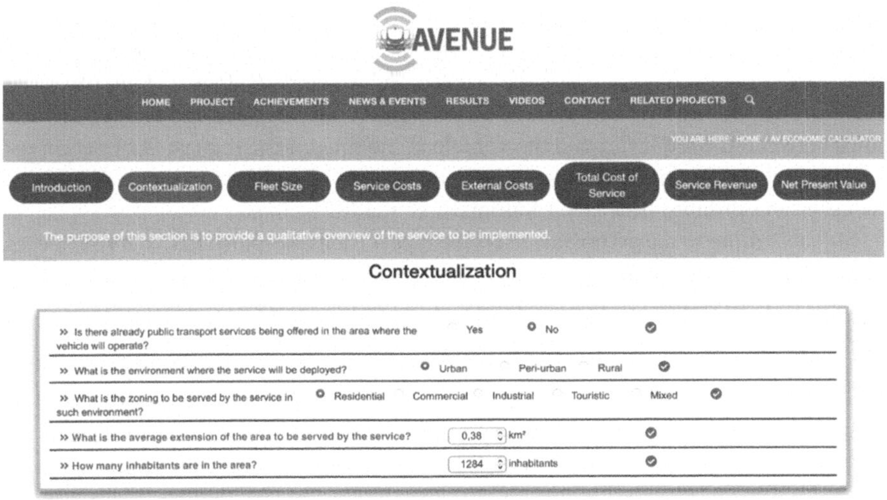


**Fig. 12.1** EASI-AV web application data entry—example of an AVENUE testing site. Source: prepared by the authors

#### 12.3.2.2 Part 2: Fleet Size Dimensioning

EASI-AV proposes four alternatives for the fleet size calculation (Table 12.1) that are guided by two main drivers: (1) service type (supply-push or demand-pull), and (2) road environment (fixed-routes or geofenced-on-demand). The tool allows the fleet size to be calculated for all combinations of service type/route environments. Once the category is selected, decision-makers enter data on selected cells if they work with the spreadsheet tool or ask for data collection online if they work on the web application (Antonialli et al., 2021).

As explained by Antonialli et al. (2021), each scenario prompts the algorithms to calculate both the local service costs (CAPEX and OPEX) and well as the local external costs (such as accidents, pollution, noise, and congestion). In addition, each scenario allows the simulation of different revenue models, not only based on ticketing or subsidies (as the current public transport offerings) but also innovative tariffs for on-demand with different revenue sources, such as custom commute to schools, hospitals, and private companies with different onboard services integrated into a mobility app, allowing new revenue streams, and a different economic balance.

For the scenario with option type 1 (fixed-route), the fleet size dimensioning is based on traditional fleet size calculations. Besides the usual general parameters characterizing the territory (route length, average speed, layover time, capacity, etc.) and specific parameters characterizing local mobility uses (percentage of public transport users in the area or numbers of operating hours per day), we considered some other specific for parameters as a way of leading to a finer calculation, such as the average operational speed (taking into account the idle time on each stop), as well as the battery autonomy and its charging time (which allows us to make a time differential to integrate in the calculation for how long a vehicle will be out of service to recharge). Simple algorithms compute these data and propose an optimum fleet size. The scenario with option type 2 (geofenced on-demand) is more complex since the algorithms must evaluate how many kilometres the vehicle may drive across the serviced area to comply with any user's demand for any direction at any time. Key elements of calculation in that option are the passenger waiting time (i.e. how long should a requester wait before a vehicle arrives) and the maximum distance between the requester and the vehicle at the time of the request. After computing these elements in addition to all elements considered for option 1, EASI-AV proposes an optimum fleet size (Antonialli et al., 2021).

Still according to the authors, for service type 1 (demand-pull), EASI-AV proposes calculations via the demand side, that is, for the cases where the demand for mobility

**Table 12.1** Service scenarios encompassed by EASI-AV; Source: adapted from Antonialli et al. (2021)

| | Service 1: demand-pull | Service 2: supply-push |
|---|---|---|
| Option 1: Fixed-route | Copenhagen (Nordhavn)<br>Luxembourg (Contern) | Lyon (Groupama Stadium)<br>Geneva (Meyrin)<br>Luxembourg (Pfaffenthal) |
| Option 2: Geofenced on-demand | Lyon(Groupama Stadium) | Geneva (Belle-Idée) |

is known (i.e. areas where public transport is already in place). Three calculation scenarios are proposed depending on the degree of knowledge of data concerning the existing transport demand (the number of passengers or the expected percentage of passengers during the peak and off-peak hours, etc.). The objective is to offer a flexible, modular tool depending on the transport demand and/or the future transport service offer. It is worth noting that from an economic standpoint, incentives will certainly be different depending on the type of vehicle chosen (e.g. one large bus versus several small buses), thereby, the input demand variables for this service type proposal must always be adjusted accordingly. For service type 2 (supply-push), the tool offers calculations via supply (i.e. areas where public transport is not yet available); thus, where demand on public transport is unknown or the service will be offered as a new transport offering in a supply-pushed strategy (Antonialli et al., 2021).

With the tool being tested with data from several testing sites in the AVENUE project, the EASI-AV-simulated presented results are consistent with the field data. For instance, the fleet size simulations carried out on EASI-AV for the Groupama Stadium testing site in Lyon (KEOLIS), the Nordhavn testing site in Copenhagen (Holo), and the Ormøya testing site in Oslo (Holo) yield the same fleet size and the real number of shuttles used for the operators. Figure 12.2 illustrates the results for one of the AVENUE testing sites yielded by the EASI-AV's web application.

#### 12.3.2.3 Part 3: Local Service Cost Assessment

In the scope of the Total Cost of Mobility as presented by Nemoto et al. (2021), the service cost assessment can be used as the follow-up of part 2 (fleet size dimensioning), or for cases where the fleet size is already known, it can be started by entering the current fleet size the users seek to evaluate.

For this part, information about the lifetime of the vehicles as well as the number of onboard safety drivers and off-board supervisors are requested. The former will allow for the calculation of vehicle depreciation, while the latter two will allow a better characterization of operating costs and possible economies of scale in terms of required staff.

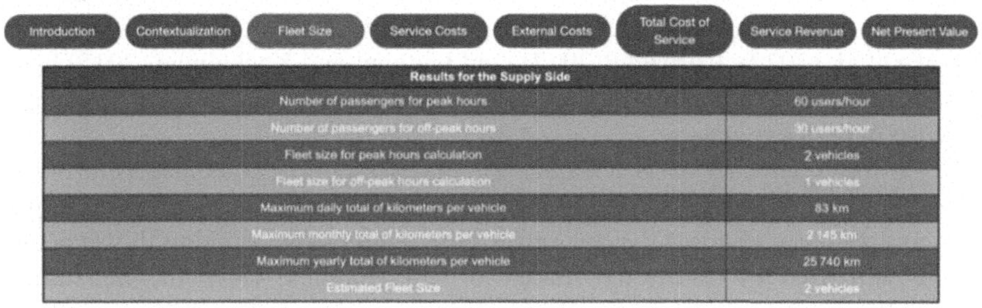


**Fig. 12.2** EASI-AV web application fleet size results—example of an AVENUE testing site. Source: prepared by the authors

The main internal costs are investment costs (or capital expenditures (CAPEX)) and operations expenditures (OPEX), both must be determined. Once all cost sources are registered, EASI-AV calculates useful KPIs such as the costs per passenger/km and per vehicle/km as well as other indicators. These ratios will be used afterwards for a detailed comparison between transport modes (Antonialli et al., 2021).

To help the user of the tool, based on an extensive literature review and benchmark with the PTOs, a list was created that explains the most relevant CAPEX and OPEX cost sources on a specific side document and via drop-down menus for the web application. In order to integrate economies of scale, the user can choose if the cost applies to a single vehicle or to the entire fleet (e.g. infrastructure works is CAPEX applied to the whole fleet, whereas acquisition is a per vehicle cost). In case the user does not know the exact cost values for the automated shuttle, the calculator also provides the option of using the standard costs (determined based on the average results obtained in the AVENUE project).

As shown in Fig. 12.3, with the summary results for one of the testing sites on the AVENUE project (anonymized due to confidentiality agreements), currently services with automated minibuses are not economically viable when compared to the baseline vehicle (6 meters human-driven bus). Since the fleet size service for this site was comprised of a single shuttle, the capital expenditures are 67% higher, and the operating costs are 29% higher as for CAPEX and the main cost differences between the Automated Vehicle and the baseline vehicle lie in commissioning costs (1567% higher) and in the purchase price of the vehicle (73% higher), in addition, it is worth mentioning the economies of scales that exist for the baseline vehicles due to their higher number on the roads.

As Konstantas (2021) explains, due to current regulatory limitations in some countries (such as Switzerland and Denmark) as well as to the experimental nature of the current projects of Automated Mobility, the commissioning costs for automated vehicles, besides being expensive (the extreme difference in commissioning costs depends very much on the test character and the innovativeness of the mobility approach), are imperatively carried out individually for each vehicle in the fleet, while for traditional human-operated vehicles, the legislation allows such costs to be applied to the whole fleet. As for the acquisition cost, it is evident that automated vehicles are currently more expensive (mainly due to the onboard automation technology and the electric powertrain and batteries). However, as Heineke et al. (2022), Cortright (2017), and Fagnant and Kockelman (2015) have pointed out, the expectation is that the costs of these vehicles will gradually decrease in the next years.

Regarding operating costs (OPEX), the major factor to be considered (in the current state of affairs) is the salary of onboard safety drivers. Since legislation still requires them inside AVs, this raises personnel costs, bringing this expense in line with that of traditional human-driven vehicles. Once the legislation waives the requirement for such professionals, this expense will be eliminated for automated vehicles, rendering them more cost-competitive (more details on these aspects are presented in Session 5). Furthermore, it should be noted that currently, the taxes and fees and maintenance costs are higher than the ones of

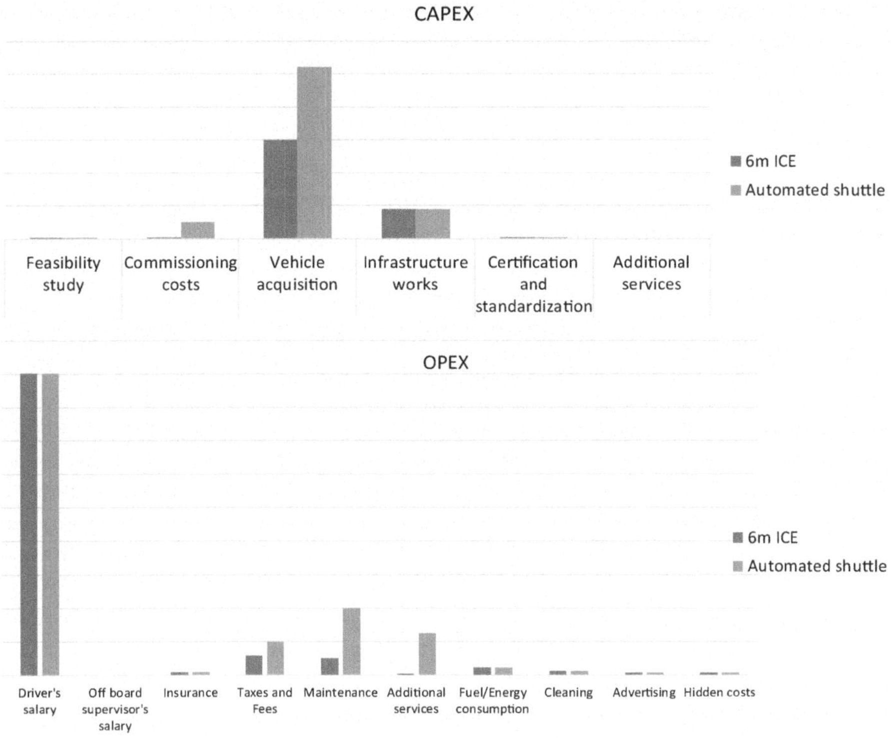


| | 6m ICE | Automated Vehicle | % Change |
|---|---|---|---|
| INTERNAL COST per passenger/km | 0.83 € | 2.51 € | 202% |
| INTERNAL COST per vehicle/km | 24.89 € | 37.60 € | 51% |
| ONE-WAY COST per passenger | 1.74 € | 5.26 € | 202% |
| ONE-WAY COST per vehicle | 52.26 € | 78.96 € | 51% |

**Fig. 12.3** Service cost assessment results—example of an AVENUE testing site. Source: prepared by the authors on EASI-AV

the baseline vehicle (73% and 296%, respectively). Finally, there is a high value of additional services (3.025% higher for AVs), which is due to the choice of the transport operator to offer the testing of additional services to the user (such as the Follow-my-Kid application of the start-up and partner of the Mobile Thinking project).

The KPIs on cost per passenger/km and cost per vehicle/km also corroborate the current economic infeasibility of the services with AVs, being 201% higher in terms of cost per passenger/km compared to the baseline vehicle and 51% higher considering cost per vehicle/km.

#### 12.3.2.4 Part 4: Local External Cost Assessment

At both local and global scales, public actions are considered in terms of sustainability. In this regard, the transport sector is no exception (Bulteau, 2016). The objective of policymakers is to reduce negative externalities of transport for the community, such as congestion, environmental pollution, and accidents. Therefore, the economic assessment takes into account externalities generated by the transport service implemented in the area, as well as at the macro-city level.

On EASI-AV, several sources of external costs for the cities are considered: congestion, accidents, air pollution (NOx and particulate matter), and noise. The monetarized values of these externalities come from the handbook of the externalities of transport (European Commission, 2019) being adjusted for inflation for the year 2020 and adapted to fit AVs. To get the results for externalities valuation, all that needs to be done is to select the country of where the shuttle will be deployed. Everything else is automatically calculated. A comparative analysis is provided between the external costs generated by the fleet size of shuttles and the chosen baseline vehicle (Antonialli et al., 2021).

It is worth noting that since this assessment is based on secondary data from the handbook of externalities of transport (which data was compiled for the year of 2016), the analysis is an approximation and only available for the countries listed in the handbook. However, the results do already provide an overview of the possible impacts (and eventual gains) of implementing AVCTs services.

Figure 12.4 exemplifies the local externalities gains for the same AVENUE testing site exemplified on Fig. 12.3. In comparison with the 6-meter internal combustion engine baseline bus, a 69% local externality cost reduction is observed in all

| | 6m ICE | Autonomous Shuttle | % Change |
|---|---|---|---|
| EXTERNAL COST per passenger/km | 0.04 € | 0.01 € | -69% |
| EXTERNAL COST per vehicle/km | 0.83 € | 0.26 € | -69% |

| | 6m ICE | Autonomous Shuttle | % Change |
|---|---|---|---|
| ONE WAY COST per passenger | 0.09 € | 0.03 € | -69% |
| ONE WAY COST per vehicle | 1.73 € | 0.54 € | -69% |

| | 6m ICE | Autonomous Shuttle | % Change |
|---|---|---|---|
| Vehicle DAILY cost | 28.42 € | 8.78 € | -69% |
| Vehicle MONTHLY cost | 625.23 € | 193.19 € | -69% |
| Vehicle YEARLY cost | 7,502.74 € | 2,318.31 € | -69% |

| | 6m ICE | Autonomous Shuttle | % Change |
|---|---|---|---|
| Fleet DAILY cost | 28.42 € | 8.78 € | -69% |
| Fleet MONTHLY cost | 625.23 € | 193.19 € | -69% |
| Fleet YEARLY cost | 7,502.74 € | 2,318.31 € | -69% |

**Fig. 12.4** Local external costs assessment results—example of an AVENUE testing site. Source: prepared by the authors on EASI-AV

| | Luxembourg (Sales-Lentz) | | Copenhagen/Oslo (Holo) | | Geneva (TPG) | Lyon (Keolis) | AVENUE average | Baseline vehicle 6m-EV (AVENUE average) |
|---|---|---|---|---|---|---|---|---|
| | Pfaffenthal (2 vehicles) | Contern (2 vehicles) | Nordhavn (1 vehicle) | Ormøya (2 vehicles) | Meyrin (1 vehicle) | Groupama (2 vehicles) | | |
| KPIs | | | | | | | | |
| Cost per passager/km | 0.73 € | 0.79 € | 1.35 € | 1.01 € | 2.52 € | 1.22 € | 1.27 € | 0.63 € |
| Cost per vehicle/km | 11.01 € | 11.78 € | 20.33 € | 15.21 € | 37.86 € | 18.29 € | 19.08 € | 9.48 € |

NOTE: detailed figures on CAPEX and OPEX have been omitted due to confidentiality requirements.

**Fig. 12.5** Total cost of services (CAPEX and OPEX)—AVENUE demonstrator sites. Source: prepared by the authors on EASI-AV

listed KPIs. This results in significant long-term savings for local taxpayers, especially regarding accident costs (94% lower for AVs), local pollution costs (99.7% lower), and noise costs (100% reduction).

The next section details the global results obtained for each testing site of the AVENUE project for which data was provided by the PTOs. Finally, an analysis of scenarios considering service offerings without onboard safety drivers and with commissioning costs applied to the entire AVs fleet is also performed.

## 12.4 Global Impact Evaluation and Scenarios

### *12.4.1 Overall AVENUE Results: KPIS*

Figure 12.5 depicts the global results for the AVENUE demonstrator sites based on the simulation outcomes from EASI-AV. The presented results are from the demonstrator sites where the operators were able to provide detailed data (results have been adjusted for inflation (December 2022).

The variations in the CAPEX values among the operating sites differ due to individual prices and levels of investments needed mainly on feasibility studies, commissioning costs, infrastructure works, and certification and standardization for each country. Those values vary according to the specificity of each site as well as based on local legislation.

The variations seen in the OPEX values are mainly due to the costs of personnel. The average salary paid for the operators and supervisors in each country varied as well as the number of operators needed for the daily operation of a single shuttle. For Lyon and Switzerland, the reported average annual salary for the safety drivers is about 90.000,00 euros, while for Luxembourg the values are around 43.133,48 euros, for Denmark around 48.000,00 euros, and for Sweden 55.700,00 euros.

It is worth noting that the KPIs calculations (Fig. 12.5) are based on the OPEX for the service and on the maximum daily mileage that a shuttle can run, which is dependent on the route length, operating hours, and frequency of the service. Consequently, the KPIs (cost passenger/km and cost shuttle/km) vary accordingly. For instance, in the Meyrin site and in Pfaffenthal the shuttles are operated much less than on the other sites (an average of 31.66 km/day and 43,27 km/day respectively versus the average of 72.73 km for the other four testing sites).

At the current state of affairs, it is safe to say that services with AVs are not yet cost-effective. The KPIs for all demonstrator sites, as well as the averages for the project (column in green on Fig. 12.5), have higher costs than the 6 m EV human-driven baseline vehicle[3] used here to provide a global comparison to the AVENUE services (column in red on Fig. 12.5). The costs of the automated vehicles used in the project are 50.39% higher in terms of cost per passenger/km and 50.31% regarding the costs per vehicle/km, considering an average occupancy of 15 passengers. In the following subsection, different scenarios are presented to show the potential economic viability of the services once certain regulatory barriers are overcome.

## *12.4.2 Simulation of Scenarios*

### 12.4.2.1 CAPEX Savings

As for possible CAPEX reductions, a possible evolution in the regulatory framework can lead to a reduction in costs for certification and standardization, feasibility studies, as well as commissioning costs. As pointed out by Konstantas (2021) feasibility studies and commissioning costs deserve special attention according to the local legislation of each country. According to the author, in Denmark, due to the lack of clear legal requirements, feasibility studies were not only time-consuming but also more cost-intensive than planned. In Switzerland, on the other hand, current legislation requires commissioning costs to be incurred for each vehicle individually and not for the fleet as a whole, leading to higher capital expenditures (particularly for the Belle Idée site, where two shuttles were used and not just one as in Meyrin). Secondly, as pointed out by Fagnant and Kockelman (2015), not only do the acquisition costs of these vehicles tend to decrease in the next years, but also their life cycle is assumed to decrease due to more modern sensors and cameras. Thus, it is possible to consider an important reduction in fleet acquisition costs.

Overall, the current average CAPEX values for the AVENUE project presented in the green column in Fig. 12.5 (both for a single AV and for the fleet) are not disproportionately higher than the average calculated for the human-driven baseline vehicle (red column in Fig. 12.5), being only 8.41% higher for a single vehicle and 21.71% higher for the total fleet.

In fact, by analysing each demonstrator site individually, (apart from Groupama in Lyon—due to its particularities that rendered the CAPEX higher than the AVENUE average), all other demonstrators already present—for a single vehicle—CAPEX values lower than the average calculated for the baseline vehicle. Therefore, even with a

[3] Values for the baseline vehicle were calculated based on the baseline average for the AVENUE demonstrator sites, considering a 6-meter Bolloré electric bus (https://www.bluebus.fr/bluebus-6-metres) with an average occupancy of 15 passengers (maximum capacity 30 passengers)

potential margin for investment cost reductions, CAPEX is not the main elements that make the current offering of services with AVs unfeasible, these fall on operating costs. The following subsection describes the possible ways to reduce these costs.

#### 12.4.2.2 OPEX Savings

Undoubtedly, the main competitive advantage in terms of operating costs for AVs when compared to traditional vehicles is the absence of drivers. However, the current regulatory framework in most countries does not allow services and projects without the presence of an onboard safety driver. Thus, the constant presence of this professional inside the vehicle suppresses their competitive promise in terms of cost reduction.

However, advances are being made in the legal frameworks, and in some countries, legislation is starting to allow trials of these services without the presence of a full-time onboard safety driver. In these cases, an off-board supervisor in a control room is responsible for ensuring the safety and operation of the service for the AV fleet. Within the scope of the AVENUE project, the first tests of this type began in Geneva at the Belle Idée site and were scheduled to begin at the Lyon site. In a medium to long-term horizon, considering technological and regulatory advances, one can imagine a 100% automated service (SAE level 5), where the vehicle would not require either an onboard safety driver or off-board supervisors (SAE, 2016).

Figure 12.6 presents the OPEX and KPIs results from the AVENUE project as well as simulation results for a (1) short-term scenario without an onboard safety driver and with an off-board supervisor and (2) a medium to the long-term scenario without any human intervention (either inside or outside the vehicle) for the operation of the service. It is worth noting that for both scenarios, all other operating costs were kept unchanged, meaning that further savings can still be made when

| | | Luxembourg (Sales-Lentz) | | Copenhagen/Oslo (Holo) | | Geneva (TPG) | Lyon (Keolis) | AVENUE average | Baseline vehicle 6m-EV (AVENUE average) |
|---|---|---|---|---|---|---|---|---|---|
| | | Pfaffenthal (2 vehicles) | Contern (1 vehicles) | Nordhavn (1 vehicle) | Ormøya (2 vehicles) | Meyrin (1 vehicle) | Groupama (2 vehicles) | | |
| AVENUE results | KPIs | | | | | | | | |
| | Cost per passager/km | 0.73 € | 0.79 € | 1.35 € | 1.01 € | 2.52 € | 1.22 € | 1.27 € | 0.63 € |
| | Cost per vehicle/km | 11.01 € | 11.78 € | 20.33 € | 15.21 € | 37.86 € | 18.29 € | 19.08 € | 9.48 € |
| Scenario 1 | KPIs | | | | | | | | |
| | Cost per passager/km | 0.49 € | 0.79 € | 1.18 € | 0.63 € | 1.86 € | 0.82 € | 0.96 € | 0.63 € |
| | Cost per vehicle/km | 7.39 € | 11.78 € | 17.81 € | 9.22 € | 27.95 € | 12.37 € | 14.42 € | 9.48 € |
| Scenario 2 | KPIs | | | | | | | | |
| | Cost per passager/km | 0.41 € | 0.57 € | 0.84 € | 0.48 € | 1.20 € | 0.69 € | 0.70 € | 0.63 € |
| | Cost per vehicle/km | 6.19 € | 8.66 € | 12.75 € | 7.23 € | 18.04 € | 10.29 € | 10.53 € | 9.48 € |

NOTE: detailed figures on CAPEX and OPEX have been omitted due to confidentiality requirements.

**Fig. 12.6** OPEX and KPIs—simulation of scenarios Source: prepared by the authors on EASI-AV considering future reductions in fees and charges, maintenance, additional services, advertising, etc.

Considering the average values for the AVENUE project (green column in Fig. 12.6), it is noted that for scenario 1 (with an onboard safety driver and without an off-board supervisor), there is a 24.41% reduction in the cost per passenger/km (dropping from an average value of 1.27€ to 0.96€) and a similar 24.42% reduction in the cost per vehicle/km (dropping from 19.08€ to 14.42€). However, these average values are still above the average values found for the baseline vehicle (red column in Fig. 12.6).

On the other hand, this scenario is already economically feasible for two of the six demonstration sites studied. In Pfaffenthal (Luxembourg) and Ormøya (Olso), the replacement of onboard safety drivers with off-board supervisors has resulted in lower costs per passenger/km and vehicle/km than the average values for the baseline vehicle (even without considering a reduction in the other listed operating costs). This represents a 32.87% reduction for both costs per passenger/km and vehicle/km in Pfaffenthal. In Ormøya a 37.62% reduction in costs per passenger/km and 39.38% reduction in costs per vehicle/km was recorded.

The results observed for scenario 2 (no onboard safety drivers and no off-board supervisors) bring the operational costs to competitive values when compared to the baseline vehicle. Although the overall average costs within the project are still above the values observed for the baseline vehicle (0.70€ versus 0.63€ for the cost per passenger/km, and 10.53€ versus 9.48€ for the vehicle/km cost), it is worth noting that in half of the demonstration sites studied, the service would prove to be economically competitive.

Both sites in Luxembourg show average values lower than those of the baseline vehicle. For Pfaffental, the cost per passenger/km would drop to 0.41€ and per vehicle/km to 6.19€, a respective reduction of 34.91% and 34.70% compared to the baseline vehicle. In Contern, the reductions would be 9.52% regarding the cost per passenger/km and 8.64% for vehicle/km. When comparing the figures to the actual results of the project, in Pfaffenthal the potential for reductions in cost per passenger/km would drop from 0.73€ to 0.41€ (a reduction of 43.83%) and from 11.01€ to 6.19€ in cost per vehicle/km (reduction of 43.77%). In Contern, the reductions would be 27.84% for the cost per passenger/km and 26.48% for the cost per vehicle/km.

The Ormøya site in Olso also provided results that would make it feasible to implement the service with AVs. With respect to the baseline vehicle, the cost per passenger/km would be 0.48€ (23.8% less than the baseline vehicle) and the cost per vehicle/km would be 7.23€ versus 9.48€ for the baseline EV (23.73% reduction). When comparing the results with the actual values obtained in the AVENUE project, the savings in terms of cost per passenger/km would be 52.47% (falling from 1.01€ to 0.48€) and 52.46% regarding the cost per vehicle/km (falling from 15.21€ to 7.23€).

It is important to emphasize the following for the three other project sites considered in the analysis: in Nordhavn (Copenhagen), Meyrin (Geneva), and Groupama Stadium (Lyon), high operating costs such as insurance, taxes and fees, maintenance, and additional services have pushed the average costs up, causing the KPIs to yield higher values than those obtained for the baseline vehicle (0.84€ and 12.75€ for Nordhavn; 1.20€ and 18.04€ for Meyrin; and 0.69€ and 10.29€ for Groupama Stadium).

Nevertheless, when comparing the results of scenario 2 for these sites with the actual results obtained in the AVENUE project, a significant reduction in costs can also be observed. For the Nordhavn site, costs per passenger/km would fall from 1.35€ to 0.84€ (a 37.77% reduction) and from 20.33€ to 12.75€ for costs per vehicle/km (37.28% reduction). At Meyrin, the reductions would be 52.38% for the cost per passenger/km and 52.35% for the cost per vehicle/km. Finally, for Groupama Stadium the reductions would be 43.44% for costs per passenger/km and 43.73% for costs per vehicle/km.

Therefore, there is a high potential in both the short and medium to long term for the economic viability of AV services. Once technological advances (allowing reductions in vehicle prices, maintenance costs, insurance, and fees) and regulatory framework advances (eliminating the need for safety-driver and/or off-board supervisor, as well as better accommodating feasibility studies and commissioning costs) are accomplished, the potential gains for passengers, transport operators, and consequently for cities are undeniable.

The main expected results would then be an important increase in the number of passengers commuting in the automated vehicles, thus rendering the services not only viable but attractive to the general public. On the other hand, if none in-vehicle services and reliable safeguard systems are provided, privacy and security issues can lower the number of passengers willing to use the service.

## 12.5 Conclusion

The large-scale deployment of automated collective vehicles, combined with online services, user profiling, and dynamic itinerary optimization, will have a disruption effect on today's public transport. The disappearance of drivers will allow transport operators to deploy more vehicles, resulting in a scaling down of vehicles. This, in turn, will allow vehicles to divert from the predefined itineraries and start offering on-demand, door-to-door services (based on dynamic online reservations and optimization), transforming public transport into a personalized transit service.

This transformation will require a high level of investments. Anticipating the economic impact of investments is a usual task for any decision-maker or investor. Therefore, the tool was developed in collaboration with transport operators and local city governments to enable an economic assessment of the implementation of automated vehicles into their transport network and the valorisation of deployment scenarios.

With public transport being a complex ecosystem, not only transport operators, passengers, and policymakers are included but a lot of different stakeholders such as software providers, mobility platforms, vehicle manufacturers, insurance companies, telecom companies, infrastructure construction companies, maintenance companies, and data provider companies. Each stakeholder may facilitate or hinder the deployment of automated collective transport. Therefore, they should be able to analyse scenarios from their own economic viewpoint. EASI-AV was successfully tested on the AVENUE experimentation sites in four European cities and proved to be a relevant support tool for decision-making in designing new mobility solutions.

**Acknowledgements** The authors would like to thank the AVENUE project consortium, in special CentraleSupélec and Professor Danielle Attias scientific coordinator of the AVENUE project, the Automated Mobility OEM Navya, and the transport operators Keolis, TPG, Sales-Lentz, and Holo. A special thank you also goes to the colleagues from HSPF Guy Fournier and from Siemens Michael Thalhofer for the concept of integrating externalities, TCM and incorporation revenues into account, as well as Eliane Nemoto, Inès Jaroudi, and Adrian Boos for their precious insights and collaborative work in the AVENUE project. The AVENUE project has received funding from the European Union's Horizon 2020 research and innovation program under grant agreement No. 769033.

This project has received funding by the European Union's Horizon 2020 research and innovation program under grant agreement No. 769033.

## References

Antonialli, F., Mira-Bonnardel, S., & Bulteau, J. (2021). Economic assessment of services with intelligent autonomous vehicles: EASI-AV. In S. Mira-Bonnardel, F. Antonialli, & D. Attias (Eds.), *The robomobility revolution of urban public transport* (1st ed.). Springer Nature.

Bansal, P., & Kockelman, K. M. (2017). Forecasting Americans' long-term adoption of connected and automated vehicle technologies. *Transportation Research Part A: Policy & Practice, 95*, 49–63.

Bösch, P. M., Becker, F., Becker, H., & Axuausen, K. W. (2018). Cost-based analysis of automated mobility services. *Transport Policy, 64*, 76–91.

Bulteau, J. (2016). Revisiting the bottleneck congestion model by considering environmental costs and a modal policy. *International Journal of Sustainable Transportation, 10*(3), 180–192.

Cortright, J. (2017). *What price for autonomous vehicles?* City Observatory. Available at: http://cityobservatory.org/what-price_autonomous_vehicles. Accessed 2 May 2023.

European Commission. (2019). Handbook on the external costs of transport. *Publications Office of the European Union [online]*. https://op.europa.eu/en/publication-detail/-/publication/9781f65f-8448-11ea-bf12-01aa75ed71a1. Accessed 15 Jan 2021.

Fagnant, D. J., & Kockelman, K. (2015). Preparing a nation for automated vehicles: Opportunities, barriers and policy recommendations. *Transportation Research Part A: Policy & Practice, 7*, 167–181.

Fournier, G., Boos, A., Wörner, R., Jaroudi, I., Morozova, I., & Nemoto, E. H. (2020). Substituting individual mobility by mobility on demand using autonomous vehicles - A sustainable assessment simulation of Berlin and Stuttgart. *International Journal of Automotive Technology and Management, 20*(4), 369–407.

Heineke, K., Heuss, R., Kampshoof, P., Kelkar, A., & Kellner, M. (2022). *The road to affordable autonomous mobility*. McKinsey & Company. Available at: https://www.mckinsey.com/industries/automotive-and-assembly/our-insights/the-road-to-affordable-autonomous-mobility#/. Accessed 2 May 2023.

Henderson, H., Veder, A., Li, L., & Johnson, M. (2017). *Feasibility of electric automated vehicles on Ohio State University Campus*. Available online: www.kb.osu.edu/bitstream/han-

dle/1811/81632/1/ENRAEDE4567_ElectricAutomatedVehicles_au2017.pdf. Accessed 14 Apr 2020.
Jaroudi, I. (2021). What is the potential impact of the transition from traditional transport to new mobility models (electric automated minibuses) in European cities? In *Proceedings of the 29th international colloquium of Gerpisa: The transformations of the global auto industry*. Available at: https://www.gerpisa.org/node/6370. Accessed 2 May 2023.
Kalakuntla, S. P. (2017). *Adopting automated bus to a transit agency: a cost-benefit analysis*. Available online: www.academia.edu/34636124/ADOPTING_AUTOMATED_BUS_TO_A_TRANSIT_AGENCY_A_COST-BENEFIT_ANALYSIS. Accessed 14 Apr 2020.
Konstantas, D. (2021). From demonstrator to public service: The AVENUE experience. In S. Mira-Bonnardel, F. Antonialli, & D. Attias (Eds.), *The robomobility revolution of urban public transport* (1st ed.). Springer Nature.
KPMG. (2015). *Connected and automated vehicle - The UK economic opportunity*. Available online: www.smmt.co.uk/wp-content/uploads/sites/2/CRT036586F-Connected-and-Automated-Vehicles-%E2%80%93-The-UK-Economic-Opportu...1.pdf. Accessed 13 Apr 2020.
Litman, T. (2018). Automated vehicle implementation predictions. Implications for Transport Planning. Available online:. www.vtpi.org. Accessed 13 Apr 2020.
Mira-Bonnardel, S. (2021). Robomobility for collective transport: A prospective user centric view. *International Journal of Automotive Technology and Management, 21*(1/2), 99. https://doi.org/10.1504/IJATM.2021.113353
Nemoto, E. H., Issaoui, R., Korbee, D., Jaroudi, I., & Fournier, G. (2021). How to measure the impacts of shared automated electric vehicles on urban mobility. *Transportation Research Part D: Transport and Environment., 93*, 1–18.
Ongel, A., Loewer, E., Roemer, F., Sethuraman, G., Chang, F., & Lienkamp, M. (2019). Economic assessment of automated electric microtransit vehicles. *Sustainability, 11*(648), 1–18.
PwC. (2019). The 2019 strategy & digital auto report: Time to get real: Opportunities in a transforming market. Available online: www.strategyand.pwc.com/gx/en/insights/2019/digital-auto-report/digital-auto-report-2019.pdf. Accessed 2 May 2023.
SAE. (2016). Taxonomy and Definitions for Terms Related to Driving Automation Systems for On-Road Motor Vehicles J3016_201609, Available online: www.sae.org/standards/content/j3016_201609/. Accessed 14 Apr 2020.

# Chapter 13
# Environmental Impact Assessment: Automated Minibuses for Public Transport

**Tobias Viere, Adrian Boos, Nicole van den Boom, Meriem Benyahya, Maher Ben Moussa, and Guy Fournier**

**Abstract** This chapter examines the environmental impacts of integrating automated minibuses into public transport systems, focusing on the trade-offs between energy savings and increased energy demand due to connectivity and automation features. A comprehensive life cycle assessment (LCA) is used to analyse the potential environmental benefits and problems of automated minibuses. The results show that the environmental benefits of these vehicles depend significantly on the electricity mix for charging, passenger occupancy, and vehicle utilisation throughout their lifetime. In particular, scenarios with renewable energy sources, high passenger occupancy, and optimal vehicle utilisation show the significant potential of automated minibuses to contribute to a more sustainable transport system. This study highlights the importance of holistic considerations, including technological, operational, and infrastructural factors, in maximising the environmental benefits of automated minibuses in urban mobility. It provides a nuanced understanding of the conditions under which automated minibuses can act as a catalyst for the transition to greener public transport.

## 13.1 Introduction

The following chapter presents an environmental impact assessment concerning the deployment of automated minibuses in public transport systems. Therefore, we present a study of the potential energy demand and savings of automated driving and an environmental life cycle assessment for automated electric minibuses in two

T. Viere (✉) · A. Boos
Pforzheim University, Pforzheim, Germany
e-mail: tobias.viere@hs-pforzheim.de

N. van den Boom · G. Fournier
Pforzheim University, Pforzheim, Germany

CentraleSupélec, LGI, Gif-sur-Yvette, France

M. Benyahya · M. B. Moussa
University of Geneva, Carouge, Switzerland



G. Fournier et al. (eds.), *Automated Vehicles as a Game Changer for Sustainable Mobility*, Contributions to Management Science,
https://doi.org/10.1007/978-3-031-61681-5_13

subchapters. Another subchapter summarises the results into environmental indicators, which are used for the sustainability assessment in Chap. 17 of this book.

The main findings of this chapter are:

- Predictive, adaptive, and information sharing through vehicle communication with infrastructure and other vehicles improves vehicle braking performance and consequently energy consumption. However, a highly connected vehicle means more processing required by the infrastructure, remote, or cloud servers which may outweigh the V2X sustainability.
- Based on results from the AVENUE pilots, the automated driving components for an automated minibus driving at 30 km on an 8-hour day require 82.1 Wh $km^{-1}$ (304.4 W x 8 h/30 km) or 15.6% of the total energy use of 520 Wh $km^{-1}$.
- The data transmission and energy consumption for 3GPP and 5GAA use cases were estimated for automated driving connectivity for 11 scenarios: platooning, sensor and state map sharing, remote driving, lane change, infrastructure-based perception of environment, collision avoidance, collective information sharing, see-through for passing, emergency trajectory alignment, intersection crossing, and cooperative driving.
- Significant potential in energy savings can be achieved in particular from intelligent route optimisation and velocity control.
- Data on energy saving for predictive functions are presented for selected cases based on literature. Among the functions, the eco route planning and traffic light assistant are cited for being urban scenarios that require little exchanged information between the vehicles and the infrastructure. This makes them very promising candidates for real energy savings achieved through the implementation of automated urban mobility.
- The energy efficiency for exchanging data within the automated minibuses ecosystem depends on the number of connections and the advancement of the deployed technologies.
- The cooperative V2X is undoubtedly the key sustainable communication mode and plays an important role in energy demand.
- A fixed route or a mature on-demand service would have different energy consumption due to different numbers of involved servers.

The life cycle assessment (LCA) of the automated minibus shows that the automated technologies in the automated minibuses, as deployed in the AVENUE pilots, are around 5% of the total energy used. When considering the near-future use case, the study points that 59% of the automated minibuses impact stems from the use phase, while component production accounts for 39%. The use phase climate impacts are mostly due to the burning of fossil fuels to produce the electricity required for driving the automated minibus. The global warming potential for each pkm is 78 g CO2eq.

The assessment of the automated minibus based on the environmental indicators shows that at the current stage, the automated minibuses face challenges to be deployed as an environmentally friendly mode of transport. These results are confirmed by the LCA study, pointing that the automated minibus at the current deployment does not show significant environmental benefits, but future use cases are likely to improve substantially. In addition, the automated minibuses qualification as environmentally friendly depends on many factors such as occupancy, vehicle speed, mileage, and lifetime. Taking into consideration the perspective on the mobility system, the automated minibuses are seen as a complementary service in public transport. In combination with door-to-door, on-demand, and driverless services, the automated minibuses are expected to improve and strengthen public transport, hence bring benefits by reinforcing shared, multi- and intermodal mobility as well.

## 13.2 Assessment of Energy Demand of CAV/AV

Automated driving technologies are likely to reduce energy demand for driving compared to traditional vehicles, e.g. by functions such as platooning and eco-driving. However, the increased demand for data transmission and processing might increase energy demand. This part elaborates on both sides, the energy demand side of automated driving technologies and the potential energy savings of automated driving. Such an assessment is crucial from an environmental perspective as energy increase or decrease effect the overall environmental performance of automated vehicles significantly. In this chapter, we would be mainly referring to automated vehicles as CAV (connected automated vehicles) instead of AV, as we aim to address various aspects of connectivity in automated driving.

### *13.2.1 Energy Demand of Automated Driving Technologies*

Throughout the scientific literature, researchers debated the energy demand of CAV to be either greedy or efficient depending on the implemented automation units, Internet technologies, and deployed services, though they clearly agreed on the direct impact of the data exchange on the overall energy consumption. This section sheds the light on the key publications that showcased the vehicle connectivity and data transfer impact on energy consumption. It also reported the efforts from literature in translating the exchanged bytes and bits into energy units.

Noroozi et al. (2023) conducted a systematic literature review on the automation impact on energy consumption. They review recent literature focusing on how energy consumption of automated vehicles is influenced by advancements in powertrain operation and driving patterns. The papers they found were organised based on various factors affecting power consumption, but it's noted that existing studies vary greatly in their design and implementation, suggesting a need for more specific and comprehensive research to accurately benchmark the energy consumption impact of vehicle automation.

Liu, Tan, et al. (2019) provided a quantitative study on the negative effects of smart vehicles on energy consumption. The authors draw attention to the fact that automated and intelligent vehicles are equipped with computing devices, advanced sensors, controllers, and actuators, in combination with connecting communication technologies, resulting in higher energy consumption compared to conventional vehicles. The authors suggest that computing platform performance, connection strength, and radar performance are the three main factors impacting the energy consumption of CAV. Their study led to the assessment of fuel consumption per 100 km for different levels of automation—primary, intermediate, and advanced intelligence (corresponding to SAE levels 3, 4, and 5 accordingly)—and the identification of different factors that potentially influence vehicle's consumption costs. The study has shown an increase in energy consumption depending on the level of intelligence with 0.78 L/100 km in primary intelligent vehicles, 1.58 L/100 km in intermediate intelligent vehicles, and 1.86 L/100 km in advanced intelligent vehicles.

Chen et al. (2021) explore how vehicle automation affects fuel consumption for electric and gasoline powered vehicles with automation levels 0, 2, and 5 across different scenarios, by analysing design changes and performance impacts. Results suggest energy savings of 4–8% in an optimistic level 5 scenario and a 10–15% increase in fuel consumption in a pessimistic scenario. It also notes that power demand varies between urban and highway driving, with inertial power dominating in urban conditions and aerodynamic power in highway conditions.

Song et al. (2023) analyse the impacts of level 2 automation on traffic efficiency and energy consumption of expressways, considering travel time, road capacity, and energy consumption. The study proposes a benefit evaluation framework and uses microscopic traffic simulation software for experiments. Different market penetration rates and traffic flow statuses are considered, along with dedicated lanes for CAVs. Results show that they generally save travel time and reduce energy consumption per vehicle, with positive economic benefits increasing with larger market penetration rates.

Gawron et al. (2018) present a life cycle assessment (LCA) of CAV sensing and computing hardware with SAE level 4 of automation exploring the potential energy and greenhouse gas (GHG) emission impacts of CAV based on six scenarios. Three of the scenarios simulate sensing and computing hardware configurations of Tesla Model S, Ford Fusion (AV test vehicles), and Waymo's Chrysler Pacifica respectively integrated into an internal combustion engine vehicle (ICEV), and the other three scenarios simulate the hardware configuration on a battery electric vehicle (BEV). They reported that the additional hardware resulted in an increase of 3–20% of energy consumption compared to conventional vehicles. However, when considering the automated driving functions (e.g. eco-driving, platooning, and intersection connectivity) facilitated by the additional hardware, the net result is up to 9% of energy (and emission) reduction based on the Tesla and Ford hardware configuration. The authors claim that data transmission is one of the four factors contributing to an increase of energy consumption. Their research studied data transmission over 4G wireless networks, which was estimated to 1.4 MB/mile and to a consumption of 1.25 MJ/GB.

Figure 13.1 depicts Gawron et al. (2018) life cycle energy estimation for a medium CAV.

Pihkola et al. (2018) evaluated the environmental impact of mobile access networks and sustainability of services within the IoT (Internet of Things) ecosystem using the LCA methodology. In their study, the authors constructed a trend of kWh per transferred gigabyte where they linked the network electricity consumption to the transferred data within the network. However, their computations were limited to the 4G mobile network consumption in Finland that can be extended to any IoT model.

Greenblatt and Shaheen (2015) focused their research on environmental impacts of CAV's on-demand services, which reduce the vehicle ownership, the number of households owning a car, and the vehicle miles (kilometres) travelled.

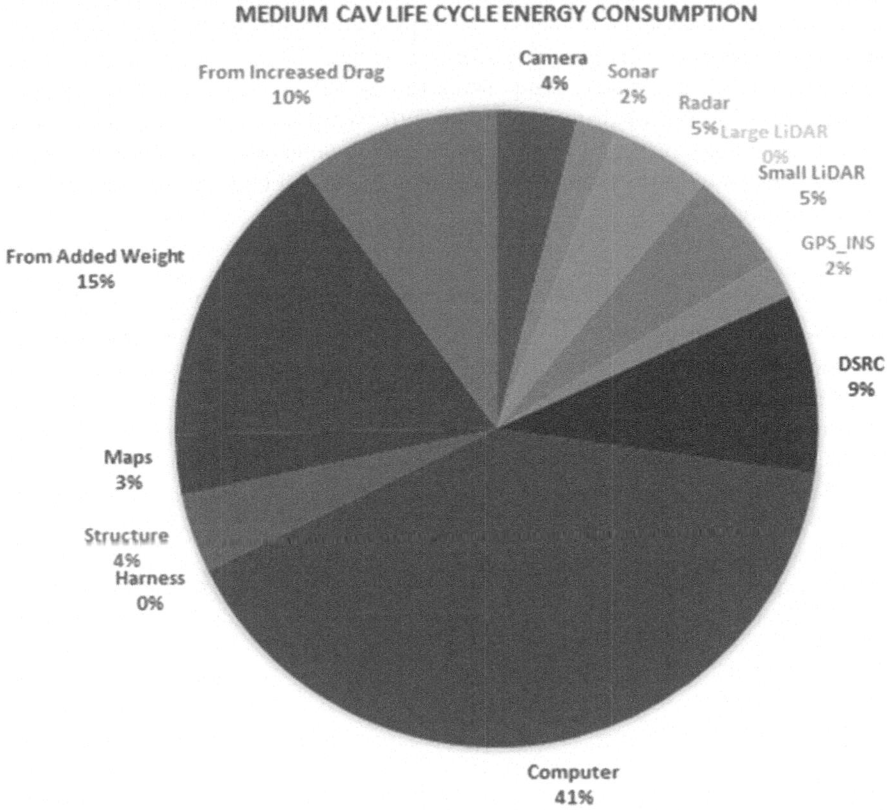


**Fig. 13.1** Medium CAV life cycle energy consumption (reprinted (adapted) with permission from (Gawron, James H./Keoleian, Gregory A./Kleine, Robert D. de/Wallington, Timothy J./Kim, Hyung Chul (2018). Life Cycle Assessment of Connected and Automated Vehicles: Sensing and Computing Subsystem and Vehicle Level Effects. Environmental science & technology 52 (5), 3249–3256. https://doi.org/10.1021/acs.est.7b04576}. Copyright {2024} American Chemical Society)

Based on some environmental background data from Gawron et al. (2018) and Baxter et al. (2018), we calculated the specific energy usage of fully automated minibuses used within AVENUE. The vehicle components required for automated driving are of particular interest and listed in Table 13.1. For each component, reference technologies and nominal power figures have been derived from the component manufacturer's information. In total, automated driving components in automated minibuses demand about 300 W. According to Gawron et al. (2018), the additional power required for a medium-sized, automated vehicle sums up to 240 W, while Baxter et al. (2018) state that 200 W is caused by the sensor layout for a mid-sized vehicle. The higher value of this study might be explained by a more detailed list of components in comparison to the studies by Gawron et al. (2018) and Baxter et al. (2018), which focus on primary hardware technology, such as sensors, radars, cameras, LiDARS, computers, and location detection.

In the current trial mode of the AVENUE project, the distance-weighted average speed of all sites equals 11.4 km/h (this calculation is based on Table 13.4). Assuming that all automated components run at full nominal power, the energy demand for 11.4 km of driving is 304.4 Wh, which equals 26.7 Wh/km. The distance-weighted average of the trial site's energy demand is 554 Wh/km. Hence only 4.8% of total energy demand is caused by the use of components required for automated driving.

**Table 13.1** Nominal power of automated driving components installed in one automated minibus

| Automated driving component | Number of components | Nominal power (W) |
|---|---|---|
| 180° Mono-Layer LiDARS | 6 | 48.0 |
| 360° Multi-Layer LiDARS | 2 | 24.0 |
| Computer | 2 | 160.0 |
| Module GNSS | 1 | 5.6 |
| Inertial Unit | 1 | 0.2 |
| World Shuttle Router | 1 | 25.5 |
| Front/Rear Cameras | 4 | 4.0 |
| Wheel Encoder | 4 | 0.6 |
| 3G & Ethernet Router | 2 | 12.0 |
| 15″ Touchscreen | 1 | 15.0 |
| Steering Encoder | 2 | 1.2 |
| Radio Modul GNSS | 1 | 0.2 |
| 4G Antenna | 1 | 5.0 |
| GPS Antenna | 2 | 3.2 |
| *Total power consumption (watt)* | | *304.4* |

### *13.2.2 Energy Demand of CAV Connectivity*

To compute CAV's energy consumption, it is important to consider the connection operations and strength. Such calculations depend on the connection hardware and its related power and time, the vehicle automation level, and the amount of exchanged data (Liu, Tan, et al., 2019). The following subsections present an in-depth classification of the different sources of consumption related to the vehicle data exchange. It addresses the energy consumption of the vehicle communication to external servers or devices, including the vehicle communication to other vehicles (V2V), to the infrastructure (V2I), to the cloud (V2C), to pedestrians (V2P), and to the grid (V2G).

### *13.2.3 CAV Connectivity Technologies*

The CAV's connectivity is built through multiple channels (El-Rewini et al., 2020): radio (AM/FM/DAB/RFID), Wi-F- (IEEE 802.11), Bluetooth, cellular (3/4/5G), and bidirectional communication (IEEE 802.11p, DSRC, WAVE), or using IoT networks (IEEE 802.15.4, Zigbee). With the presence of wireless connections, virtual ad hoc networks (VANET) can be spontaneously created among CAVs, leading to V2V communication. With the increase of modern concepts, infrastructure (V2I) and additional devices (V2X) are required to assist the VANETs for data storage and data transmission for long distances (El-Rewini et al., 2020; Lee & Atkison, 2021). V2X also compasses cloud (V2C) and grid (V2G) communication in addition to any further devices or peripherals interacting with the vehicle such as smartphones (V2P), car keys or Bluetooth devices (Domínguez et al., 2019). Being hyper-connected by nature, CAVs sustainability embeds the inherited energy consumption of data transfer technologies through their communication networks. Furthermore, the V2X network topology requires further processing either at the RSU (roadside unit) or at a remote server, which would cost additional computation resources, cause delays, and hence increase the energy consumption (Belogaev et al., 2020).

In academic literature, multiple approximations and estimations are used to quantify wireless cellular networks' energy consumption. In 2020, the 4G energy consumption was assessed to be around 0.1 kWh/GB (Andrae & Edler, 2015; El-Rewini et al., 2020; Pihkola et al., 2018), including the network, data centre computations, and data storage. Masoudi et al. (2019) added that 5G networks promise higher efficiencies (up 1 Mbit/J) to the energy consumption within the IoT ecosystem. Further research recommends new IoT technologies such as Zigbee for efficient energy consumption (Gheorghiu & Minea, 2016).

To improve the vehicular energy consumption related to data transmission, researchers have studied some protocols for higher energy efficiency. Pihkola et al. (2018) highlight that new efficiency measures that have been deployed within the last decade lower the energy consumption of Internet data transmission to 0.1 kWh/GB in 2020 instead of 12.35 kWh/GB in 2010. Dong et al. (2016) proposed an optimum cluster management method to reduce the V2V transmission power while using DSRC and LTE. Passafiume et al. (2020) proposed a battery-less transponder plugged to an RSU supporting the V2I communication.

### *13.2.4 Data Transmissions of CAV*

In the last decades, various consortia have been active in defining and studying various V2X data transmission technologies and protocols and their application in real-world use cases of automated driving with 3GPP and 5GAA alliances have been the most active in this domain. An early 3GPP report (3GPP, 2015) defined data transmission use cases using 4G-based long-term evolution-vehicle (LTE-V) and 3GPP (2018), 3GPP (2019), as well as 5GAA (2020) and defined data transmission use cases using 5G based C-V2X technologies. More advanced real-world scenarios with real traffic situations are being evaluated in various EU-funded telecommunication projects such as METIS 2020 and METIS 2020-II and V2X projects such as 5GCAR, 5GCroCo, and 5G CARMEN. It is not clear, though, whether these projects measure data transmission levels and whether such measures are accessible. Furthermore, simulation platforms such as Fraunhofer's Simulation Platform for Cellular V2X Fraunhofer IIS (2021) may also be useful as tools to collect data about real-world traffic situations.

Based on 3GPP and 5GAA use cases and data and the summarisation of the use cases as conducted by Kanavos et al. (2021), we calculate the energy consumption for V2X for various automated driving functions. We employed the 4G energy consumption measure of 0.1 kWh/GB (Andrae & Edler, 2015; El-Rewini et al., 2020; Pihkola et al., 2018) to calculate the average energy usage of automated driving connectivity for each of the eleven scenarios.

Table 13.2 presents the estimated data transmission rates for each use case based on 5GAA and 3GPP estimations along with our calculated energy consumption for each use case. Kanavos et al. (2021) point out that 3GPP data transmission estimations are on the higher level to prepare the telecom networks to support extreme cases of each use case, while 5GAA has more modest values based on more efficient implementation of the services. We believe more in the modest values of 5GAA as we believe that every market-conscious car manufacturer would strive to lower their vehicle's energy consumption and is more inclined to develop energy-efficient implementation of their automated driving services as long as it does not affect the driving security.

**Table 13.2** Data transmission and energy consumption for different automated functions

| | | Data transmission | | Energy consumption | |
|---|---|---|---|---|---|
| UC | Title | 5GAA | 3GPP | 5GAA | 3GPP |
| 1 | Platooning | 8–48 kbps | 1.5–325 kbps | 0.0001–0.0006 W | 0.00002–0.004 W |
| 2 | Sensor and state map sharing | 4–47 mbps | 25 mbps | 0.05–0.59 W | 0.31 W |
| 3 | Remote driving | 400 kbps–36 mbps | 1–20 mbps | 0.005–0.45 W | 0.013–0.25 W |
| 4 | Lane change | 120 kbps | – | 0.0015 W | – |
| 5 | Infrastructure-based perception of environment | 4–155 mbps | 1 Gbps | 0.05–1.9 W | 12.5 W |
| 6 | Collision avoidance | 10 mbps | – | 0.125 W | – |
| 7 | Collective information sharing | 120 kbps | 50 mbps | 0.0015 W | 0.625 W |
| 8 | See through for passing | 8 mbps | 10–700 mbps | 0.1 W | 0.125–8.75 W |
| 9 | Emergency trajectory alignment | 48 kbps | 30 mbps | 0.0006 W | 0.375 W |
| 10 | Intersection crossing | 8–25 kbps | 50 mbps | 0.0001 W | 0.625 W |
| 11 | Cooperative driving | – | 384 kbps | – | 0.0048 W |

### *13.2.5 Potential Energy Savings Through Predictive Driving Functions*

Automated and connected driving functions do not only control the perception, decision-making, and driving command execution to move the vehicle in a safe and convenient way, but they also enable energy savings by optimising the driving route selection, motion planning, and powertrain operation. This is an important aspect when considering the connectivity options to be implemented since a few additional data provided by the infrastructure plus a few megabytes of additional bandwidth needed might pay off significantly in terms of energy saved.

Hu et al. (2017) and Connor et al. (2021) discussed the direct correlation between the vehicle connectivity, velocity, and the battery consumption and their impacts to the environment. The authors studied real-world driving scenarios for electric buses using V2I and V2V technologies. According to their findings, the V2I and V2V communications provide energy savings that are up to 27% of battery cost reduction. Bo et al. (2019) also asserted the beneficial impact of V2I to have an optimal energy control. The US Department of Energy, through the NREL study (Stephens, 2016), reported 2–6% fuel savings by adopting the V2I smart intersections.

The predictive functions typically combine models of the vehicle and its powertrain with external data such as the upcoming driving route characteristics and traffic conditions to predictively control the vehicle. Examples of such functions are provided in Table 13.3 below, including published data on corresponding energy savings. Considering the literature for energy-saving potentials indicated in Table 13.3, it is noteworthy that:

**Table 13.3** Predictive automated driving (AD) functions and their energy-saving potential

| Function | Description | Main application area | Potential energy savings | Source | Required infrastructure or data |
|---|---|---|---|---|---|
| Eco route planning | Identifies routes with lowest predicted energy consumption based on upcoming routes and traffic conditions, optionally including charging point selection | Routes with multiple paths and varying traffic conditions | Average 12.5% Up to 48% | (Fiori et al., 2018; Kubicka et al., 2016) | Communication with off-board digital maps including route topography, road network, and live traffic speed data provided by vehicle fleet |
| Traffic light assistant (TLA) Or Green light optimal speed advisory (GLOSA) | Predictively adjusts velocity to reach upcoming traffic lights at the start of their expected green phases | City driving with traffic lights | Average 23%–36% | (EU Horizon 2020 EVC1000, 2018) | V2I communication for upcoming signal phase and timing (SPAT) information, e.g. provided by traffic light roadside units |
| Predictive adaptive cruise control | Predictively adjust velocity to maintain appropriate headway distances to a preceding target vehicle | Motorways driving with multiple connected vehicles | Average 13%–15% | (EU Horizon 2020 EVC1000, 2018) | Onboard sensors for target object and motion detection, optionally V2V |
| Platooning | Cooperatively maintains efficient headway distances between multiple connected vehicles reducing accelerations and aerodynamic drag | Motorway driving with multiple connected vehicles | Average 4%[a] | (Bichiou & Rakha, 2020) (Hussein & Rakha, 2020) | V2V with low latency |
| Predictive cruise control or eco-driving | Predictively adjust velocity based on upcoming hills, speed limits, and/or curves, including efficient gears election in case of multi-ratio gearboxes | Hilly motorways without traffic ahead | Average 5%[b] | (Volvo, 2021) | Onboard navigation system maps or electronic horizon system with cloud communication for time-varying data such as dynamic speed limits |

**Table 13.3** (continued)

| Function | Description | Main application area | Potential energy savings | Source | Required infrastructure or data |
|---|---|---|---|---|---|
| Predictive thermal management | Predictively adapts thermal management of battery, e-motor, and inverter using future velocity and ambient conditions | Driving routes with cold or hot ambient conditions | Up to 11%[c] | (Auer et al., 2015) | Communication with off-board digital maps including route topography and traffic speed data for the driven route |
| Predictive hybrid and fuel cell control | Predictively plans when to charge or discharge the battery over the route, as well as the power distribution between powertrain components (ICE, electric motor, fuel cell) | Hybrid, range extender, and fuel cell vehicles | Up to 5% | (Huss et al., 2021) | Communication with off-board digital maps including route topography and traffic speed data for the driven route |

[a]Results for conventional (ICE-driven) car at 100 km/h cruising speed
[b]Results for conventional (ICE-driven) commercial vehicles, smaller savings expected for light vehicles with electric recuperation potential
[c]Results assuming full preview of route velocity available and excluding energy consumption required for component preconditioning, i.e. energy provided by the electrical grid before driving

- The achievable energy savings are generally heavily dependent on the defined vehicle and use cases, resulting in wide ranges of savings typically being published for similar functions by different authors.
- The energy savings also strongly depend on the particular baseline to which they are calculated, which often consists of different types of human drivers or non-predictive control algorithms.

Nevertheless, the available published results demonstrate that a significant potential in energy savings can be achieved in particular from intelligent route optimisation and velocity control.

### *13.2.6 Discussion and Implications for Environmental Impact Assessment*

The nexus of data processing and exchange within the automated driving landscape raises challenges to consider while assessing CAV's sustainability. To this end, the energy consumption related to CAV's data transfer depends on wireless technologies, the cooperative communication modes, and the implemented services. Automated minibuses may support different types of internet connections which result in large differences in energy consumption. The driverless wireless network can vary from 4G and 5G to DSRC, which definitely impacts the amount of exchanged data and hence the vehicle energy consumption (Masoudi et al. (2019). Cooperative V2X is expected to be the favourable communication mode with regard to energy use, according to Bo et al. (2019) and Stephens (2016). Predictive, adaptive information sharing provided through vehicle communication with infrastructure and other vehicles improved vehicle's braking performance and consequently its energy consumption. However, a highly connected vehicle also requires more processing within its infrastructure, remote, or cloud servers. This may outweigh the V2X sustainability. A fixed route or a mature on-demand service would not have comparable energy consumption as the number of involved servers and processing will not be proportionate (Greenblatt & Shaheen, 2015). As with every exchanged data within the automated minibus ecosystem, the energy efficiency can fall over to either high or low energy demand depending on the number of connections and the advancement of the deployed technologies.

Although the reported potential energy savings through predictive driving functions differ between studies, it seems evident that the savings are likely to counterbalance and even overcompensate the energy costs associated with the communication modules. It is 'common sense' in the European research community that it will be impossible to implement large-scale automated urban mobility in a safe way without infrastructure support; both in-vehicle and infrastructure communication equipment will already be there regardless of the use of predictive functionalities. Consequently, the additional cost for employing predictive functionalities is merely the additional bandwidth used by the additional data which need to be transmitted. Especially the first two functions mentioned in Table 13.3, *eco route planning* and *traffic light assistant*, apply to urban scenarios and require only a little information to be exchanged between the vehicles and the infrastructure, which makes them very promising candidates for real energy savings achieved through the implementation of automated urban mobility.

## 13.3 LCA (Life Cycle Assessment) Model

This part summarises the final results of a life cycle assessment (LCA) study on automated minibuses, which was also published as an analysis in Transportation Research Part D (Huber et al., 2022).

The LCA study investigated the environmental impacts of automated minibuses to be integrated into the public transport of cities, guided by the following research questions: (1) Which environmental impacts are associated with the operation of an automated minibus? (2) What are the main drivers of these impacts, and how can these be reduced? (3) What conclusions can be drawn from these findings for the role of automated minibuses in future public transportation systems?

### 13.3.1 Goal and Scope of the LCA Study

The common functional unit of a passenger kilometre (pkm) enables the comparison between automated minibuses and other modes of transport.

The automated minibus under investigation is 4.75 m long, 2.11 m wide, and 2.65 m high, weighs 2400 kg, and can carry 15 passengers (11 seated and 4 standing) with a maximum speed of 25 km/h.

The automated minibuses operation involved fixed routes for public transport, with trials on on-demand, door-to-door, and pooling options for passengers to request the vehicle to come to a designated pick-up location through the use of a mobile application (Navya, 2018). The automated minibuses are intended for use in a public transport system, not as a replacement for individual vehicles.

The assessment adopts a cradle-to-grave approach, encompassing the main life cycle phases recommended by Duce et al. (2013), namely, component production, vehicle assembly, use, and end-of-life treatment. The production of components has been divided into three categories: battery manufacturing, the production of automated driving components, and the production of other bus components. In order to accurately assess the environmental impact of these components throughout their life cycle, it is necessary to account for the material and energy inputs and outputs at each stage of the LCA. The following six impact categories were employed: acidification, climate change, eutrophication, ozone depletion, photochemical ozone formation, and resource depletion. A control calculation showed that the five remaining environmental impact categories presented similar results in terms of their impact.

### 13.3.2 Life Cycle Inventory and Data Collection

The life cycle inventory encompasses all environmental impacts associated with the system under investigation, including the inflow and outflow of materials and energy. A preliminary generic automated minibus model has been constructed using existing literature data (Gawron et al., 2018; Hawkins et al., 2013; Majeau-Bettez et al., 2011). This model was then refined by incorporating primary data obtained from the automated minibus manufacturer and public transport operators involved in the AVENUE project.

The primary data for this study were obtained from the demonstrator sites of the AVENUE project, where an automated minibus manufacturer provided information on vehicle components (such as weight, functions, and nominal power) and transport operators provided data on the use of these vehicles in public transportation. The data collection took place between 2019 and early 2021 in an iterative manner.

The sources of information regarding component production, including batteries and automated driving components, are Majeau-Bettez et al. (2011), Hawkins et al. (2013), Gawron et al. (2018), and Moreno Ruiz et al. (2020). The manufacturer supplied all the components and their respective weights for the vehicles in this study. The total weight of all the automated minibus components used is over 99% of the total automated minibus weight, which meets the requirements of standard

LCA. The assembly of these components was based on data from Majeau-Bettez et al. (2011), Hawkins et al. (2013), and Gawron et al. (2018) which represents industry-scale assembly of electric passenger cars. Additional secondary data was sourced from well-known LCA databases, chiefly ecoinvent 3.7 (Moreno Ruiz et al., 2020). A comprehensive overview of all components and materials required for vehicle assembly, along with their life cycle inventories (LCI) and details on primary and secondary sources, is provided in the study's supplementary material.

Table 13.4 presents data on the use phase of automated minibus service on fixed-route buses in five trial sites from September 2018 to January 2021. The data reveals differences in the average speed, expected annual mileage per shuttle, average

**Table 13.4** Automated minibus use phase data

| Site | Lyon, Groupama (France) | Contern (Luxembourg) | Luxembourg, Pfaffenthal (Luxembourg) | Copenhagen Nordhavn (Denmark) | Oslo, Ormoya (Norway) |
|---|---|---|---|---|---|
| Data collection period | November 2019–January 2021 | September 2018–January 2021 | September 2018–January 2021 | September 2020–January 2021 | December 2019–January 2021 |
| Route length [km] | 1.3 | 2.2 | 1.2 | 1.3 | 1.6 |
| Average driving speed [km/h] | 10 | 15 | 17 | 8 | 10 |
| Total mileage during data collection [km] | 12,492 | 1900 | 9000 | 2000 | 23,000 |
| Annual mileage during data collection [km] | 9994 | 786 | 3724 | 4800 | 19,714 |
| Total passengers | 5545 | 650 | 25,060 | 1300 | 6600 |
| Average passenger trip length [km] | 1.3 | 2.2 | 1.0 | 0.4 | 0.8 |
| Average vehicle occupancy | 0.6 | 0.8 | 2.8 | 0.3 | 0.2 |
| Average energy demand [Wh/km] | 480 | 780 | 510 | 590 | 590 |
| Minimum energy demand [Wh/km] | No data | 480 | 350 | 300 | 300 |

vehicle occupancy, and average and minimum energy demand. It should be noted that the extremely low average occupancy is due to the experimental nature of the trial sites, where vehicles are also used for functional and technical testing purposes.

Umberto® software was utilised to create a product life cycle model and analyse the results. The comprehensive automated minibus LCA model comprises 198 processes and 42 subnets across 4 hierarchical levels. A portion of the overall model is depicted in Fig. 13.2.

### *13.3.3 Scenario Setting*

Using primary data gathered from trial sites, we have generated a near-future use case and worst- and best-case values for scenario analysis. These values can also be combined to form an ideal use case. The relevant parameters for these scenarios include the automated minibuses expected lifetime, annual mileage, average passenger occupancy, energy demand, energy source, and the used battery LCA data. The parameter settings are described in detail and summarised in Table 13.5.

According to the manufacturer and transport operators involved in this study, the battery lifetime can be used as a proxy for the overall automated minibus lifetime. If one charging process occurs per day and the automated minibus operates for 5 days per week, the estimated battery lifetime is 7.7 years. This has been rounded up to 7 years to account for probable losses and reduced efficiency as the battery ages (Hadjipaschalis et al., 2009; Oliveira et al., 2015). The high cost of batteries for electric vehicles has led to the belief that the lifespan of an automated minibus is the same as the lifespan of its battery, although it is acknowledged that the rapid advancement of automated minibus technology may result in obsolescence and decommissioning before the end of the battery's lifespan. Conversely, some studies on batteries for electric vehicles suggest longer lifetimes, such as 10 years (Deng et al., 2017). The average lifetime of 7 years is obtained through scenario analysis, which ranges from 3 to 10 years in some cases.

The near-future use case is based on an annual mileage of 20,000 km, while the scenario analysis considers a range of mileages from 5000 km (which is roughly the average of other trial sites in Table 13.5) to 36,500 km, assuming a daily operating distance of 100 km.

The expected average occupancy for the near-future use case is five passengers, which is above the current trial data but aligns with transport operators' economic feasibility expectations. The scenario analysis considers a worst-case value of one passenger on average and a best-case value of ten passengers on average.

The energy demand for the near-future use case is 554 Wh/km, which is the distance-weighted average of the trial site's typical energy consumption. This is within the manufacturer's specifications of 520 Wh/km, which was measured with one person on board, traveling at an average speed of 6.6 km/h, and an outside temperature of 30 degrees Celsius, while the vehicle's interior was cooled down to 16 °C. The energy demand for the autonomous mobility system encompasses all automated components,

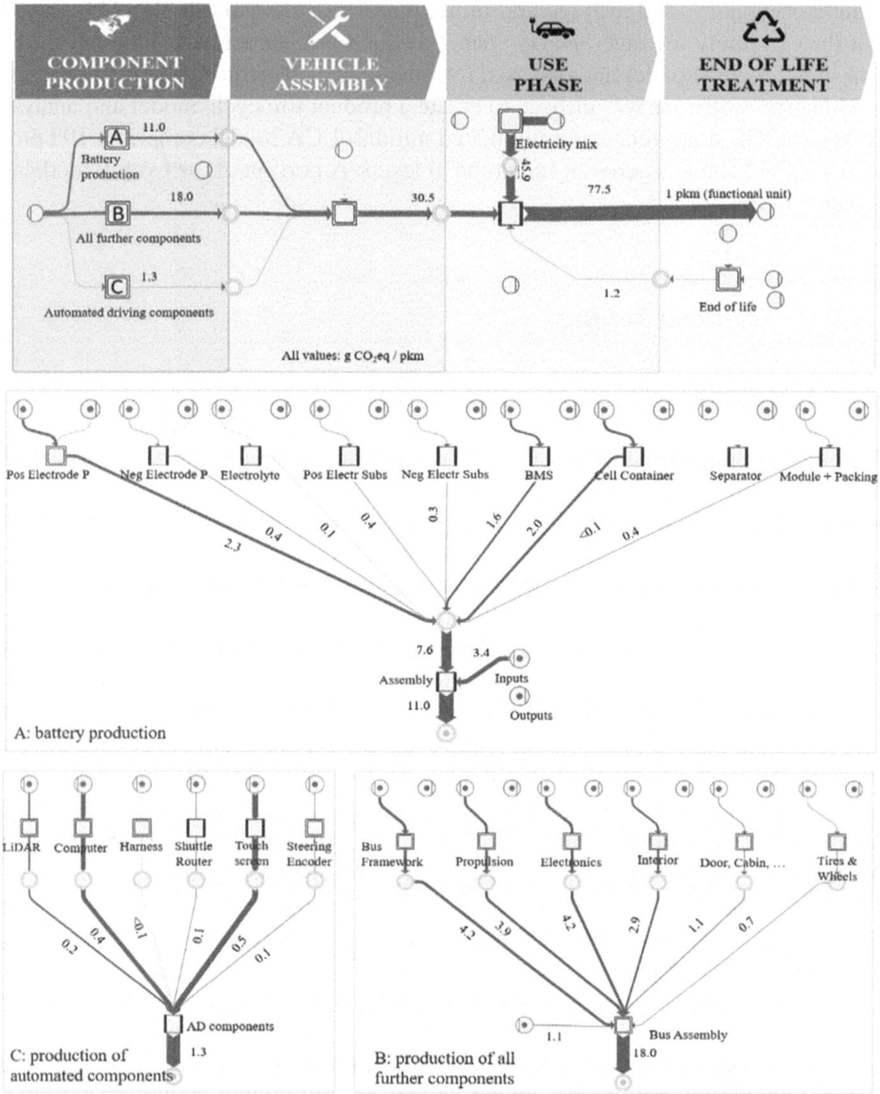


**Fig. 13.2** Automated minibus life cycle model main model with subnets for battery production (**a**), production of all further components (**b**), and production of automated components (**c**). Sankey diagrams depict global warming potential in g CO2eq per pkm for the automated minibus near-future use case. In the Petri-net-based material flow network approach underlying the Umberto LCA software, blue squares represent processes or subnetworks, and circles represent input points (green), output points (red), and connection points (yellow). All values have been rounded to one digit

**Table 13.5** Parameter setting for near-future use case and scenario analysis

| Parameters | Near-future use case | Worst case | Best case |
|---|---|---|---|
| Expected lifetime [years] | 7 | 3 | 10 |
| Annual mileage [km] | 20,000 | 5000 | 36,500 |
| Average passenger occupancy | 5 | 1 | 10 |
| Energy demand [Wh/km] | 554 | 780 | 332 |
| Energy source | Electricity mix Europe | Fossil energy mix (using Poland as approximation) | 100% renewable energy mix (using Norway as approximation) |

passenger interaction components, and the electric driving components. Given that various factors such as speed, temperature, weight, and others can impact automated minibus energy consumption, it is crucial to conduct scenario analysis. The distance-weighted average of the minimum energy demand at the trial site (332 Wh/km) represents the best-case scenario, while the highest average energy demand at a trial site (780 Wh/km) represents the worst-case scenario for scenario analysis.

For the near-future use case, we have assumed a European electricity mix with 418 grams of CO2 equivalent per kilowatt-hour (kWh) (ecoinvent 3.7 databases with low voltage electricity market datasets for Europe, Poland, and Norway). In contrast, the worst-case scenario for scenario analysis is a mostly fossil electricity mix with 1037 grams of CO2 equivalent per kWh, while the best-case scenario is an almost entirely renewable electricity mix with 23 grams of CO2 equivalent per kWh. Battery production has been modelled using detailed data from the literature.

## 13.4 Results

This section will provide an analysis of the environmental impacts of additive manufacturing, as well as the results of scenario simulations and assessments of the automated components. Additionally, a comparison of automated minibuses with other modes of transportation will be presented to provide a more comprehensive understanding of the findings.

### *13.4.1 Life Cycle Impacts of Automated Minibuses*

Table 13.6 displays the environmental impact per passenger kilometre for the selected impact categories, disaggregated into the life cycle phase components of production (separated into battery, automated, and all other bus components), vehicle assembly, use, and end of life.

**Table 13.6** Environmental impacts for 1 pkm of automated minibuses in a near-future use case and in comparison to an ideal use case; climate change (global warming potential 100 years) measured in g CO2eq (carbon dioxide equivalents); acidification in mol H + eq (proton equivalents); eutrophication in kg Peq (phosphorous equivalents), ozone depletion in kg CFC-11 eq (trichlorofluoromethane equivalents), photochemical ozone formation (POCP) in NMVOCeq (non-methane volatile organic compound equivalents), resource depletion in kg Sbeq (antimony equivalents)

| | Climate change | | Acidification | | Freshwater eutrophication | | Ozone layer depletion | | Photochemical ozone creation | | Resource depletion | |
|---|---|---|---|---|---|---|---|---|---|---|---|---|
| | (g $CO_2$eq) | | (mol H + eq) | | (kg Peq) | | (kg CFC-11 eq) | | (kg NMVOCeq) | | (kg Sbeq) | |
| Automated driving components | 1.25 | 2% | 9.11E-06 | 2% | 1.08E-06 | 1% | 7.30E-11 | 0% | 4.96E-06 | 2% | 2.32E-07 | 2% |
| Battery production | 11.00 | 14% | 6.99E-05 | 15% | 9.59E-06 | 9% | 4.39E-08 | 92% | 3.43E-05 | 14% | 1.44E-06 | 11% |
| Further bus components | 18.00 | 23% | 1.30E-04 | 28% | 4.68E-05 | 45% | 1.27E-09 | 3% | 9.32E-05 | 39% | 1.06E-05 | 83% |
| Component production | 30.30 | 39% | 2.09E-04 | 45% | 5.75E-05 | 55% | 4.53E-08 | 95% | 1.32E-04 | 55% | 1.23E-05 | 97% |
| Final assembly | 0.19 | 0% | 8.51E-07 | 0% | 8.03E-08 | 0% | 9.22E-12 | 0% | 4.56E-07 | 0% | 4.65E-10 | 0% |
| Use (driving) | 45.90 | 59% | 2.51E-04 | 54% | 4.59E-05 | 44% | 2.34E-09 | 5% | 1.05E-04 | 44% | 4.19E-07 | 3% |
| End of life | 1.17 | 2% | 3.77E-06 | 1% | 3.69E-07 | 0% | 2.47E-11 | 0% | 2.53E-06 | 1% | 2.80E-09 | 0% |
| Total near-future use case | 77.50 | 100% | 4.65E-04 | 100% | 1.04E-04 | 100% | 4.77E-08 | 100% | 2.41E-04 | 100% | 1.27E-05 | 100% |
| Total ideal use case | 6.92 | 9% | 4.48E-05 | 10% | 1.18E-05 | 11% | 8.84E-09 | 19% | 2.84E-05 | 12% | 2.49E-06 | 20% |

In the near-future use case, each pkm has a global warming potential of 78 g $CO_2eq$ (as shown in Fig. 13.2 with a Sankey visualisation of the global warming potential within the automated minibus life cycle model). The majority of this impact, at 59%, stems from the use phase, while component production accounts for 39%. The primary climate impacts resulting from the use phase are primarily caused by the combustion of fossil fuels to power the operation of the automated minibus. For this reason, the use phase also accounts for 54% of the overall acidification potential. In all other environmental impact categories, the production of components either dominates moderately (eutrophication potential and photochemical ozone depletion potential) or by a wide margin (ozone layer depletion potential and resource depletion potential). The assembly and end-of-life phases of the product have no significant impact in any of the chosen environmental impact categories, which is why their modelling is based on average literature data.

The aforementioned Table 13.5 outlines the optimal parameter settings for the best possible scenario. By utilising these settings, a practical example with high automated minibus lifetime mileage, high passenger occupancy, low energy demand, and renewable energy supply can be modelled and evaluated. The total environmental impacts of this ideal scenario are presented in Table 13.6, revealing substantial reductions in environmental impact across all categories, with 80–91% reduction in resource depletion and climate change impact.

### 13.4.2 Scenario Analysis

The scenario analyses have been executed for all the parameters listed in Table 13.5 and for every environmental impact category under consideration. The findings of the scenario analyses, depicted in Fig. 13.3, demonstrate the paramount importance of non-technical parameters in determining the overall environmental impact of

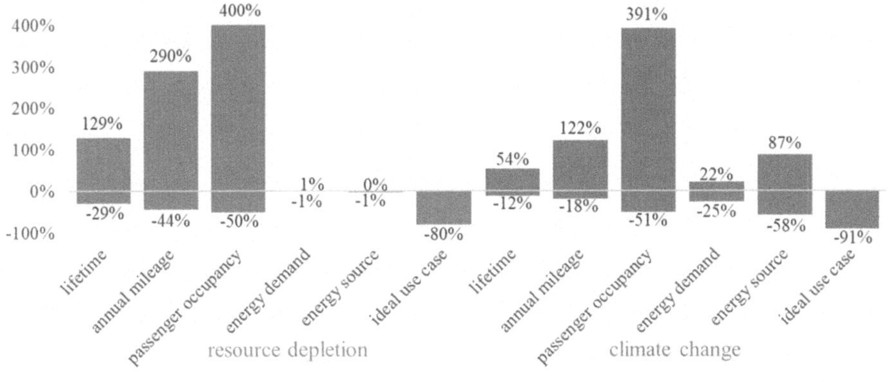


**Fig. 13.3** Scenario analyses for resource depletion potential and global warming potential (near-future use case = 0%)

automated minibuses in terms of resource depletion and global warming potential. The depletion of resources and the impact of climate change are significantly affected by low passenger occupancies and low annual mileage, whereas the energy demand of the vehicle has only a limited impact. Additionally, the energy source used to charge the automated minibus battery is the most important factor in reducing the climate impact. Operating the automated minibus within a fully renewable energy system can reduce the global warming potential by 58%.

The environmental impact category of climate change demonstrates the disparate effects of different parameters on the production and use phases (see Fig. 13.4). While the vehicle's lifetime and mileage only affect the production phase, the energy demand and energy mix influence the use phase. Occupancy affects both the production and use phases equally. The relationship between the use phase and the production phase, as shown in Fig. 13.4, exhibits strong fluctuations. In extreme cases, such as very high mileage and a high fossil energy mix, the use phase dominates. Conversely, when renewable energies are used, the importance of the use phase decreases from a climate perspective, and the importance of production for the overall result increases.

### 13.4.3 Impact of Automated Components

The focus of the study lies in the components necessary for automated driving. Table 13.6 indicates that the production of these components has a minimal impact on environmental performance, accounting for less than 2% in all impact categories. Additionally, the energy consumption of these components is of importance, despite the shift towards renewable electricity sources. The use phase's significance is reduced in the long run, but energy demand and energy mix remain critical factors in overall performance in the present. Table 13.7 provides a list of reference technologies and their corresponding nominal power figures for all the components used in automated manual driving (for a

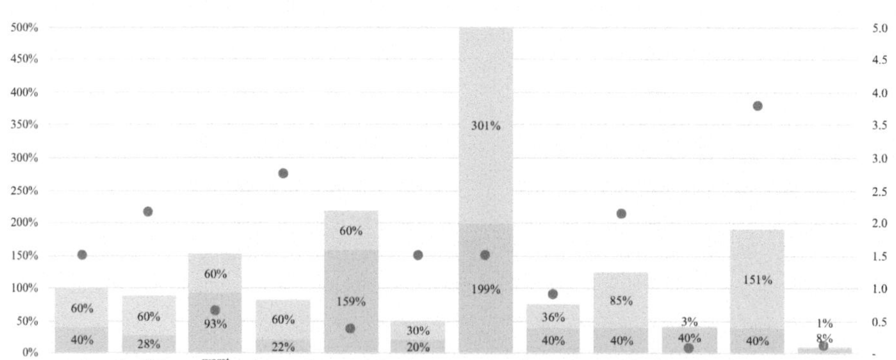


**Fig. 13.4** Scenario analysis (global warming potential) comparing automated minibus production and use phase (near-future use case = 100%)

**Table 13.7** Nominal power of automated driving components installed in one automated minibus (Light detection and ranging sensors (LiDARS), GNSS, GPS)

| Automated driving component | Number of components | Nominal power (W) |
|---|---|---|
| 180° monolayer LiDARS | 6 | 48.0 |
| 360° multi-layer LiDARS | 2 | 24.0 |
| Computer | 2 | 160.0 |
| Module GNSS | 1 | 5.6 |
| Inertial unit | 1 | 0.2 |
| World shuttle router | 1 | 25.5 |
| Front/rear cameras | 4 | 4.0 |
| Wheel encoder | 4 | 0.6 |
| 3G and Ethernet router | 2 | 12.0 |
| 15″ touchscreen | 1 | 15.0 |
| Steering encoder | 2 | 1.2 |
| Radio module GNSS | 1 | 0.2 |
| 4G antenna | 1 | 5.0 |
| GPS antenna | 2 | 3.2 |
| Total | 304.5 | |

more detailed table that includes manufacturers, models, and Internet sources, please refer to the supplementary material). In summary, the components used in automated manual driving require approximately 300 W of power in total.

According to Gawron et al. (2018) and Baxter et al. (2018) the power requirements for a medium-sized, automated vehicle are estimated to be 240 W and 200 W, respectively. The higher value of the present study might be explained by a more comprehensive list of components compared to Gawron et al. (2018) and Baxter et al. (2018), which primarily focus on primary hardware technology, such as sensors, radars, cameras, light detection and ranging sensors (LiDARS), computers, and location detection.

Table 13.7 presents the average speed at the various trial sites. The distance-weighted average speed of all sites is 11.4 km/h. If all automated components are running at their full nominal power, the energy consumption for a distance of 11.4 km is 304.5 Wh, which translates to 26.7 Wh/km. The average energy demand of the trial sites is 554 Wh/km. Therefore, only 4.8% of the total energy demand is due to the use of components required for autonomous driving.

### *13.4.4 Contextualisation*

With a high degree of confidence, the environmental impact per pkm of near-future and ideal future applications of automated minibuses is significantly lower than the current trial cases. To compare its performance with other modes of transportation, the automated minibuses' climate change impact per km is contrasted against literature values of other vehicles (Table 13.8). For all vehicles, the impact is calculated for off-peak, average, and peak operation. The average number of passengers per vehicle used in the calculation is 1.58, based on a study by Chester and Horvath (2009). Figure 13.5 shows the climate change impacts of all transportation modes, including the automated minibuses near-future and ideal use case.

**Table 13.8** Climate impacts, lifetime mileages, and passenger occupancies for various individual and public transportation vehicles (based on [1] Puig-Samper Naranjo et al., 2021; [2] Gawron et al., 2018; [3] Kemp et al., 2020; [4] Nordelöf et al., 2019; [5] this paper): abbreviations: *Ind.* individual, *ICEV* internal combustion engine vehicle, *HEV* hybrid electric vehicle, *BEV* battery electric vehicle, *BECAV* battery electric connected and automated vehicle, *ICECAV* internal combustion engine connected and automated vehicle, *SUV* sports utility vehicle, *BEB* battery electric bus, *PHEB* plug-in hybrid electric bus, *HEB* hybrid electric bus, *AM NF* automated minibus near-future use case, *AM ideal* automated minibus ideal use case

| | Peak operation | Off-peak operation | Average operation | Peak occupancy | Off-peak occupancy | Average occupancy | Lifetime mileage |
|---|---|---|---|---|---|---|---|
| Unit | g $CO_2$eq/pkm | g $CO_2$eq/pkm | g $CO_2$eq/pkm | No. of passengers | No. of passengers | No. of passengers | km |
| Ind. ICEV petrol [1] | 52 | 261 | 131 | 5 | 1 | 1.58 | 150,000 |
| Ind. ICEV diesel [1] | 48 | 241 | 121 | 5 | 1 | 1.58 | 150,000 |
| Ind. HEV, EU electricity [1] | 44 | 222 | 111 | 5 | 1 | 1.58 | 150,000 |
| Ind. BEV, EU electricity [1] | 27 | 135 | 68 | 5 | 1 | 1.58 | 150,000 |
| Ind. small BECAV, US electricity [2] | 44 | 221 | 140 | 5 | 1 | 1.58 | 257,494 |
| Ind. medium BECAV, US electricity [2] | 45 | 223 | 141 | 5 | 1 | 1.58 | 257,494 |
| Ind. large BECAV, US electricity [2] | 50 | 250 | 158 | 5 | 1 | 1.58 | 257,494 |
| Ind. small ICECAV [2] | 74 | 372 | 235 | 5 | 1 | 1.58 | 257,494 |

**Table 13.8** (continued)

| Unit | Peak operation g $CO_2$eq/pkm | Off-peak operation g $CO_2$eq/pkm | Average operation g $CO_2$eq/pkm | Peak occupancy No. of passengers | Off-peak occupancy No. of passengers | Average occupancy No. of passengers | Lifetime mileage km |
|---|---|---|---|---|---|---|---|
| Ind. medium ICECAV [2] | 75 | 375 | 237 | 5 | 1 | 1.58 | 257,494 |
| Ind. large ICECAV [2] | 86 | 431 | 273 | 5 | 1 | 1.58 | 257,494 |
| Ind. BECAV SUV, US electricity [3] | 27 | 134 | 85 | 5 | 1 | 1.58 | 321,868 |
| Ind. ICECAV van [3] | 60 | 301 | 85 | 5 | 1 | 1.58 | 321,868 |
| Public BEB, EU electricity [4] | 7 | 154 | 48 | 105 | 5 | 16 | 780,000 |
| Public PHEB, diesel [4] | 13 | 202 | 63 | 80 | 5 | 16 | 780,000 |
| Public HEB, diesel [4] | 10 | 211 | 66 | 102 | 5 | 16 | 780,000 |
| Public diesel bus [4] | 16 | 304 | 95 | 95 | 5 | 16 | 780,000 |
| Public–AM NF [5] | 39 | 387 | 77 | 10 | 1 | 5 | 140,000 |
| Public–AM ideal [5] | 7 | 69 | 14 | 10 | 1 | 5 | 365,000 |

Compared to other forms of public transportation, the near-future use case of automated minibuses has higher climate impacts per pkm, with the exception of the comparison to the average operation of diesel buses. However, it should be noted that all other forms of public transportation use larger buses with higher passenger numbers for peak, average, and off-peak operation. The ideal use case of automated minibuses performs better than any other form of transportation, demonstrating the significant potential for EMA to improve the environment further and optimise it. It should be viewed with caution, as no ideal use case was calculated for the other forms of transportation, and a renewable energy mix would also have a positive effect on all other battery-electric and hybrid vehicles.

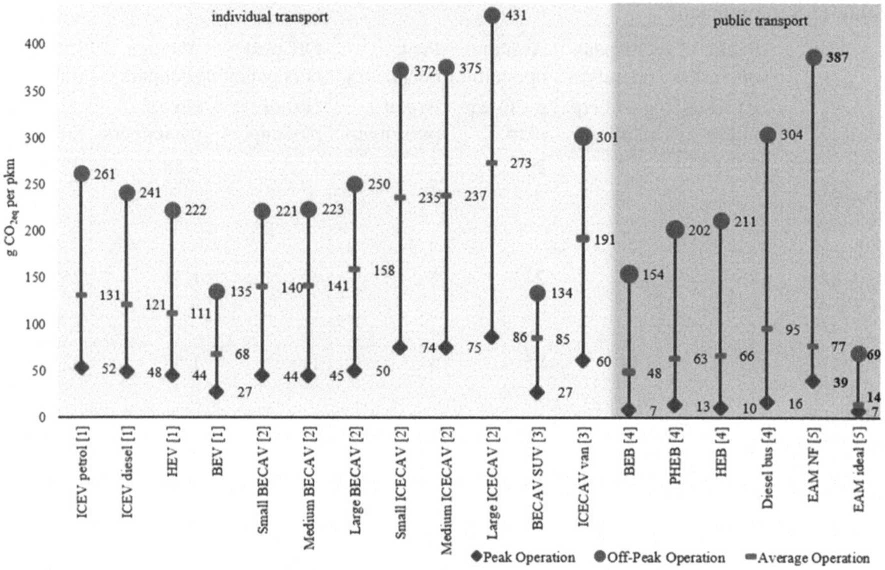


Fig. 13.5 Climate impact of different transportation modes in g $CO_2$eq per pkm (own compilation, based on [1] Puig-Samper Naranjo et al., 2021; [2] Gawron et al., 2018; [3] Kemp et al., 2020; [4] Nordelöf et al., 2019; [5] this study, all abbreviations are detailed in Table 13.8)

## *13.4.5 Discussion and Limitations*

The present LCA identifies the key factors that influence the environmental impact of automated minibuses. Similar to electric vehicles in general (as noted in Helmers et al., 2017), the energy consumption and electricity mix used for charging batteries have a significant impact on the vehicle's climate performance. Moreover, systemic factors, such as the utilisation of the vehicle in terms of annual mileage and passenger occupancy, also play a crucial role. Therefore, whether automated minibuses can be considered environmentally friendly depends on various factors. While an infrequently used automated minibus with few passengers is highly unlikely to have a favourable environmental impact, a heavily used automated minibus that is also fully utilised in terms of its passengers can outperform other transportation methods in terms of environmental benefits. Based on measurement data from operating automated minibuses, a near-future use case has been defined that is considered to be very achievable in the coming years. This use case already shows a very good environmental performance under the aforementioned framework conditions.

The use case calculated in an optimal manner also highlights the potential environmental benefits of automated minibuses, as long as all conditions and parameters are optimal. The most environmentally intensive phases of an automated minibuses life cycle are the production of components and driving, and the components required for automated driving play a minor role, contributing less than 2% to production-related climate impacts and approximately 5% to driving-related climate impacts.

The LCA results presented here are promising, particularly because they indicate the potential for achieving higher mileages in the short term, and there is a positive attitude (goodwill) expressed by potential users towards the automated minibus (Korbee et al., 2024). This study also has some limitations that are important to consider. For example, the data used was collected from pilot sites provided by public transport operators and vehicle manufacturers. These sites have an experimental nature and were affected by COVID restrictions, which limited the amount of data collected and the number of passengers in the vehicles. As data accuracy is a recurrent concern in LCAs of emerging technologies (Arvidsson et al., 2018; Hetherington et al., 2014), future research could focus on utilising longer time-series of data collected from regular operations rather than relying on data from demonstration and trial site operations. To address the potential consequences of innovative technology data, this study suggests utilising scenario development. The purpose of these findings is not to present specific numbers for inclusion in product declarations but rather to offer valuable insights on the key factors affecting the future environmental impact of automated minibuses.

The low relevance of automated driving components is comparable to studies of automated vehicles (Gawron et al., 2018).

Although other studies forecast a much greater effect of automated components on overall energy consumption (Brown et al., 2013; Gawron et al., 2018; Gonder et al., 2016; Grisoni & Madelenat, 2021; inria, 2019; Krail, 2021; Pihkola et al., 2018; Saujot et al., 2017; Wadud et al., 2016).

The deployment of automated vehicles necessitates adaptations in both physical and digital infrastructure, as pointed out by Noussan & Tagliapietra, 2020, and the implementation of vehicle-to-everything (V2X) technologies requires additional technical infrastructure.

The following out-of-vehicle technical infrastructure is necessary for the operation of automated vehicles: roads, sensors to detect special signals, and a long-range wireless network (Liu, Tight, et al., 2019). The examined automated minibus neither transmits large amounts of data to the outside nor requires extensive additional technical infrastructure, which may change in the future. Notably, the study did not consider the long-term benefits of automated driving compared to human driving. Research suggests that connectivity and cooperative technologies will lead to better traffic anticipation, modulated driving, better manoeuvring (Fagnant & Kockelman, 2015), and better ride-matching capacity (Shaheen & Bouzaghrane, 2019), which could further reduce energy consumption. However, the use of automated vehicles may simultaneously reduce energy demand through efficient driving and increase it through additional data processing and transmission (see, e.g. Stephens et al., 2019). The current state of research does not allow for a definite conclusion, and the impact of data processing and transmission on the automated minibus under investigation is currently minor. The implementation of innovation and efficiency improvements in batteries is expected to decrease the negative impact on the environment caused by automated minibuses.

The automated minibus energy demand of 554 Wh per driven km in comparison to other electric vehicles is quite high, as reported by Puig-Samper Naranjo et al.,

2021 and Bauer et al., 2015. Since the components for automated driving do not significantly influence this consumption, other factors must be responsible for this high level of energy consumption. Specifically, the heating and cooling of the vehicles, as well as their low speed, are noteworthy. The entire interior of the vehicle is constantly cooled on warm days and warmed up on cold days, which results in high energy consumption. To further reduce the overall energy consumption and environmental impacts, higher speeds, reduced heating and cooling behaviour, and additional energy efficiency measures on the vehicle could be implemented, but these are not within the scope of this study.

### *13.4.6 Conclusion and Consequences for Future Mobility Systems*

This LCA study highlights the potential of automated minibuses as part of public transport. If automated minibuses are utilised extensively in terms of mileage and regularly used by multiple passengers, they offer significant environmental benefits and perform similarly or better than other public transport vehicles. However, it becomes clear that automated minibuses play a specific role in the overall mobility system and cannot replace all other modes of transport. Currently, the performance of automated minibuses, which features low speed, low passenger capacity, door-to-door service, on-demand service, and driverless operation, is seen as complementary to public transport. For example, automated minibuses can cover the 'first and last mile' or take over off-peak operations of regular buses, increasing the availability, flexibility, efficiency, and reliability of local public transport, which brings significant environmental benefits, especially when replacing individual motorised transportation. However, there may be some rebound effects if automated minibuses replace walking or biking or lead to more travel due to their convenience and comfort (Grisoni & Madelenat, 2021; inria, 2019; Saujot et al., 2017).

Taking into account the environmental and sustainability science perspective, this study highlights the limitations of LCA studies that solely concentrate on individual vehicles and vehicle types. The environmental benefits of automated minibuses are influenced by the individual vehicle's performance but are largely determined by the vehicle's utilisation, occupancy, and integration within a comprehensive transportation system. As a result, comparing different automated minibus types or brands is relatively inconsequential in this context.

The findings of this research are relevant to decision-makers at various levels of policy and public transport operators. For the latter, the study offers clear understanding of the environmental benefits and drawbacks of the automated minibus implementation. For policymakers, the study underscores the importance of developing plans and frameworks for the deployment of autonomous vehicles in a timely manner to maximise environmental benefits. A public transport system that integrates automated minibuses at strategic points and is multimodal and flexible appears to be a promising approach.

## 13.5 Final Environmental Indicators for Sustainability Assessment of Pilot Sites

The assessment of the pilot sites finalises the analysis presented in the second iteration of the environmental deliverable, and it focuses on the environmental indicators. The assessment presents the data collected from the pilot sites as well as the recent updates of the methodology and results. It serves as background information and data for the final sustainability deliverable 8.12.

The objective of this section is to investigate the environmental performance of the automated minibus through mobility indicators. Sustainability indicators are a powerful tool to simplify, quantify, analyse, and communicate complex information (Innamaa & Salla, 2018; KEI, 2005; Singh et al., 2009). In addition, urban sustainability indicators are fundamental to support target setting and performance reviews and enable communication among policymakers, experts, and the general public (Shen et al., 2011; Verbruggen & Kuik, 1991).

The environmental indicators and respective units of assessment are presented in Table 13.9.

In addition to the environmental indicators, the indicators for the sustainability assessment comprehend the social, economic, governance, and system performance of the automated minibus (Nemoto et al., 2021).

Each indicator requires a specific methodology (refer to APPENDIX A), and the value of the indicators is represented on a scale of 1–5, with 5 being considered the best score. For each indicator we:

1. Defined a parameter.
2. Defined a scale, with minimum and maximum values considering the environmental impacts of main urban modes of transport, e.g. walking, cycling, small and big cars, and bus (freight transport and air transport were not comprehended, for example).
3. Calculated the indicator value for the automated minibus according to the demonstrator site.

**Table 13.9** The environmental indicators and units of assessment

| Environmental indicators | Unit of assessment |
|---|---|
| Energy efficiency | Automated minibus energy consumed for passenger per km (kWh/pkm) |
| Use of renewable energy | Automated minibus use phase, energy source, and percentage of renewable energy sources (%) |
| Noise pollution | Automated minibus traffic noise (dB) |
| Air pollution | Automated minibus emissions of air pollutants, PM levels (ug/m3), NOx, CO emissions |
| Climate change | Automated minibus GHG emissions: $CO_2$eq/pkm |

The results are presented on a spider chart, providing a disaggregated overview of the indicators. This allows for identifying the weaknesses and strengths of each indicator (WBCSD, 2015), also for a comparison between the pilot sites.

The limitations of the assessment concern the innovativeness of the automated minibus. The technology is still in a test and development phase. Hence, the main limitations concern the fact that the pilot projects are restricted to a local/neighbourhood area, and the automated minibuses drive in mixed traffic area at a low average speed (10–18 km/h). The automated minibuses drive on a fixed route (with the exception of 'Belle Idée' test site, where on-demand service has been tested), and the safety driver on board the automated minibus is required in case human intervention is required, as well as to report the performance of the automated minibuses in general.

These limitations reduce the performance and usability of the automated minibuses. In addition, the demonstrator sites have been facing constraints due to the COVID-19 pandemic. As a result, there have been interruptions in the pilot tests, and some transport companies have limited the maximum number of passengers to four during certain periods. This factor has a negative impact on the automated minibuses performance assessed by the environmental indicators.

The next section presents the results for the environmental indicators for five AVENUE demonstrator sites: Pfaffenthal and Contern (Luxembourg City), Groupama Stadium (Lyon, France), Ormøya (Oslo, Norway), and Nordhavn (Copenhagen, Denmark). The transport operators provided primary data; therefore, the results and analysis rely on the data presented in Table 13.4 (Chap. 3), which also provided inputs for the LCA study.

An overview of the pilot trials is present in Table 13.10, and the results per site are illustrated in Fig. 13.6, followed by analysis and conclusions.

The indicators addressing 'local air pollution' and 'local noise pollution' do not vary from site to site because they are assessed according to the vehicle. As an electric vehicle, the automated minibus has a good score on local air pollution. It is explained by the fact that BEVs in their use phase have zero exhaust emissions, e.g.

**Table 13.10** Overview of the AVENUE demonstrator sites

| City | Pilot | Characteristics of route | Type of passenger |
|---|---|---|---|
| Lyon | Groupama Stadium | Fixed route with stops 1.3 km Will become an on-demand, door-to-station service | Regular workers, people with reduced mobility (medical Centre nearby) |
| Copenhagen | Nordhavn | Fixed route with stops, 1,2 km, will become an on-demand, door-to-door service | Residents of the area, tourists |
| Oslo | Ormøya | Fixed route with stops, 3,6 km, | Residents of the area |
| Luxembourg | Contern | Fixed route with stops, on-demand. 2.2 km | Employees working at campus Contern |
| | Pfaffenthal | Fixed route with stops, on-demand 1.2 km | Workers, tourists, residents, and visitors of Luxembourg city |

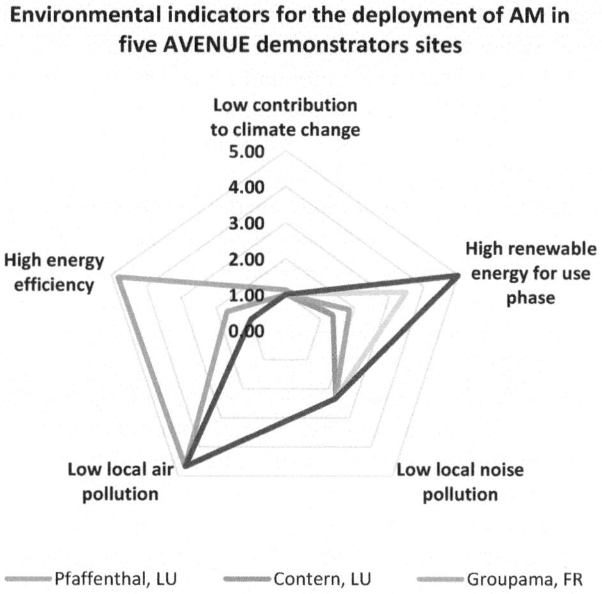


**Fig. 13.6** Environmental performance of the automated minibuses in the demonstrator sites. The scale ranges from 1 to 5, with 5 as the best score and 1 worst score

NOx and PM, and they just emit PM locally from road and tyre and brake wear, like other motor vehicles (European Environment Agency, 2018). The air pollutant emissions for the electricity generation to charge BEV batteries occur in power stations and tend to impact less densely populated areas (ibid). For this reason, the local air pollution emissions are assessed here for the use phase, as they affect cities (more densely populated areas) and consequently cause greater human exposure and potential health damage.

For local noise pollution, the automated minibus as an EV do not differ significantly from ICEV in the usual traffic and from 30 km/h speed. This is due to the fact that 'the tyre/road noise increases more with increasing speed than the propulsion noise, and therefore the tyre/road noise dominates the propulsion noise at high speeds' (Marbjerg, 2013). Therefore, the automated minibuses as an EV play a role to avoid local noise pollution for urban traffic during the night in low-speed areas (Jochem et al., 2016). Since the automated minibuses currently run at a low speed of 11–18 km/h, their noise pollution is slightly lower than ICEVs and lower than regular buses.

In all the pilot sites, the automated minibuses scored poorly for 'low contribution to Climate Change'. This indicator is highly affected by the low occupancy of the automated minibuses, due to the characteristics of the pilots, such as temporary and new services, the newness of the technology, as well as the interruptions of the trials due to the constraints of the COVID-19 pandemic and the reduction in mobility and in the use of public transport. Further, the climate change indicator is affected by the vehicle lifetime, total mileage, and electricity mix, as pointed by the LCA study (Chap. 3). From all the sites, Pfaffenthal (Luxembourg) presents a better performance due to the higher average of vehicle occupancy, while Ormøya (Oslo) and

Nordhavn (Copenhagen) present the lowest performance for climate change due to the very low average occupancy and low mileage in the case of Nordhavn.

Likewise, the energy efficiency indicator is directly impacted by the average occupancy of the automated minibus. Therefore Pfaffenthal (Luxembourg) presents a good score (with an average occupancy of 2 or 8 passengers), in contrast to the other sites.

The indicator of renewable energy for the use phase varies according to the share of energy from renewable sources in gross electricity consumption in each country. In this case, Nordhavn (Copenhagen) and Ormøya (Oslo) present a good score since Denmark and Norway have a share of energy from renewable sources in gross electricity consumption of 62% and 100%, respectively, in contrast to 9% in Luxembourg and 21% in France.

In relation to Chap. 2, it is worth noting that the pilot trial at Groupama Stadium (Lyon) comprehends V2I communication, meaning that three traffic light junctions operate in communication with the automated minibus. The V2X communications were not taken into account for the environmental indicators at this stage. However, on a larger scale, vehicle communications and connectivity could contribute to reducing energy impacts in mobility. Lee and Kockelman (2019), for example, pointed out that energy savings resulting from vehicle-to-infrastructure connectivity and smart intersections range from 6% to 30%, thanks to improvements in traffic interactions and better fuel-efficient driving (see more in Chap. 2).

The assessment of automated minibuses based on the environmental indicators point out that the automated minibuses face challenges to be deployed as an environmentally friendly mode of transport at the current stage.

A key factor targeting 'low contribution to climate change' is primarily to increase the vehicle occupancy and secondly through technology development, to increase the vehicle speed, mileage, and lifetime. Likewise, by aiming at a better energy efficiency, it is crucial to increase vehicle occupancy. Therefore, it is important that the automated minibuses are deployed in routes in order to cover real gaps in mobility, with more permanent services and good acceptance. And as mentioned previously, the average occupancy of the automated minibuses was also affected by the COVID pandemic, interruptions in the trials, and mobility restrictions.

The automated minibuses, as a BEV, can highly contribute to the reduction in local air pollution, and while targeting the reduction of local noise pollution, the automated minibuses present limited advantages in comparison to regular cars and buses, reducing noise during the night and at low-speed areas.

In the future, the automated minibus has the potential to be deployed as environmentally friendly mobility taking into account technological improvements, better social acceptance and usability, better integration into urban mobility as part of intermodal and MaaS systems, as well as shared and electric mobility.

The sustainability assessment study aims to set goals for the future deployment of the automated minibuses and therefore monitor the progress of the environmental and remaining sustainability indicators towards a more sustainable operation.

## 13.6 General Discussion and Conclusion

The different assessments in this deliverable reach interesting results on the current and future performance of automated minibuses. The energy demand analysis of the automated components in the automated minibus shows that the energy efficiency depends on the wireless technologies, the cooperative communication modes, and the implemented services. The automated technologies in the automated minibus, as deployed in the AVENUE pilots, are around 5% of the total energy used. The potential evolution of the energy demand depends on the different types of Internet connections. The driverless wireless network impacts the transmitted data and eventually the overall energy consumption or savings. Moreover, predictive and adaptive driving functions and information sharing in the automated minibus are likely to improve the acceleration and braking processes and hence contribute to overall energy savings. The energy demand also differs if the automated minibus provides an on-demand service. Overall, the energy-saving potential from predictive driving functions is highly likely to outweigh the energy consumption from data transmission energy.

Going beyond the energy analysis related to driving itself, the presented LCA study focuses on environmentally relevant energy flow and materials throughout the life cycle of the vehicle within the public transportation system. The study reiterates findings from other studies that energy consumption and the electricity mix used for charging have significant climate performance impacts. It also shows a low relevance of automated driving components in current deployment circumstances. The LCA study was used to identify factors that contribute to the environmental impact of automated minibuses. The environmental benefits of automated minibuses rely on the utilisation rate and occupancy factor and thus on their integration in the overall transportation system.

The current deployment of automated minibuses does not show significant environmental benefits, but future use cases are likely to improve substantially. The development and assessment of environmental indicators for an overall sustainability assessment corroborate the LCA study's conclusions. Occupancy, vehicle speed, mileage, and lifetime play an important role in reducing environmental impacts. Automated minibuses are particularly beneficial if they can be deployed to close former public mobility gaps, which would otherwise lead to the use of individual cars.

**Acknowledgements** This project has received funding by the European Union's Horizon 2020 research and innovation program under grant agreement No. 769033.

## References

3GPP. (2015). *Technical specification group services and system. Telecommunication management* (13th ed.). Valbonne. 3GPP TS 32.155.

3GPP (Ed.). (2018). *Technical specification group services and system. Aspects study on enhancement of 3GPP support for 5G V2X services* (16th ed.). 3GPP TR 22.886.

3GPP (Ed.). (2019). *Technical specification group services and system aspects. Enhancement of 3GPP support for V2X scenarios; stage 1* (16th ed.). 3GPP TS 22.186.

5GAA. (2020). *Working Groupd STandards and Spectrum. Study of spectrum needs for safety related intelligent transportation systems* ((1st). ed.). München. 5GAA S-200137.

Andrae, Anders/Edler, Tomas (2015). On global electricity usage of communication technology: Trends to 2030. Challenges 6 (1), 117–157. https://doi.org/10.3390/challe6010117.

Arvidsson, Rickard/Tillman, Anne-Marie/Sandén, Björn A./Janssen, Matty/Nordelöf, Anders/ Kushnir, Duncan/Molander, Sverker (2018). Environmental assessment of emerging technologies: Recommendations for prospective LCA. Journal of Industrial Ecology 22 (6), 1286–1294. https://doi.org/10.1111/jiec.12690.

Auer, Markus/Wiedemann, Jochen/Widdecke, Nils/Kuthada, Timo (2015). Increase of range of battery electric vehicles through thermal management. ATZ Worldwide 117 (7–8), 64–71. https://doi.org/10.1007/s38311-015-0037-6.

Bauer, Christian/Hofer, Johannes/Althaus, Hans-Jörg/Del Duce, Andrea/Simons, Andrew (2015). The environmental performance of current and future passenger vehicles: Life cycle assessment based on a novel scenario analysis framework. Applied Energy 157 (3), 871–883. https://doi.org/10.1016/j.apenergy.2015.01.019.

Baxter, Jared A., Merced, Daniel A./Costinett, Daniel J./Tolbert, Leon M./Ozpineci, Burak (2018). Review of electrical architectures and power requirements for automated vehicles, 2018 IEEE transportation electrification conference and expo (ITEC), Long Beach, CA, USA, pp. 944–949. https://doi.org/10.1109/ITEC.2018.8449961.

Belogaev, Andrey/Elokhin, Alexey/Krasilov, Artem/Khorov, Evgeny/Akyildiz, Ian F. (2020). Cost-effective V2X task offloading in MEC-assisted intelligent transportation systems. IEEE Access 8, 169010–169023. https://doi.org/10.1109/ACCESS.2020.3023263.

Bichiou, Y., & Rakha, H. (2020). Vehicle platooning: An energy consumption perspective. In *2020 IEEE 23rd international conference on intelligent transportation systems (ITSC)* (pp. 1–6). https://doi.org/10.1109/ITSC45102.2020.9294525

Bo, Zhu/Di, Wu/MinYi, Zheng/Nong, Zhang/Lin, He (2019). Electric vehicle energy predictive optimal control by V2I communication. Advances in Mechanical Engineering 11 (2), 168781401882152. https://doi.org/10.1177/1687814018821523.

Brown, A., Repac, B., & Gonder, J. (2013). *Autonomous vehicles have a wide range of possible energy impacts. NREL/PO-6A20-59210*. National Renewable Energy Lab. Available online at https://www.nrel.gov/docs/fy13osti/59210.pdf

Chen, Yuche/Sun, Ruixiao/Wu, Xuanke (2021). Estimating bounds of aerodynamic, mass, and auxiliary load impacts on autonomous vehicles: A powertrain simulation approach. Sustainability 13 (22), 12405. https://doi.org/10.3390/su132212405.

Chester, Mikhail V./Horvath, Arpad (2009). Environmental assessment of passenger transportation should include infrastructure and supply chains. Environmental Research Letters 4 (2), 24008. https://doi.org/10.1088/1748-9326/4/2/024008.

Connor, Wesley D./Wang, Yongqiang/Malikopoulos, Andreas A./Advani, Suresh G./Prasad, Ajay K. (2021). Impact of connectivity on energy consumption and battery life for electric vehicles. IEEE Transactions on Intelligent Vehicles 6 (1), 14–23. https://doi.org/10.1109/TIV.2020.3032642.

Deng, Yelin/Li, Jianyang/Li, Tonghui/Zhang, Jingyi/Yang, Fan/Yuan, Chris (2017). Life cycle assessment of high capacity molybdenum disulfide lithium-ion battery for electric vehicles. Energy 123, 77–88. https://doi.org/10.1016/j.energy.2017.01.096.

Lozano Domínguez, José Manuel/Mateo Sanguino, Tomás Jesús (2019). Review on V2X, I2X, and P2X communications and their applications: A comprehensive analysis over time. Sensors (Basel, Switzerland) 19 (12), 2756. https://doi.org/10.3390/s19122756.

Dong, P., Du, X., Sun, J., & Zhang, H. (2016). Energy-efficient cluster management in heterogeneous vehicular networks. In *2016 IEEE conference on computer communications workshops (INFOCOM WKSHPS)* (pp. 644–649). https://doi.org/10.1109/INFCOMW.2016.7562156

Duce, A. D., Egede, P., Öhlschläger, G., Dettmer, T., Hans-Jörg, A., Bütler, T., & Szczechowicz, E. (2013). *eLCAr: Guidelines for the LCA of electric vehicles*. https://doi.org/10.13140/RG.2.1.2782.8244

El-Rewini, Zeinab/Sadatsharan, Karthikeyan/Selvaraj, Daisy Flora/Plathottam, Siby Jose/Ranganathan, Prakash (2020). Cybersecurity challenges in vehicular communications. Vehicular Communications 23, 100214. https://doi.org/10.1016/j.vehcom.2019.100214.

EU Horizon 2020 EVC1000. (2018). Predictive control strategy based on enhanced connectivity level. D5.2. No. 824250.

European Environment Agency. (2018). Electric vehicles from life cycle and circular economy perspectives - TERM 2018. Accessed 18 Dec 2018.

Fagnant, D. J., & Kockelman, K. (2015). Preparing a nation for autonomous vehicles: Opportunities, barriers and policy recommendations. *Transportation Research Part A: Policy and Practice, 77*, 167–181. https://doi.org/10.1016/j.tra.2015.04.003

Fiori, Chiara/Ahn, Kyoungho/Rakha, Hesham A. (2018). Optimum routing of battery electric vehicles: Insights using empirical data and microsimulation. Transportation Research Part D: Transport and Environment 64, 262–272. https://doi.org/10.1016/j.trd.2017.08.007.

Fraunhofer, I. I. S. (2021). *Simulation platform for cellular V2X. C-V2XSim*. Institute for Integrated Circuits IIS. Available online at https://www.iis.fraunhofer.de/en/ff/kom/mobile-kom/simulation.html

Gawron, James H./Keoleian, Gregory A./Kleine, Robert D. de/Wallington, Timothy J./Kim, Hyung Chul (2018). Life cycle assessment of connected and automated vehicles: Sensing and computing subsystem and vehicle level effects. Environmental Science & Technology 52 (5), 3249–3256. https://doi.org/10.1021/acs.est.7b04576.

Gheorghiu, Razvan Andrei/Minea, Marius (2016). Energy-efficient solution for vehicle prioritisation employing ZigBee V2I communications. 2016 international conference on applied and theoretical electricity (ICATE), 1–6. https://doi.org/10.1109/ICATE.2016.7754691.

Gonder, Stephens/Lin, Chen/Gohlke, Liu (2016). Estimated bounds and important factors for fuel use and consumer costs of connected and automated vehicles. NREL/TP-5400-67216. Golden. National Renewable Energy Lab.. Available online at https://www.nrel.gov/docs/fy17osti/67216.pdf.

Greenblatt, Jeffery B./Shaheen, Susan (2015). Automated vehicles, on-demand mobility, and environmental impacts. Current Sustainable/Renewable Energy Reports 2 (3), 74–81. https://doi.org/10.1007/s40518-015-0038-5.

Grisoni, Anahita/Madelenat, Jill (2021). Le véhicule autonome: quel rôle dans la transition écologique des mobilités ? LA FABRIQUE ECOLOGIQUE.. Available online at https://www.lafabriqueecologique.fr/app/uploads/2020/02/Rapport-Complet_Ve%CC%81hicule-autonome-et-Transition e%CC%81cologique_La Fabrique Ecologique-Forum-Vies-Mobiles-1.pdf.

Hadjipaschalis, Ioannis/Poullikkas, Andreas/Efthimiou, Venizelos (2009). Overview of current and future energy storage technologies for electric power applications. Renewable and Sustainable Energy Reviews 13 (6–7), 1513–1522. https://doi.org/10.1016/j.rser.2008.09.028.

Hawkins, Troy R./Singh, Bhawna/Majeau-Bettez, Guillaume/Strømman, Anders Hammer (2013). Comparative environmental life cycle assessment of conventional and electric vehicles. Journal of Industrial Ecology 17 (1), 53–64. https://doi.org/10.1111/j.1530-9290.2012.00532.x.

Helmers, Eckard/Dietz, Johannes/Hartard, Susanne (2017). Electric car life cycle assessment based on real-world mileage and the electric conversion scenario. The International Journal of Life Cycle Assessment 22 (1), 15–30. https://doi.org/10.1007/s11367-015-0934-3.

Hetherington, Alexandra C./Borrion, Aiduan Li/Griffiths, Owen Glyn/McManus, Marcelle C. (2014). Use of LCA as a development tool within early research: Challenges and issues across different sectors. The International Journal of Life Cycle Assessment 19 (1), 130–143. https://doi.org/10.1007/s11367-013-0627-8.

Hu, Xiaosong/Wang, Hong/Tang, Xiaolin (2017). Cyber-physical control for energy-saving vehicle following with connectivity. IEEE Transactions on Industrial Electronics 64 (11), 8578–8587. https://doi.org/10.1109/TIE.2017.2703673.

Huber, Dominik/Viere, Tobias/Horschutz Nemoto, E., Jaroudi, I., Korbee, D., Fournier, G. (2022). Climate and environmental impacts of automated minibuses in future public transportation. Transportation Research Part D: Transport and Environment 102, 103160. https://doi.org/10.1016/j.trd.2021.103160.

Huss, A., Boksebeld, S., Wikström, N., Teuschl, G., & Jones, S. (Eds.). (2021). *Real-time predictive control utilizing route preview for hybrid & range extender vehicles*. FISITA World Congress.

Hussein, A. A./Rakha, H. A. (2020). Vehicle platooning impact on drag coefficients and energy/fuel saving implications. ArXiv.

Innamaa, S./Salla, K. (2018). Key performance indicators for assessing the impacts of automation in road transportation. Results of the trilateral key performance indicator survey. VTT.. Available online at https://cris.vtt.fi/en/publications/key-performance-indicators-for-assessing-the-impacts-of-automatio. Accessed 17 Jan 2020.

inria. (2019). Véhicules autonomes et connectés. Les défis actuels et les voies de recherche (2nd ed.). inria.

Jochem, Patrick/Doll, Claus/Fichtner, Wolf (2016). External costs of electric vehicles. Transportation Research Part D: Transport and Environment 42, 60–76. https://doi.org/10.1016/j.trd.2015.09.022.

Kanavos, A., Fragkos, D., & Kaloxylos, A. (2021). V2X communication over cellular networks: capabilities and challenges. *Telecom, 2*(1), 1–26. https://doi.org/10.3390/telecom2010001

KEI. (2005). Knowledge economy indicators, Work package 7. State of the Art Report on Simulation and Indicators.

Kemp, Nicholas J./Keoleian, Gregory A./He, Xiaoyi/Kasliwal, Akshat (2020). Life cycle greenhouse gas impacts of a connected and automated SUV and van. Transportation Research Part D: Transport and Environment 83, 102375. https://doi.org/10.1016/j.trd.2020.102375.

Korbee, D., Naderer, G., Nemoto, E. H., Frank, L., & Fournier, G. (2024). Who will accept travelling with automated shuttle buses? Defining target groups for potential users. *Transportation Research Part F; Travel Psychology and Behaviour*. (in press).

Krail, M. (2021). *On autopilot to a more efficient future? How data processing by connected and autonomous vehicles will impact energy consumption*. Fraunhofer. Available online at https://www.agora-verkehrswende.de/fileadmin/Projekte/2020/Automatisiertes_Fahren/Agora-Verkehrswende_On-Autopilot-to-More-Efficient-Future.pdf

Kubicka, M., Klusacek, J., Sciarretta, A., Cela, A., Mounier, H., Thibault, L., & Niculescu, S. I. (2016). Performance of current eco-routing methods. In *2016 IEEE intelligent vehicles symposium (IV)* (pp. 472–477). https://doi.org/10.1109/IVS.2016.7535428

Lee, Michael/Atkison, Travis (2021). VANET applications: Past, present, and future. Vehicular Communications 28 (9), 100310. https://doi.org/10.1016/j.vehcom.2020.100310.

Lee, J./Kockelman, K. M. (2019). Energy implications of self-driving vehicles. 98th annual meeting of the Transportation Research Board in Washington DC.. Available online at https://www.google.com/url?sa=t&rct=j&q=&esrc=s&source=web&cd=&ved=2ahUKEwj7vNGb2pvuAhUJDGMBHSZRDTwQFjABegQIBBAC&url=http%3A%2F%2Fwww.caee.utexas.edu%2Fprof%2Fkockelman%2Fpublic_html%2FTRB19EnergyAndEmissions.pdf&usg=AOvVaw1e17rb1ozoFRJsM8-uo1Do.

Liu, Z., Tan, H., Kuang, X., Hao, H., & Zhao, F. (2019). The negative impact of vehicular intelligence on energy consumption. *Journal of Advanced Transportation*, 1–11. https://doi.org/10.1155/2019/1521928

Liu, Y., Tight, M., Sun, Q., & Kang, R. (Eds.) (2019). A systematic review: Road infrastructure requirement for Connected and Autonomous Vehicles (CAVs).

Majeau-Bettez, Guillaume/Hawkins, Troy R./Strømman, Anders Hammer (2011). Life cycle environmental assessment of lithium-ion and nickel metal hydride batteries for plug-in hybrid and battery electric vehicles. Environmental Science & Technology 45 (10), 4548–4554. https://doi.org/10.1021/es103607c.

Marbjerg, G. (2013). Noise from electric. A literature survey. Vejdirektoratet. Available online at https://www.google.com/url?sa=t&rct=j&q=&esrc=s&source=web&cd=&ved=2ahUKEwiG37W-1LHtAhUJhRoKHT1KDQ4QFjADegQIBRAC&url=https%3A%2F%2Fwww.vejdirektoratet.dk%2Fapi%2Fdrupal%2Fsites%2Fdefault%2Ffiles%2Fpublications%2Fnoise_from_electric_vehicles_0.pdf&usg=AOvVaw1ksLoAa2OmmfiYsj1WTR_O

Masoudi, Meysam/Khafagy, Mohammad Galal/Conte, Alberto/El-Amine, Ali/Francoise, Brian/Nadjahi, Chayan/Salem, Fatma Ezzahra/Labidi, Wael/Sural, Altug/Gati, Azeddine/Bodere, Dominique/Arikan, Erdal/Aklamanu, Fred/Louahlia-Gualous, Hasna/Lallet, Julien/Pareek, Kuldeep/Nuaymi, Loutfi/Meunier, Luc/Silva, Paulo/Almeida, Nuno T./Chahed, Tijani/Sjolund, Tord/Cavdar, Cicek (2019). Green mobile networks for 5G and beyond. IEEE Access 7, 107270–107299. https://doi.org/10.1109/ACCESS.2019.2932777.

Moreno Ruiz, E., Valsasina, L., FitzGerald, D., Symeonidis, A., Turner, D., Müller, J., Minas, N., Bourgault, G., Vadenbo, C., Loannidou, D., & Wernet, G. (2020). *Documentation of changes implemented in the ecoinvent database v3.7 & v3.7.1 (2020.12.17)*. Ecoinvent Assoc.

Navya. (2018). AUTONOM® SHUTTLE. For more fluid mobility. Available online at https://navya.tech/wp-content/uploads/documents/Brochure_Shuttle_EN

Nemoto, Eliane Horschutz/Issaoui, Roukaya/Korbee, Dorien/Jaroudi, Ines/Fournier, Guy (2021). How to measure the impacts of shared automated electric vehicles on urban mobility. Transportation Research Part D: Transport and Environment 93, 102766. https://doi.org/10.1016/j.trd.2021.102766.

Nordelöf, Anders/Romare, Mia/Tivander, Johan (2019). Life cycle assessment of city buses powered by electricity, hydrogenated vegetable oil or diesel. Transportation Research Part D: Transport and Environment 75 (2), 211–222. https://doi.org/10.1016/j.trd.2019.08.019.

Noroozi, Mohammad/Moghaddam, Hanieh Rastegar/Shah, Ankit/Charkhgard, Hadi/Sarkar, Sudeep/Das, Tapas K./Pohland, Timothy (2023). An AI-assisted systematic literature review of the impact of vehicle automation on energy consumption. IEEE Transactions on Intelligent Vehicles 8 (6), 3572–3592. https://doi.org/10.1109/TIV.2023.3268300.

Noussan, Michel/Tagliapietra, Simone (2020). The effect of digitalization in the energy consumption of passenger transport: An analysis of future scenarios for Europe. Journal of Cleaner Production 258 (2), 120926. https://doi.org/10.1016/j.jclepro.2020.120926.

Oliveira, Luis/Messagie, Maarten/Rangaraju, Surendraprabu/Sanfelix, Javier/Hernandez Rivas, M., van Mierlo, J. (2015). Key issues of lithium-ion batteries—From resource depletion to environmental performance indicators. Journal of Cleaner Production 108, 354–362. https://doi.org/10.1016/j.jclepro.2015.06.021.

Passafiume, Marco/Collodi, Giovanni/Cidronali, Alessandro (2020). Design principles of batteryless transponder for vehicular DSRC at 5.8 GHz. IEEE Journal of Radio Frequency Identification 4 (4), 491–505. https://doi.org/10.1109/JRFID.2020.3020398.

Pihkola, Hanna/Hongisto, Mikko/Apilo, Olli/Lasanen, Mika (2018). Evaluating the energy consumption of mobile data transfer—From technology development to consumer behaviour and life cycle thinking. Sustainability 10 (7), 2494. https://doi.org/10.3390/su10072494.

Puig-Samper Naranjo, G., Bolonio, D., Ortega, M. F., & García-Martínez, M.-J. (2021). Comparative life cycle assessment of conventional, electric and hybrid passenger vehicles in Spain. *Journal of Cleaner Production, 291*, 125883. https://doi.org/10.1016/j.jclepro.2021.125883

Saujot, Mathieu/Sartor, Oliver/Brimont, Laura (2017). Comment accélérer la mobilité durable avec le véhicule autonome ? Iddri.. Available online at https://www.iddri.org/sites/default/files/import/publications/ib0217_veicule-autonome_ms-os-lb.pdf.

Shaheen, Susan/Bouzaghrane, Mohamed Amine (2019). Mobility and energy impacts of shared automated vehicles: A review of recent literature. Current Sustainable/Renewable Energy Reports 6 (4), 193–200. https://doi.org/10.1007/s40518-019-00135-2.

Shen, Li-Yin/Jorge Ochoa, J./Shah, Mona N./Zhang, Xiaoling (2011). The application of urban sustainability indicators—A comparison between various practices. Habitat International 35 (1), 17–29. https://doi.org/10.1016/j.habitatint.2010.03.006.

Singh, Rajesh Kumar/Murty, H. R./Gupta, S. K./Dikshit, A. K. (2009). An overview of sustainability assessment methodologies. Ecological Indicators 9 (2), 189–212. https://doi.org/10.1016/j.ecolind.2008.05.011.

Song, Haokun/Zhao, Fuquan/Zhu, Guangyu/Liu, Zongwei (2023). Impacts of connected and autonomous vehicles with level 2 automation on traffic efficiency and energy consumption. Journal of Advanced Transportation 2023, 1–15. https://doi.org/10.1155/2023/6348778.

Stephens, T. S. (2016). *Estimated bounds and important factors for fuel use and estimated bounds and important factors for fuel use and consumer costs of connected and automated vehicles*. National Renewable Energy Lab.

Stephens, T. S., Auld, J., Chen, Y., Gonder, J., Kontou, E., Lin, Z., Xie, F., Mohammadian, A., Shabanpour, R., & Gohlke, D. (2019). *Assessing energy impacts of connected and automated vehicles at the U.S. national level—Preliminary bounds and proposed methods*. In G. Meyer & S. Beiker (Eds.), *Road Vehicle Automation 5* (Lecture Notes in Mobility). Springer. https://doi.org/10.1007/978-3-319-94896-6_10

Verbruggen, H., & Kuik, O. (1991). *Indicators of sustainable development: An overview*. Springer.

Volvo. (2021). *Volvo FH. Product guide*. Volvo.

Wadud, Zia/MacKenzie, Don/Leiby, Paul (2016). Help or hindrance? The travel, energy and carbon impacts of highly automated vehicles. Transportation Research Part A: Policy and Practice 86, 1–18. https://doi.org/10.1016/j.tra.2015.12.001.

WBCSD. (2015). Methodology and indicator calculation method for sustainable urban mobility. Available online at https://www.eltis.org/sites/default/files/trainingmaterials/smp2.0_sustainable-mobility-indicators_2ndedition.pdf

# Chapter 14
# Environmental Impact Assessment: Externalities of Automated Electric Vehicles for Public Transport

**Ines Jaroudi, Adrian Boos, Tobias Viere, and Guy Fournier**

**Abstract** This chapter provides a comprehensive analysis of the economic and environmental externalities associated with the use of automated minibuses in public transport, using a mixture of scenario planning and an externality model in the context of the AVENUE project cities. By analysing six different deployment scenarios, including the substitution of automated minibuses for buses and private cars, this study sheds light on the potential shifts in external costs and benefits. This chapter carefully assesses the impact of the deployment of automated minibuses on reducing external costs, taking into account factors such as energy efficiency, connectivity, automation features passenger numbers and vehicle utilisation rates. The results show that the environmental and economic outcomes of deploying automated minibuses depend significantly on the specific deployment strategies, highlighting scenarios in which automated minibuses could either reduce or exacerbate external costs. Through a detailed assessment of these scenarios, the chapter provides a nuanced understanding of how the strategic integration of automated minibuses into urban transport systems can influence the broader goals of economic sustainability and environmental protection. The study emphasises the importance of aligning automated minibus deployment strategies with city-specific goals and the broader sustainability agenda and provides valuable insights for policymakers, urban planners and transport stakeholders.

I. Jaroudi (✉)
Fraunhofer ISI, Karlsruhe, Germany

Pforzheim University, Pforzheim, Germany

CentraleSupélec, LGI, Gif-sur-Yvette, France
e-mail: ines.jaroudi@isi.fraunhofer.de

A. Boos · T. Viere
Pforzheim University, Pforzheim, Germany

G. Fournier
Pforzheim University, Pforzheim, Germany

CentraleSupélec, LGI, Gif-sur-Yvette, France



G. Fournier et al. (eds.), *Automated Vehicles as a Game Changer for Sustainable Mobility*, Contributions to Management Science,
https://doi.org/10.1007/978-3-031-61681-5_14

## 14.1 Introduction

The following assessment addresses six scenarios for potential deployment in AVENUE cities and the resulting external costs. It focuses on the potential increase or decrease in externalities as an indicator for the scenarios in which mobility should evolve. The summary of categories, impacts and methods using the work of van Essen et al. (2019), Jochem et al. (2016), Héran and Ravalet (2008), Fagnant and Kockelman (2018), and Shalkamy et al. (2015) is shown in Table 14.1.

In this chapter, the scenarios planning methodology (intuitive logic approach, key factors and driving forces) as well as the six selected scenarios (replace all buses, replace all cars, expand the network, targeted expansion of the network, robotaxis, AM in MaaS) are explained and the externality categories are introduced. First, the externalities and scenarios model is applied in detail to the case study of Geneva, where the mobility behaviour of the city, the results in terms of savings or losses per scenario and their implications are presented. Then, the results are discussed, such as the limitations of the analysis, comparison between the scenarios, potential policy recommendations, and rebound effects. Afterwards, it is applied to the other three cities of AVENUE (Luxembourg, Lyon and Copenhagen). In a final stage, a comparative analysis is conducted between the cities.

**Table 14.1** Recapitulation of the categories, effects and methods of externalities

| Categories | Impacts | Methods |
|---|---|---|
| Tank to wheel (ttw) air pollution (PM2.5, PM10, NOx, SO2 and NMVOC) | Health effects, crop losses, building damage and biodiversity loss | Damage cost estimation |
| Tank to wheel (ttw) climate change (CO2, N2O and CH4) | Sea-level rise, biodiversity loss, water management difficulties, extreme weather conditions and crop failures | Avoidance cost approach |
| Well-to-tank (wtt) aggregated emissions | Similar to TTW | Similar to TTW |
| Noise | Annoyances, health effects caused by road traffic noise | Willingness-to-pay, burden of disease approach |
| Accidents | Material damage, production losses and administrative, medical and human costs | Damage cost approach |
| Congestion | Not meeting passengers' mobility demand due to the temporary scarcity of infrastructure (traffic flow reaches its capacity) | Delay cost and deadweight loss approaches |
| Parking space | Urban space in km2 | Based on Héran and Ravalet (2008) |
| Production phase | Climate change emissions | Avoidance cost approach |

## 14.2 Scenarios

### *14.2.1 Methodology*

Scenario planning is a way to imagine potential paths in the future (Derbyshire & Wright, 2017). The scenarios are built using the intuitive logic approach (ILA). ILA defines driving forces (political, economic, technological, ecological, social and legal) as well as key factors that help structure the scenarios. These factors are either quantitative and predictable, such as demographics, while others are qualitative and less predictable, such as user acceptance and policies (Huss & Honton, 1987; Lindgren & Bandhold, 2009). The ILA strength lies in its flexibility (Zmud et al., 2015).

For this study, the driving forces help answer how and why the AM might be deployed in each scenario; they are defined, based on Townsend (2014) and Milakis et al. (2017) scenarios, as follows:

Technology advancement (automation technology and digital services).
Urban policy (political agenda for mobility and sustainability).
Transportation offer (trends of use and modes available).
The users (most likely to use the AM in the scenario).

As for the key factors, they were determined through a deliberative process within the AVENUE team. The trial sites, the interdisciplinary nature of the project, and the work of (Beukers, 2019; Korbee et al., 2021; Viere et al., 2021) were used to select the following key factors:

Whether the AM are replacing one mode of transport or multiple modes
Whether AM supports or competes with public transport (PT) (replace buses modal share or not)

These two factors help estimate the modal shift within each scenario, which is crucial to estimating the future transportation performance in person-km (pkm) based on mobility censuses and the consequent externality estimations.

The setting of key forces and key factors is combined with the system boundaries of each scenario. Figure 14.1 shows the classification of the scenarios based on the key factors.

The boundaries are defined by the key parameters of the vehicles, circulation specifications and AM service. Thus, the analysis is continued as a qualitative assessment that studies the direct and indirect consequences of the modal shift from each scenario using observations from AVENUE pilot studies and previous research on AV deployment. The overall structure is presented in Fig. 14.3, and it is the basis for the scenario description. The methodology helps limit the uncertainty of the scenarios by building plausible stories (Lindgren & Bandhold, 2009; Amer et al., 2013). See Fig. 14.2 for the steps of the scenario description and Fig. 14.3 for the methodology:

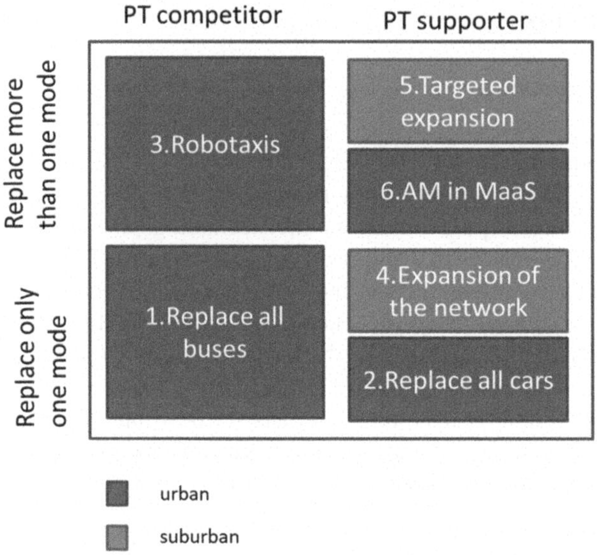


**Fig. 14.1** The classification of scenarios based on key factors

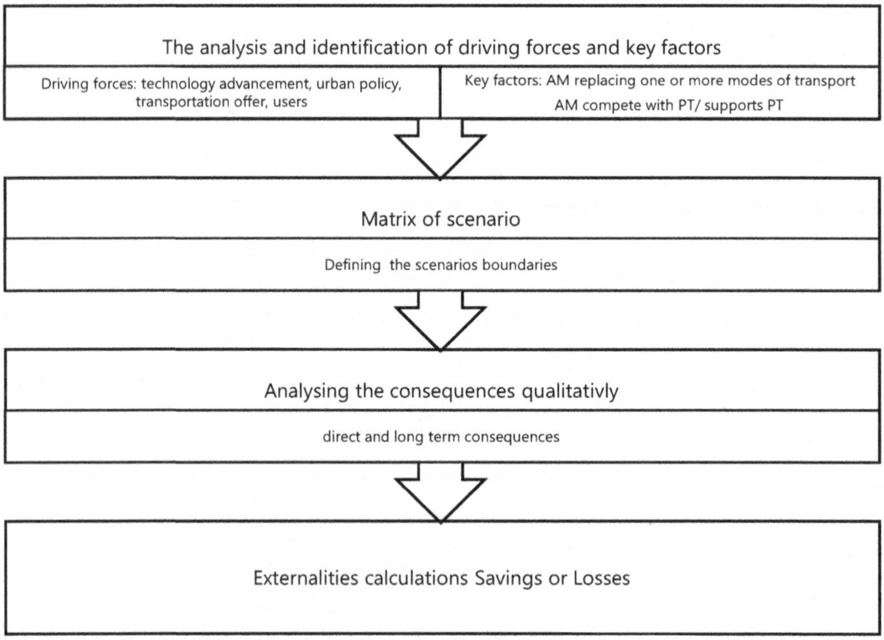


**Fig. 14.2** Steps of scenario building

Methodology framework

**System Boundaries (assumptions)**

Area urban / suburban level

**Vehicle and circulation specification**

Capacity factors

Speed limits in the area

Traffic situation: average congestion on roads

Vehicles lifetime

**AM service**

Intermodal and seamless, first/last mile, on-demand and door-to-door

**Scenario development (circumstances): Key factors (to define the scenario in the transportation system)**

It supports public transport or competes with public transport

Replaces only one mode of transport, or it replaces more than one modal share

**Scenario development (circumstances) Driving forces (to frame the scenario):**

Technology development (level of automation SAE, Connectivity V2X, digital services)

Regulatory conditions (policies: supportive or restrictive or other)

Transportation offer (description of other modes and dominance of private or public stakeholders)

Users (profile of AM users)

**Modal shift**

The predicted AM modal share from absorbing other modes of transport modal share(based on the survey and literature)

**The qualitative assessment: Direct consequences**

**Environmental and Societal impacts**

Emissions

Accidents

Congestion

Noise

**Impacts on mobility**

Travel demand - Increased demand due to on-demand service,

Road capacity-levels of congestion during off/on peak hours, the value of time)

**Impacts on infrastructure**

Parking

**The qualitative assessment: Long term consequences (rebound effects)**

**Impacts on mobility**

On the transportation system (other secondary modal shifts mobility behaviour, induced demand-due to the convenience of service)

**Impacts on infrastructure (urban development)**

Roads, city plans, jobs density, demographics

**Fig. 14.3** Methodology of scenario description

## 14.2.2 *The Scenario Description*

Based on the methodology described, the analysis focuses on six scenarios that concern the deployment of AM in cities. At this level, the scenarios occur in a standard EU city (to be applied to Geneva, Lyon, Luxembourg and Copenhagen) and in the next decade, which is a safe estimation of when is it most likely to have AV technology on the roads (Milakis et al., 2017).

Out of the six scenarios, five focus on different modal shifts caused by the deployment of AM, while one focuses on robotaxis. Among the overall scenarios, four are set in the city centre, while two are in smaller urban dwellings (characterised by smaller urban densities) surrounding urban areas like villages, towns and suburban areas.

The scenario description using the proposed methodology helps pinpoint the advantages and disadvantages of each deployment strategy. It also seeks to learn about potential obstacles and catalysers to reduce the environmental footprint of introducing AV in urban areas. It is reinforced by a deliberative process with the AVENUE experts (Antonialli et al., 2021), where the most plausible scenarios are drafted based on the key factors and driving forces.

### 14.2.2.1 Scenario 1: Replace All Buses (Sc1)

This scenario occurs in the city centre. AM are implemented on a fixed schedule, yet they operate on flexible routes within a geofenced area.

According to Hansen et al. (2021), geofencing involves the utilisation of GPS or similar location-based technologies to define virtual boundaries or zones within which AM can function. These boundaries are typically determined by either the service provider or regulatory authorities. The minibuses are meant to support the public transport network, mainly rail transport, by replacing diesel buses. The AM are more flexible and less costly (Milakis et al., 2017).

Technological advancements such as achieving level 4/5 automation, improved sensory capabilities and AV platooning support the replacement (Wadud et al., 2016). The transportation system at the time of introducing the AM does not differ significantly from the status quo. The electrification process of the bus fleet is limited; motorised individual mobility is dominant.

The replacement might lead to a ripple effect that reduces car use and increases train ridership in the long term. Integrating electric powertrains in AM would lead to a decrease in both air pollution and greenhouse gas emissions compared to the typical buses. It is also wise to predict a change in the building environment, where more roads and train stations could be redesigned to assimilate the change. The city would assign more pick-up and drop-off points around the stations. Eventually, modernising the public transport and integrating AM in ITS will make the city centre more attractive, thus attracting more jobs and increasing the population (Narayanan et al., 2020).

**Table 14.2** Occupancy factors (van Essen et al., 2019)

| Vehicle | Occupancy factors |
|---|---|
| Car | 1.6 |
| Bus | 19.0 |
| AM | 6.0 |
| Robotaxi | 1.2 |

#### 14.2.2.2 Scenario 2: Replace All Cars (Sc2)

This scenario is also set in a high-density urban area such as a city centre. The AM are deployed to replace all cars trips. The service of the AM is on-demand, door-to-door, and running on flexible routes. The AM would have on average six people per vehicle; see Table 14.2. We assume no restrictions on waiting time (for a waiting time of more than 4 min, there is no need for an increase in fleet size to meet an increase in travel demand as well as empty runs based on Fagnant and Kockelman (2018).

It is also set to support public transport and only replace cars by introducing policies restricting the circulation of internal combustion engine vehicles (ICEV). The scenario is supported by policies that mirror the current environmental agenda of sustainable cities, such as the zero-emission strategy or carbon tax, presenting the ultimate goal for sustainable mobility (Fournier et al., 2020; ITF, 2015; ITF, 2017; ITF, 2020).

The replacement of cars with automated vehicles such as the AM is bound to have a significant effect on accidents rates and traffic congestion, as explained by Childress et al., 2015; Alazzawi et al., 2018; Auld et al., 2017; Litman, 2021. Less AM are needed to meet the travel demand. This leads to an increase in road capacity and thus a decrease in road traffic, especially during peak hours (Wadud et al., 2016). On the one hand, the AM would be integrated better with other long-distance transportation such as trains as AM are used more by travellers. On the other hand, they might reduce the active transportation modal share because the AM provides on-demand and door-to-door short-distance trips (Milakis et al., 2017).

#### 14.2.2.3 Scenario 3: Robotaxis (Sc3)

Robotaxis are described as shared automated vehicles in numerous studies (Alazzawi et al., 2018; Fagnant & Kockelman, 2018; Jones & Leibowicz, 2019; Litman, 2021). Although they might be comparable to the AM in terms of services (on-demand, door-to-door), they differ in the occupancy factors, speed, vehicle size and integration with PT (Merlin, 2017). The AM is a bigger vehicle that could carry up to 15 passengers, it rarely provides single-ride trips and it requires longer waiting times for pick-up. On the contrary, the robotaxis are destined mostly for single ridership. They are operated by private stakeholders, they could drive faster and they have reduced waiting times. They are convenient, especially if the passenger privileges their privacy (UITP, 2017).

For this scenario, the robotaxis serve the city centre as well. They do offer door-to-door and on-demand trips but no ridesharing services. They are competing with public transport, replacing more than one mode of transport (cars, buses and walking).

The regulatory conditions are described best as a "Laissez-faire" outcome. This means that there are no policies to regulate the AV market. Private stakeholders are seeking to maximise their profit which would have unpredictable consequences on sustainable mobility and people's welfare (Niles, 2019). The regulatory conditions also translate into a deteriorating public transport offer that manifest in a high dependency on individual motorised mobility. A laissez-faire approach is likely to result in a rise in traffic volume and a potential increase in pkm. Concurrently, anticipated enhancements in network efficiency, often assumed, may not materialise without effective government intervention (Cohen & Cavoli, 2018).

The trend of AV markets driven by shareholders' interest would create a race to optimise the services: higher speeds, less waiting times and more vehicles; therefore an increase in overall vehicle travelled km "VKM" is expected.

The long-term consequences of this scenario on the mobility system are the modal shift from active modes of transport and public transportation. Thus, this scenario is expected to cause induced demand as a rebound effect, and it could even reduce public transport ridership as it is very convenient (Litman, 2022; Niles, 2019; UITP, 2017). Furthermore, urban planning follows car-centric strategies, where the building environment is designed to accommodate private vehicles rather than the people. The spread of AV means new roadway design features such as improved lane markings, signs designed to be read electronically and wireless repeaters in tunnels to provide internet access (Childress et al., 2015).

#### 14.2.2.4 Scenario 4: Expand the Network (Sc4)

In this scenario, the AM are deployed to support public transportation in low-density urban areas such as small villages and suburban areas where the PT offer is limited compared to more urbanised areas. The service provided by the AM is seamless intermodal trips. The modal shift is to replace one mode of transportation, mainly cars. The technological development is similar to previous scenarios. The policies in place that enable the expansion of the transportation network with AM are drafted to increase accessibility, reduce reliance on individual vehicles, and develop surrounding areas to cities. The modal shift is partially from cars to AM. The direct repercussions of the scenario, in terms of environmental and societal impacts, should be similar to the "replace all cars" scenario. The direct effect on mobility is that the AM is more susceptible to making empty runs and carrying fewer passengers on board. The AM are meant to attract car users. Thus they should provide services that could compete with the comfort and convenience of an individual vehicle.

For the indirect consequences, we predict that the AM would increase overall PT ridership. However, it is difficult to determine the effect on walking and biking, but as it offers short trips, it could also replace some walking and biking. The improvement in the public transport network outside of densely populated areas would make them more attractive to inhabitants.

#### 14.2.2.5 Scenario 5: Targeted Expansion of the Public Transport Network (Sc5)

The targeted expansion scenario is similar to scenario 4 "expand the network". In this scenario, the AM are supporting the public transport network. It also partially replaces car modal share. The prominent feature in this scenario is that it also replaces a share of the buses. Specifically, the AM are deployed to replace night buses and low occupancy buses. The bus service is at capacity (or even over capacity) during peak hours but underutilised during the day (Pyddoke, 2020).

Moreover, it runs frequently empty in areas with high car ownership (Adra et al. (2004). Hence, substituting these buses with AM could reduce the environmental impact. This scenario's direct and indirect consequences are similar to scenario 4.

#### 14.2.2.6 Scenario 6: AM in MaaS (Sc6)

The AM are deployed within MaaS to better provide on-demand services that bridge the first and last mile and provide seamless and intermodal trips. They are deployed in highly dense areas such as city centres. Their introduction aims to support PT. They are positioned to influence more than one mode of transport (cars, walking and biking). The technological innovations in AV are similar to the previous scenarios. However, there are significant advancements in digital on-demand services, interoperability, ticketing, utilising mobile apps, the cloud, ride-pooling and routing algorithms (Plested, 2021).

The regulations to support this deployment strategy rely on public and private collaboration for MaaS services, platform management, open API, and data sharing. Other regulations in place are similar to scenario 2 "replacing all cars", where the city is seeking to implement more sustainable practices in line with the sustainable urban mobility plan (SUMP) and smart city initiatives. They adopt fuel and parking measures and push and pull regulations (in line with transport demand management (TDM)) to prevent the use of ICEV in the city centre and reduce the environmental and societal impact. The public transportation offer is efficient and reliable. However, there are gaps connecting travellers to mobility hubs (e.g. tram and metro stations). Thus, the AM seeks to capture first- and last-mile travellers that would have driven to reach a train or a tram station. The modal shift to be studied in this scenario concerns the share of journeys within an intermodal trip that connects to or from a train/tram station. The deployment of AM would reduce emissions.

We consider that for the AM to meet the travel demand and remain competitive the waiting time is less than 4 min. Hence, there is an increase in VKM due to pooling and rerouting to pick-up and drop-off passengers (ITF, 2020; Jones & Leibowicz,

2019; Milakis et al., 2017; Moreno et al., 2018). Reduction in parking space is expected (Zakharenko, 2016).

The long-run consequences of this scenario is an increase in PT ridership, as the AM provide seamless intermodal and last-mile trips and is considered as a mobility gap filler. However, the convenience of the service could replace more short-distance trips from walking and biking (Shen et al., 2018).

#### 14.2.2.7 General Comments

It is important to note that the speed in all scenarios is limited to 30 km/h for all road transportation. The average occupancy factors are those of the Delft report (van Essen et al., 2019) for traditional road transportation. Also, we follow the Huber et al. (2022) assessment where the AM occupancy factor is five passengers. We assume an occupancy factor of 1.2 for the robotaxis which is more comparable to a normal taxi where individual trips are significant. These factors vary in suburban scenarios to translate its deployment particularities. Table 14.2 presents the different occupancy factors.

### *14.2.3 Modal Shifts*

The modal shift is essential to determine the direct and long-term consequences in the qualitative description above. Moreover, it is a key component to the externality estimations. Thus, some assumptions were needed to determine the AM modal share in each scenario. The modal shift explanation and resources for each scenario are in Table 14.3.

**Table 14.3** Modal shift for the scenarios

| | | | |
|---|---|---|---|
| Sc1 | Scenario 1 | Replace bus share | – |
| Sc2 | Scenario 2 | Replace car share | – |
| Sc3 | Scenario 3 | −20% from cars<br>−6% from buses<br>−13% walking | Ward et al. (2019); Clewlow & Gouri, (2017); Heineke et al. (2019) |
| Sc4 | Scenario 4 | Depends on willingness question | |
| Sc5 | Scenario 5 | Depends on willingness question<br>4% from buses | Mancret-Taylor and Boichon (2015), Adra et al. (2004) |
| Sc6 | Scenario 6 | Depends on willingness question<br>Based on journey to a train station in intermodal trips<br>3% from walking<br>1.7% from biking | Paydar et al. (2020), Giansoldati et al. (2020), Gebhardt et al. (2016) |

### *14.2.4 The Representative Survey*

The social impact assessment is based on surveys to study the potential acceptance of users and the trends related to AM use. In this assessment, we focus on the willingness questions that were part of the representative surveys. These specific questions help determine the potential modal shift in some scenarios.

Scenario: AM in MaaS: The modal share of cars to be absorbed by the AM is determined by using the question: "how willing are you to give up your car if AM offers a service that bridges the first and last mile?". The modal share corresponds to the percentage of respondents who were very willing, who consider their residential area as a city centre and who use their cars daily.

Scenario: Expand the network and targeted expansion of the network: the modal share of cars to be absorbed by the AM is determined by using the question: "How willing are you to give up your car if AM is part of a seamless, intermodal trip?". The modal share corresponds to the percentage of respondents who were very willing and who consider the area they live in to be either a big town or a small to medium village and who use their cars daily.

## 14.3 The Externality Methodology

### *14.3.1 Production Phase*

The manufacturing of vehicles is a complex process; it accounts for steps from the extraction of raw materials to the production of the components (Chester, 2008; Pero et al., 2018).

The assessment focuses on the climate change impact on the production phase. The emissions are CO2, CH4, and N2O; it is based on climate avoidance costs. According to van Essen et al. (2019), the avoidance costs are determined by averaging values from the literature for the short and medium term (up to 2030) and the long term (2040–2060). The values used in this estimation are central and short- to medium-run climate avoidance which is 100 €/tCO2e.

The marginal cost is estimated as follows:

$$\text{Marginal cost for production}\left(\text{¢}-\frac{\text{cent}}{\text{pkm}}\right) \equiv \frac{\begin{array}{c}\text{total emissions of CO2e}\\ \text{during production}\left(\text{tCO2e}\right)\end{array}}{\begin{array}{c}\left(\text{expected lifetime mileage}\left(\text{km}\right)\\ \times\ \text{the average occupancy}\\ \times 100 \times \text{central short run value}\left(\text{¢}\right)\end{array}}$$

**Table 14.4** Marginal cost for the production phase

| Vehicle | Bus | Car | Electric car | AM |
|---|---|---|---|---|
| Marginal cost in €-cent/pkm | 0.18 | 0.38 | 0.64 | 0.3 |

The total emissions, as well as the expected lifetime mileage, were determined using the ecoinvent database for the car and electric car, while for the bus, the values were taken from Chester and Horvath (2009). Finally, Huber et al. (2022) as well as Viere et al. (2021) study on the LCA of the AM were used. More details about the estimations are found in the environmental impact assessment chapter. The marginal costs are presented in Table 14.4. These values are applicable at the EU level.

### *14.3.2 Parking Space*

In the analysis, only the savings from parking spots for private vehicles are considered. The buses are usually stored and maintained in dedicated bus garages or depots, and they do not interfere with daily traffic. Although it raises an issue for land-use and transportation planning, it presents less nuisance to cities compared with private vehicles and thus, it is not covered in this assessment, similar to the AM (Lai et al., 2013).

To determine the overall parking space, the fleet size for cars and the AM is needed in each scenario. Thus, the fleet size calculator that is presented in the previous chapter is used. Moreover, we assume a linear relation between the parking space reduced and the modal share of car trips reduced.

## 14.4 Assumptions and Boundaries

To be able to calculate the external costs and build on work presented in Antonialli (2019) and Jaroudi et al. (2021) assessment, we assume the following:

The AM deployment could affect motorised mobility, public transport and active mobility (Janasz, 2018). The AM are introduced in mixed traffic (with no prior presence of automated technology on the roads).

Every additional person-kilometre travelled on the AM is travelled less on the other transportation modes. Therefore, the total transport performance remains identical compared to the reference scenario (Bubeck et al., 2014).

Active mobility's negative externalities as walking and biking are considered negligible (Keall et al., 2018).

The study does not account for the increase in transport performance due to population increase and other sociodemographic changes; we are focused on the effect of unpredictable factors such as the integration within the transportation system and policies. That is why we chose to simplify the calculations by omitting their effects (Huss & Honton, 1987).

Intermodal trips consist of two or more modes of transport (car/bus/walking/biking+train) (Fraedrich et al., 2015), while monomodal or unimodal trips are considered as one mode.

Walking trips accounted as a mode of transport for intermodal trips from 600 m or more (more than 5 min) (Gebhardt et al., 2016).

The average speed of circulation for all vehicles ranges from 25 to 30 km/h.

The AM operate on "other urban roads" or "other interurban roads" based on the scenario, and we estimate that the average traffic flow[1] is near capacity.

## 14.5 Application of the Externalities Model to Geneva

In the following part, the assessment applies this model by adjusting the marginal costs (€-cent per pkm) to the context of the city and estimating the future transportation performance in person-km (PKM). It considers the four cities of AVENUE as a case study. First, Geneva is studied in detail to showcase the specifics of the scenarios and the externalities calculations. Second, the results are presented for Luxembourg, Lyon and Copenhagen, applying the steps shown in Fig. 14.4.

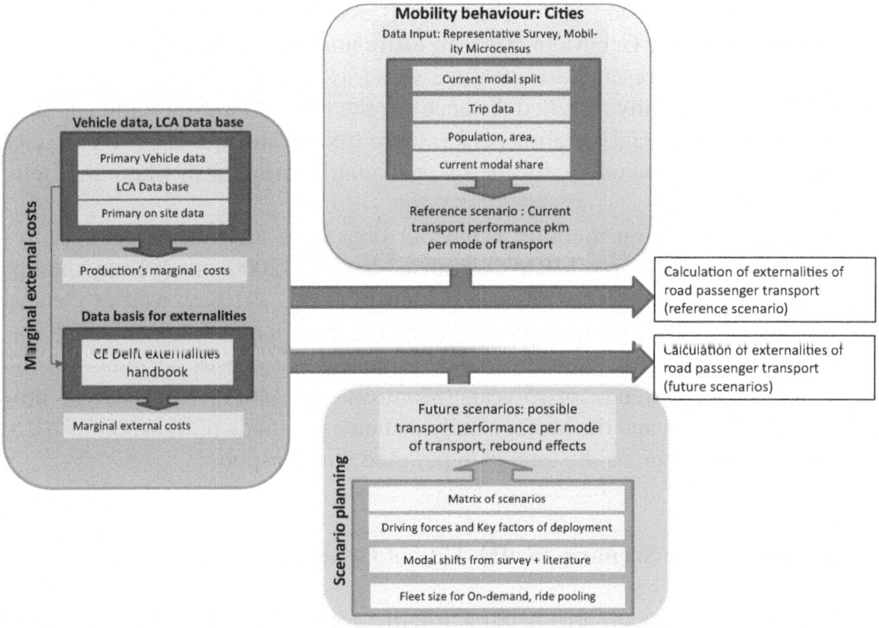


**Fig. 14.4** Externalities calculations and data methodology

---

[1] Traffic flow = volume of traffic/capacity of a traffic on a link ("near capacity" is v/c between 0.8 and 1, "congested" between 1 and 1.2, "over capacity" is above 1.2)

### 14.5.1 The Mobility Behaviour Profile in Geneva (National Census Data)

Geneva is the second most populated city in Switzerland, with 197,376 inhabitants. The "Mobilités 2030" long-term strategy aims to address mobility challenges in Geneva Canton, such as underdeveloped public transport in low-density areas, narrow city centre streets and the absence of tangential routes. The promotion of low-carbon mobility is part of the Climate Plan 2018–2022 in Geneva Canton. Automated driving could replace motorised individual mobility, promote active modes of transport and make the transportation system more robust against crises. The legal framework for autonomous vehicles is being developed, and the "Digital Switzerland" strategy will address the necessary digital key factors for autonomous driving implementation (Etat de Genève, 2013; Jenelius & Cebecauer, 2020; OFCOM, 2018).

### 14.5.2 Mobility Behaviour: The Reference Scenarios

To calculate the externalities, we refer to the current status with no AV on the roads, as mentioned in assumptions and boundaries. Based on the data available, we used the mobility census for Geneva in 2015. To distinguish between our urban and suburban scenarios, we present two reference scenarios; each is characterised by the modal split, average daily travelled distance per the mode of transportation, area in km2, population, and transportation performance (pkm). The first one is for the city of Geneva, while the second is for the second suburban ring, also where the Belle Idee pilot is operated.

The analysis relies on mobility behaviour data from a census (microrecensements mobilité et transports (MRMT)) realised between 2000 and 2015 by the statistical office of the Canton of Geneva (Montfort et al., 2019) as well as Geneva Public Transport-TPG (2016). The MRMT is an extension of a Suisse national census comprising the participation of 4500 residents. The census accounts for the evolution of mobility trends for 15 years (Montfort et al. (2019)). The analysis provides trip distribution and the average daily distance per mode of transport in 2015. These values help estimate the total pkm per mode of transport.

#### 14.5.2.1 Reference Scenario (Sc01): City of Geneva

This scenario is the comparison point for the scenarios "replace all cars" and "replace all buses", "Robotaxis" and "AM in MaaS". The trip distribution and the average daily distance per the mode of transport in 2015 are computed in Table 14.5. The population is 197,376 in the city of Geneva (Rietschin, 2015). The area in the city of Geneve is 15.93 km2 (Service de la mensuration officielle, 2005). The overall daily trips amount to 711,000 trips (Montfort et al., 2019). The modal split is estimated using Montfort et al. (2019) and TPG (2016) analysis for

**Table 14.5** Mobility behaviour and transport performance in the city of Geneva 2015

| | Mode of transport | Total transport performance in million pkm = Transport performance * population | Transport performance (pkm) | Average daily distance (km) | Modal share (trips) |
|---|---|---|---|---|---|
| Private mode | Private cars | 972.57 | 4927.50 | 13.50 | 22.60% |
| Public mode | Bus | 389.03 | 1971.00 | 5.40 | 12.30% |
| | Tram metro | – | – | 0 | 6.40% |
| Active mobility | Biking | 72.04 | 365.00 | 1 | 6.80% |
| | Walking | 187.31 | 949.00 | 2.6 | 47.60% |
| Other | | 72.04 | 365.00 | 1 | 4.30% |

**Table 14.6** Marginal and total costs in €-cent per pkm in city of Geneva

| Externality category | Car (marginal) | Bus (marginal) | Car (total) | Bus (total) |
|---|---|---|---|---|
| Air pollution | 0.63 | 0.76 | 6.09 | 2.94 |
| Climate change | 1.31 | 0.44 | 12.72 | 1.70 |
| Wtt | 0.42 | 0.19 | 4.09 | 0.74 |
| Noise | 1.92 | 0.84 | 18.67 | 3.27 |
| Accidents | 1.35 | 1.62 | 13.13 | 6.30 |
| Congestion | 39.30 | 6.50 | 382.22 | 25.29 |
| Production | 0.38 | 0.18 | 3.68 | 0.71 |
| Total externalities per mode of transport | | | 440.61 | 40.95 |
| Total externalities—Reference scenario | 481.56 | | | |

the mobility in Geneva. These values are also used to estimate the new transport performance for the scenarios. The estimated transport performance and modal split is in Table 14.5.

Using annual transport performance from Table 14.5 and the marginal costs for buses and cars for Switzerland from Table 14.6, we estimate the total external costs for road passenger transport for Geneva in 2015.

The total external costs in 2015 are important because they represent the reference point that is used to compare the impacts of the AM introduction and calculate the potential increase or decrease in external costs (before and after the deployment of the AM).

#### 14.5.2.2 Reference Scenario (Sc02): Second Couronne

This area has an overall population of 26,102 inhabitants and an area of 162 km2 (Federal Statistical Office, 2013; Service de la mensuration officielle, 2005). Based on Montfort et al. (2019), we estimate around 94,000 daily trips for all the

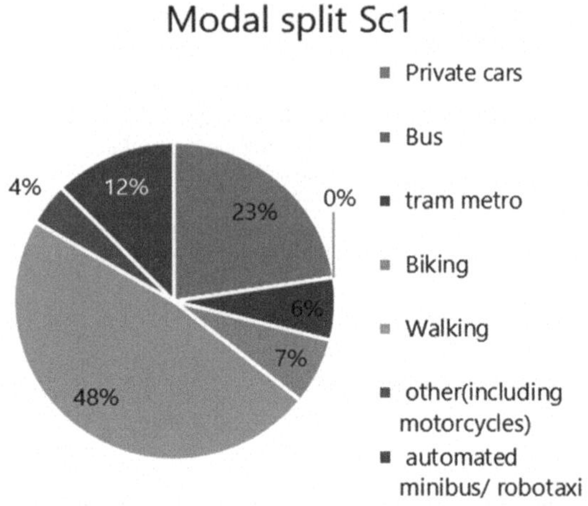


**Fig. 14.5** Modal shift for the scenario: replacing all buses in Geneva

**Table 14.7** The mobility behaviour and transport performance for the second suburban ring in Geneva, 2015

| Mode of transport | | Total transport performance in million pkm | Transport performance (pkm) | Average daily distance in km | Modal share |
|---|---|---|---|---|---|
| Private mode | Private cars | 240.1 | 9198.00 | 25.2 | 43.9% |
| Public mode | Bus | 39.1 | 1496.50 | 4.1 | 9.504% |
| | Tram metro | 19.1 | 730.00 | 2 | 4.896% |
| Active mobility | Biking | 7.6 | 292.00 | 0.8 | 4.5% |
| | Wa lking | 21.0 | 803.00 | 2.2 | 31.5% |
| | Other | 20.0 | 766.50 | 2.1 | 5.7% |

inhabitants in the second couronne consisting of 11 municipalities. The modal split is in Fig. 14.5.

The mobility behaviour and the transportation performance in Table 14.7 present the values used to estimate the new pkm per mode of transportation used in the scenarios.

In this scenario, to better reflect the reduced public transportation offer, we adjust the occupancy factors. The bus has an average ten passengers on board. The estimation is based on the higher dependency on cars and a lower modal share of public transport, in general, in suburban areas compared to city centres (high-density population areas): The car modal share is 44% compared to 22.3% in the city centre. And the buses' modal share is 10%, while it is 15% in the city of Geneva. We assume that the AM are circulating on interurban other roads. The marginal costs are adjusted accordingly and included in Table 14.8, as well as the total external costs (Total external costs (2015), $Y = \sum_i \sum_j xij$, (2015)).

**Table 14.8** Marginal and total costs in €-cent per pkm in second suburban ring in Geneva

| Externality category | Car (marginal) | Bus (marginal) | Car (total) | Bus (total) |
|---|---|---|---|---|
| Air pollution | 0.63 | 1.43 | 1.50 | 0.56 |
| Climate change | 1.31 | 0.83 | 3.14 | 0.33 |
| Wtt | 0.42 | 0.36 | 1.01 | 0.14 |
| Noise | 1.92 | 1.60 | 4.61 | 0.62 |
| Accidents | 1.35 | 3.08 | 3.24 | 1.20 |
| Congestion | 25.50 | 7.98 | 61.22 | 3.12 |
| Production | | | 0.91 | 0.14 |
| Total externalities per mode of transport | | | 75.64 | 6.11 |
| Total externalities | 81.74 | | | |

These values are used to estimate the increase or decrease in external costs in the suburban scenarios: "expand the network" and "targeted expansion of the network".

In the following part, the scenarios in the city centre are analysed. Afterwards, those in suburban parts are presented. The assessment relies on data from the marginal costs in Table 14.4, Table 14.6, and Table 14.8, scenario description and the reference scenarios (see Fig. 14.4) by applying the steps from the externalities model.

## 14.5.3 Scenarios in City Centre

### 14.5.3.1 Replace All Buses (Sc1)

Following, the scenario "replace all buses" is tested in the city of Geneva. First, the marginal costs for the vehicles are presented in Table 14.9 based on Table 14.6 and the marginal costs for the AM. Second, the AM are deployed to replace the 12.3% bus trips share. The estimated new modal share is shown in Fig. 14.5.

Similarly, we assume that the AM conduct the trips of the buses with the same average distance of the bus trips from the reference scenario. This led to a prediction of the transport performance and, consequently, the total externalities.

Scenario (Sc1) leads to an increase in external costs compared to (Sc01), mostly due to congestion. The increase in externalities equals around +12 million euros; see Fig. 14.6. If the congestion is omitted, the scenario shows a decrease in externalities of −11 million euros. The wtt costs more in terms of external costs, but all the other environmental impacts are considered positive if the buses are replaced with AM (see Fig. 14.6). However, this is still offset by almost +23 million lost due to congestion. The congestion is also explained by the occupancy of an AM being

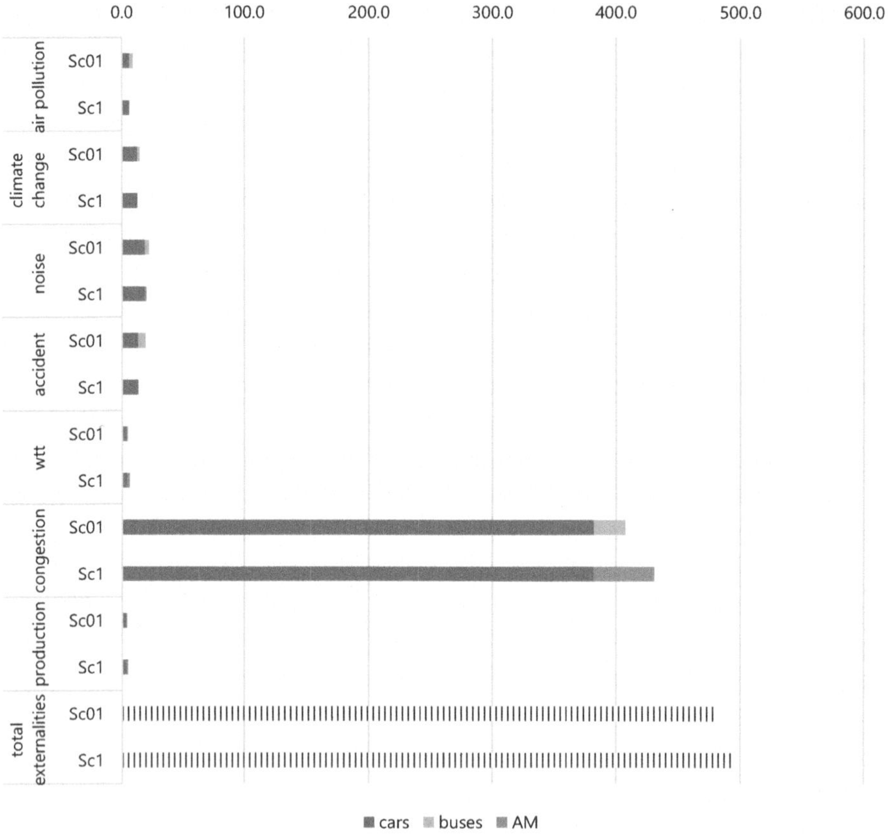


**Fig. 14.6** The increases or decreases in externalities with congestion in million euros in Sc1

lower than that of a bus, as well as the difference in the sizes of the vehicles (both factors interfere in the estimation of the marginal cost of congestion). In this scenario, there is no estimation for parking space saved since it focuses on bus replacement (Fig. 14.7).

#### 14.5.3.2 Replace all Cars (Sc2)

The marginal costs are those in Table 14.9 (from Sc1). We replace the car modal share in the city of Geneva (i.e. 22.6%) with the traffic circumstances of 2015 (Fig. 14.8).

Replacing all cars with AM leads to a reduction in all external cost categories except the wtt costs, where there is an increase of +3.4 million euros. The wtt emissions increase could be justified by reliance on electricity generated from fossil fuels (Kasten et al., 2016). In general, this scenario could save up to 308 million

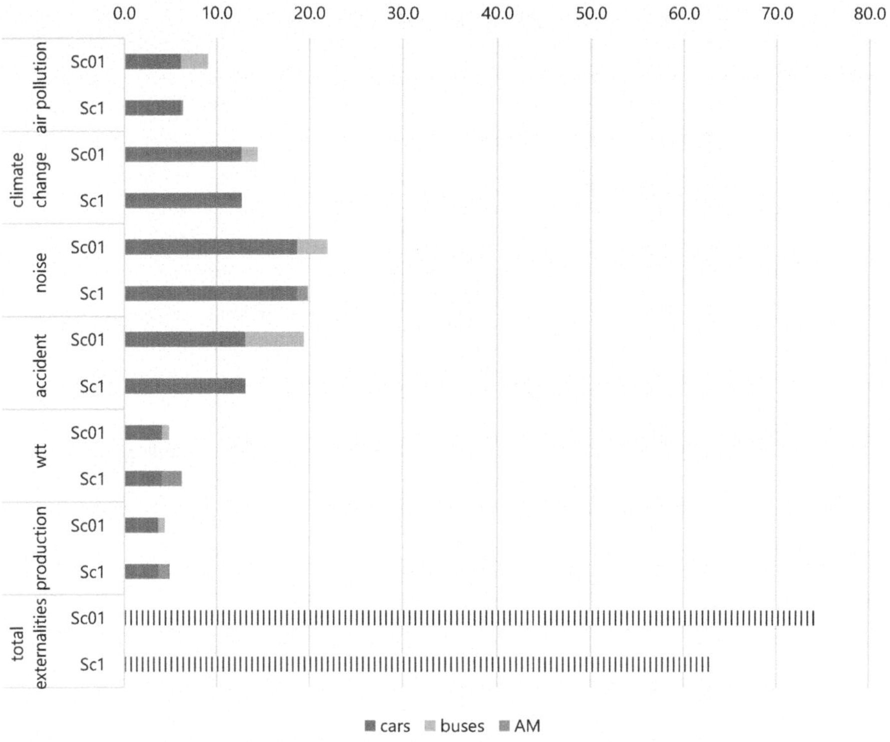


**Fig. 14.7** The increases or decreases in externalities without congestion in million euros in Sc1

**Table 14.9** Marginal costs (car, bus, and AM) in €-cent per pkm in Geneva city

| Externality category | Car | Bus | AM |
|---|---|---|---|
| Air pollution | 0.63 | 0.76 | 0.05 |
| Climate change | 1.31 | 0,44 | – |
| Wtt | 0.42 | 0.19 | 0.54 |
| Noise | 1.92 | 0.84 | 0.30 |
| Accidents | 1.35 | 1.62 | – |
| Congestion | 39.30 | 6.50 | 13.10 |

euros, with the majority coming from the congestion savings (−210 million euros); see Figs. 14.9 and 14.10.

The environmental impact of replacing all cars with the AM is reflected in gains of around 40 million euros. The findings are aligned with most studies that support a shift from individual motorised mobility toward electric and shared transportation. Using the fleet calculator, the estimated fleet size needed to replace all car trips in the city of Geneva with AM is around 1380 minibuses for a waiting time superior to

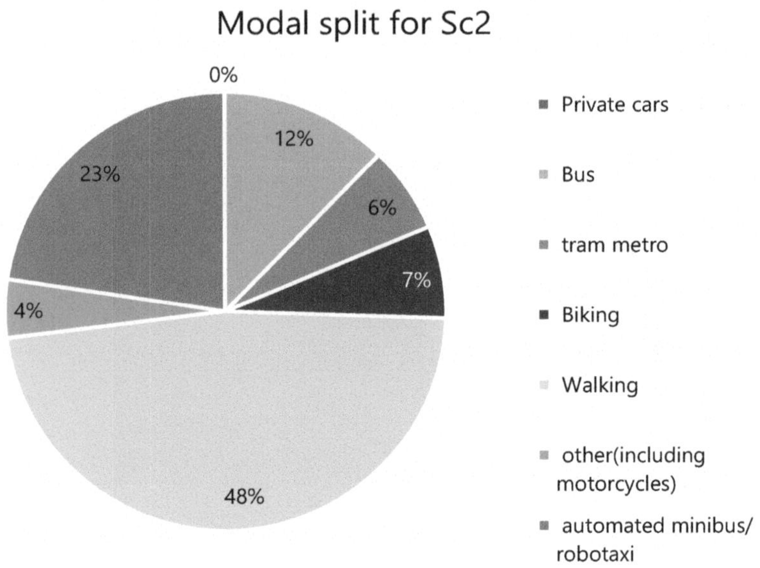


**Fig. 14.8** Modal shift for scenario Sc2: replacing all cars in Geneva

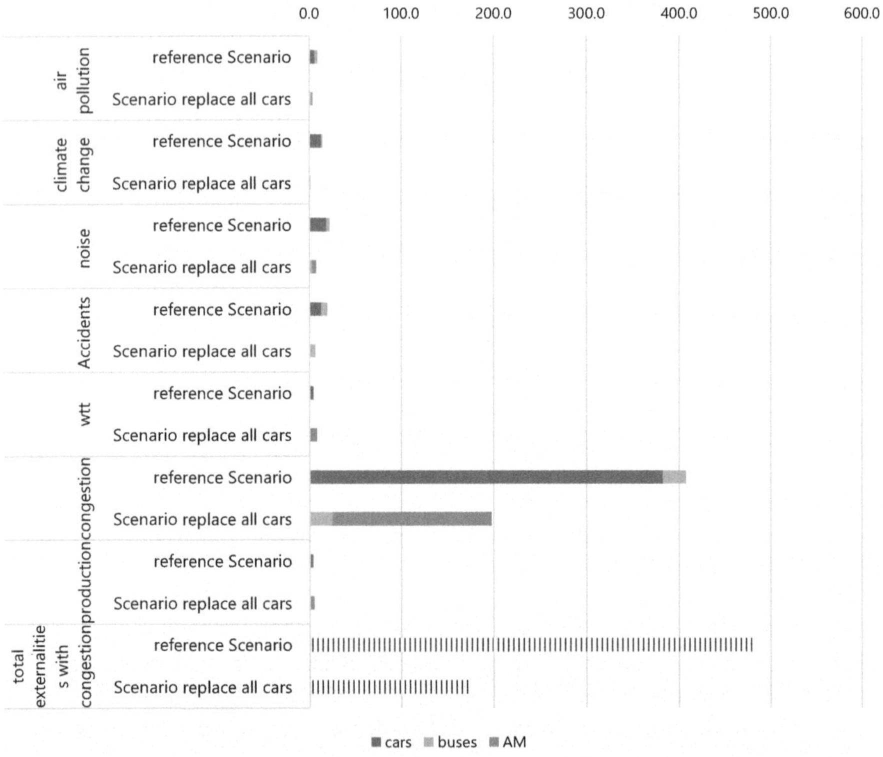


**Fig. 14.9** The increases or decreases in externalities with congestion in million euros in Sc2

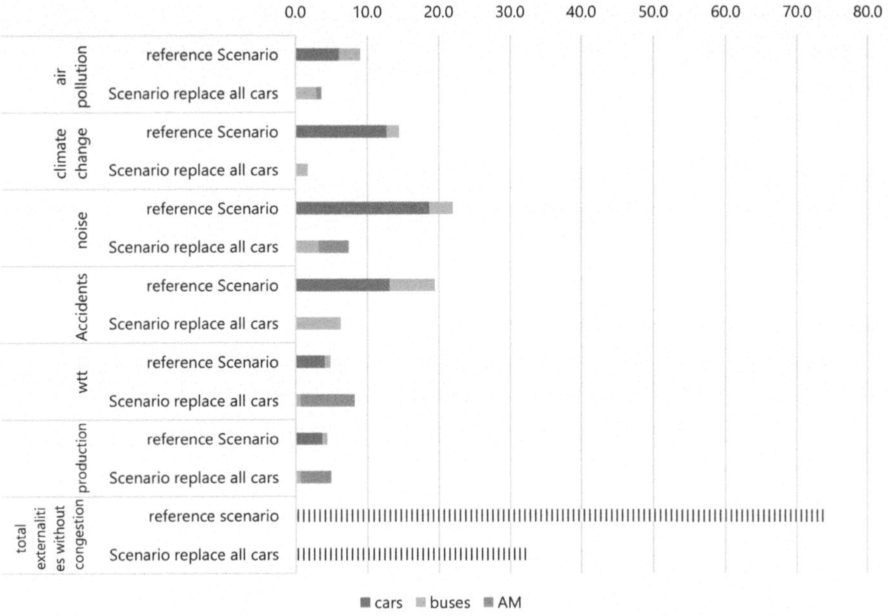


**Fig. 14.10** The increases or decreases in externalities without congestion in million euros in Sc2

4 min. Finally, the savings in parking space is 0.65 km2, the equivalent of 64,824 parking spots.

#### 14.5.3.3 Scenario 3: Robotaxis (Sc3)

This scenario focuses on AV's integration in the transportation network in the form of robotaxis. The marginal costs for the robotaxis depend on van Essen et al. (2019) estimations for electric vehicles (EV). In this scenario, we estimate an occupancy factor for the robotaxi of 1.2. The marginal costs in the Delft assessment for the electric vehicle use an estimation of an occupancy factor of 1.6 (as well as for an ICEV, Table 14.2). Thus, the marginal costs for the different external costs categories are adjusted for an occupancy factor of 1.2 (new marginal cost for robotaxi = marginal cost for EV x 1.6/1.2).

The robotaxis' marginal costs are presented in Table 14.10. Furthermore, to better translate the effect of the low occupancy factors, we assume that the robotaxis are operating on congested roads.

The robotaxis modal share is explained in Table 14.11, the new modal split of the scenario in Fig. 14.11. 20% of car modal share (~22%) means the AM will absorb around 5% of car trips and, similarly, a 6% reduction in bus share means around 1% of trips replaced by AM. Finally, for the walking trips, we assume the new modal share of walking is 37% (~48%–13% = 37%).

The deployment of robotaxis in the city centre with a laissez-faire outcome causes a significant increase in external costs compared to Sc01. In fact, it is estimated to cost around +162 million euros in terms of environmental and societal

**Table 14.10** Marginal costs for scenario robotaxis in €-cent per pkm

| Externality category | Car | Bus | Robotaxis |
|---|---|---|---|
| Air pollution | 0.63 | 0.76 | 0.07 |
| Climate change | 1.31 | 0.44 | – |
| Wtt | 0.42 | 0.19 | 1.11 |
| Noise | 1.92 | 0.84 | 1.92 |
| Accidents | 1.35 | 1.62 | – |
| Congestion | 48.50 | 8.00 | 64.67 |

**Table 14.11** Robotaxis modal share composition

| | | | |
|---|---|---|---|
| Robotaxi modal share 18.2% | 4.78 | % (reduced from cars) | Ward et al. (2019); Zhong et al. (2020) |
| | 13.0 | % (reduced from walking) | Fournier et al. (2020) |
| | 0.79 | % (reduced from buses) | Mishra et al. (2015) |

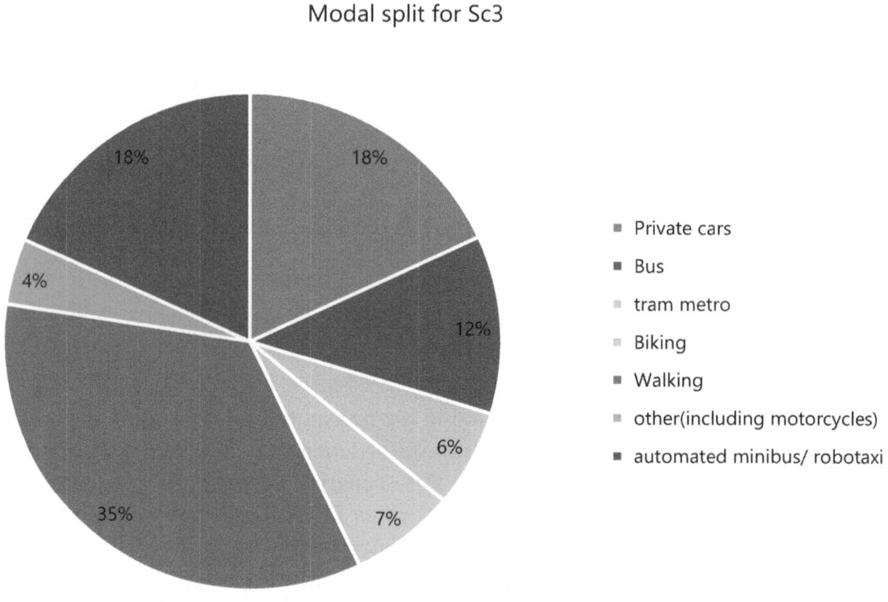


**Fig. 14.11** Modal shift for scenario Sc3: robotaxis in Geneva

impacts (see Fig. 14.12). Figure 14.13 is used to gauge the environmental impacts. The scenario also leads to decreased gains in the environmental and accidents categories except for the noise and wtt categories. The overall difference in external costs compared to Sc01 without the congestion is around one million euros.

Accounting for the increased travel demand, the fleet calculator by Fournier et al. (2020) shows a fleet of 1058 robotaxis for an on-demand service with a waiting time of 1 min. The 1058 robotaxis could cover 18.2% of all daily trips within

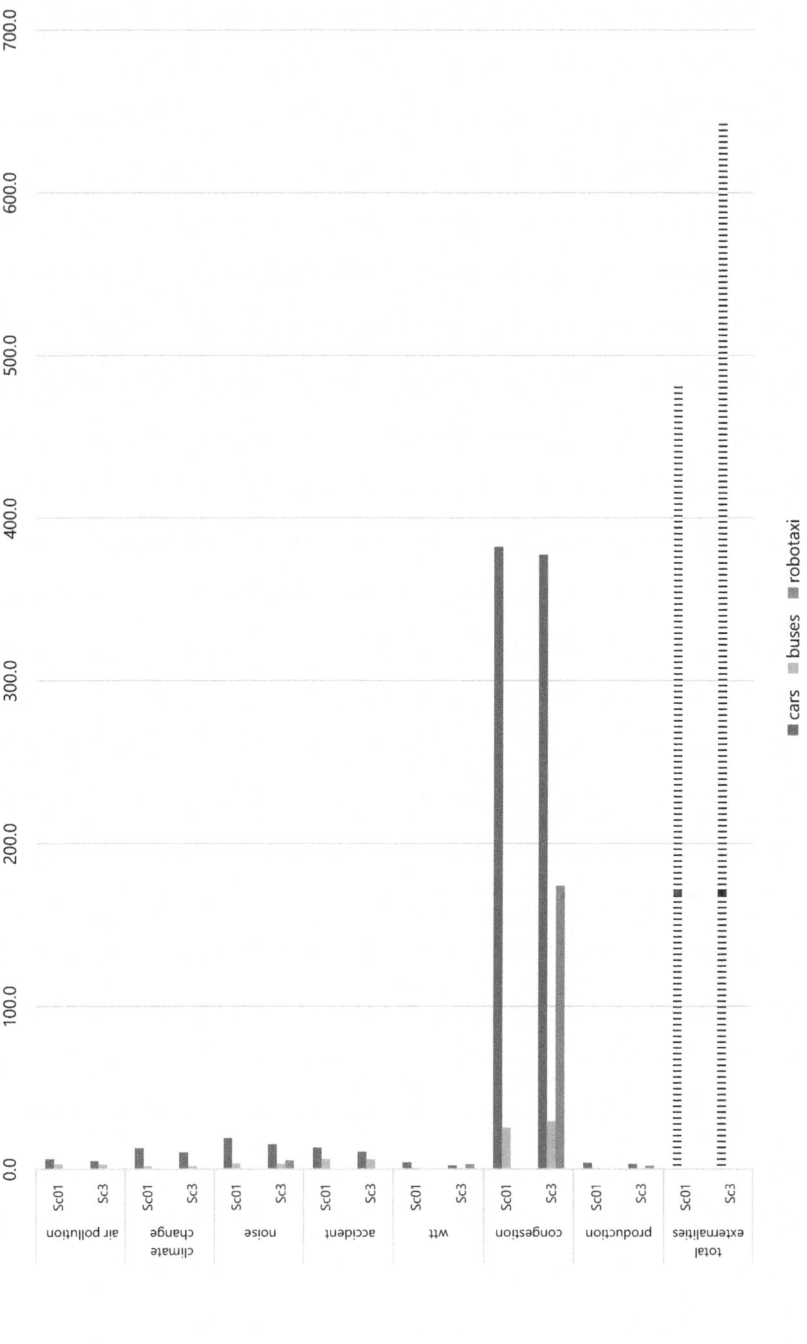


**Fig. 14.12** The increases or decreases in externalities with congestion in million euros in Sc3

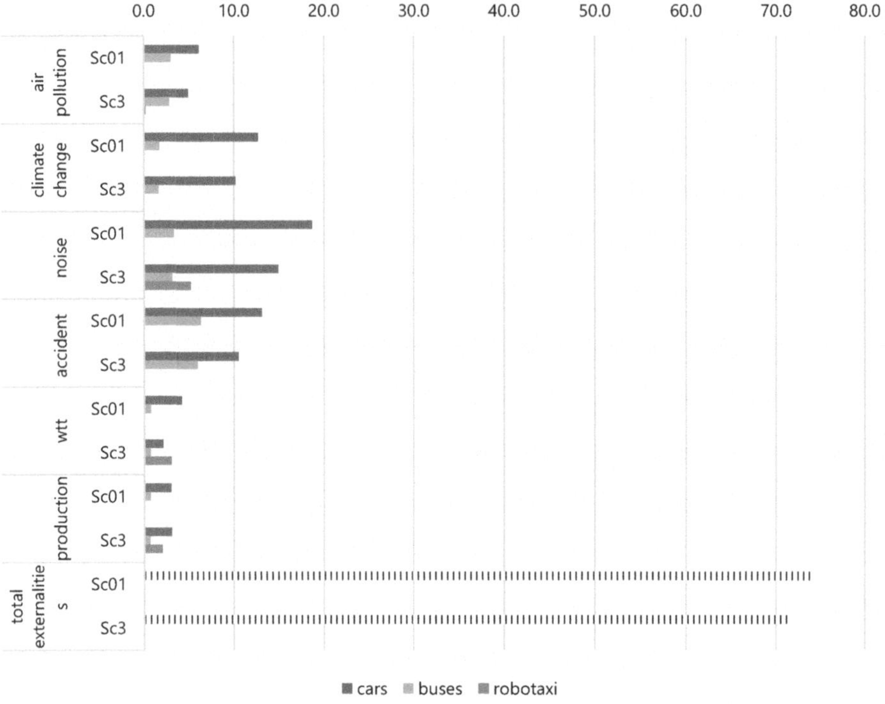


**Fig. 14.13** The increases or decreases in externalities without congestion in million euros in Sc3

a geofenced area of 15.93 km2. We follow the simulation results from (Bischoff & Maciejewski, 2016) since the simulation has similar conditions to Sc3; thus, one robotaxi replaces ten cars. Hence the new car fleet is 54,624 cars compared to the original 64,824 cars (Bischoff & Maciejewski, 2016). The 10,200 parking spots saved result in an additional 0.1 km2 of available space. We do not account for parking space for robotaxis based on numerous studies that assume the robotaxis are to be stored off roads in special facilities (AV parks) to be maintained and charged or would be circulating all day (Hayes, 2011; ITF, 2015; Nourinejad et al., 2018; Zhang & Guhathakurta, 2017). Moreover, these garages are not part of the public domain since they belong to the private stakeholders (the robotaxis are operated by private stakeholders; see 5.2.2, scenario 3 robotaxis); hence, they would not affect public infrastructure.

Finally, we test the results for when the robotaxis are a part of a pooling service. Based on Alazzawi et al. (2018), we account for an occupancy factor of 2.4 instead of 1.2. The subscenario with ridesharing registers an increase in externalities that accounts for around 76 million euros. Alternatively, if we consider Mosquet et al. (2015) estimation for an average of four people per vehicle, the increase is around 41 million euros. See Table 14.12 below.

This shows a correlation between the occupancy factors and the externalities. Hence, ridesharing could contribute to a better environmental and societal impact when deploying robotaxis.

#### 14.5.3.4 Scenario 6: AM in MaaS (Sc6)

Table 14.9 includes the marginal costs used to calculate the externalities of integrating the AM in MaaS in the city centre of Geneva.

The modal split of Sc6 is explained in Table 14.3. The willingness to give up the cars for daily car users in the city centre of Geneva is used to estimate the modal share of cars that is absorbed by the AM. It is important to note that the sample for daily car users in the city of Geneva was limited, with only 34 respondents.

After conducting a social assessment, it was found that the level of willingness to use AM in Geneva was similar to that of Lyon. As a result, the sample used for

**Table 14.12** Scenario 3 robotaxis applied with and without ridesharing service in Geneva increase in externalities

| Service | Individual | Ridesharing | |
|---|---|---|---|
| Occupancy factor (average person per vehicle) | 1.2 | 2.4 | 4 |
| Increase in externalities in million euros | 163.2 | 75.8 | 41.8 |

**Table 14.13** AM in MaaS modal share composition

| | | |
|---|---|---|
| The AM modal share = 11.12% | 6.42 | % of cars |
| | 3.00 | % of walking |
| | 1.70 | % of biking |

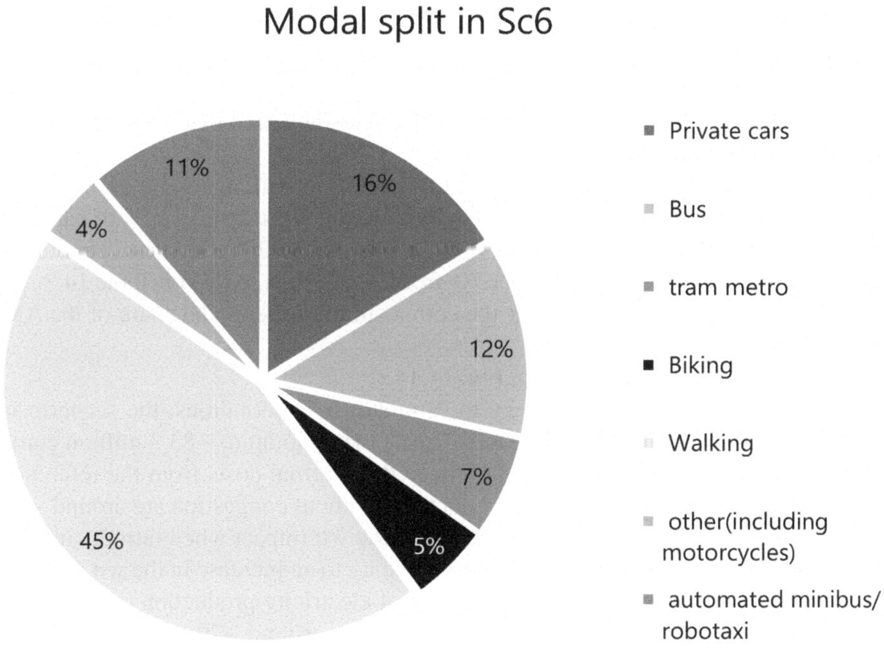


**Fig. 14.14** Modal shift for scenario Sc6: AM in MaaS in Geneva

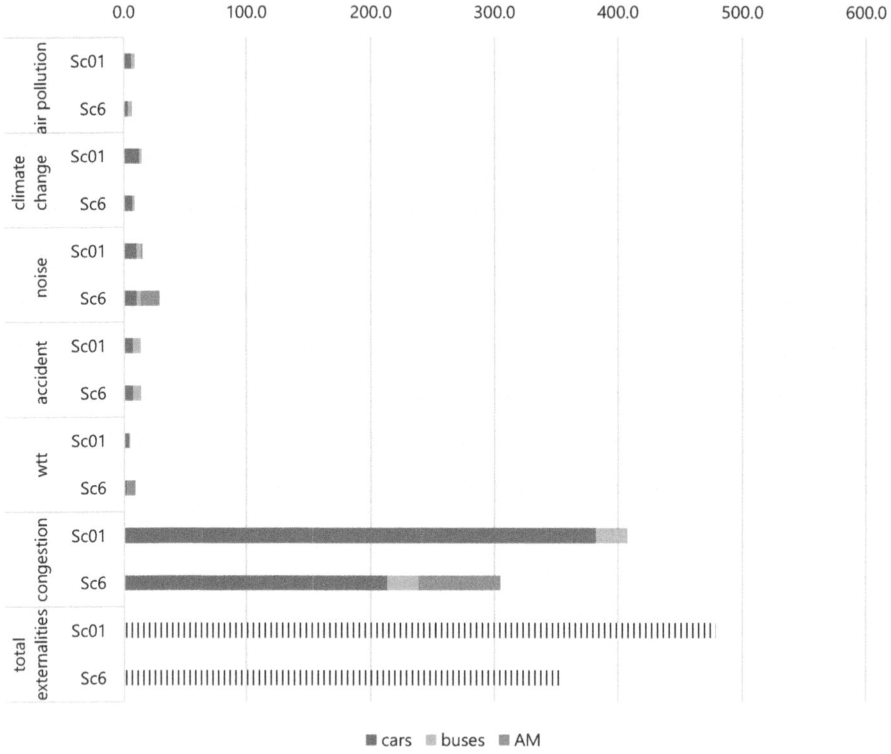


**Fig. 14.15** The increases or decreases in externalities with congestion in million euros in Sc6

the study included 81 individuals who use their cars daily and live in the city centre of both Lyon and Geneva combined.

From this sample, 23 respondents said they were very willing to give up their cars if the AM bridges first and last miles. Eventually, we consider that 28% of car trips could be replaced by the AM in MaaS, the equivalent of 6.42% of all trips. The AM is also replacing 3% from walking and 1.7% from biking trips, based on Table 14.3.

Table 14.13 and Fig. 14.14 show the composition of the modal share of the AM in this scenario.

The new modal split is shown in Fig. 14.14.

Following the same methodology to estimate the externalities, the scenario is predicted to lead to reductions in external costs that amount to −83.3 million euros compared to Sc01 (AM in MaaS will reduce the external costs from the reference scenario by 83.3 million euros). The reductions without congestion are around −13 million euros. Similarly to previous scenarios, the wtt impact when introducing the AM in densely populated areas is negative (leading to an increase in the wtt external costs). This is explained mostly by the effect of electricity production on air pollution. Congestion presents the bulk of the reductions, around −70 million euros. We estimate around −four million euros in savings for the categories of climate change, accidents and noise. See Fig. 14.15 for the results with congestion and Fig. 14.16 for results without congestion.

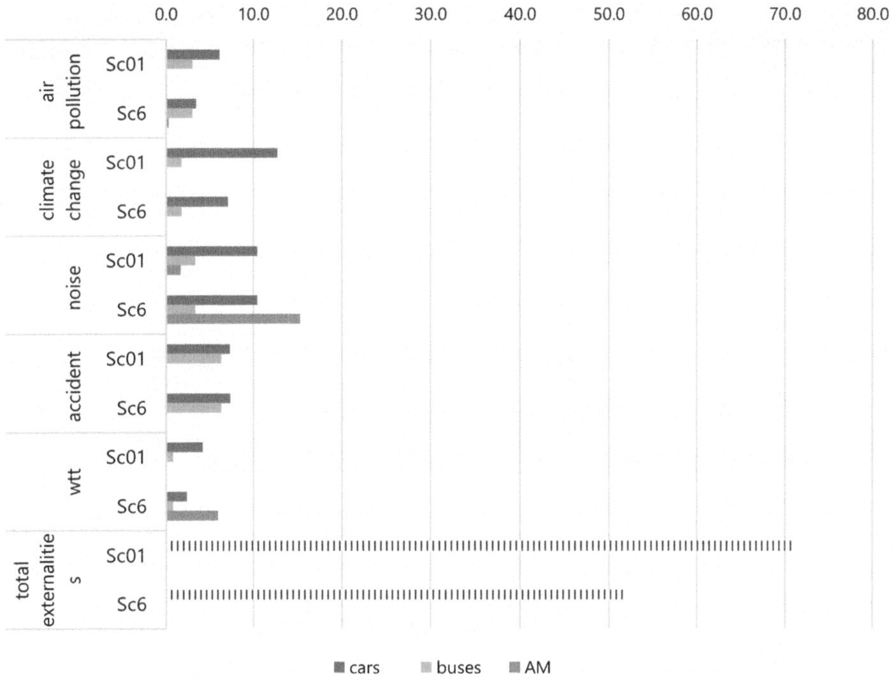


**Fig. 14.16** The increases or decreases in externalities without congestion in million euros in Sc6

This contributes to savings in parking spaces of 0.04 km2 or around 4160 parking spots. Finally, we consider the results if the AM in MaaS in deployed in suburban areas. This could be interesting to see the difference in applying the same scenario in different settings (urban and suburban), and it highlights the importance of targeted and context-based deployment strategy. The analysis for suburban scenarios is applied in this case.

Based on the social survey, we can anticipate a replacement rate of cars by AM of 12.3%. The AM will also absorb 3% of walking and 1.7% of biking, similar to the analysis of the urban version of the scenario described previously. The decrease in externalities is estimated to be around 12 million euros. Without accounting for the congestion, we have a decrease of two million euros. See the comparative Table 14.14 below.

### *14.5.4 Suburban Scenarios*

For our two suburban scenarios, we use the mobility behaviour of the second suburban ring in Geneva (Sc02).

**Table 14.14** Scenario 6 AM in MaaS applied in urban and suburban Geneva

| Decrease in externalities in million euros | Setting for AM in MaaS | |
|---|---|---|
| | urban | Suburban |
| With congestion | 83.38 | 11.95 |
| Without congestion | 12.92 | 1,85 |

**Table 14.15** Marginal costs (car, bus and AM) in €-cent per pkm in second suburban ring of Geneva

| Externality category | Car | Bus | AM |
|---|---|---|---|
| Air pollution | 0.63 | 1.43 | 0.04 |
| Climate change | 1.31 | 0.83 | – |
| Wtt | 0.42 | 0.36 | 0.41 |
| Noise | 1.92 | 1.60 | 0.23 |
| Accidents | 1.35 | 3.08 | – |
| Congestion | 25.50 | 7.98 | 6.38 |

Table 14.8 is used to calculate the total externalities in the scenarios. The AM marginal costs are also adjusted to take into account that the AM will compensate for the shortage of public transportation in these areas. The occupancy factor is 8, and the marginal costs estimated in D8.4 have been adjusted accordingly in the Table 14.15 below.

#### 14.5.4.1 Scenario 4: Expand the Network (Sc4)

To expand the network in suburban areas, the AM are deployed to replace a percentage of cars' modal share. Thus, we rely on the explanation from Table 14.3 as well as why we used the sample of Lyon and Geneva from scenario 6.

Finally, the modal share of the AM equals 11.2%, absorbed from the cars' modal share as presented in Fig. 14.17.

This modal shift leads to a decrease in externalities. We account for around −13 million euros. Consistent with the previous scenarios, congestion accounts for the highest impact, around −12.3 million euros; see Fig. 14.18. We register a slight reduction in the WTT category in this scenario. In general, there are reductions in all externalities categories. Without congestion, the reductions are around −three million euros; see Fig. 14.19.

The scenario presents reductions in external costs of −18 million euros for the city. If the congestion is not accounted for, the targeted expansion still leads to reductions of around −five million euros.

The fleet size for AM is estimated as 220, with a waiting time of 6 min and exit and entry time of 3 min (Fournier et al., 2020).

In conclusion, the savings in parking space equals 0.01 km2, around 1367 parking spots.

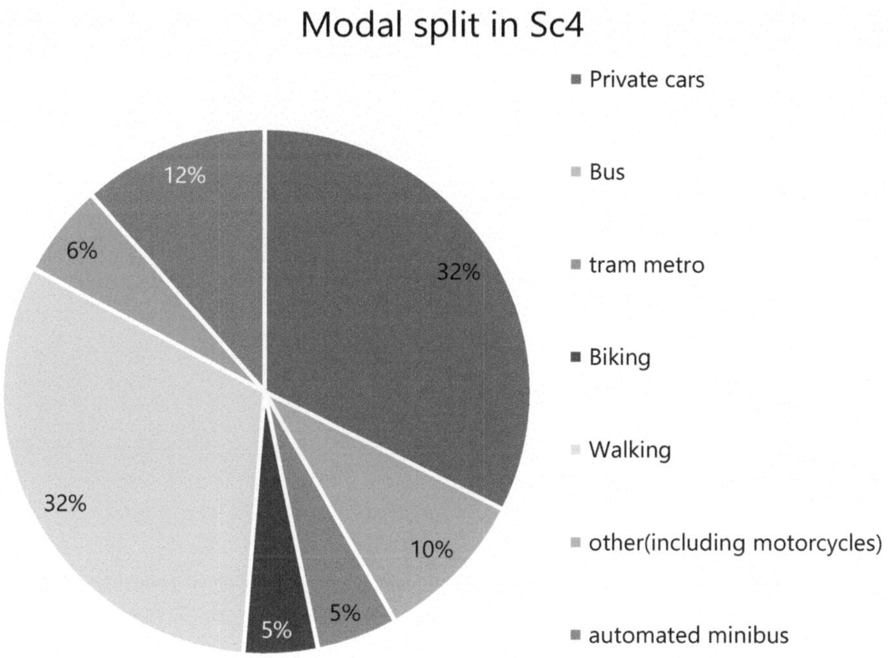


**Fig. 14.17** Modal shift of Sc4: expand the network compared to Sc02

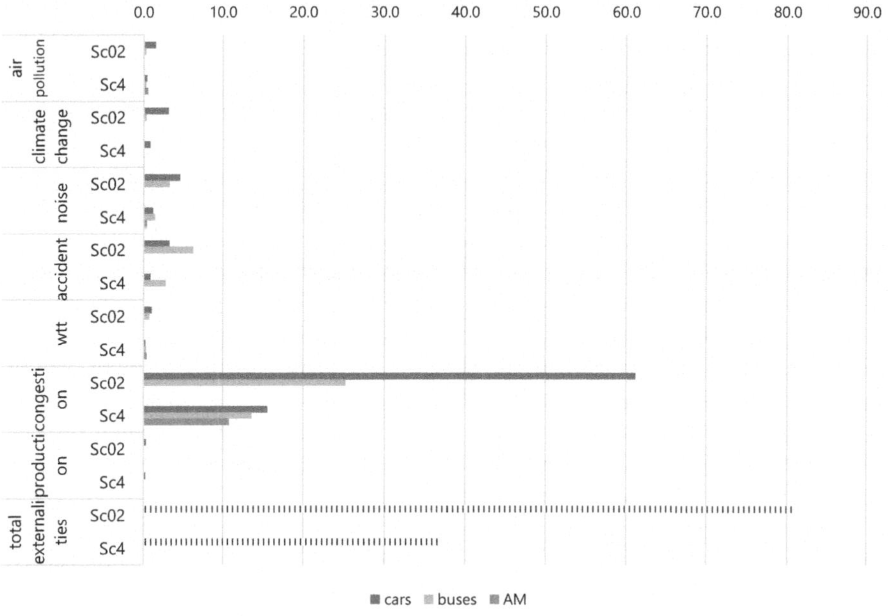


**Fig. 14.18** The increases or decreases in externalities with congestion in million euros in Sc4

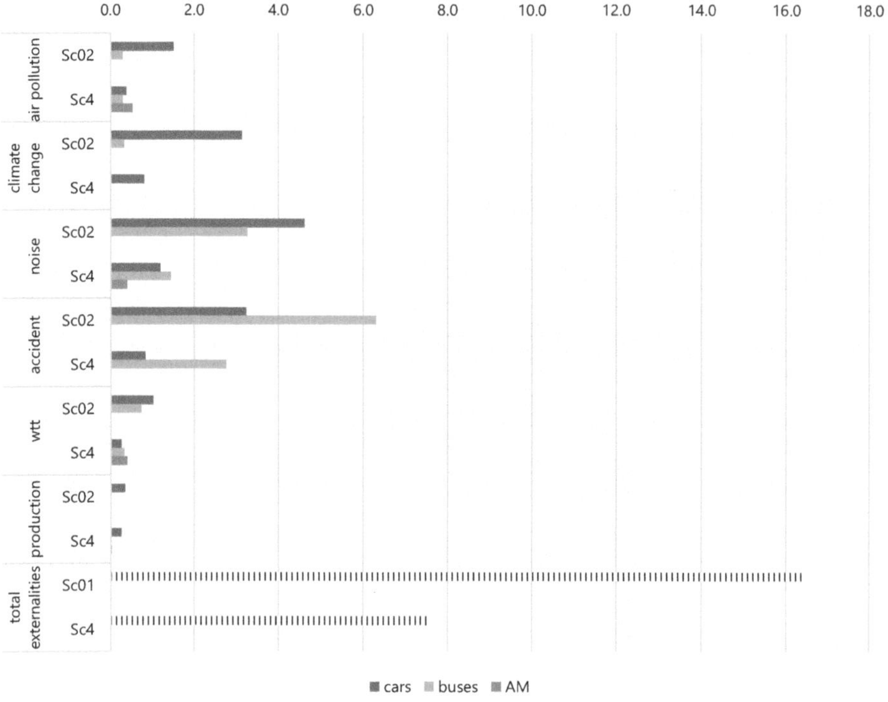


**Fig. 14.19** The increases or decreases in externalities without congestion in million euros in Sc4

#### 14.5.4.2 Scenario 5: Targeted Expansion (Sc5)

The targeted expansion scenario builds on the previous scenario. The same modal shift from cars to AM remains around 11.2%. However, in this case, we consider that the AM are targeted to also replace low occupancy buses and night buses. The demand for public transport fluctuates during the day and during its route. The notion of welfare optimisation justifies maximising the capacity of the bus (the bigger the bus, the better the ability to meet the peak time travel demands). Thus, public transport operators would justify running big buses (fitting up to 60 passengers). First, it is difficult to predict exactly when a bus is running empty (even during off-peak time, there might be exceptional demand). Second, operating another fleet of smaller vehicles would require more drivers and thus more costs than just keeping the big buses. However, as drivers are no longer necessary in an automated vehicle, these costs are eliminated. Furthermore, the on-demand feature makes the service customisable according to the demand (Pyddoke, 2020).

Determining the specific ratio of buses with low ridership is a complex process since the number varies unpredictably. That is why we consider Adra et al. (2004) study. In their analysis, empty running km for buses is around 11% based on data from Paris public transport system (RATP). Moreover, Mancret-Taylor and Boichon (2015) report stipulates around 4% of bus trips are taken between midnight and

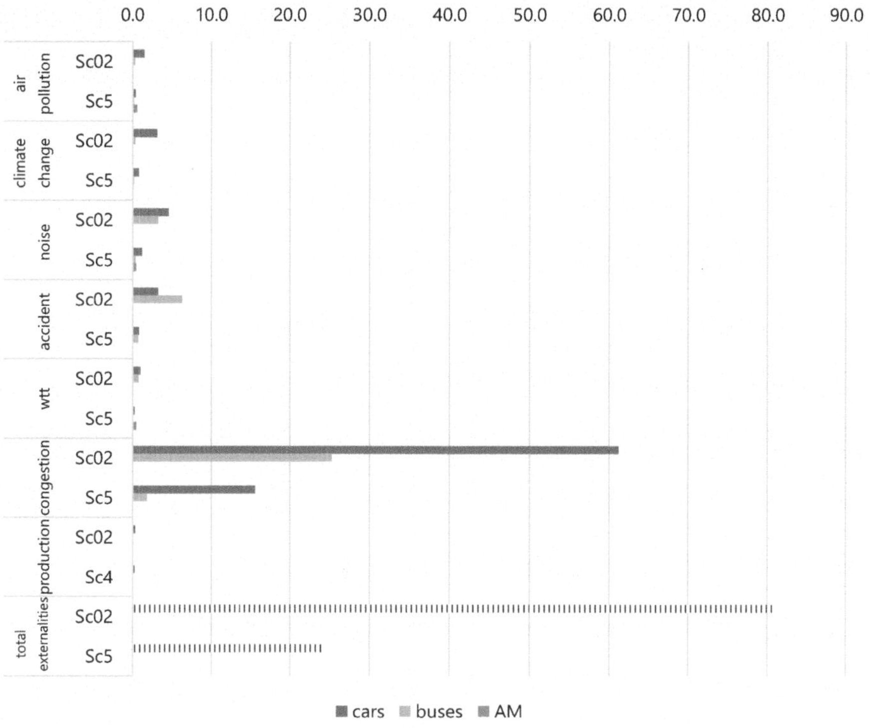


**Fig. 14.20** The increases or decreases in externalities with congestion in million euros in Sc5

5 a.m. Hence, we assume that more than 12% of all bus trips in suburban areas are replaced by AM trips. Finally, the modal share of AM in this scenario is 13.4%. Sc5 also registers reductions in external costs of around −15 million euros and −4 million euros without counting the external costs from congestions. The parking savings are similar to the previous scenario, "expand the network"; we estimate savings of 0.1 km2 (Figs. 14.20 and 14.21).

## 14.6 Discussion

To better demonstrate the scenarios and compare the results, Table 14.16 includes increases or decreases in external costs per scenario for each of the six scenarios analysed.

The two scenarios that incurred increases in external costs were "robotaxis" and "replacing all buses". The first is set to compete with public transport, while the second is set to update the bus service. However, both demonstrate that replacing traditional public transportation without a planned strategy (or with a laissez-faire

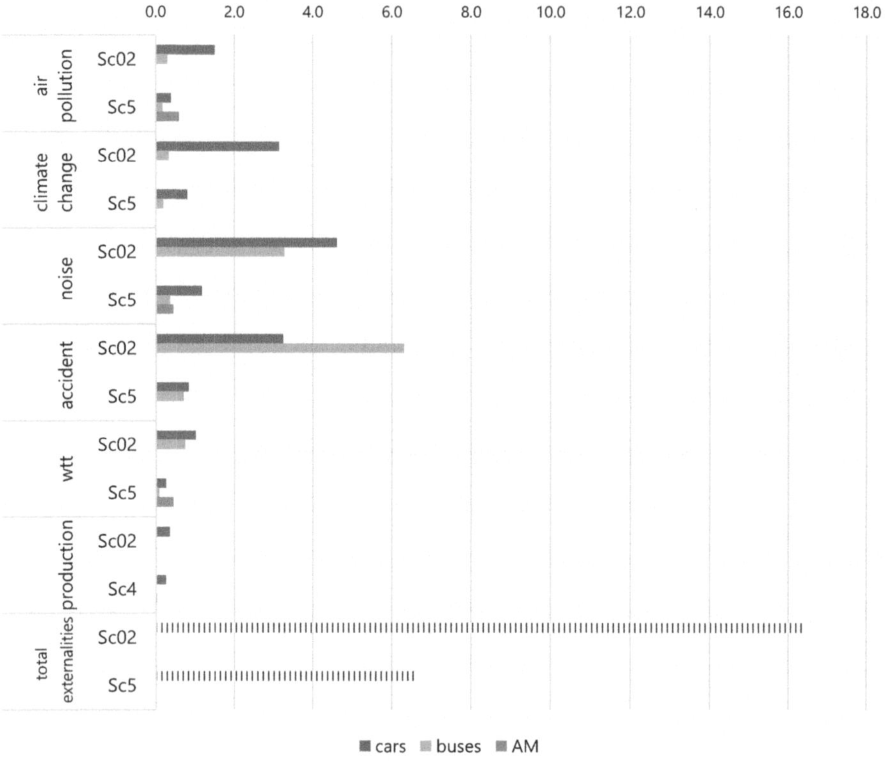


**Fig. 14.21** The increases or decreases in externalities without congestion in million euros in Sc5

outcome) would worsen the impact of deploying AV in cities and diminish their potential benefits on the environment and society.

Replacing the bus scenario results matches the simulation of replacing buses with shared AV in Helsinki (ITF, 2017). Thus, it is not recommended to replace all the buses with AM but rather target the off-peak hours of bus trips like in scenario 5 “targeted expansion”, where the bus service is mixed with AM on-demand. The path to update the bus service would be better by replacing the fleet with electric buses.

The implementation of robotaxis exacerbates the shortcomings of the transportation system, as it perpetuates a model of individual mobility that has proven to be detrimental to the city’s environment.

Replacing all cars offers the greatest potential for savings, which aligns with many urban initiatives aimed at reducing car usage. This approach is in line with Geneva’s efforts to promote sustainable transportation, such as soft modes and public transit. Although replacing all cars clearly results in cost savings, it may be more difficult to implement in suburban areas, limiting its overall impact and potential (Duarte & Ratti, 2018; ITF, 2015; ITF, 2017). Moreover, it is more challenging to realise as it depends strongly on people’s acceptance of the technology and their willingness to get rid of their cars.

**Table 14.16** The increases (+) and decreases (–) in external costs in the six scenarios of AM deployment in the Canton de Geneve

| | Scenario | Setting | The modal share of AV | The scenario modal shift | Sources used for modal shift | Decrease (–) or increase (+) in external costs in million euros | Savings in Parking space in km2 | Equivalent number of parking spaces |
|---|---|---|---|---|---|---|---|---|
| 1 | Replacing all buses | Urban | 12% | – Replace all bus trips in city Centre = > 12% of all trips | | +12.11 | – | – |
| 2 | Replace all cars | Urban | 23% | – Replace all car trips in city Centre = > 23% of all trips | | –307.95 | 0.65 | 64,824 |
| 3 | Robotaxis | Urban | 18% | – Replace 7% of bus trips= > 1% of all trips<br>– Replace 20% of car trips = > 4.5% of all trips<br>– Replace 13% of all trips (from walking modal share) | Ward et al. (2019), Clewlow & Gouri, (2017), Heineke et al. (2019) | + 161.8 | 0.16 | 15,800 |
| 4 | Expand network | Suburban | 12% | – Replace the car trips from daily car users who would give up their car –26% of car trips = > 11.5% of all trips | – Representative survey | –12.94 | 0.01 | 1367 |
| 5 | Targeted expansion | Suburban | 13% | – Same for car trips from Sc4 = > 11.5% from all trips<br>– Replace all trips on night buses and empty-running buses-12% from all bus trips = > 1.2% from all trips | – Representative survey Mancret-Taylor and Boichon (2015), Adra et al. (2004) | –14.86 | 0.01 | 1367 |
| 6 | AM in MaaS | Urban | 11% | – Same for car trips from Sc4<br>– Replace car last/first mile trips to connect to a rail station = > 6.5% of all trips<br>– Replace walking last/first mile trips to connect to a rail station = > 3% of all trips<br>– Replace biking last/first mile trips to connect to a rail station = > 1.7% of all trips | – representative survey Paydar et al. (2020), Giansoldati et al. (2020), Gebhardt et al. (2016) | –83.38 | 0.04 | 4160 |

On a larger scale, MaaS could reduce environmental deterioration and increase the efficiency of the shuttles even if it absorbs some active mobility means. It is the most realistic and could provide better results. It could be part of an on-demand and door-to-door service across the canton of Geneva to reduce the long-distance trips conducted with cars and improve connectivity to mobility hubs.

Moreover, the suburban scenarios show optimistic results, which justifies the need to strengthen the transportation network where there are shortcomings of public transport. The gains might be limited compared to other scenarios gains (AM in MaaS or replacing all cars). Nevertheless, these deployment strategies would strengthen the overall transportation offer across the canton and encourage passengers to use trains more. This would lead to further gains in suburban and urban areas alike. Concurrently, high-density population areas usually contain more short-distance trips in smaller areas, as opposed to suburban and interurban areas. In this case study, the city of Geneva (one commune with a population density of 12,500 inhabitants/km2) accounts for 711,000 daily trips, while the second suburban couronne (11 communes with a population density of 161 inhabitants/km2) accounts for 94,000 daily trips. Hence, this makes the comparison between the two sets of scenarios biased. In order to maximise the savings, the AM could cover longer distances and connects communes (rather than operate exclusively within the commune).

If we focus on the externalities categories, we notice that the congestion presents the factor that leads to the biggest reduction in external costs (Sc 2, 4, 5 and 6) or the biggest increases (Sc 1 and Sc 3). It reflects the transport pricing and value of time. Replacing buses that usually operate within specific lanes with AM would potentially slow down the traffic flow, whereas deploying more individual vehicles like robotaxis would affect the traffic congestion.

This is incremental for policymakers as it showcases the perils of traffic jams as it worsens the traffic flow, which affects daily life and air pollution and GHG emissions. Reducing congestion is a leading cause of externalities gains in our scenarios. Dominating congestion externalities are aligned with Jochem et al. (2016) and van Essen et al.'s (2019) results for road traffic congestion.

The categories of accidents, air pollution, production and climate change show savings across the six scenarios. Hence, any introduction of AV, whether AM or robotaxis, will have positive impacts on accident rates and air pollution GHG emissions. The air pollution and GHG emission externalities during the wtt phase, on the other hand, show negative results for all urban scenarios. This is explained by the fact that the production of electricity for battery charging is strongly energy-intensive, and it involves air emissions, thus causing a not negligible environmental burden, as proved by Pero et al. (2018). Furthermore, the Robotaxis scenario is the only case where it causes a negative noise externality which is understandable since the marginal cost is the same as that of an ICEV. The decision not to distinguish between an EV and ICEV is due to the similarity in speed and the occupancy factors for both vehicles (Jochem et al., 2016).

Also, the introduction of AV in the city will, in general, lead to savings in parking spaces. The more cars are replaced, the more urban area is free. Free urban space means reshaping the built environment to be more green and liveable, which again aligns with the urban strategy of Geneva (Etat de Genève, 2013).

The direct costs of the deployment on a microscale for the pilots were analysed previously in Antonialli et al. (2021). Building on this analysis, we note that it is important to account for the effects of letting go of the drivers. The AM are operating with safety drivers on board in the AVENUE pilots. However, in the future, they will operate without human interference on board. This used a major advantage for replacing traditional public transport with AM as it will significantly cut the costs for the public transport operators, which would make it easier for them to adopt the technology and make their public transport more attractive and competitive. However, the layoffs mean an increase in the unemployment rate. For instance, Transport Public de Geneve (TPG), the public transport operator in Geneva, hires almost 1300 bus drivers. In Sc1, eliminating all buses might seem disadvantageous for the city in terms of environmental and societal impacts. It could reduce around 50 million euros of labour costs for TPG. However, this would have a more negative impact on the society by decreasing 1300 jobs, even if we account for the creation of new jobs: off-site safety operator positions are limited and would not compensate around 1000 posts, which would create labour market disruption and have a significant effect on the economy as well (Nikitas et al., 2021; Sousa et al., 2018).

To conclude, the externalities calculations supported most of the initial assumptions for the consequences of the deployment in the scenarios. It also helps clarify the ambiguity of the direct environmental effects of the AM. The analysis provides insights that would help policymakers decide on how to deploy the AM (or robotaxis) to support the prosperity of their cities.

## 14.7 The Model Applied to Other Cities

In this part, we apply selected scenarios to the AVENUE cities. First, we use the representative survey to filter which scenarios are interesting to be studied based on the sample of respondents that are using their cars daily in the geographic areas of the city centre, large towns and small to medium villages. For Luxembourg, the sample of respondents that are daily car users and live in the city was too small. Thus, we select the scenarios in the suburban and preurban parts. For Lyon, we follow the same analysis as for Geneva; we combine the samples for Lyon and Geneva. Finally, for Copenhagen, we focus on the urban scenarios (replacing all cars, replacing all buses, robotaxis and AM in MaaS) since the sample of respondents is significantly larger than those from small to medium villages.

### 14.7.1 Copenhagen

For this study, we focus on the municipality of Copenhagen for the four urban scenarios. The data used is based on the Danish national census for 2019 (Christiansen & Baescu, 2021). The modal share for the urban scenarios is from 2015 based on Kayser (2017). This represents the most recent data we could find that represents the mobility behaviour of the municipality of Copenhagen. The modal share is 26% private cars, 5% bus, 10% train, 36% biking, 22% walking and 1% other (including motorcycles). This is combined with the average distances per the mode of transport from the greater area of Copenhagen based on Christiansen and Baescu (2021).

Copenhagen municipality has around 623,000 residents and an area of 86.4 km2 (Statistics Denmark, 2021). The average daily trips per person are around 2.8. We account for a fleet of 252,600 cars and around 1000 buses (Movia, 2019).

The scenarios applied in this case are: "replacing all cars", "replacing all buses", "robotaxis", and "AM in MaaS". By using the same analysis and methods described for Geneva, we obtain the potential increase or decrease in external costs as presented in Table 14.17.

The robotaxis scenario remains the one with the highest potential increase in external costs compared to the status quo, while "replacing all cars" is the one with the highest reductions in external costs.

### 14.7.2 Luxembourg

The scenario that we consider for Luxembourg is the communes of the Luxembourg canton without the city of Luxembourg. The area is 187 km2, and the population is 58,079 inhabitants (Ville de Luxembourg, 2021). The mobility behaviour data was limited. Therefore, we used averages for modal share from Luxembourg and average distances on EU level.

**Table 14.17** The potential increases or decreases in external costs in Copenhagen for Sc1, Sc2, Sc3 and Sc6

| Scenario | The modal share of AV (AM or robotaxi) | The difference in externalities (million euros) | The decrease in external costs without congestion |
|---|---|---|---|
| Replacing all buses | 5% | +(8.28) | −1.33 |
| Replace all cars | 26% | −964.96 | −94.27 |
| Robotaxis | 18% | +(452.07) | −3.28 |
| AM in MaaS | 7% | −86.54 | −8.37 |

**Table 14.18** The decrease in external costs in Luxembourg for Sc4 and Sc5

| Scenario | Setting | AM modal share | Decrease in external costs in million euros |
|---|---|---|---|
| Targeted expansion | Suburban | 12% | −27.11 |

The externalities results following the estimations in marginal costs for Luxembourg, the transport performance and the predicted modal shift are presented in Table 14.18.

The results show that Luxembourg canton would reduce its external costs if it deploys the AM in the suburban parts.

### *14.7.3 Lyon*

We focus on the area of the Greater Lyon (Metropole de Lyon) composed of the areas of Lyon-Villeurbanne (the centre Lyon) as the city centre and the ring of the Lyon Metropole (Couronne de la Metropole de Lyon). Thus, the reference scenarios are applied to these areas: Sc01 is the scenario of Lyon-Villeurbanne, and Sc02 is the scenario of the ring of the Lyon Metropole, both in 2015.

The transport performance and the mobility behaviour data are computed using a mobility survey for 2015, conducted by Lyon (Sytral, 2016).

The City of Lyon is one of the most populated cities in France, with a population of around 500,000 inhabitants and a surface of 47.9 km2. As for the mobility behaviour, using the average number of 3.3 daily trips for each inhabitant, we could estimate a sum of 1.65 million overall trips a day. In the city, the motorisation rate is 414 cars per 1000 residents. Thus, we count 207,000 cars. The modal split is 26% private cars, 9% bus, 16% tram metro, 3% biking, 45% walking and 1% other (including motorcycles)..

For the suburban ring, the population is around 885,927 inhabitants, and the surface is 486 km2.

The average number of daily trips is 3.43 for an average distance of 20 km compared to the 13 km average for the City of Lyon. In this area, there are around 30 million daily trips. The car fleet accounts for 575,853 cars. The modal split is 57% private cars, 5% bus, 9% tram metro, 1% biking, 27% walking and 1% other (including motorcycle).

We apply the scenarios and the externalities model as it was applied for Geneva.

The results are presented in Table 14.19 below.

The results here differ slightly from previous findings. In this case, updating the city's bus fleet by replacing the vehicles with AM would lead to decreases in the external costs.

**Table 14.19** The potential increase or decrease in external costs in million euros Lyon for all six scenarios

| Scenario | Setting | The modal share of AV | Increase + or decrease—in externalities | Increase + or decrease—in externalities—without congestion |
|---|---|---|---|---|
| Sc1-replacing all buses | Urban | 9% | −6.41 | 2.7 |
| Sc2-replace all cars | Urban | 26% | −319.65 | 41.86 |
| Sc3-Robotaxis | Urban | 19% | +133.75 | 4.98 |
| Sc4-expand network | Suburban | 15% | −170.74 | 30.32 |
| Sc5-targeted expansion | Suburban | 16% | −174.21 | 32.82 |
| Sc6-AM in MaaS | Urban | 12% | −88.89 | 11.6 |

**Table 14.20** Summary of the results for AVENUE cities, increase (+) or decrease (−) in external costs for each scenario in million euros

| | | Geneva | | Lyon | | Copenhagen | | Luxembourg | |
|---|---|---|---|---|---|---|---|---|---|
| Scenario | Setting | Modal share AV | Increase decrease million euros | Modal share AV | Increase or decrease in million euros | Modal share AV | Increase or decrease in million euros | Modal share AV | Increase or decrease in million euros |
| Replacing all buses | Urban | 12% | +12.11 | 9% | +6.41 | 5% | + (8.28) | – | – |
| Replace all cars | Urban | 23% | −251.94 | 26% | −319.65 | 26% | −964.96 | – | – |
| Robotaxis | Urban | 18% | +161.8 | 19% | +133.75 | 18% | + (452.07) | – | – |
| Expand network | Suburban | 12% | −15.42 | 15% | −170.74 | – | – | 10% | −23.89 |
| Targeted expansion | Suburban | 13% | −17.63 | 16% | −174.21 | – | – | 12% | −27.11 |
| AM in MaaS | Urban | 11% | −83.38 | 12% | −88.89 | 7% | −86.54 | – | – |

## 14.8 Conclusion

In general, the robotaxis scenario will always lead to increases in external costs when deployed in competition with public transport. A deployment with ridesharing could improve the balance. The environmental and accident categories record a slight decrease in externalities, but the decrease in external costs (without congestion) remains the lowest out of all the urban scenarios. The AM in MaaS, replacing all cars and the suburban scenarios consistently show decreases in external costs. Replacing all cars leads to the highest reduction in externalities (Table 14.20). While

**Table 14.21** Summary of the results for AVENUE without congestion

| | | Geneva | | Lyon | | Copenhagen | | Luxembourg | |
|---|---|---|---|---|---|---|---|---|---|
| Scenario | Setting | Modal share AV | Decrease million euros | Modal share AV | Decrease million euros | Modal share AV | Decrease million euros | Modal share AV | Decrease million euros |
| Replacing all buses | Urban | 12% | 11.02 | 9% | 2.7 | 5% | 1.33 | – | – |
| Replace all cars | Urban | 23% | 46.77 | 26% | 41.86 | 26% | 94.27 | – | – |
| Robotaxis | Urban | 18% | 2.57 | 19% | 4.98 | 18% | 3.28 | – | – |
| Expand network | Suburban | 12% | 3.24 | 15% | 30.32 | – | – | 10% | 3.29 |
| Targeted expansion | Suburban | 13% | 4.75 | 16% | 32.82 | – | – | 12% | 5.5 |
| AM in MaaS | Urban | 11% | 12.92 | 12% | 11.66 | 7% | 8.37 | – | – |

AM in MaaS is a user-centric approach pull strategy, replacing all cars is a typical push strategy which forbids cars in the city. Thus replacing all cars is more difficult to realise due to low social acceptance. For Copenhagen, the most appealing scenario seems to be to replace all car trips by the AM (see Table 14.21). The number of the city's residents and the high average daily distance for car trips would justify this potential reduction.

Lyon shows that the emphasis should be on strengthening its public transport in suburban parts by deploying the AM serving seamless and intermodal services and also in the city, where the AM could be deployed within MaaS to bridge first and last-mile gaps and enhance connections to the rail stations.

Geneva would benefit from the AM deployment in the urban areas, mostly as seen in scenarios 2 and 6 since it has the highest reduction in externalities. It could reduce car access to the city centre and introduce the AM to support public transport and replace all car trips in these areas. Relying on the users' acceptance, the most realistic approach is to introduce the AM gradually, first as part of a MaaS service. Where AM are filling existing mobility gaps rather than completely replacing individual mobility, as the users' acceptance increases, passengers would switch more and more to the AM instead of relying on their cars. If this introduction is accompanied by urban policies such as road pricing and no-car zones, it will further deter citizens from using their cars in the city centre, which would increase the modal shift to the AM, active mobility and public transport.

Deploying the AM in suburban areas in Luxembourg also shows a reduction in externalities. This strategy fits with the canton's plans to reduce car use and improve connections to train stations to better serve cross-border travellers.

The scenarios with the externalities calculations provide indicators on the recommended deployment strategy that would fit with a European city's sustainable development. In conclusion, AM in MaaS presents a significant reduction in externalities, and it is a scenario supported by social acceptance as shown in the scenario description. This scenario also supports the public transport network, which makes it more sustainable. In contrast, the robotaxis scenario would increase externalities and compete with public transport. It is more convenient but also would lead to more congestion and pollution compared to the AM. However, the model has some limitations that are described in the following part.

### *14.8.1 Limitations of the Model*

It is important to present the limitations of the model and where it falls short. It is a given that the accuracy of the results is dependent on the accuracy of the input data. For instance, Luxembourg and Copenhagen case studies would benefit greatly from updating the mobility behaviour data to better reflect the areas in question (city centre or suburban ring).

This study provides important results on AM as a new force capable of transforming the transportation sector. It foresees scenarios of embedding AM in societal and political contexts and their impacts.

It has the potential to be more robust if the following limitations are addressed:

The external noise costs depend on the population density, traffic status and time of day. However, the data on a country level for these specific contexts is limited. This requires more fine-tuning to reflect the contextual impact.

Similarly, monetising congestion is complicated as it depends on an EU-level study to reach national marginal costs (Jochem et al., 2016). The meta-analysis requires inputs of speed-flow functions, demand curves and value of time (VOT) at a large scale and poses a challenge to downscale to smaller cities.

The use of national-level marginal costs for city-level assessment is a limitation. For instance, air pollution national external costs might underestimate those on an urban level (PM emissions differ between rural and urban areas). This could be addressed using the European values for urban and rural parts in a sensitivity analysis.

We should also consider whether the use of AM will increase the use of traditional public transport and estimate this increase in modal share. Notably, the study of the induced demand requires more analysis since it is difficult to predict the exact numbers for this rebound effect.

An optimised assessment should also account for the potential demographic changes and their effect on VKM (in Geneva, between 2000 and 2015, there was an increase in travel demand by 10% attributed to the increase in the population, for example).

The reliance only on one source to estimate the marginal costs (the Delft handbook of external costs by van Essen et al. (2019) gives one version of the analysis: a sensitivity analysis using different marginal costs would make the model more robust.

A further limitation is accident marginal costs estimation when AV interacts with human driving. The methodology of calculation is still evolving with growing knowledge.

### 14.8.2 Rebound Effects

As mentioned in the limitations, the deployment of the scenarios should account for potential rebound effects. Since this analysis tries to imagine potential future scenarios based on current data and observations, it is limited to what could be extrapolated quantitatively. The study of rebound effects thus is better addressed qualitatively.

Hymel et al. (2010) describe the rebound effect as an increase in vehicle usage due to an unintended effect of raising fuel efficiency and decreasing the cost of car usage through policies and technological advancements. They also mention induced demand as a rebound effect to road transportation where an increase in infrastructure capacity attracts new traffic and causes it to reach its capacity, which was not the intended goal (Hymel et al., 2010). The AM would provide a convenient, affordable and safe option to passengers (Onat et al., 2023). Thus, it could lead to more trips, as it provides more trips to people who were not commuting with vehicles in the first place, such as children and the elderly. In addition, it might cause a secondary modal shift after its implementation. It could reduce active mobility and public transportation shares even further than first predicted (Childress et al., 2015; Fagnant & Kockelman, 2018; Zmud & Sener, 2017). This might lead to a vicious circle of deploying more vehicles to meet the new demand. Then, as an unintended effect, more people shift to using the AM. Thus, the operators will need to deploy even more vehicles to meet the increasing demand. Thus, the AM would aggravate the traffic congestion and increase the environmental footprint. Hence, the rebound effect undermines the gains from reducing the use of individual mobility by causing new external costs because of reducing walking, biking and public transport trips. Ergo, it is crucial that the deployment is accompanied by a regulatory framework to monitor the introduction of the AV in the transportation system and reduce potential rebound effects (Möller et al., 2019).

### 14.8.3 Policy Recommendations

To summarise replacing all cars has the best impact for reducing external costs but due to the expected problems with the citizen acceptance a political problem for implementing would probably raise. AM in MaaS seems thus to be the best solution for the cities as external costs can decrease substantially. At the same time, positive externalities through network effects can further be expected making the transport system more efficient for the TPOs and reliable for the citizen. The combination of AM and MaaS would enable a change of the mobility paradigm enabling sustainable mobility and a reliable and improved transport system. Last but not least, AM in MaaS would consequently enable SUMP without dos and don'ts. Traditional (unpopular) push and pull strategies (see, e.g. (TUMI, 2018)) could so become obsolete: the results of a representative survey among 1816 citizens (of which 1526 have privately owned vehicles) in Lyon, Copenhagen, Luxembourg and Geneva confirm that 45% of car drivers are "willing" (22%) or even "very willing" (23%) to give up using their own car to use AM to bridge the first and the last mile if this were available. If the service is on-demand and door-to-door, the acceptance could be even higher (Korbee et al., 2022).

**Acknowledgements** This project has received funding by the European Union's Horizon 2020 research and innovation program under grant agreement No. 769033.

## References

Adra, N., Michaux, J.-L., André, Michel (2004). Analysis of the load factor and the empty running rate for road transport. Artemis - Assessment and reliability of transport emission models and inventory systems.

Alazzawi, S., Hummel, M., Kordt, P., Sickenberger, T., Wieseotte, C., & Wohak, O. (2018). Simulating the impact of shared, autonomous vehicles on urban mobility—a case study of Milan. *Simulating Autonomous and Intermodal Transport Systems, 2*, 94–76.

Amer, M., Daim, T. U., & Jetter, A. (2013). A review of scenario planning. *Futures, 46*, 23–40. https://doi.org/10.1016/j.futures.2012.10.003

Antonialli, F. (2019). International benchmark on experimentations with Autonomous Shuttles for Collective Transport. Available online at https://hal-centralesupelec.archives-ouvertes.fr/hal-02489797/document

Antonialli, F., Fournier, G., Jaroudi, I., Mira-Bonnardel, S., & Thalhofer, M. (2021). D8.4 second iteration economic impact. Available online at https://cordis.europa.eu/project/id/769033/results.

Auld, J., Sokolov, V., & Stephens, T. S. (2017). Analysis of the effects of connected–automated vehicle technologies on travel demand. *Transportation Research Record: Journal of the Transportation Research Board, 2625*(1), 1–8. https://doi.org/10.3141/2625-01

Beukers, J. (2019). *First iteration Geneva large scale pilot use case demonstration report*. TPG. Available online at https://cordis.europa.eu/project/id/769033/results

Bischoff, J., & Maciejewski, M. (2016). Simulation of city-wide replacement of private cars with autonomous taxis in Berlin. *Procedia Computer Science, 83*(4), 237–244. https://doi.org/10.1016/j.procs.2016.04.121

Bubeck, S., Tomaschek, J., & Fahl, U. (2014). Potential for mitigating greenhouse gases through expanding public transport services: A case study for Gauteng Province, South Africa. *Transportation Research Part D: Transport and Environment, 32*, 57–69.

Chester, M. V. (2008). Life-cycle environmental inventory of passenger transportation in the United States. Dissertation, Berkeley, University of California. Available online at http://repositories.cdlib.org/its/ds/UCB-ITS-DS-2008-1

Chester, M. V., & Horvath, A. (2009). Environmental assessment of passenger transportation should include infrastructure and supply chains. *Environmental Research Letters, 4*(2), 24008. https://doi.org/10.1088/1748-9326/4/2/024008

Childress, S., Nichols, B., Charlton, B., & Coe, S. (2015). Using an activity-based model to explore the potential impacts of automated vehicles. *Transportation Research Record: Journal of the Transportation Research Board, 2493*(1), 99–106. https://doi.org/10.3141/2493-11

Christiansen, H., Baescu, O. (2021). The Danish National Travel Survey - Annual statistical report for Copenhagen Area. DTU orbit. TU 0620v1. Available online at orbit.dtu.dk

Clewlow, R.R. and Mishra, G.S., (2017). Disruptive transportation: The adoption, utilization, and impacts of ride-hailing in the United States

Cohen, T., & Cavoli, C. (2018). Automated vehicles: Exploring possible consequences of government (non)intervention for congestion and accessibility. *Transport Reviews, 39*(1), 129.

Derbyshire, J., & Wright, G. (2017). Augmenting the intuitive logics scenario planning method for a more comprehensive analysis of causation. *International Journal of Forecasting, 33*(1), 254–266. https://doi.org/10.1016/j.ijforecast.2016.01.004

Duarte, F., & Ratti, C. (2018). The impact of autonomous vehicles on cities: A review. *Journal of Urban Technology, 25*(4), 3–18. https://doi.org/10.1080/10630732.2018.1493883

Etat de Genève. (2013). Mobilités 2030. Stratégie multimodale pour Genève. Accessed 3 July 2019.

Fagnant, D. J., & Kockelman, K. M. (2018). Dynamic ride-sharing and fleet sizing for a system of shared autonomous vehicles in Austin, Texas. *Transportation, 45*(1), 143–158. https://doi.org/10.1007/s11116-016-9729-z

Federal Statistical Office. (2013). Données communales (NOAS04, NOLC04, NOLU04).

Fournier, G., Boos, A., Wörner, R., Jaroudi, I., Morozova, I., & Nemoto, E. H. (2020). Substituting individual mobility by mobility on demand using autonomous vehicles—a sustainable assessment simulation of Berlin and Stuttgart. IJATM, 20, 369.

Fraedrich, E., Beiker, S., & Lenz, B. (2015). Transition pathways to fully automated driving and its implications for the sociotechnical system of automobility. *European Journal of Futures Research, 3*(1), 91. https://doi.org/10.1007/s40309-015-0067-8

Gebhardt, L., Krajzewicz, D., Oostendorp, R., Goletz, M., Greger, K., Klötzke, M., Wagner, P., & Heinrichs, D. (2016). Intermodal urban mobility: Users, uses, and use cases. *Transportation Research Procedia, 14*(2), 1183–1192. https://doi.org/10.1016/j.trpro.2016.05.189

Giansoldati, M., Danielis, R., & Rotaris, L. (2020). Train-feeder modes in Italy. Is there a role for active mobility? *Research in Transportation Economics, 2247*(1), 100990. https://doi.org/10.1016/j.retrec.2020.100990

Hansen, L., Arnesen, P., Graupner, S.-T., Lindkvist, H., Leonardi, J., Al Fahel, R., & Andersson, K. (2021). Current state of the art and use case description on geofencing for traffic. *GeoSence*.. https://sintef.brage.unit.no/sintef-xmlui/handle/11250/2826636

Hayes, B. (2011). Leave the driving to it. *American Scientist, 99*(5), 362. https://doi.org/10.1511/2011.92.362

Heineke, K., Kampshoff, P., Kellner, M., & Benedikt, K. (2019). *Change vehicles: How robo-taxis and shuttles will reinvent mobility*. McKinsey & Company.

Héran, F., & Ravalet, E. (2008). *La consommation d'espace-temps des divers modes de déplacement en milieu urbain*. Application au cas de l'Ile de France.

Huber, D., Viere, T., Nemoto, E. H., Jaroudi, I., Korbee, D., & Fournier, G. (2022). Climate and environmental impacts of automated minibuses in future public transportation. *Transportation Research Part D: Transport and Environment, 102*, 103160. https://doi.org/10.1016/j.trd.2021.103160

Huss, W. R., & Honton, E. J. (1987). Scenario planning—What style should you use? *Long Range Planning, 20*(4), 21–29. https://doi.org/10.1016/0024-6301(87)90152-X

Hymel, K. M., Small, K. A., & van Dender, K. (2010). Induced demand and rebound effects in road transport. *Transportation Research Part B: Methodological, 44*(10), 1220–1241. https://doi.org/10.1016/j.trb.2010.02.007

ITF. (2015). *Urban mobility system upgrade. How shared self-driving cars could change city traffic*. OECD.

ITF. (2017). *Shared mobility simulations for Helsinki. Case-specific policy analysis*. OECD.

ITF. (2020). *Shared mobility simulations for Lyon*. OECD.

Janasz, T. (2018). *Paradigm shift in urban mobility. Towards factor 10 of automobility*. Springer.

Jaroudi, Ines, Nemoto, E. H., Korbee, D., Bulteau, Julie, Viere, T., & Fournier, G. (Eds.) (2021). What is the potential impact of the transition from traditional transport to new mobility models (electric automated minibuses) in European cities? *The transformations of the global automotive industry: Digitalisation, ecological transition and the impact of the Covid 19 crisis*, Paris, 8 June.

Jenelius, E., & Cebecauer, M. (2020). Impacts of COVID-19 on public transport ridership in Sweden: Analysis of ticket validations, sales and passenger counts. *Transportation Research Interdisciplinary Perspectives, 8*, 100242. https://doi.org/10.1016/j.trip.2020.100242

Jochem, P., Doll, C., & Fichtner, W. (2016). External costs of electric vehicles. *Transportation Research Part D: Transport and Environment, 42*, 60–76. https://doi.org/10.1016/j.trd.2015.09.022

Jones, E. C., & Leibowicz, B. D. (2019). Contributions of shared autonomous vehicles to climate change mitigation. *Transportation Research Part D: Transport and Environment, 72*, 279–298. https://doi.org/10.1016/j.trd.2019.05.005

Kasten, P., Bracker, J., Haller, M., & Purwanto, J. (2016). Electric mobility in Europe—Future impact on the emissions and the energy systems. Final report of task 2 - Assessing the status of electrification of road transport passenger vehicles and potential future implications for the environment and European energy system. In *EEA project: Assessing the status of electrification of the road transport*. https://www.oeko.de/fileadmin/oekodoc/Assessing-the-status-of-electrification-of-the-road-transport-passenger-vehicles.pdf (passenger vehicles and potential future implications for the environment and European energy system)

Kayser, A. (2017). *Copenhagen: The cycling city. Technical workshop*. City of Copenhagen, Create.

Keall, M. D., Shaw, C., Chapman, R., & Howden-Chapman, P. (2018). Reductions in carbon dioxide emissions from an intervention to promote cycling and walking: A case study from New Zealand. *Transportation Research Part D: Transport and Environment, 65*(3), 687–696. https://doi.org/10.1016/j.trd.2018.10.004

Korbee, D., Naderer, G., Dubielzig, M., Mathe, L., Helfer, L. (2021). Deliverable 8.7 Second iteration social impact assessment. AVENUE project. Available online at https://h2020-avenue.eu/public-delivrables/. Accessed 22 Mar 2021.

Korbee, D., Naderer, G., & Fournier, G. (2022). The potential of automated minibuses in the socio-technical transformation of the transport system. In *Transport research arena (TRA) conference, Lisbon*.

Lai, L. W. C., Chau, K. W., Ching, K. S. T., Kwong, J. W. Y., Cheung, P. A. C. W., & Lorne, F. T. (2013). Acquisition and redevelopment of bus depots: A Hong Kong land policy and planning case study. *Habitat International, 39*(1), 75–84. https://doi.org/10.1016/j.habitatint.2012.10.002

Lindgren, M., & Bandhold, H. (2009). *Scenario planning - Revised and updated. The link between future and strategy*. Palgrave Macmillan.

Litman, T. (2021). *Autonomous vehicle implementation predictions implications for transport planning*. Victoria Transport Policy Institute.

Litman, T. (2022). *Autonomous vehicle implementation predictions. Implications for transport planning*. Victoria Transport Policy Institute. Available online at https://www.vtpi.org/avip.pdf

Mancret-Taylor, V., & Boichon, N. (2015). *La Réponse Des Transports Collectifs À La Mobilité Nocturne*. L'Institut D'Aménagement Et D'Urbanisme. Note Rapid 692.

Merlin, L. (2017). Comparing automated shared taxis and conventional bus transit for a small city. *JPT, 20*, 19–39.

Milakis, D., Snelder, M., Van Arem, B., Van Wee, B., & de Almeida Correia, G. H. (2017). Development and transport implications of automated vehicles in The Netherlands: Scenarios for 2030 and 2050. *European Journal of Transport and Infrastructure Research, 17*(1). https://doi.org/10.18757/ejtir.2017.17.1.3180

Mishra, G. S., Clewlow, R. R., Mokhtarian, P. L., & Widaman, K. F. (2015). The effect of carsharing on vehicle holdings and travel behavior: A propensity score and causal mediation analysis of the San Francisco Bay Area. *Research in Transportation Economics, 52*, 46–55. https://doi.org/10.1016/j.retrec.2015.10.010

Möller, J., Daschkovska, K., & Bogaschewsky, R. (2019). Sustainable city logistics: Rebound effects from self-driving vehicles. In C. Jahn, W. Kersten, & C. M. Ringle (Eds.), *Digital transformation in maritime and city logistics: Smart solutions for logistics. Proceedings of the Hamburg international conference of logistics (HICL)* (Vol. 28, pp. 299–337). epubli GmbH. https://doi.org/10.15480/882.2501

Montfort, H., De Faveri, P., & Bisso, S. (2019). La mobilite des habitants du canton de Geneve. Résultats Des Microrecensements Mobilité et Transports (MRMT) 2000 À 2015. Republic et Canton de Geneve.

Moreno, A. T., Michalski, A., Llorca, C., & Moeckel, R. (2018). Shared autonomous vehicles effect on vehicle-km traveled and average trip duration. *Journal of Advanced Transportation, 2018*(1), 1–10. https://doi.org/10.1155/2018/8969353

Mosquet, X., Dauner, T., Lang, N., Rüßmann, M., Agrawal, R., Schmieg, F., & Mei-Pochtler, A. (2015). *Robo-taxis and the new mobility*. BCG. Available online at https://www.bcg.com/publications/2015/automotive-consumer-insight-robo-taxis-new-mobility

Movia. (2019). Movia arsrapport. Available online at https://www.moviatrafik.dk/media/7737/movia-aarsrapport-2019.pdf

Narayanan, S., Chaniotakis, E., & Antoniou, C. (2020). Shared autonomous vehicle services: A comprehensive review. *Transportation Research Part C: Emerging Technologies, 111*, 255–293.

Nikitas, A., Vitel, A.-E., & Cotet, C. (2021). Autonomous vehicles and employment: An urban futures revolution or catastrophe? *Cities, 114*(4), 103203. https://doi.org/10.1016/j.cities.2021.103203

Niles, J. (2019). *Full potential of future robotaxis achievable with trip-based subsidies and fees applied to the for-hire vehicles of today*. Mineta Transportation Institute Publications.

Nourinejad, M., Bahrami, S., & Roorda, M. J. (2018). Designing parking facilities for autonomous vehicles. *Transportation Research Part B: Methodological, 109*, 110–127. https://doi.org/10.1016/j.trb.2017.12.017

OFCOM. (2018). Digital Switzerland strategy. Available online at https://www.bakom.admin.ch/bakom/en/homepage/digital-switzerland-and-internet/strategie-digitale-schweiz/strategy.html. Accessed 7 Mar 2019.

Onat, N. C., Mandouri, J., Kucukvar, M., Sen, B., Abbasi, S. A., Alhajyaseen, W., Kutty, A. A., Jabbar, R., Contestabile, M., & Hamouda, A. M. (2023). Rebound effects undermine carbon footprint reduction potential of autonomous electric vehicles. *Nature Communications, 14*, 6258. https://www.nature.com/articles/s41467-023-41992-2

Paydar, M., Fard, A. K., & Khaghani, M. M. (2020). Walking toward metro stations: The contribution of distance, attitudes, and perceived built environment. *Sustainability, 12*(24), 10291. https://doi.org/10.3390/su122410291

Pero, F. D., Delogu, M., & Pierini, M. (2018). Life cycle assessment in the automotive sector: A comparative case study of internal combustion engine (ICE) and electric car. *Procedia Structural Integrity, 12*(11), 521–537. https://doi.org/10.1016/j.prostr.2018.11.066

Plested, D. (2021). *The future of urban mobility. Flexible, scalable, on-demand transit*. Deloitte.

Pyddoke, R. (2020). *Analysis of daily variation in bus occupancy rates for city-buses in Uppsala and optimal supply*. Wedish National Road & Transport Research Institute. Working papers 8.

Rietschin, R. (2015). *Memento Statistique du canton de Geneve*. Republic et Canto de Geneve.

Service de la mensuration officielle. (2005). Statistiques cantonales. les 45 communes genevoises

Shalkamy, A., Said, D., & Radwan, L. (2015). Influence of carriageway width and horizontal curve radius on passenger car unit values of two-lane two-way rural roads. *International Knowledge Sharing Platform, 7*(3), 157.

Shen, Y., Zhang, H., & Zhao, J. (2018). Integrating shared autonomous vehicle in public transportation system: A supply-side simulation of the first-mile service in Singapore. *Transportation Research Part A: Policy and Practice, 113*, 125–136.

Sousa, N., Almeida, A., Coutinho-Rodrigues, J., & Natividade-Jesus, E. (2018). Dawn of autonomous vehicles: Review and challenges ahead. *Proceedings of the Institution of Civil Engineers - Municipal Engineer, 171*(1), 3–14. https://doi.org/10.1680/jmuen.16.00063

Statistics Denmark. (2021). Population figures from the census by nationa part and time. Available online at https://statbank.dk/statbank5a/default.asp?w=1280

Sytral. (2016). *Enquête déplacements 2015 de l'aire métropolitaine lyonnaise. Principaux résultats*. Syndicat mixte des transports pour le Rhône et l'agglomération Lyonnaise.

Townsend, A. (2014). *Re-programming mobility. The digital transformation of transportation in the United States*. Rudin Center for Transportation Policy & Management NYU.

TPG. (2016). *Transports publics genevois*. Rapport annuel de gestion.

TUMI. (2018). *Transport demand management. Push and pull.* TUMI. Available online at https://www.transformative-mobility.org/publications/transport-demand-management-push-and-pull

UITP. (2017). Policy brief. Autonomous vehicles: A potential game changer for urban mobility. Brussels. Available online at https://www.uitp.org/publications/autonomous-vehicles-a-potential-game-changer-for-urban-mobility/

van Essen, H., van Wijngaarden, L., Schroten, A., Sutter, D., Bieler, C., Maffii, S., Brambilla, M., Fiorello, D., Fermi, F., Parolin, R., & El Beyrouty, K. (2019). *Handbook on the external costs of transport* (2019th ed., p. 68). Publications Office of the European Union.

Viere, T., Nemoto, E., Jaroudi, I., & Korbee, D. (2021). *Deliverable 8.2 second iteration environmental impact assessment.* AVENUE project. Available online at https://h2020-avenue.eu/public-delivrables/

Ville de Luxembourg. (2021). Le portail des statistiques. Available online at https://statistiques.public.lu/fr/index.html

Wadud, Z., MacKenzie, D., & Leiby, P. (2016). Help or hindrance? The travel, energy and carbon impacts of highly automated vehicles. *Transportation Research Part A: Policy and Practice, 86*(6), 1–18. https://doi.org/10.1016/j.tra.2015.12.001

Ward, J. W., Michalek, J. J., Azevedo, I. L., Samaras, C., & Ferreira, P. (2019). Effects of on-demand ridesourcing on vehicle ownership, fuel consumption, vehicle miles traveled, and emissions per capita in U.S. states. *Transportation Research Part C: Emerging Technologies, 108*(3), 289–301. https://doi.org/10.1016/j.trc.2019.07.026

Zakharenko, R. (2016). Self-driving cars will change cities. *Regional Science and Urban Economics, 61*, 26–37.

Zhang, W., & Guhathakurta, S. (2017). Parking spaces in the age of shared autonomous vehicles: How much parking will we need and where? *Transportation Research Record: Journal of the Transportation Research Board, 2651*(1), 80–91. https://doi.org/10.3141/2651-09

Zhong, H., Li, W., Burris, M. W., Talebpour, A., & Sinha, K. C. (2020). Will autonomous vehicles change auto commuters' value of travel time? *Transportation Research Part D: Transport and Environment, 83*, 102303.

Zmud, J., Tooley, M., Baker, T., & Wagner, J. (2015). *Paths of automated and connected vehicle deployment: Strategic roadmap for state and local transportation agencies.* A&M Transportation Institute. https://static.tti.tamu.edu/tti.tamu.edu/documents/161504-1.pdf. Accessed 11 March 2022

Zmud, J. P., & Sener, I. N. (2017). Towards an understanding of the travel behavior impact of autonomous vehicles. *Transportation Research Procedia, 25*, 2500–2519. https://doi.org/10.1016/j.trpro.2017.05.281

# Chapter 15
# Social Impact Assessment: Changing Mobility Behaviour by Understanding Customer Needs and Attitudes

**Dorien Duffner-Korbee, Gabriele Naderer, Niklas Liebhauser, and Guy Fournier**

**Abstract** This chapter presents results from the AVENUE social impact assessment. The assessment focuses on the social impacts of the deployment of automated minibuses in the four official AVENUE cities: Luxembourg, Copenhagen, Geneva, and Lyon. The following studies have been conducted: (1) a qualitative study on mobility needs, mobility gaps, and expectations on automated minibuses in the pilot site Nordhavn, in Copenhagen, Denmark, as well as in a potential replicator site in Singen, Germany; (2) a quantitative, representative study on mobility behaviour, attitudes on automated minibuses, and social acceptance of automated minibuses in the four AVENUE cities; and (3) a study on user experiences of passengers of the automated minibuses service in Nordhavn, Copenhagen. This social impact assessment revealed that the majority of citizens have not yet adopted a clear position towards automated minibuses (AM) but rather a positive, open-minded (benevolent) attitude. As there are no stable attitudes yet, there is a possibility to influence (nudge) preferences through well-targeted communication campaigns. The willingness to use the automated minibus service is higher when it is offered on-demand, door-to-door, integrated into a public transport system. The automated minibus is considered as an alternative to the use of their private car, with 45% of car drivers 'willing' (22%) or even 'very willing' (23%) to give up using their own car to use automated minibuses to bridge the first and the last mile. In addition, users of the automated minibus are generally satisfied with this experience, which even leads to increasing acceptance and a reduction of perceived risks compared to non-users. An important

---

D. Duffner-Korbee (✉)
Pforzheim University, Pforzheim, Germany

Fraunhofer ISI, Karlsruhe, Germany
e-mail: dorien.duffner-korbee@isi.fraunhofer.de

G. Naderer · N. Liebhauser
Pforzheim University, Pforzheim, Germany

G. Fournier
Pforzheim University, Pforzheim, Germany

CentraleSupélec, LGI, Gif-sur-Yvette, France



G. Fournier et al. (eds.), *Automated Vehicles as a Game Changer for Sustainable Mobility*, Contributions to Management Science,
https://doi.org/10.1007/978-3-031-61681-5_15

prerequisite for users to shift from using their private car to an automated minibus service is to increase flexibility of use by providing an on-demand, door-to-door service. If an automated minibus is integrated into the urban public and private transport system, it has the potential to become a real game changer for urban mobility. The integration of automated minibuses therefore represents a real system innovation that takes into account the mobility needs of potential users.

## 15.1 Introduction

The current private car-based mobility paradigm is facing severe challenges, such as meeting climate goals, reducing congestion and air pollution (Banister, 2008; Fournier et al., 2020; Geels et al., 2017; Kuss & Nicholas, 2022; Pribyl et al., 2020). To respond to these challenges, several trends have emerged, including electrification, automation, low-carbon transition, shared transportation, and intermodal transport (Fagnant & Kockelman, 2015; Geels, 2012; Hirschhorn et al., 2019). Innovative mobility modes and systems are developing, such as mobility as a service (MaaS), car-sharing systems, and automated vehicles (Firnkorn & Müller, 2011; Pangbourne et al., 2020). Of particular interest in this chapter is the automated minibus. Automated minibuses are compact, electric-powered, pod-like vehicles deployed in public transport systems (Huber et al., 2022; Nemoto et al., 2021; Nordhoff et al., 2018). This chapter presents the AVENUE social impact assessment, focusing on social acceptance as the capability to attract and retain users.

Social acceptance of an innovation is a prerequisite for its use. Acceptance does not inevitably lead to the adoption (use) of the innovation but rather signifies/represents a positive intentional attitude towards the innovation (Kilian-Yasin et al., 2016). Assessing social acceptance is thus crucial for predicting whether, and under which circumstances, an innovation will be accepted and used. One consideration that is often forgotten in studies on social acceptance is the previously unmet needs that impact social acceptance.

Therefore, in this chapter, an approach based on needs and attitudes is applied to assess social acceptance. Attitudes towards automated minibuses are comprised of the subsequent dimensions: perceived risks and benefits, performance expectations, trustworthiness of the technology, effectiveness, accessibility, price, and willingness to use (Banister, 2008; Hirschhorn et al., 2019; Kyriakidis et al., 2015; Nordhoff et al., 2018). Recent studies demonstrate that attitudes on automated minibuses, such as willingness to use, are strongly influenced by previous experiences with the automated minibuses (Zoellick et al., 2019). These prior experiences are also a vital predictor to evaluate behavioural intentions. Hence, to assess the potential of automated minibuses, it is important to incorporate insights on user experiences.

The following research questions are guiding the research:

What mobility needs does the customer have, and how can these be met by automated minibuses?

What attitude do customers have towards automated minibuses, and how does that influence acceptance?

And finally, how does the specific user experience affect the attitude towards and acceptance of automated minibuses?

## 15.2 Research Approach

To answer these research questions, a multi-method study was designed, combining qualitative and quantitative methods. Table 15.1 provides an overview of the studies. The mobility needs assessment (RQ1) consists of 20 in-depth, semi-structured interviews with citizens; 8 citizens live in a pilot site and 12 citizens live in an area with potential automated minibus service.

To evaluate the attitudes towards automated minibuses and mobility in general (RQ2), a quantitative, representative survey among 1.816 potential users of four cities (Copenhagen, Geneva, Luxembourg, and Lyon) that host pilot sites of the AVENUE project was conducted. User experiences (RQ3) were evaluated through a survey among 68 users in the AVENUE pilot site in Copenhagen.

### *15.2.1 Mobility Needs, Mobility Gaps*

Two qualitative studies were conducted to identify mobility requirements and their associated deficiencies. One of the studies was conducted in Germany, specifically in Singen, a small village situated between Karlsruhe and Stuttgart, while the other

**Table 15.1** Overview studies conducted for the social impact assessment

| Objective | Study | Target group | Method | Sample size |
|---|---|---|---|---|
| Mobility needs assessment (RQ1) | Mobility needs, mobility gaps and expectations of automated minibuses | Citizens | Qualitative interviews | $n = 12$ |
| | Mobility behaviour at pilot site level | Residents of the pilot site | Qualitative, longitudinal study interviews | $n = 8$ |
| Attitudes towards automated minibuses (RQ2) | Expectations, attitudes, and acceptance of automated minibuses | Potential users | Quantitative, online survey | $n = 1816$ |
| User experiences (RQ3) | The user experience of actual users | Users of the automated minibus | Quantitative, online questionnaires | $n = 68$ |

was conducted in Copenhagen's Nordhavn area. The objective of the study conducted in Singen was to gain an understanding of the current situation, problems, and needs of the residents regarding their mobility. Furthermore, the study aimed to analyse expectations and concerns regarding possible future mobility solutions, such as the automated minibus. The objective of the study conducted in Copenhagen was to gain insight into the mobility needs, attitudes, and mobility behaviour of citizens living near an AVENUE pilot site. Moreover, our objective was to determine whether the presence of an automated minibus in their neighbourhood had an impact on their mobility needs, attitudes, and mobility behaviour.

Participants were selected based on a theory-based sampling method. A qualitative approach does not strive for statistical representativeness but rather psychosocial representativeness. Therefore, theoretically relevant criteria were defined for the selection of the respondents: age, gender, PRM, place of residence, perceived mobility gaps, satisfaction with PT, as well as current mobility behaviour. Twelve citizens were selected from Singen and eight from Copenhagen. Qualitative, semi-structured interviews were conducted in Singen. In Copenhagen, the data collection was divided into different steps. First, exploratory interviews were conducted to gain an understanding of the participants' attitudes towards mobility and their mobility behaviour. In order to assess a change in attitude, participants were sent six online surveys over the course of 6 weeks. After the completion of the six questionnaires, a concluding interview was conducted to evaluate the alterations in perceived needs, attitudes, and mobility behaviour.

### *15.2.2 Attitudes and Social Acceptance*

To understand how automated minibuses are perceived and accepted in cities across Europe, a large-scale, representative online survey was launched in the four AVENUE cities in 2021. The online questionnaire was fully structured and programmed in 'Questback/Tivian'. A total of 1816 citizens participated in the survey across the four AVENUE demonstrator cities. The survey invitation was disseminated through a sample procured from a market research institute. The sample was representative based on gender and age distribution. The sample structure was diverse, reaching all age groups, female and male potential users, employees, as well as students and households with or without children (see Table 15.2). For data analysis, the statistical software SPSS was used.

A three-step approach was followed in developing the questionnaire. First, the main parameters and indictors were assessed based on theoretical insights and literature. Second, a repository of questions used in previous questionnaires on automated driving was created (based on Bernauer & Wicki, 2018; Keolis Downer, 2018; Kilian-Yasin et al., 2016; Schoettle & Sivak, 2014). The questions were grouped according to our main parameters and indicators. Where necessary, additional questions were formulated. Third, experts, such as public transport operators in the cities of focus, provided feedback on the questions.

**Table 15.2** Sample structure representative online survey

| | Total respondents | 1816 |
|---|---|---|
| City | Lyon | 501 (28%) |
| | Copenhagen | 491 (27%) |
| | Geneva | 284 (16%) |
| | Luxembourg | 540 (30%) |
| Age | Younger than 36 years | 25% |
| | Older than 36 years | 75% |
| Gender | Male | 47% |
| | Female | 53% |

For the majority of the questions, we applied a 5-point Likert scale (Brosius et al., 2012). Only the endpoints are named with (1) describing the negative configuration, e.g. 'fully disagree' or 'not important at all', and (5) describing the positive form, e.g. 'fully agree' or 'very important'. To not overstrain respondents, an additional answer option (6) 'I can't judge' or 'I don't know' was added in specific questions. To avoid cancelling none of the questions was set as mandatory.

### 15.2.3 *User Experience*

User surveys were conducted to acquire a more in-depth understanding of the users' interaction with the new service and their perceptions of it. In the user survey, users were asked to reflect on their experience using the automated minibus and to evaluate its usability. The survey consisted of two blocks. A first block of the survey was dedicated to the evaluation of the service. A second block of the survey addressed more general questions about attitudes towards automated minibuses. The second block contained questions that were also part of the large-scale representative survey.

A total of 68 respondents in Copenhagen completed the survey. The survey consisted of two blocks. These participants were asked to conduct the survey, while using the automated minibus. Therefore, the decision to use the service was not influenced by the researchers. The user survey was conducted online, using Questback/Tivian software. Passengers received a flyer from the safety operators with concise information and a link to the online survey. Since the questionnaire could also be completed via mobile devices, passengers had the choice whether they wanted to answer the questionnaire during their ride with the automated minibus or afterwards.

Additionally, respondents were recruited through the aforementioned large-scale representative survey (see Sect. 2.2). In this survey, respondents were asked whether they had previously used an automated minibus. If they had, they were subsequently requested to respond to a set of questions from the user survey. A total of 126 respondents of the large-scale representative survey indicated that they had previously used an automated minibus service and completed the additional user survey questions. In the presentation of the results of the user survey (Sect. 5.1), this group is not mixed with the users of the Nordhavn site. In comparing users to non-users (Sect. 5.2), these respondents are treated as a separate group.

**Table 15.3** Mobility needs

| | Mobility needs | Reliefs |
|---|---|---|
| Mobility needs | Needs for effectiveness | Spatial flexibility |
| | | Temporal flexibility |
| | | Reliability |
| | Needs for cognitive relief | Ease of use |
| | | Convenience |
| | Needs for emotional relief | Safety |
| | | Security |
| | Needs for physical relief | Comfort |
| | Needs for environmental relief | Impacts on climate change |
| | Needs for financial relief | Cost of mobility |

## 15.3 Mobility Needs

A crucial element for the successful implementation of innovations is their compatibility with current, actual, and latent needs. The first step in understanding how new mobility systems like automated minibuses can meet these needs is to understand current, actual, latent, and unmet needs in both prospective and actual AVENUE pilot sites. Therefore, the main aim of this needs assessment is to understand what the relevant needs are. In the analysis and in presenting the results, we divide between the mobility needs and the aspects that are required or are in place to fulfil these needs. The latter is termed 'reliefs' in Table 15.3.

According to Schmitt and Altstötter-Gleich (2010), attitude is defined as the 'degree to which a person likes or dislikes an object' (2010, p. 146). The attitude thus always refers to a specific attitude object. In the context of this study, the attitude object is the 'integration of automated minibuses', which is expressed by the perceived suitability of the automated minibus to best meet personal mobility needs.

### *15.3.1 Needs Assessment: Singen and Nordhavn*

Singen has a population of approximately 3800 and is approximately 10 square kilometres in size. The village is characterized by a significant difference in altitude. Due to the difference in altitude, the village is divided into two distinct sections, namely 'Unterdorf', which is located in the lower part, and 'Oberdorf', located in the upper part. The majority of the stores, as well as the train station, are situated on the outskirts of Unterdorf. The village is well connected to major roads, such as the B10. Additionally, there is a bus route that connects Singen to other nearby villages. Cars are the most used transportation mode among our participants, followed by walking and cycling, while trains and buses are used only rarely or not at all (Fig. 15.1).

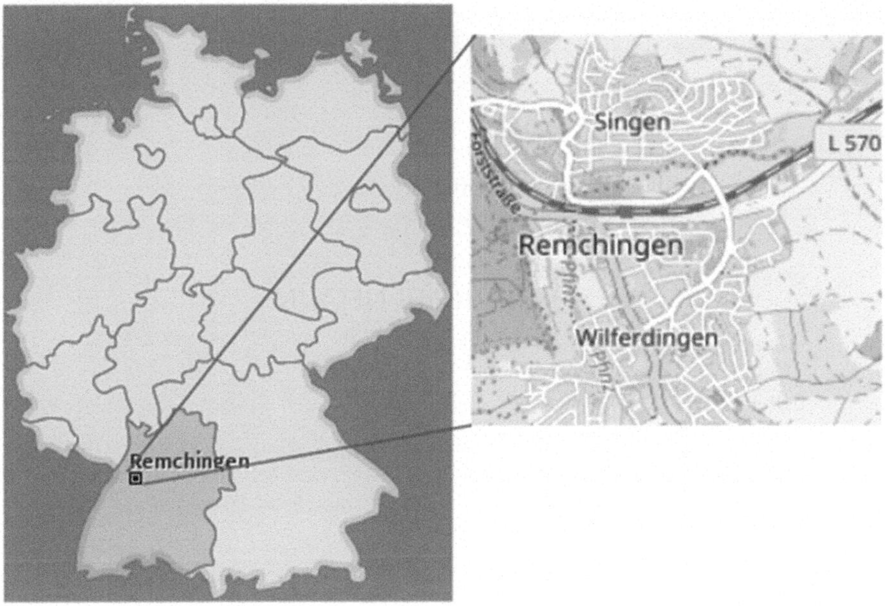


**Fig. 15.1** Location of Singen, Remchingen

Nordhavn is a dynamic industrial port that is currently being transformed into Copenhagen's new international waterfront district, accommodating both residential and commercial structures. Upon completion of Nordhavn's development, the area will accommodate more than 40,000 residents and 40,000 employees. The vision is for Nordhavn to become an environmentally friendly neighbourhood and contribute to Copenhagen's image as an environmental metropolis. The introduction of new energy sources and eco-friendly modes of transportation will help establish Nordhavn as a leading example of eco-friendly construction and design. Currently, the Nordhavn area is served by a nearby S-train station and bus stops. However, there are no direct buses or trains operating in the area. This presents a significant opportunity for automated vehicles to serve as a novel public transport (PT) solution and enhance the connectivity of the region beyond its current status. The automated minibus route is located in the Århusgadekvarteret area.

Our participants use all available public transportation options. They consider the metro to be an integral part of their daily mobility, alongside walking and cycling. At the beginning of the study, participants expressed a strong preference for walking. Private cars were less attractive to the study respondents, and in some cases, even a strong aversion against their use was identified (Fig. 15.2).

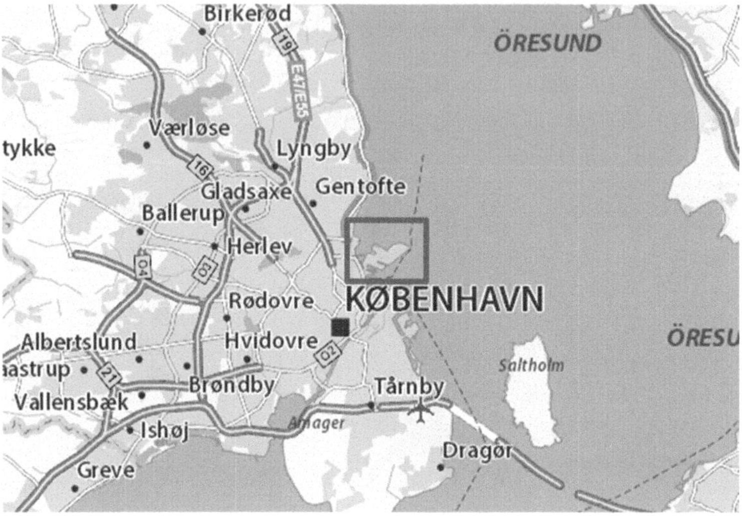


**Fig. 15.2** Location of Nordhavn

## *15.3.2 Needs for Effectiveness*

The analysis indicates that effectiveness is the primary need for mobility. The ability to travel swift and uninterrupted from point A to point B is essential for effective mobility. In order to achieve effectiveness, both spatial and temporal flexibility are essential. A combination of spatial and temporal flexibility allows users to access and use the transport mode whenever and wherever they want, providing users with a certain degree of freedom.

Respondents in Singen stress the importance of being able to reach and leave the train station at any given moment to catch a scheduled train and then return to the station. This mobility need is not being met at this time. The perception of impracticality is based on factors such as the distance of bus stops and departure times, as well as the frequency of bus stops. Hence, travelling by bus is currently perceived as ineffective and time-consuming.

> The bus stop is so far from my house. Until I have walked to the bus stop, I may as well walk right down to the station. (Participant from Singen)

In Copenhagen, participants expressed satisfaction with the spatial flexibility of the transportation system, owing to the tightly knit network of stations that can be quickly accessed from any location. Furthermore, metro and train can bypass traffic congestion on the streets and provide a direct connection between two locations. Nonetheless, the requirement for temporal flexibility freedom is not always fully fulfilled, as trains and metros in Copenhagen operate with reduced frequency in certain areas or at specific times of the day, such as at night. The advantages of the bus network in Copenhagen are seen in the accessibility of the bus stops. However, some bus lines and stations have recently been eliminated due to the extension of the metro. Therefore, some participants complained about longer travel times or poorer access to the bus stops. Due to the substantial traffic volume in Copenhagen, private cars and buses are not regarded as flexible and effective mobility alternatives.

### *15.3.3 Cognitive Relief*

The cognitive needs are those mobility needs that involve stress reduction and psychological well-being, such as being able to travel spontaneously without requiring detailed planning and not having to change the means of transport several times on one trip. Using bicycles provides cognitive relief because no planning is required, but it can also lead to other stressful problems, such as fear that the bike will be stolen. Copenhagen has a relatively flexible public transportation system, which provides a certain degree of cognitive relief. Nonetheless, certain issues remain, such as the positioning of bus stops, as well as the frequency and network of the public transportation system. As for cycling, the well-developed cycling paths provide an alternative to the congested streets, thus offering cognitive relief from the stress of being stuck in traffic with the car. On the contrary, the situation is sometimes referred to as 'cycling chaos', which may also result in stress and diminish the cognitive relief derived from cycling (see also the section on safety).

### *15.3.4 Reliability*

The need for reliability involves a need for the mode of transport to arrive punctually and sufficient information supply. To attain a high level of effectiveness, a high level of reliability of the respective means of transport is required.

> I always ask myself, which is the fastest transport? There are two things I'm considering, firstly with which mode of transport I'm reaching my destination faster and also in time regarding the punctuality. So that I can trust it. (Participant from Copenhagen, a man without handicap, age 55)

The bus system in Singen fails to meet the needs for reliability and punctuality due to delays and an inadequate information policy. Therefore, the need for reliability can be better fulfilled through using the car. In Copenhagen, the need for reliability is also considered to be inadequately met by the metro and train. At times, the trains are subject to delays owing to disruptions, maintenance, or train failures. Hence, the reliability of the public transportation system is compromised and fails to provide adequate freedom, flexibility, and cognitive relief, thereby compromising its effectiveness.

### *15.3.5 Needs for Physical Relief*

Needs for physical reliefs relate needs that are associated with the physical environment, such as the need for comfort. This need is particularly acute in Singen due to its steep, hilly roads. These areas are easily accessible by car but can be challenging when walking, using a bike, or taking the bus. The need for physical relief is especially acute if larger purchases must be transported.

> … And then there is again the question of how to get up the hill with the purchase? That's why I use the car again. (Participant from Singen)

In addition to the topography, poor road conditions (few bike lanes, parked cars, detours over railroad tracks) and bad weather are problems for physical relief when using the bicycle. Thus, bicycles provide less physical relief in comparison to a car.

In the Copenhagen study, we identified several elements of physical relief, including the ability to transport purchases, provide protection from adverse weather conditions, and maintain high standards of hygiene. A specific need is expressed by respondents with reduced mobility (PRM) stating a need for physical relief due to the physical burden of their handicap. The reduction in the number of bus lines and stations in Copenhagen has led to a decrease in the accessibility of bus stops, which has resulted in a reduction in physical relief.

### *15.3.6 Needs for Emotional Relief*

The need for emotional relief is multifaceted. Among other aspects, it encompasses perceptions of safety and security. In general, all forms of transportation should be safe and secure. In Singen, a special need for safety when returning home from the train station at night was explicitly addressed. The area around the train station, including the underpass, has been called a particularly dangerous area, so using the car is considered a better way to ensure emotional well-being. In Copenhagen, public transport doesn't always meet the need for emotional comfort and security because trains and metros run at a lower frequency in certain areas or at certain times of the day. In general, female participants expressed a sense of discomfort while awaiting trains at night.

> When it is really late in the night I never use busses, to be honest. And also, the trains are only every 20 min, so I do not feel comfortable waiting at the station or even being in the s-train. (Participant from Copenhagen, woman, age 40)

Moreover, not all participants in Copenhagen felt safe when riding the bicycle. Several participants even referred to the current situation in Copenhagen as 'cycling chaos', which results in a sense of unease. Fun, entertainment, or other emotional experiences while travelling were not mentioned as a major need. This need only exists for long-distance trips and is not relevant for shorter trips.

### *15.3.7 Needs for Environmental Relief*

Environmental relief and impacts on climate change are a mobility need that is not important to the majority of respondents in Singen. Only participants who already have a strong awareness of environmental protection value this need highly.

> To be honest, the environmental aspect was not a reason for me. Or just like everyone tries to be a bit resource-conserving, but that wasn't really a motive for me. (Participant from Singen)

In Copenhagen, two groups of respondents can be distinguished. One group consists of 'engaged environmental advocates', who are highly involved in climate change and sustainability. These individuals disapprove of the fact that the majority of buses in Copenhagen utilize fossil fuels. Nonetheless, they express optimism regarding the public transportation system, which is regarded as an environmentally friendly system, wherein resources are utilized effectively. The other group are 'environmental friends' who are aware of climate change but for whom sustainable mobility is less relevant. Thus, the current transportation system meets their need for sustainable transportation.

### *15.3.8 Needs for Financial Relief*

Price was not a mobility need for the participants in case the price is seen as reasonable for the mode of transportation.

> The costs are important when thinking of which mode of transport. But it is more that it should be a justified price, like when I get a high standard of service and good quality, I'm willing to pay more. (Participant from Copenhagen, a man without handicap, age 55)

In Singen, for example, the high bus fares were criticized. The degree of current need fulfilment is—in the view of the citizens—disproportionate to the ticket prices.

### *15.3.9 Wrap Up*

The needs identified among the respondents show one thing above all: there is no acute need for 'automated' minibuses. Neither the need for being electrically powered in the sense of greater environmental friendliness nor for being automated in the sense that no driver is needed has priority. Other needs are the focus. The main needs identified are needs for a high level of effectiveness, gained by a high level of flexibility and reliability and specific types of relief: cognitive relief, emotional relief, and physical relief. Environmental relief and financial relief are secondary needs.

## 15.4 Attitudes and Acceptance

The following chapter presents the results of the representative, large-scale quantitative study on attitudes and acceptance of automated minibuses by the citizens of the four AVENUE pilot cities.

### *15.4.1 Awareness and Willingness to Use Automated Minibuses*

A slight majority of respondents (55%) knew about the existence of automated minibuses before the survey. In 2021, the awareness of automated minibuses was greatest in Luxembourg (72%) and lowest in Copenhagen (40%). The difference in awareness is most likely due to difference in visibility of the pilot sites. The site in Luxembourg was situated in a central area, while the site in Copenhagen was located in a new neighbourhood. There is a wide range of information sources through which participants had gained awareness of the automated minibuses: newspaper (37%), radio/tv (50%), the Internet (33%), social media (15%), seen on the test site (22%), or via friends (13%). Only a small number of respondents (7% overall) have used the automated minibus service before (Table 15.4). That could be due to the small portion of the cities covered by the automated minibus service, and in addition, the Corona pandemic decreased the willingness to use PT in general.

To gain insight into the acceptance of citizens to use the automated minibus services, respondents were asked to think about their willingness to use the automated minibus. The consensus in academic literature (see among others Nordhoff et al., 2020; Hirschhorn et al., 2019; Zoellick et al., 2019) is that there is generally a high level of acceptance and willingness to use automated minibuses. We inquired for willingness to use the automated minibuses in three different scenarios of deployment to obtain a more nuanced picture of the willingness to use automated minibuses. The respondents were presented with information on these three deployment scenarios before the questions; see Box 1.

**Table 15.4** Awareness of automated shuttles

| | Copenhagen ($n = 491$) | Geneva ($n = 284$) | Luxembourg ($n = 540$) | Lyon ($n = 501$) | Overall ($n = 1816$) |
|---|---|---|---|---|---|
| Awareness of automated shuttles[a] | 197 | 141 | 387 | 278 | 1003 |
| | 40% | 50% | 72% | 56% | 55% |
| Experience with automated shuttles | 21 | 22 | 40 | 43 | 126 |
| | 4% | 8% | 8% | 8% | 7% |

No statistical dependence between city and experience with automated minibuses

[a]Statistical dependence between city and awareness of automated minibuses, (chi-square (3) = 107.917, $p < 0.001$)

**Box 1 Explanation Provided to the Respondents on the Three Automated Minibus Modes of Deployment**

The automated minibuses can be integrated in the public transport system to bridge the distance between your place of departure or destination to a public transport station. This is called a service that bridges the first and last mile.

The automated minibuses can provide a service that is not based on a fixed timetable. The automated minibus comes at a time requested by you, to a location specified by you. This location can be the nearest bus stop, or even your doorstep. This is called an on-demand, door-to-door service.

The automated minibuses service can be integrated in a system together with other means of transportation. This means that you can easily combine a trip with the automated minibus with a trip on the train, bus or taxi. As the automated minibuses will operate on-demand, the system will allow you to plan your trip seamlessly. This is called a seamless, intermodal trip.

The willingness to use is high. Irrespective of the scenario, more than 50% 'would be willing' or even be 'very willing' to use automated minibuses, even two-thirds (63%) in case an on-demand service is offered (see Table 15.4). The willingness to use the automated minibuses as an on-demand service is high in all four cities, with the highest level observed in Geneva, where a generally higher use of buses is observed. Willingness is lowest in Copenhagen. Here, potential users believe that automated minibuses are best suited for short to medium distances. For these distances, however, bicycles already offer an adequate alternative due to the topography of Copenhagen (Fig. 15.3).

Likewise, the willingness to use the automated minibuses as a service that links the first and last mile is high in all four cities, once again highest in Geneva, where a higher usage of buses is observed, followed by Lyon. By contrast, it is lower in Copenhagen and Luxembourg, but differences are only partially significant (Fig. 15.4).

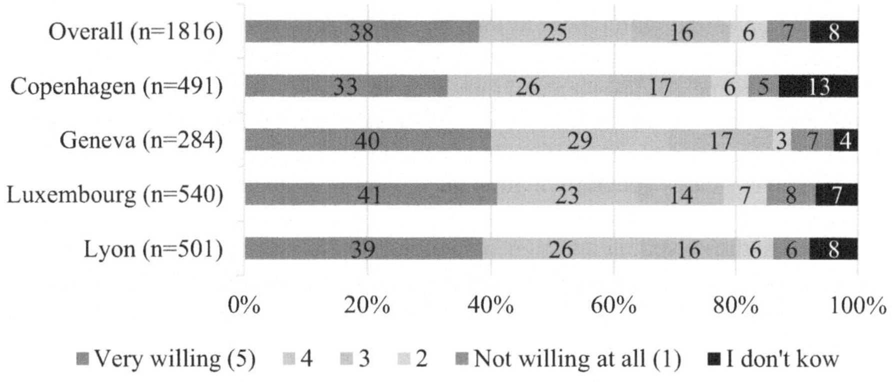


**Fig. 15.3** Willingness to use automated minibus per city if the automated minibus offers an on-demand, door-to-door service (in %, $n = 1816$)

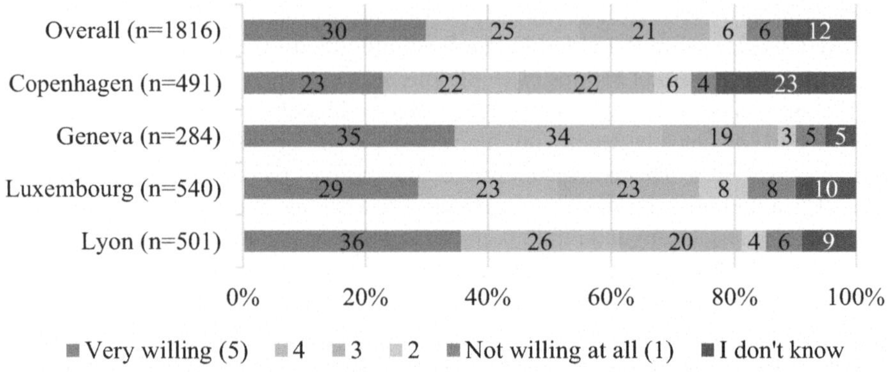


**Fig. 15.4** Willingness to use automated minibus per city if the automated minibus offers a service that bridges the first and last mile (in %, $n = 1816$)

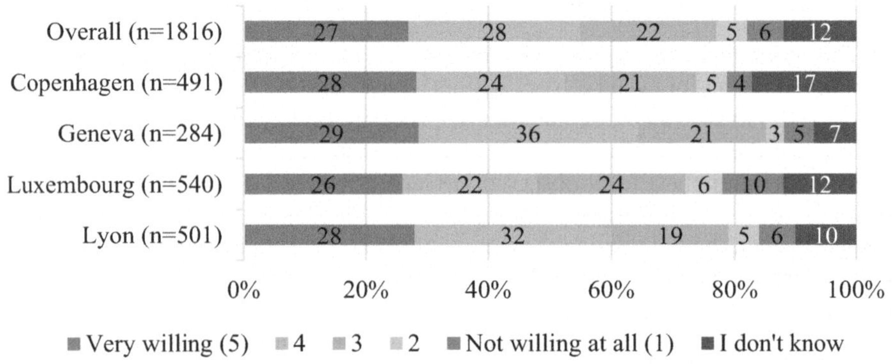


**Fig. 15.5** Willingness to use automated minibus per city if the automated minibus is part of a seamless, intermodal trip (in %, $n = 1816$)

Similar effects can be seen in the scenario where the service is offered as part of a seamless trip (Fig. 15.5).

In addition to the willingness to use, the respondents were also asked to think about their willingness to reduce their car use. Of interest are those respondents that actually (field time 2021) own a car; therefore, the results exclude respondents who do not own a car. Hence, the presented figures are based on a slightly smaller sample, as only 1526 out of the 1816 respondents own a car. The willingness to reduce the use of one's car is generally high in all scenarios. In the case of an on-demand, even more than 50% would be 'willing to reduce use of own car' (Fig. 15.6).

Similar to the preceding question, respondents were asked to consider their willingness to give up the use of their car completely. Approximately 50% 'would be willing' to give up the use of their own car, in case of an on-demand service. If the automated minibus service would bridge the first and the last mile, 45% of car drivers are 'willing' (22%) or even 'very willing' (23%) to give up using their own car to use automated minibuses (Fig. 15.7).

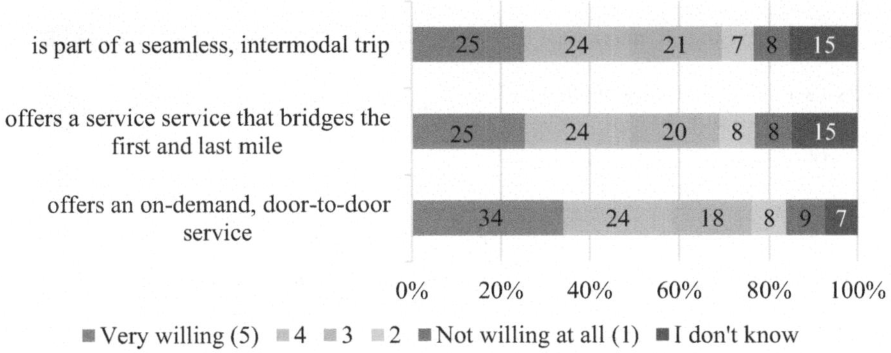


**Fig. 15.6** Willingness to reduce the use of your car if the automated minibus… (in %, $n = 1526$*). *Reduced base, as non-car owners are excluded

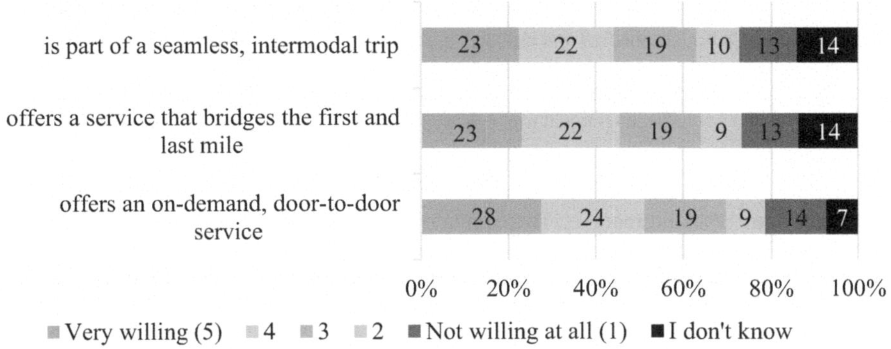


**Fig. 15.7** Willingness to give up the use of your car if the automated minibus… (in %, $n = 1526$*) *Reduced base, as non-car owners are excluded

The respondents were also asked to reflect on their willingness to increase the use of public transport if automated minibuses offer an on-demand service. At least 50% 'would be willing' to increase the use of public transport if the automated minibus offers an on-demand service (Fig. 15.8).

The respondents were also asked to think about their willingness to pay for the automated minibus service. The willingness to pay more for an automated minibus, compared to current public transport, is low—see Fig. 15.9. About 6% are very willing, and 30% are willing to pay at least a bit more if the service would be on-demand door-to-door. The automated minibus is not regarded as a truly inspiring innovation (see two-factor model, Herzberg 1968) but rather as a service that removes existing deficits in terms of effectiveness and flexibility (hygiene factor), i.e. it merely reduces the existing dissatisfaction. This could explain why most people are not willing to pay more. The subjective measure might even be that respondents feel that they are paying too much for the current, rather unsatisfactory service.

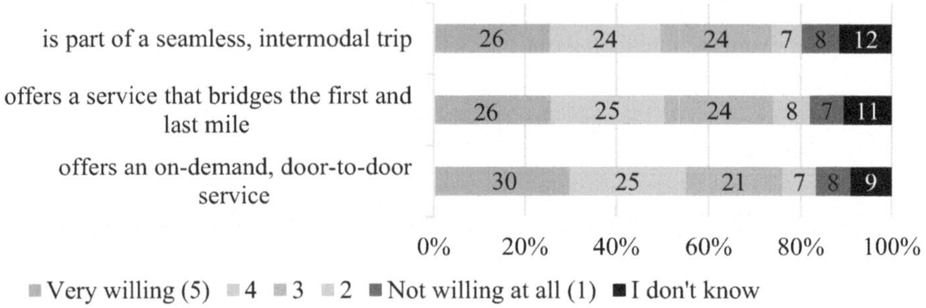


**Fig. 15.8** Willingness to increase the use of public transport systems (including current offers), if the automated minibus... (in %, $n = 1816$)

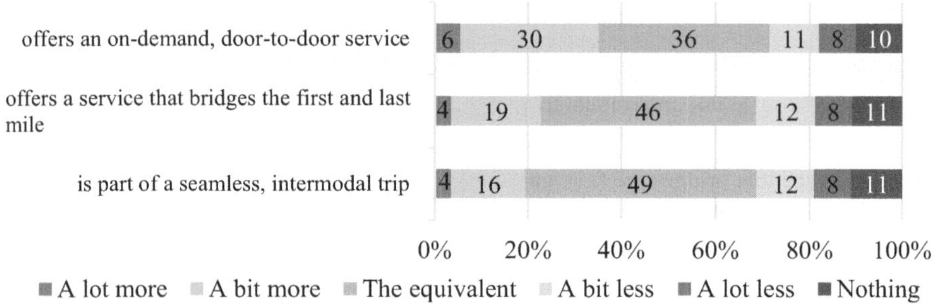


**Fig. 15.9** Willingness to pay to use the automated minibus, if the automated minibus... (in %, $n = 1273$*). *This question was not asked in Luxembourg, as Luxembourg offers free public transport

In addition to the different scenarios, respondents were asked to select a mobility option for a specific journey. The options differ in terms of flexibility and duration, cost, and environmental friendliness, see Fig. 15.10. Two of the options included the automated minibus. The results show that 25% prefer a fast but expensive option, either by private car or robotaxi, 26% prefer a cheap, environmentally friendly option, even if this implies a longer travel time, and 28% choose a service that incorporates the automated minibus. The latter seems to be mostly driven by the travel time. These results are in line with identified factors for selecting prefered mode of transport (see Fig. 15.11). Here, 59% of the respondents indicated that speed and travel are ranked first, second or third most important factor in selecting their prefered mode of transport.

Q: Which of the presented mobility options would you prefer? (overall n=1816)

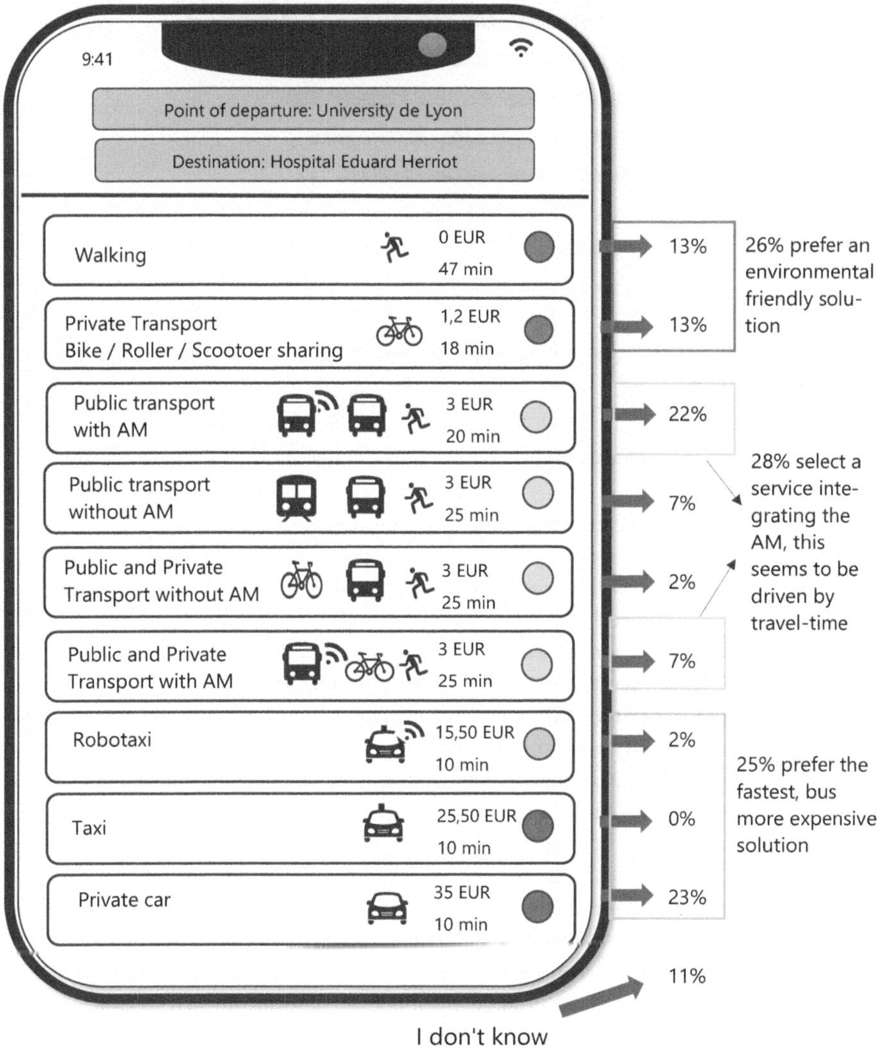


**Fig. 15.10** Preference for mobility options

## *15.4.2 Advantages and Concerns*

The general impression is that the respondents have a positive perception of the advantages of automated minibuses. About two-thirds expect the automated minibuses to ('agree' or 'fully agree') provide more freedom for people with reduced mobility (64%) that they will increase flexibility by providing a greater number of

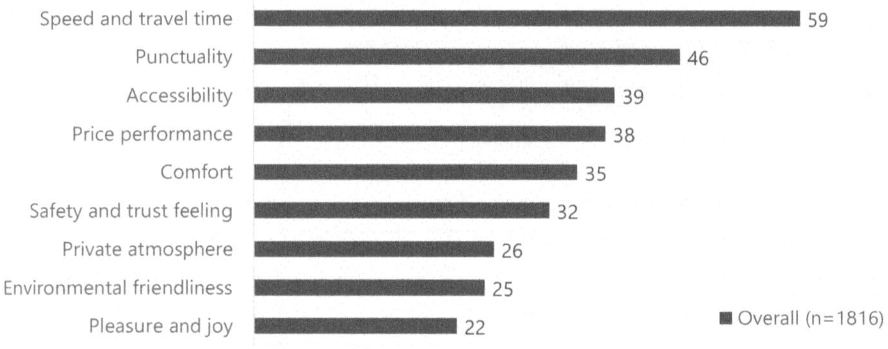


**Fig. 15.11** Importance of factors for selecting the preferred mode of transport, divided by the market segments (in %)

start and stop locations (64%) and by increasing the frequency of PT offers (63%)—see Fig. 15.12.

Additionally, a positive environmental impact is expected, e.g. reducing the negative impact on the environment (60%) and congestion (49%). A pleasant and comfortable journey (54%) with guaranteed security due to installed video cameras inside the buses is expected (52%) by every second respondent. More prudent are the positive expectations concerning the question of whether the automated minibus will truly lead to fewer accidents by eliminating human errors.

When it comes to perceived concerns, the picture is slightly more nuanced. Respondents show a high agreement with concerns regarding the operation of the automated minibus (this includes the interaction with motorized (49%) and nonmotorized traffic (46%), its reaction to unforeseen situations (54%), and issues of liability in the case of an accident (44%). Social concerns are also pertinent to respondents. 41% of the respondents are in agreement with the statement 'jobs get lost'. These concerns are consistent with the results of our social media analysis. Here, more than 1200 comments were analysed. The comments were associated with different online textual contents, images, or video clips from around the world associated with the topic of automated public transit stemming from a variety of channels ranging from the online press to internet communities and blogposts through the multiple social networks. The study suggests that 'social concerns' is undoubtedly the most represented of the previously identified categories (concerning 43% of the comments). In particular, the negative consequences for the labour market (14.9% of the occurrences) seem to be a concern (Fig. 15.13).

Respondents are also worried that the software could be hacked, as 44% agree that this could happen. 43% would not feel safe in the event of harassment and assault, and 41% also express concern that the technology may not yet be fit to drive. These concerns also match the results of our social media study. Following social concerns, 'security concerns' (18% of the comments) are the most important concerns for people. Still, there seems to be a certain level of trust in 'the system'. Only a small share of the respondents are worried that the system may not be secure

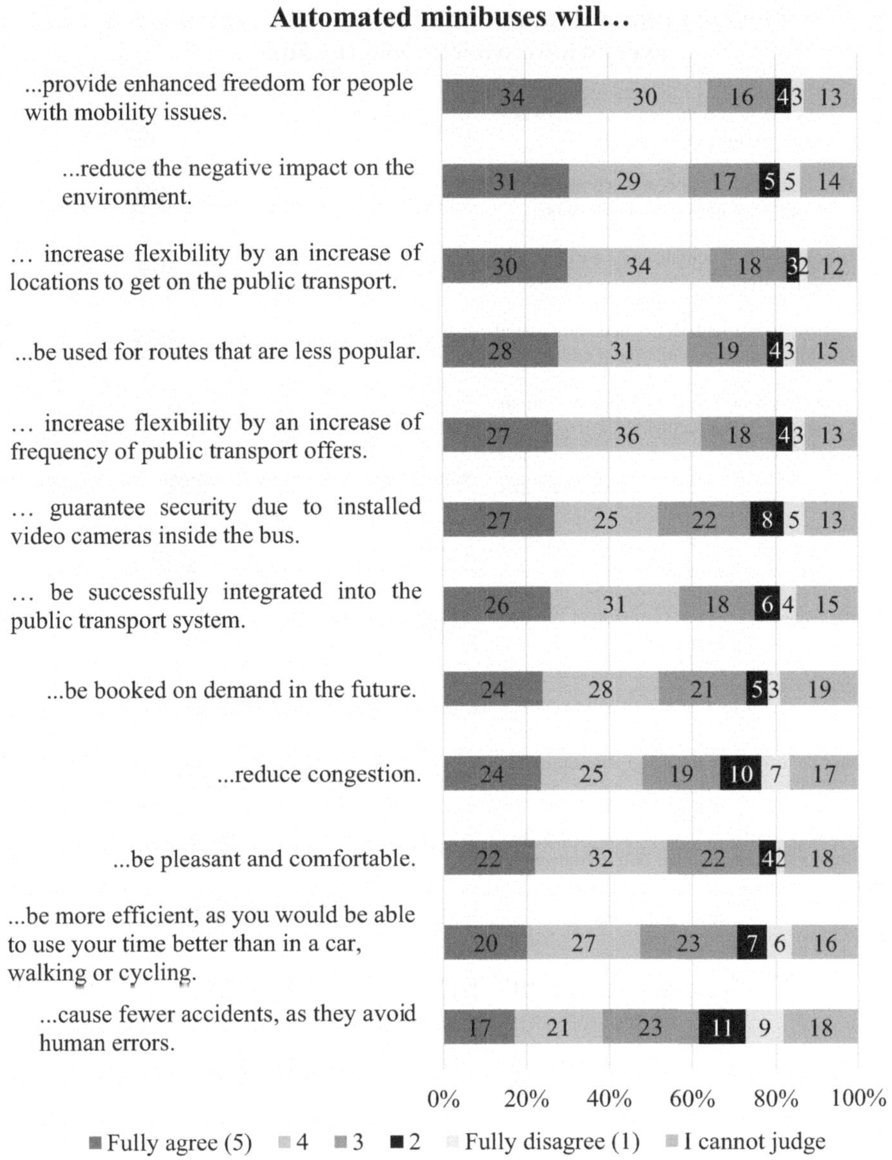


**Fig. 15.12** Perceived advantages of automated minibuses (in %, $n = 1816$)

(29%) or is not reliable (24%). Since many respondents still believe that it is unclear how the automated minibuses will react in critical situations, the presence of a safety operator is still important for the majority of participants.

Generally, criticisms and concerns towards the automated public transit system remain high, especially worries about safety, security, and society. These findings can be seen both in the large-scale, representative online survey and in our social

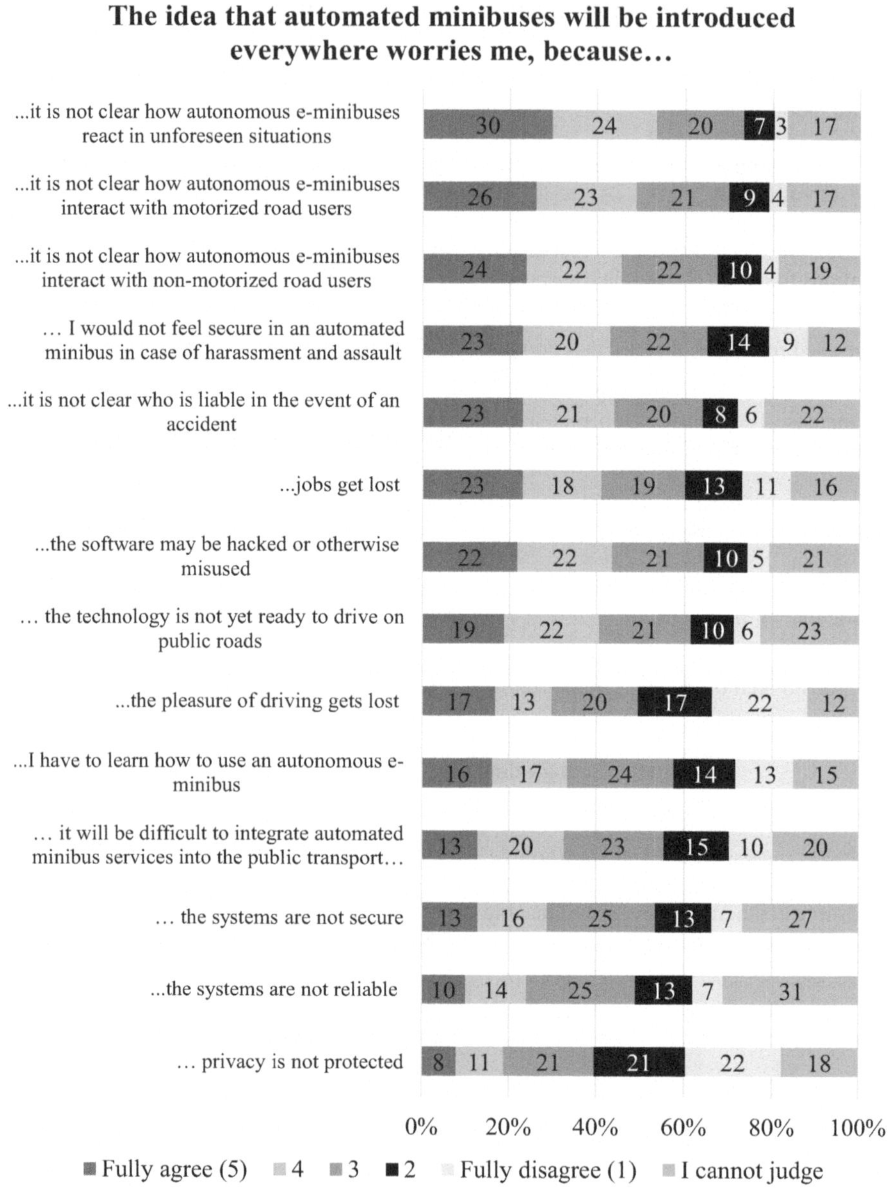


**Fig. 15.13** Perceived concerns over automated minibuses (in %, $n = 1816$)

media analysis. Nonetheless, both studies reveal that people are aware of the benefits of automated minibuses. This means that while the attitude towards the automated minibuses is still not exclusively positive, a change in attitude is observable. For example, the social media study described in Chap. 7.3 in Korbee et al. (2021) found that the most quoted comment is that automated PT is needed and its introduction is very welcome (17.6% of the comments). Moreover, the share of comments expressing that accidents with automated PT will be decreased has risen by 142.9% indicating that trust in automated PT also seems to have improved. At the same time, the part of negative remarks has mostly decreased in comparison to the first monitoring. For instance, the share of remarks asserting that automated public transports are dangerous has decreased (−27.8%), same for those claiming that automated minibuses are unreliable (−71.4%) or susceptible to being hacked, bugged, or used for terrorist attacks (−39.4%).

## 15.5 User Experience

A user survey is a widely accepted research tool for gaining valuable insights into the experiences of individuals using specific services and for investigating the usability of the new service (Korbee et al., 2019). The participants in this survey were approached by a safety operator while travelling with the automated minibus, i.e. they weren't travelling with the automated minibus for the sole purpose of taking part in a user survey. In the online user survey, they were asked to reflect on their experience and assess the usability, their overall satisfaction, as well as their satisfaction with specific aspects.

### *15.5.1 User Experience and Acceptance*

There is a wide range of information sources through which users became aware of the automated minibus: the most important sources are the internet (28%), direct contact on test site (28%), or informal sources such as word of mouth (17%) and friends (10%). Only one in five users read about it in newspapers (18%) and social media (16%) or heard a news item on the radio (16%).

The analysis of the motivation behind the use of the automated minibus reveals that the users did not plan on using the automated minibus but rather made a spontaneous decision, following emotions such as curiosity (88%), or they cannot cite a specific reason at all (43%). This demonstrates that the barriers to use the automated minibuses are quite low, but it also indicates that there is still a lack of real conviction for this new mobility system. Positive attitudes that could result in planned use (see Theory of Planned Behaviour (Ajzen, 1991)) have not yet been established. Only 4% report that they use the automated minibus due to previous positive experiences. The analysis shows that use is presently not motivated by the market presence in the sense of concrete knowledge and conviction but is mainly driven by spontaneous interest.

Approximately half of the respondents (46%) utilized the automated minibuses as a substitute for cycling, while 17% substituted walking (as depicted in Fig. 15.14). A high substitution of active mobility could be due to the length and location of the line and the high modal share of cycling in Copenhagen.

**Fig. 15.14** Means of transport replaced by the automated minibus (in %, $n = 66$)

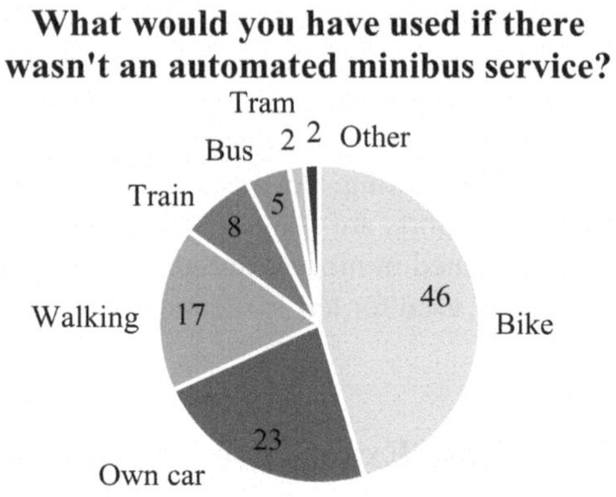


Users were asked to reflect on their satisfaction with their last ride. In general, the users expressed a high level of satisfaction with their previous ride, with a mean score of 4.5 out of a possible 5, with 5 being the highest level of satisfaction. Almost two-thirds (59%) of the users expressed a high level of satisfaction. An analysis of the aspects that users were most satisfied with indicates that especially the overall atmosphere in the automated minibus was experienced as pleasant: temperature (4.6), cleanliness (4.6), noise level in the bus (4.5), atmosphere in the bus (4.4), and security from outside the bus (4.4). Items that are more difficult to evaluate are related to the quality of service, such as punctuality, due to the lack of experience of the users. Most of the respondents had experienced the automated minibus for the first time. Consequently, more than 42% of the respondents did not answer these questions. If they were able to assess the quality of service, the subsequent features were satisfying:

Location of stops (4.3)
Punctuality (4.2)
Reliability (4.1)
Waiting time (4.1)

Compared to the highly satisfying aspects, the following aspects were rated slightly worse on average:

Speed/travel time (3.9)
Frequency of service (3.9)
Connection to other transport means (3.0)

The information offered on the bus is again satisfying (4.5); other information offers are rarely evaluated. This again underscores the fact that current users did not schedule their ride and therefore did not require more detailed information in advance.

Users appear to experience the ride in a relaxed atmosphere. This is confirmed by the results of the question 'what describes your feeling/emotions towards automated minibuses best?' These feelings are reported as being characterized by optimism (74%) and curiosity (68%). Suspicious or anxious feelings occur extremely rarely (less than 5%). These positive feelings are reinforced by the low incidence of

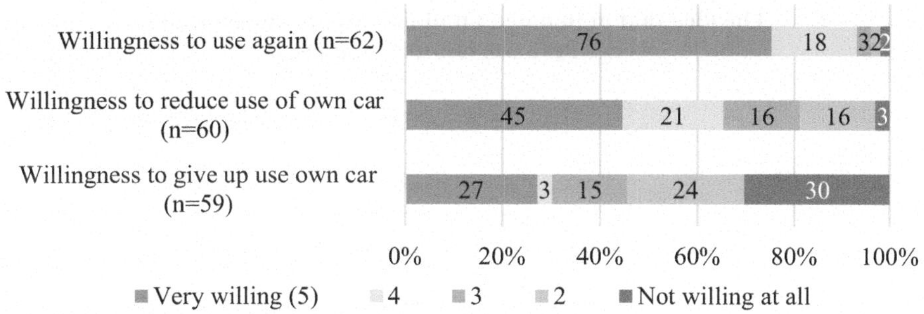


**Fig. 15.15** Willingness to use again (in %)

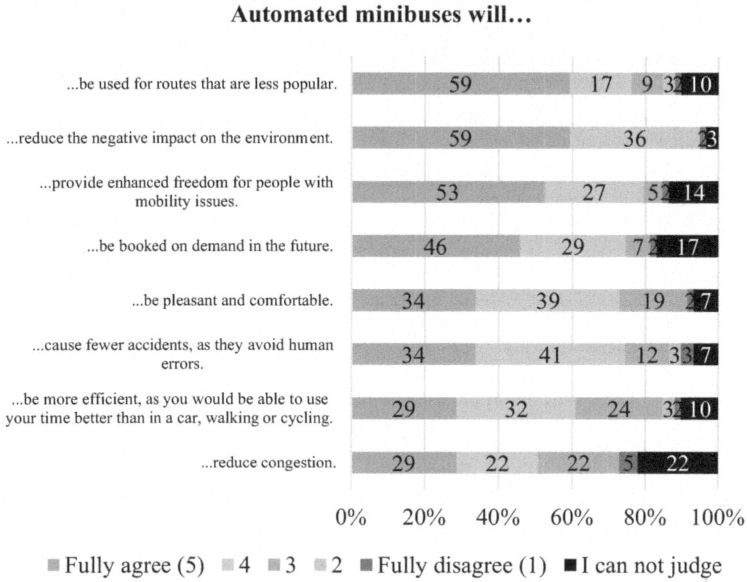


**Fig. 15.16** Perceived benefits of automated minibuses (in %, $n = 57$)

problems. When asked explicitly, 26% of the respondents (16 users out of 62 users that answered this question) encountered problems during their trip: unplanned stops (7 users), conflicts with other road members (8 users), long interruptions (5 users), or sudden braking manoeuvre (8 users).

The willingness to use the automated minibuses again is very high; 76% are very willing to use the automated minibus again, as shown in the first bar of Fig. 15.15. Only 2% of the users are hesitant to use it again, and another 2% of the users indicate that they are not willing to use it again.

In order to assess the acceptance of automated minibuses, it is essential to gain a more profound understanding of the perceived benefits (Fig. 15.16) and concerns (Fig. 15.17). Overall, the impression is that the respondents have a positive

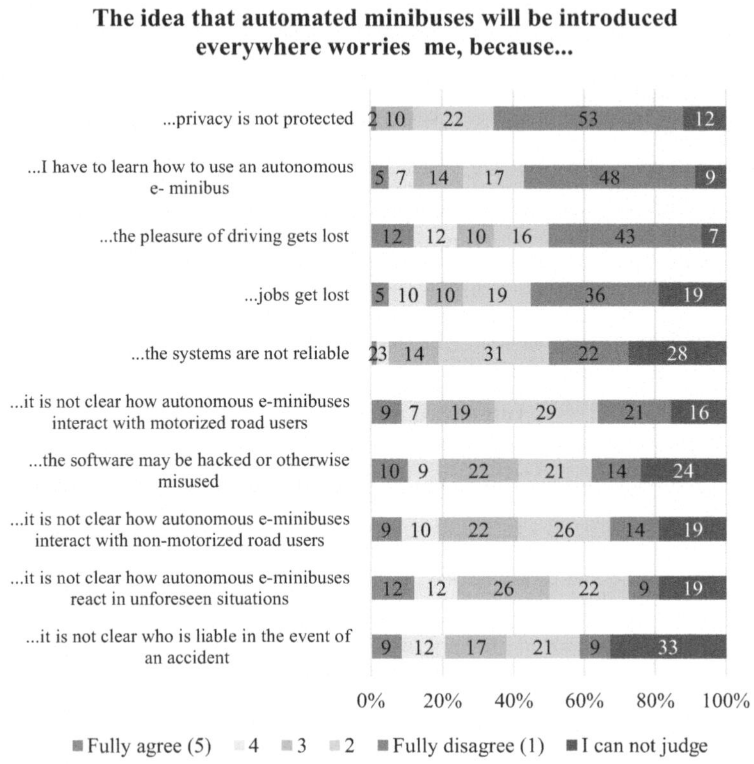


**Fig. 15.17** Perceived concerns about automated minibuses (in %, $n = 54$)

perception about the benefits of automated minibuses. All items asked had a minimum agreement of 50% for the perceived benefit that automated minibuses can reduce congestion and a maximum agreement of 95% that automated minibuses can be used for routes that are less popular. The perceived benefit that automated minibuses can reduce congestion received the most critical assessment, with only a small majority of 51% agreeing. These figures are similar to those in the original survey of Keolis Downer in Australia (Keolis Downer August, 2018). The potential benefits were scored on a 10-point scale, ranging from 8.1 for more efficient use of time to 8.9 for enhanced freedom for people with reduced mobility (PRM).

When it comes to the perceived concerns, the picture is somewhat more differentiated. Five out of ten concerns are not classified as such by most users. Hence, over 50% of the users disagree with the concerns about privacy, the pleasure of driving, learning to use the new system, job losses, and the reliability of the systems. A smaller percentage of users disagree with apprehensions regarding the functioning of the automated minibus. This includes the interaction with motorized and nonmotorized traffic, its reaction to unforeseen situations, and issues of liability in the case of an accident.

AM development also includes creating services beyond transportation. An argument made in support of the development of such services is that it can increase the number of users and add an additional benefit to the system compared to other transport systems. One service discussed is the presence of a safety operator. Opinions differ on whether a safety operator is needed on board the automated minibus. For a smaller share of users (18%), the presence of a safety operator is (very) important. A larger proportion of users (53%) believe that a safety operator is not important (at all).

Suggested services, such as providing feedback via QR codes, in-vehicle entertainment, access to WIFI, and an app to help with trip planning, are generally not considered as important by the respondents of this survey (Fig. 15.18). In the answer category 'other', respondents suggested the following services:

Feedback from the vehicle concerning its decision-making and status
Screen indicating the journey status (number of stops, estimated time of arrival, etc.)
More handles in the bus, primarily for the passenger seat in the middle
A plan indicating the location of the bus stops

The price of the automated minibuses is an essential selection criterion for the users for using the automated minibus. A majority of 58% state that the price is (very) important for their decision to use the automated minibus. No one indicated that price is not important at all. To explore the opportunities for future use and business cases, a question was integrated to compare the willingness to pay for the automated minibus in comparison to other means of transport. Only 24% of the users refuse to pay the same or an even higher price. However, the fact that only 11% of users are

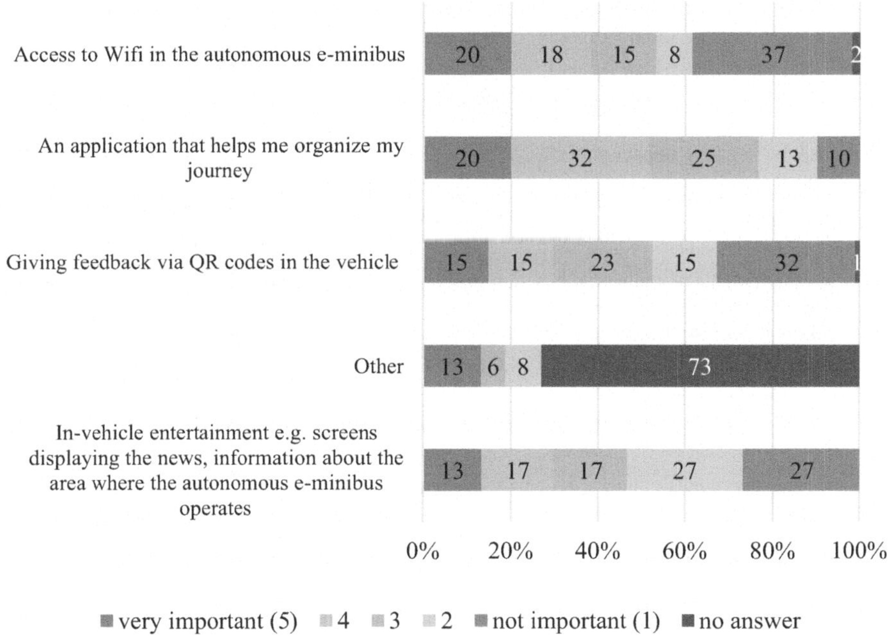


**Fig. 15.18** Evaluation of services (in %, $n = 60$)

willing to pay more indicates that users do not see such a large improvement that it would justify a higher willingness to pay. Again, this can be interpreted to imply that the automated minibus cannot entirely replace other systems but that it is seen as an obvious necessity that does not justify the additional cost for the user.

### *15.5.2 Users vs Non-users*

An important question is whether respondents who have used an automated minibus are more willing to use the automated minibus in different scenarios and whether an experience with the automated minibuses decreased the perception of risks and improved the perception of benefits. In this section, therefore, we compare responses of non-users with those of users. In designing the questionnaires, we took care to keep the questions the same for the different target groups. Table 15.5 and Table 15.6 show a comparison between the responses of citizens who have never used an automated minibus (gathered through the survey in the four AVENUE cities, 2021) with citizens who have experience with an automated minibus service. For the second group, we present the results in two groups: the users from the survey in the four AVENUE cities (2021) and the users of the Copenhagen site (2020). Overall, users in Copenhagen appear to be most convinced. Slightly less but still partially more

**Table 15.5** Perceived benefits of automated minibus services in non-users and users

| Mean[a] | Top boxes 4 and 5 (agree) / Low boxes 1 and 2 (do not agree) | Non-users survey 2021 ($n$ = 1690) | | Users survey 2021 ($n$ = 126) | | Users Copenhagen ($n$ = 57) | |
|---|---|---|---|---|---|---|---|
| … reduce the negative impact on the environment | | 3.9 ($n$ = 1577) | 60 | 3.8 ($n$ = 123) | 66 | 4.6 ($n$ = 57) | 95 |
| | | | 10 | | 12 | | 2 |
| … reduce congestion | | 3.6 ($n$ = 1509) | 49 | 3.7 ($n$ = 120) | 61 | 3.9 ($n$ = 46) | 51 |
| | | | 17 | | 17 | | 5 |
| … be more efficient, as you would be able to use your time better than in a car, walking or cycling | | 3.6 ($n$ = 1520) | 47 | 3.7 ($n$ = 113) | 55 | 3.9 ($n$ = 53) | 61 |
| | | | 13 | | 13 | | 5 |
| … provide enhanced freedom for people with mobility issues. | | 4.0 ($n$ = 1577) | 64 | 4.0 ($n$ = 119) | 60 | 4.5 ($n$ = 51) | 80 |
| | | | 7 | | 12 | | 2 |
| … be pleasant and comfortable | | 3.8 ($n$ = 1497) | 54 | 3.9 ($n$ = 122) | 68 | 4.1 ($n$ = 55) | 73 |
| | | | 6 | | 8 | | 2 |
| … be booked on-demand in the future | | 3.8 ($n$ = 1480) | 52 | 3.7 ($n$ = 113) | 52 | 4.4 ($n$ = 49) | 75 |
| | | | 8 | | 14 | | 2 |
| … cause fewer accidents, as they avoid human errors | | 3.3 ($n$ = 1483) | 38 | 3.7 ($n$ = 114) | 56 | 4.1 ($n$ = 55) | 75 |
| | | | 20 | | 16 | | 6 |

[a]Means: 1, is fully disagree; 5, fully agree. Hence, the higher the mean, the more positive. For the calculations of the means, we excluded those respondents that selected 'I cannot judge'; therefore, the basis varies—and is depicted after every means

**Table 15.6** Perceived risks of automated minibus services in non-users and users

| Mean | Top boxes 4 and 5 (agree) in % / Low boxes 1 and 2 (do not agree) in % | Non-users survey 2021 ($n$ = 1690) | | Users survey 2021 ($n$ = 126) | | Users Copenhagen ($n$ = 57) | |
|---|---|---|---|---|---|---|---|
| … it is not clear how automated minibuses react in unforeseen situations | | 3.8 ($n$ = 1501) | 54 | 3.6 ($n$ = 117) | 31 | 3.0 ($n$ = 47) | 24 |
| | | | 10 | | 14 | | 31 |
| … it is not clear how automated minibuses interact with motorized road users | | 3.7 ($n$ = 1502) | 49 | 3.4 ($n$ = 118) | 47 | 2.4 ($n$ = 49) | 16 |
| | | | 13 | | 18 | | 50 |
| … it is not clear how automated minibuses interact with nonmotorized road users | | 3.7 ($n$ = 1477) | 46 | 3.3 ($n$ = 117) | 43 | 2.7 ($n$ = 47) | 19 |
| | | | 14 | | 25 | | 40 |
| … it is not clear who is liable in the event of an accident | | 3.6 ($n$ = 1417) | 24 | 3.3 ($n$ = 110) | 24 | 2.9 ($n$ = 39) | 21 |
| | | | 14 | | 42 | | 30 |
| … the software may be hacked or otherwise misused | | 3.6 ($n$ = 1439) | 44 | 3.4 ($n$ = 118) | 43 | 2.8 ($n$ = 44) | 19 |
| | | | 15 | | 23 | | 35 |
| … the technology is not yet ready to drive on public roads | | 3.5 ($n$ = 1402) | 31 | 3.3 ($n$ = 115) | 44 | X | x |
| | | | 16 | | 26 | | x |
| … the systems are not reliable | | 3.1 ($n$ = 1253 | 24 | 2.8 ($n$ = 109) | 19 | 2.0 ($n$ = 42) | 5 |
| | | | 20 | | 36 | | 53 |
| … the systems are not secure | | 3.2 ($n$ = 1332) | 29 | 2.9 ($n$ = 109) | 28 | X | x |
| | | | 20 | | 31 | | x |

[a]Means: 1, is fully disagree; 5, fully agree. Hence, the higher the mean, the more negative. For the calculations of the means, we excluded those respondents that selected 'I cannot judge'; therefore, the basis varies—and is depicted after every means

convinced than non-users are the users identified in the representative survey. These effects can be explained by several reasons: users may be more open-minded than non-users in the first place. Especially in Copenhagen, usage was strongly motivated by curiosity. This may also explain that these very open-minded users even show higher goodwill. Users in Copenhagen see the lowest risks and the highest benefits. This is especially the case for the statement that the automated minibuses will 'reduce the negative impacts on the environment'. 95% of users in Copenhagen approve with this statement compared to 60% of the non-users. Both user groups display higher levels of agreement with the statements that 'automated minibuses will be pleasant and comfortable', that the 'automated minibuses will cause fewer accidents', and that the 'automated minibuses will be more efficient'. However, non-users are not more likely to agree with the statement that the 'automated minibuses can reduce congestion'. A higher percentage of users agree with the assertion that 'the technology is not yet ready to drive on public roads'. Users identified in the representative survey do not differ substantially from non-users in terms of perceived risks.

## 15.6 Conclusions

The AVENUE social impact assessment demonstrates that the majority of citizens are inclined towards a positive, receptive (goodwill) attitude. More than half of the potential users ‘would be willing’ or even ‘very willing’ to use the automated minibuses irrespective of the three different scenarios tested. Of the car owners, 50% would be willing to reduce or even give up the use of their own car if the automated minibus service was offered as part of an on-demand, door-to-door service.

It is critical to understand the perspective of users and potential users. For potential users, the benefit for them (as individual users) is the most important. Potential benefits include greater temporal and spatial flexibility, reduced waiting time, high speed, and more affordable transportation. Thus, it is essential that the automated minibus can compete with other means of transport in these aspects. Since automated minibuses also compete with private cars in terms of a possible lack of private atmosphere, especially for the critical reserved and unconvinced refusers, privacy must also be the focus. Given the current state of technology, the automated minibus operates at low speed (between 8 and 20 km/h). This low speed leads to low ratings by users (Nordhoff et al., 2018). These results are corroborated by Krueger et al. (2016) who define travel time and waiting time as critical determinants for the use and acceptance of the automated minibus. In addition, it is essential for customers that automated minibuses can fulfil the need for cognitive relief. The service option that best meets respondents’ needs is the on-demand service, which has the highest willingness to use of all three scenarios, with more than 60%. There are many unfulfilled needs due to gaps in the current PT offer but also resulting from gaps experienced in using one’s own car. This creates/leads to at least latent needs which could potentially be met by the automated minibuses. In particular, the need for more flexibility could be satisfied here.

In general, the impression is that respondents have a positive perception of the benefits of automated minibuses. When it comes to the perceived concerns, the picture is somewhat more nuanced. Respondents show a high level of agreement with concerns relating to the functioning of the automated minibus. Particularly, an increase in fears about how the automated minibuses may interact with other traffic members can be observed. As long as it is still unclear to many citizens how the automated minibuses may interact with other motorized or nonmotorized traffic members, a supervisor is still required. One way to positively influence attitudes is through well-targeted communication campaigns. An additional way to shift attitudes is to bring people in contact with the automated minibuses. This is because real experience in the automated minibus generally has a positive impact on the trust in the system: a comparison of the findings of the quantitative survey with potential users and the quantitative survey with users in Nordhavn demonstrates that user experience is an essential factor in reducing the perceived concerns and increasing acceptance of the automated minibus. Nevertheless, a high level of goodwill among potential users and a high level of satisfaction among users lead to a high level of willingness to (re)use and experience the new system several times. Yet, this does not result in a regular usage or substitution of privately owned cars, which can be observed especially in Copenhagen. The explanation for this is that, according to the findings of all included studies, citizens do not perceive an acute need for a complete substitution of current public transport offers. The extent to which classical transport modes can be substituted by automated minibuses also depends heavily on the satisfaction with the existing transport options. In cities like Copenhagen, where satisfaction with the offer of public transportation is already elevated, the need for new transportation services is thus lower. Potential users do not expect the automated minibus service to completely replace all current mobility options; it is not expected that the service will replace the use of private cars or bicycles completely, but they see the potential to substitute the current bus system.

Since there is a risk that negative experiences with conventional public transportation systems may also reduce acceptance of the automated minibuses, a clear differentiation of the new system from the existing bus offers must be communicated. The automated minibuses should not only be introduced as a new bus offer but as a completely new service that combines the advantages of a car (high level of flexibility) with the advantages public transport (higher cognitive and environmental relief as well as lower costs). We concluded that automated minibuses have the potential to fundamentally transform urban mobility when integrated as an on-demand, door-to-door service in a public transport or MaaS system. That said, it is not the innovation of automated minibuses itself that is key but rather its integration in the urban public transport systems and, at a later stage, in a MaaS / ITS. This integration will fulfil the mobility needs of potential users, taking the introduction of automated minibuses from a product innovation to a system innovation. Lastly, the AVENUE social impact assessment reaches the conclusion that by integrating automated minibuses into a public transport system, public transport can become as flexible as private cars. This will lead to high acceptance. Automated minibuses also have additional potential for improvement, as they are more efficient in use (shared) and thus contribute to a more sustainable urban mobility system.

**Acknowledgements** This project has received funding by the European Union's Horizon 2020 research and innovation program under grant agreement No. 769033.

## References

Ajzen, I. (1991). The theory of planned behavior. *Organizational Behavior and Human Decision Processes, 50*(2), 179–211. https://doi.org/10.1016/0749-5978(91)90020-T

Banister, D. (2008). The sustainable mobility paradigm. *Transport Policy, 15*, 73–80. https://doi.org/10.1016/j.tranpol.2007.10.005

Bernauer, T., & Wicki, M. (2018). Die Linie 12 im öffentlichen Meinungsbild: Zwischenbericht zur ersten Umfrage zum Pilotversuch eines automatisierten Busbetriebs in Neuhausen am Rheinfall. In E. T. H. Zurich (Ed.). *Institute of Science, Technology and Policy*. Available online at https://doi.org/10.3929/ethz-b-000282612. Accessed 11 Aug 2018.

Brosius, H., Haas, A., Koschel, F. (2012). Methoden der empirischen Kommunikationsforschung. Eine Einführung. 6., durchges. Aufl. Wiesbaden: VS Verlag für Sozialwissenschaften (Studienbücher zur Kommunikations- und Medienwissenschaft).

Fagnant, D. J., & Kockelman, K. (2015). Preparing a nation for autonomous vehicles: Opportunities, barriers and policy recommendations. *Transport Research A: Policy and Practice, 77*, 167–181. https://doi.org/10.1016/j.tra.2015.04.003

Firnkorn, J., & Müller, M. (2011). What will be the environmental effects of new free-floating car-sharing systems? The case of car2go in Ulm. *Ecological Economics, 70*, 1519–1528. https://doi.org/10.1016/j.ecolecon.2011.03.014

Fournier, F., Boos, A., Wörner, R., Jaroudi, I., Morozova, I., & Nemoto, E. H. (2020). Substituting individual mobility by mobility on demand using autonomous vehicles: A sustainable assessment simulation of Berlin and Stuttgart. *The International Journal of Automotive Technology, 20*, 369–407.

Geels, F. W. (2012). A socio-technical analysis of low-carbon transitions: Introducing the multi-level perspective into transport studies. *Journal of Transport Geography, 24*, 471–482. https://doi.org/10.1016/j.jtrangeo.2012.01.021

Geels, F. W., Sovacool, B. K., Schwanen, T., & Sorrel, S. (2017). The socio-technical dynamics of low-carbon transitions. *Joule, 1*(3), 463–479. https://doi.org/10.1016/j.joule.2017.09.018

Hirschhorn, F., Paulsson, A., Sørensen, C. H., & Veeneman, W. (2019). Public transport regimes and mobility as a service: Governance approaches in Amsterdam, Birmingham, and Helsinki. *Transportation Research Part A: Policy and Practice, 130*, 178–191. https://doi.org/10.1016/j.tra.2019.09.016

Huber, D., Viere, T., Nemoto, E. H., Fournier, G., Horschutz Nemoto, E., Jaroudi, I., & Korbee, D. (2022). Environmental impacts of automated shuttles in future public transportation// Climate and environmental impacts of automated minibuses in future public transportation. *Transportation Research Part D: Transport and Environment, 102*(6), 103160. https://doi.org/10.1016/j.trd.2021.103160

Keolis Downer. (Ed.). (2018). Futur-driven Autonobus: Pilot project at la Trobe University. Available online at https://www.keolisdowner.com.au/autonobus-trial-victoria/. Accessed 8 Nov 2018.

Kilian-Yasin, K., Wöhr, M., Tangour, C., & Fournier, G. (2016). Social acceptance of alternative mobility systems in Tunis. *Transportation Research Procedia, 19*, 135–146. https://doi.org/10.1016/j.trpro.2016.12.074

Korbee, D., Naderer, G., Dubielzig, M., Mathe, L., & Helfer, L. (2021). D8.9 Social impact assessment. Available online at https://h2020-avenue.eu/public-delivrables/. Accessed 4 May 23.

Korbee, D., Naderer, G., & Nemoto, E. H. (2019). Deliverable 8.8 First report on social impact assessment.

Krueger, R., Rashidi, T. H., & Rose, J. M. (2016). Preferences for shared autonomous vehicles. *Transportation Research Part C: Emerging Technologies, 69*, 343–355. https://doi.org/10.1016/j.trc.2016.06.015

Kuss, P., & Nicholas, K. A. (2022). A dozen effective interventions to reduce car use in European cities: Lessons learned from a meta-analysis and transition management. *Case Studies on Transport Policy, 10*(3), 1494–1513. https://doi.org/10.1016/j.cstp.2022.02.001

Kyriakidis, M., Happee, R., & de Winter, J. C. F. (2015). Public opinion on automated driving: Results of an international questionnaire among 5000 respondents. *In Transportation Research Part F: Traffic Psychology and Behaviour, 32*, 127–140. https://doi.org/10.1016/j.trf.2015.04.014

Nemoto, E. H., Issaoui, R., Korbee, D., Jaroudi, I., & Fournier, G. (2021). How to measure the impacts of shared automated electric vehicles on urban mobility. *Transportation Research Part D: Transport and Environment, 93*, 102766.

Nordhoff, S., Winter, J. d., Madigan, R., Merat, N., van Arem, B., & Happee, R. (2018). User acceptance of automated shuttles in Berlin-Schöneberg: A questionnaire study. *Transportation Research Part F: Traffic Psychology and Behaviour, 58*, 843–854. https://doi.org/10.1016/j.trf.2018.06.024

Nordhoff, S., Stapel, J., van Arem, B., & Happee, R. (2020). Passenger opinions of the perceived safety and interaction with automated shuttles: A test ride study with 'hidden' safety steward. *In*

*Transportation Research Part A: Policy and Practice, 138*, 508–524. https://doi.org/10.1016/j.tra.2020.05.009

Pangbourne, K., Mladenovic, M. N., Stead, D., & Milakis, D. (2020). Questioning mobility as a service: Unanticipated implications for society and governance. *Transportation Research Part A: Policy and Practice, 131*, 35–49. https://doi.org/10.1016/j.tra.2019.09.033

Pribyl, O., Blokpoel, R., & Matowicki, M. (2020). Addressing EU climate targets: Reducing CO2 emissions using cooperative and automated vehicles. *Transportation Research Part D: Transport and Environment, 86*, 102437.

Schmitt, M., & Altstötter-Gleich, C. (2010). *Differentielle Psychologie und Persönlichkeitspsychologie*. Beltz PVU.

Schoettle, B., & Sivak, M. (2014). Public opinion about self driving vehicles in China, Indica, Japan, The U.S., The U.K., and Australia. Edited by The University of Michigan. *Transportation Research Institute (UMETRI-2014-30)*. Available online at https://deepblue.lib.umich.edu/handle/2027.42/108384. Accessed 20 June 2024.

Zoellick, J. C., Kuhlmey, A., Schenk, L., & Schindel., D., Blüher, S. (2019). Amused, accepted, and used? Attitudes and emotions towards automated vehicles, their relationships, and predictive value for usage intention. *Transportation Research Part F: Traffic Psychology and Behaviour, 65*, 68–78. https://doi.org/10.1016/j.trf.2019.07.009

# Chapter 16
# Governance Impact Assessment, Regulatory Recommendations and Challenges

**Lionel Binz**

**Abstract** This chapter critically examines the regulatory and governance challenges posed by the deployment of automated vehicles (AVs) and highlights the need for adaptable laws and governance models to accommodate AVs in public transport systems. Various governance approaches are described, including global, state, self-directed, cooperative, social, technological and financial governance, and their importance in the context of AV regulation is highlighted. The discussion highlights the need for an integrated, collaborative regulatory approach that recognises the unique characteristics of AVs and argues for the harmonisation of laws at different levels to facilitate this transition. Through a comprehensive analysis, the chapter highlights the multiple impacts of AVs on legal and societal domains and presents a set of legal recommendations for authorities. These recommendations emphasise governance that not only promotes societal benefits but also addresses critical ethical, privacy and liability concerns. By charting a path towards a governance framework that supports the safe and beneficial integration of AVs into urban mobility, this chapter contributes valuable insights into overcoming the legal and regulatory hurdles that stand in the way of realising the full potential of automated public transport systems.

## 16.1 Introduction

The impact of AV on governance is not well researched yet. The aim of this chapter is to raise first insights on how governance should be adapted to the upcoming change in the mobility paradigm based on AVs as a game changer. In this chapter, the impact of automated vehicles (AVs) on legislation will be briefly exposed, as well as the types of governance that are and could be used for their actual and future

L. Binz (✉)
Cantonal Transports Office in Geneva (Etat de Genève), Petit-Lancy, Switzerland
e-mail: Lionel.binz@protonmail.ch



G. Fournier et al. (eds.), *Automated Vehicles as a Game Changer for Sustainable Mobility*, Contributions to Management Science,
https://doi.org/10.1007/978-3-031-61681-5_16

regulation. A second part will briefly present a selection of legal recommendations to public authorities that were drafted within the AVENUE project.

### 16.1.1 The Deployment of AVs Will Impact All Areas of the Law

Automated vehicles (AVs) and the mobility services they provide will require changes in all areas of the law:

- Administrative/public law (through modifications to road traffic rules and regulations (World Economic Forum, 2019, p. 5))
- Civil law (amongst others by updating the regulations of civil liability and product liability in cases of accidents (PUNEV, 2020, p. 99))
- Criminal law (for the case of accidents with bodily injuries, misuse cases of AVs and, potentially, for infringements of road traffic rules committed by AVs (Council of Europe CDPC, 2020))

These areas of the law can be either:

1. Strongly harmonised and regulated at international and/or regional (European) levels, with small to no room for different material rules (e.g. having rules with different content) at national level
2. Mostly or completely regulated by national legislation, which means the material legal rules are defined at a national level and can therefore be different from country to country (for instance, the position of the driver's wheel on the left or ride side of the vehicle)
3. A mixture between both, with for instance international principles leaving room for national differences (UNECE, 2022)

### 16.1.2 Types of Governance

As the deployment of AVs has the potential to significantly impact all areas of the law, questions related to the best or more efficient governance and regulatory frameworks arise.

Governance can be described as the process of making and enforcing decisions within an organisation, state or society. It is the process of interactions through the laws, social norms, power (social and political) or language as structured in communication of an organised society over a social system (family, social group, formal or informal organisation, a territory under a jurisdiction or across territories). It is done by the government or the parliament of a State, by a market or by a network. It is the process of choosing the right course amongst the actors involved in a collective problem that leads to the creation, reinforcement or reproduction of acceptable conduct and social order (Möltgen-Sicking & Winter, 2019).

As regards the word "law", we will use the following terminology that makes the distinction between:

- Laws in the formal sense, i.e. norms that are adopted by the State's legislative body (e.g. the parliament) in the course of the regular legislative process (entitled "law").
- Laws in the material sense, i.e. general and abstract norms that can be adopted by other bodies than the legislative, such as the State's executive body (e.g. the government). They can be called regulations, ordinances, acts, rules and so on.

Here are some example of different approaches to governance and regulation (International Association of Public Transport, 2021):

- *Global governance:* Global governance involves the creation of international agreements, treaties and institutions to regulate the development, deployment and use of new technologies on a global scale, providing harmonisation such as seen above. This type of governance is particularly important for technologies that have cross-border implications, such as AVs. In the field of automotive, the UNECE and other bodies are already playing this role and are working on topics related to AVs (Lagrange, 2021).
- *Governmental regulation:* This type of governance involves the creation and enforcement of laws and regulations by government bodies to control the development, deployment and use of new technologies. Governmental regulation is seen as necessary to protect public safety and security, privacy and other public interests. In the case of automated vehicles, governmental regulation will be necessary for areas such as road traffic safety, cybersecurity, data protection and liability, for instance (World Economic Forum, 2019, pp. 8 et seq.). Governmental regulation at state level takes place into global governance, where it can be partly defined by binding international instruments. In the field of new technologies, the European Commission has already enacted various acts regulating amongst others the processing of data (such as the General Data Protection Regulation, the upcoming Data Act, the Data Governance Act, the Digital Markets Act, the Digital Service Act, the AI Act etc.).
- *Self-governance:* This type of governance is based on the principle of voluntary compliance by individuals or organisations that are involved in the development, deployment or use of new technologies such as AVs. Self-governance may involve the creation of industry standards, codes of conduct or best practices that guide the behaviour of the stakeholders in a particular industry. For AVs, the interests of public safety and security (amongst others) do not allow relying on self-governance alone, and governmental regulations are required. In particular fields, where safety and security are not at stake, lawmakers and governments might allow the stakeholders (and the markets) to regulate themselves, following a so-called laissez-faire strategy.
- *Collaborative governance or co-regulation:* This involves collaboration between lawmakers (e.g. governments and legislative bodies), industry, researchers and other stakeholders to develop and implement policies and standards (Law

Commission of England and Wales and the Scottish Law Commission, 2022a). This approach should be followed for AVs, especially at this early stage of AV deployment, as it helps ensuring that the interests of all stakeholders are taken into account and can promote greater acceptance of automated vehicles amongst the public. Safety, with, for instance, the definition of passive safety requirements (and the necessary trade-offs related to vehicle architecture), might be solved efficiently through collaborative governance (International Association of Public Transport, 2021).
- *Social governance:* Social governance refers to the use of non-governmental organisations, community groups and other civil society actors to shape the development and deployment of new technologies. Social governance may involve public education, advocacy or activism to promote responsible innovation and ensure that the benefits of new technologies are shared equitably. Social governance will be needed to ensure the acceptation of AVs by the public and to ensure, for instance, that PRMs get involved in the development and will benefit from AVs.
- *Technological governance:* Technological governance involves the development of technical standards and protocols that ensure the interoperability and compatibility of automated vehicles with other systems and technologies. This approach can help ensure that automated vehicles can operate safely and efficiently within existing transportation systems. In the field of AVs, AVENUE, for example, has shown the necessity to provide open datasets and application protocol interfaces (APIs) for the common good.
- *Financial governance:* Financial governance involves the development of policies and regulations related amongst others to the financing of automated vehicles, such as tax incentives or subsidies for the development and purchase of automated vehicles by public transport operators. Regulation also plays an incentive role, as "strict" regulations with high type-approval costs or triggering high costs for the deployment of automated vehicles have an important financial impact for vehicles manufacturers or public transport operators. Besides, it should be noted that an optimal use of AVs might require massive investments in public infrastructure.

During AVENUE, voices have been raising the fact that European regulations, for instance, in the field of type-approval and/or getting an approval for AV experimentations, as well as in the field of data privacy with the General Data Protection Regulation, tend to be more restrictive than other regulations (viz. in the USA or in Asian countries). According to these, from an economical point of view, this would result in a comparative disadvantage for research and vehicle manufacturers in Europe. A counter-argument is that, in the long run, these higher standards will also support achieving better products, output and results, increasing safety and security and avoiding unnecessary incidents. As of today, according to the author's opinion, the approach to vehicle automation followed in the USA has resulted in more accidents with body injuries or even death cases involving advanced driver assistance systems or automated driving systems than in Europe (Tesladeaths, n.d.).

Overall, effective governance of automated vehicles will likely require a combination of these different approaches, tailored to the specific needs and characteristics of different regions and countries but also to the specific AV domain to regulate.

### 16.1.3 Major Impacts on Legislation and on the Societal Activities they Regulate

The deployment of automated vehicles will have major impacts in many fields involving public authorities and interests, far beyond formal regulations provided by public, civil or criminal laws:

- *Legal and regulatory framework:* As mentioned before, the introduction of automated vehicles will require a re-evaluation of many existing legal and regulatory frameworks, as it will be briefly summarised in the following chapters. Lawmakers and executive bodies at different levels will need to consider new international and national regulations and instruments to address the unique characteristics of automated vehicles. AVs are part of the upcoming technological revolution that will be triggered by artificial intelligence and the internet of things which will require updates and changes in many existing regulations as well as the creation of new legal frameworks.
- *Infrastructure and urban planning:* Automated vehicles will require new infrastructure and urban planning (Martin et al., 2023) to support their safe and efficient operation. They could also completely change the landscape of our roads and cities. Some changes have already begun: vehicle manufacturers are developing the use case of "Remote Controlled Parking" systems (RCP), where an "Automatically Commanded Steering Function" (ACSF) of the vehicle performs the dynamic driving task (DDT) of the vehicle without any human intervention. In this use case, the driver and passengers get out of the vehicle, which is empty of any human driver and parks itself into a free parking lot "alone" by using the ACSF. This allows gaining parking space, as vehicles equipped with RCP can park themselves in much narrower spaces than the existing ones, where space is needed for the driver and the passengers to open the doors of the vehicle in order to enter or exit the vehicle (Huonder & Raemy, 2016).
- *Public investments in new technologies and infrastructure* will be needed, forcing lawmakers and executive bodies to define the AV environment they wish for their roads, such as the type of communication networks (vehicle-to-vehicle, vehicle-to-infrastructure etc.) and related sensors and communication channels (Nationale Plattform Zukunft der Mobilität, 2020), charging stations etc. to support the deployment of AVs. These developments will also be pushed by the fact that conventional vehicles are becoming connected vehicles, and AVs are a subcategory of connected vehicles. In this field, decision-makers will have to

proceed to technological, political and financial trade-offs, for instance, whether they want AVs that do not rely on sensors placed in the road infrastructure or whether the road infrastructure should be able to communicate with AVs (National Plattform Zukunft der Mobilität, 2021a, 2021b) to provide them information, and/or whether redundancies are needed, in particular for the sake of safety and security. An example: using its sensors, a fully automated vehicle could detect on its own whether a traffic light is red or green, but an "intelligent" or connected traffic light could also communicate through wireless communication channels if it is red or green and provide this information to the AV. These decisions will have important financial impacts on public budgets and might also impact the rhythm of AV deployment, by accelerating or slowing it down.

- *Employment and workforce:* The introduction of automated vehicles will have a significant impact on the transportation industry, leading to inevitable job losses in certain sectors such as public transportation, trucking and taxi services (International Association of Public Transport, 2017). Governments will need to consider the impact on these workers and develop policies to address the potential disruption. Driver unions have, for instance, started to protest against the development of automated vehicles in California in the beginning of 2024 (TheLastDriverLicenseHolder, 2024a), and a few cases of attacks against automated vehicles have been reported recently in San Francisco (TheLastDriverLicenseHolder, 2024b). In parallel, AVs will also create new jobs that didn't exist previously.
- *Social and economic impact*: Automated vehicles will significantly impact society and the economy, by (possibly) reducing traffic congestion, improving safety and increasing mobility for persons with reduced mobility. Governments will need to consider the potential social and economic impact of automated vehicles and develop policies to support their deployment in a way that benefits society as a whole (Smith, 2016). According to the research of the AVENUE projects, some current forms of AVs, such as robotaxis or private automated vehicles, will not benefit to society as much as shared and automated forms of public transportation, which should be sustained by public authorities.
- *Ethical considerations:* The development and deployment of automated vehicles also raises ethical considerations, such as how to program the automated driving system to take decisions in situations where competing interests are at hand, as described in the famous "trolley dilemma" (Foot, 1978) where an automated driving system would need to choose between hurting the AVs' passengers or hurting third parties such as pedestrians on the road. Governments will need to consider these ethical considerations and develop policies to ensure that automated vehicles operate in an ethical and socially responsible manner. Germany, for instance, has already tackled this question with its Ethics Commission that released guidelines for AVs in 2018 (Ethik-Kommission, 2017).

## 16.2 AVENUE Recommendations

The following sections will briefly present the most important legal recommendations for public authorities that were drafted during the AVENUE project. They enter into the categories of global governance, governmental regulation and collaborative governance or co-regulation defined above.

The first section will be dedicated to cross-sectoral recommendations, followed by another dedicated to specific topics.

### *16.2.1 General Recommendations Not Limited to Specific Areas of the Law*

#### 16.2.1.1 Prioritising the General Interest Through Encouraging Legislation

Roughly summarised and without going into the details of the SAE Levels, the current development of AVs can be categorised in three different approaches or use cases:

- Privately owned automated vehicles, e.g. conventional vehicles with a driver and a steering wheel, improved with advanced driver-assistance systems (ADAS) or automated driving systems (ADS), answering private mobility needs of individuals owning these vehicles.
- Robotaxis, which are automated vehicles that can be similar to privately owned vehicles, but are owned and deployed by commercial companies with the purpose of providing ride-hailing services and generating profits for the companies.
- Shared automated vehicles integrated in a MaaS (Mobility as a Service) or an ITS (Intelligent Transport System), often falling into the category of minibuses/shuttles, deployed by transport operators providing traditional transport services and/or used door to door and on demand in a MaaS or an ITS as mobility gap filler to complement private or public transport services enabling positive externalities (see Chap. 18 about vision).

If we look at the current regulations being adopted by public authorities in the field of automated vehicles, the current lawmaking momentum lies mostly with the regulation of the first two categories, namely, privately owned automated vehicles and/or robotaxis:

- The first AV-specific regulation enacted at international level (as a result of global governance defined above) was the regulation dedicated to the Automated Lane Keeping System (ALKS), a (SAE Level 3) highway traffic jam assistant designed for conventional, privately owned vehicles with automated driving systems. It was enacted by the United Nations Economic Commission for Europe (UNECE) in 2020 and consolidated to 130 km/h in 2022 (UNECE, 2022). Since then, in a

few European countries, ALKS has been type-approved and can be lawfully used on certain highways (KBA, n.d.).
- At national levels, most of the experimentations and legislative output are focused on the regulation of robotaxis (see, for instance, in San Francisco with Cruise and Waymo amongst others (TheLastDriverLicenseHolder, n.d.)).

This can be explained by the fact that these types of vehicles (e.g. privately owned vehicles with ADAS or ADS or robotaxis) and their use cases are promoted by private stakeholders with very important resources wishing to enter in what they consider a new promising market. These stakeholders are either "tech" companies (such as Waymo, a company curated by Alphabet/Google (Waymo, n.d.), or Mobileye, curated by Intel (Mobileye, n.d.)) or manufacturers of conventional vehicles vying to join the automation trend (such as Mercedes (The Verge, n.d.), Cruise/General Motors (Cruise, n.d.) etc.).

They can also be both type of actors, working together in the form of joint ventures and consortiums. As a consequence, the legislation enacted by legislative bodies and governments follows the technical developments proposed by these stakeholders and reacts to it with a "bottom-up" approach, similar to collaborative governance, although often the public authorities lack of the necessary knowledge and involvement.

Shared automated minibuses, which are, according to the AVENUE research results, the more desirable automated driving use cases from a societal point of view, did not stand in the focus of lawmakers until recently.

Thus, priority should be given to their regulation, as research within AVENUE (and other projects as well (Fabulos, n.d.)) has demonstrated that shared automated minibuses are the use case of AVs that will provide greater and better societal benefits (International Association of Public Transport, 2021) than privately owned automated vehicles or robotaxis, which might trigger negative externalities such as an increase in traffic congestion and weaken public transportation and accordingly generate additional costs for public transport authorities.

Therefore, governments and lawmakers should play an orientating role by encouraging the most promising use cases and by creating incentives. Accordingly, they should also regulate and mitigate use cases considered to be less desirable from a societal point of view.

#### 16.2.1.2 Cooperation and Harmonisation at European Level

For the development of automated vehicles, at European level, an aligned and holistic legal European approach towards automated vehicles is desirable (European Commission, 2020a).

As seen during the AVENUE project, the same automated minibuses went through very different approval proceedings in the four countries of experimentations, with important time differences, to receive an approval from public authorities. The cost and documentation required from each of the partners in France,

Luxembourg, Denmark and Switzerland were also very different, although the automated minibuses were of the same model and manufacturer.

Therefore, an aligned and holistic legal European approach is desirable and necessary for many reasons, starting from road safety to a commercial, single market perspective, with the promotion of the European industry and encouragement of technological leadership. Such an approach and harmonisation can be defined through European institutions and regulations, which have yet to be created and defined.

#### 16.2.1.3 Cooperation Between All Stakeholders

In order to cope with the fast evolving technologies and business models involving automated vehicles and minibuses, various governance models can be implemented, such as collaborative governance or co-regulation, where all stakeholders get involved in the definition of the regulation/policy through an inclusive approach and with the sharing of knowledge and resources (International Association of Public Transport, 2021).

Such approaches are needed as private stakeholders (vehicle manufacturers, "tech" companies, software developers etc.), public transportation operators (PTOs), public transportation authorities (PTAs) and further stakeholders (organisations in charge of accident analysis, insurers etc.) already involved in the deployment of AVs can identify their needs and legislative issues through their experience already gathered while experimenting.

Similar results might also be achieved through self-governance of non-institutional platforms and/or private associations, representing the interests of the various stakeholders. Harmonisation through public institutions or involving public institutions might be more desirable, as private associations might follow and defend the interests of their members that may differ and strive for different goals such as maximisation of profits. This might result in unsatisfying results from a societal point of view, as the potential positive societal benefits and outcomes studied in the AVENUE project might not be reached.

#### 16.2.1.4 Coordination Between Authorities

AVs will stand in the focus of different governmental authorities (Law Commission of England and Wales and the Scottish Law Commission, 2022b; Monti & De Streel, 2022), as they function based on multiple technologies supervised by different authorities: type approval authorities for the "road vehicle" components, telecommunication authorities for the use of antennas and radiocommunications, data protection authorities as regards the processing of personal data and even competition authorities in the case of potential concentration of market power by actors involved in the AV chain.

As of today, these authorities rarely work together. These authorities will also need to coordinate themselves and develop new skills and competences related to digitalisation and automation.

#### 16.2.1.5 Standardisation

As a result of self-governance, the vehicle industry often relies on proprietary standards and formats (for instance, each vehicle manufacturer uses its own format of a high definition map, which is generally not compatible with other manufacturers' maps). This creates high curation costs when data is transferred from one company to another and high costs to develop new services (Iacob et al., 2021; Easme & Cosme, 2018). Standardisation and interoperability in all domains should be encouraged as much as possible.

#### 16.2.1.6 Definitions and Vocabulary

The vocabulary used to describe automated driving technologies should not mislead the public and create false expectations. A certain level of harmonisation that would increase the public's awareness and comprehension of these new technologies is desirable (Nationale Platform Zukunft der Mobilität, 2021b).

### *16.2.2 Recommendations Related to Road Traffic Rules and Conventions*

#### 16.2.2.1 Amendments to the Existing Road Traffic Rules and Conventions

During many decades, the presence of a human driver, located inside the vehicle and performing the "dynamic driving task", was a central assumption of the international road traffic conventions such as the Vienna and Geneva Conventions, as well as for many national road traffic laws. With the deployment of AVs, a fundamental paradigm change is currently operating, and states should depart from this twentieth-century approach to driving (Smith, 2022; Vellinga, 2019).

At international level, the UNECE is currently working to deliver a regulation for ADs in 2026 (Pichereau, 2021). However, the timeframe for the final adoption for such regulation might be too slow and take too long before the effective deployment on AVs on the roads.

Each country should therefore proceed as soon as possible to an extensive assessment of its road traffic regulations, in order to prepare and amend:

- Legal or technical provisions that implicitly or explicitly relate to the presence of a human driver performing the dynamic driving task from inside the vehicle

- Legal or technical provisions that might pose interpretation problems with the introduction of automated vehicles
- Legal or technical provisions involving human negotiation, interpretation or communication in specific situations (for instance, when a policeman is regulating the traffic at a specific spot through hand signals)

It should also be noted that automated vehicles are programmed to comply strictly with the existing traffic rules and not infringe them. In the road traffic reality with human drivers, however, various situations can occur that require a trade-off between obstructing the traffic flow and infringing road traffic rules. For instance, after an incident with the damaged vehicles located on the lane, vehicles driven by human drivers will carefully cross the lane to circumvent the obstacle. This simple arbitration might be difficult to implement in the case of automated vehicles, as this solution might imply to infringe traffic rules.

Solutions to such problematic situations requiring trade-offs and arbitrations should be discussed with the various concerned authorities (such as road safety, police, traffic law, politicians etc.), the developers of automated vehicles and researchers.

### *16.2.3 Supervision of Automated Vehicles*

#### 16.2.3.1 Allow Remote Supervision of AVs

Having a safety operator located inside each automated minibus (as required by today's experimentations' approvals) generates important functioning costs and does not constitute an economically viable solution in the long term for commercial and large-scale deployment (Probst, 2021). Therefore, legislative and executive bodies should promote further research and create innovative regulations, such as "regulatory sandboxes", permitting the deployment of automated minibuses (and AVs in general) without a safety operator, the safety operator being replaced by a (human) remote supervision that monitors several automated minibuses in parallel and intervenes if/when necessary.

Enabling the remote supervision of automated minibuses and vehicles (e.g. without a safety operator located inside the vehicle) and creating a dedicated legal framework are crucial steps for PTOs to deploy automated minibuses commercially at large scale and to provide economically viable transportation solutions in the future.

Some European Member States such as France and Germany, as well as the European Implementing (European Commission, 2022) and Delegated Regulations (European Commission, 2022) issued in July and August 2022, as well as the interpretation report published in February 2024 (EU Regulation 2022/1426), are paving the way from a legal point of view in Europe, by permitting experimentations and/or by enacting legal frameworks allowing remote supervision at distance under an ordinary approval regime. In February 2023, the Law Commission of England has published its advice related to remote driving for drivers beyond line of sight (Lawcom, n.d.), which includes the question of remotely driving AVs.

#### 16.2.3.2 Qualifications of the "Safety Operators"

In parallel and during a "transition phase", the qualifications currently expected from safety operators located inside automated minibuses are often as high or even higher than for drivers of conventional public transport vehicles (Roedl, n.d.), generating important functioning costs. Therefore, the requirements for safety operators should be watered down and adapted to the reality of operating automated minibuses. After a first period of practice where the AVs' operational design domain is tailored to the physical location of their deployment, and problematic "hotspots" get identified, AVs are able to circulate with operation time over 90% in automated mode, where the automated driving system is successfully performing the automated driving task without any intervention of the safety operator.

The situation in Europe is quite behind in comparison to other countries. In 2022, Cruise and Waymo received commercial licenses to deploy small fleets of robotaxis in San Francisco, operating only with remote supervision: robotaxis are cruising the streets without any human driver inside, besides the passengers (TheLastDriverLicenseHolder, n.d.). Beginning in March 2024, Waymo received an authorisation to expand its robotaxis beyond San Francisco and south of the city to the Bay Area, as well as in parts of Los Angeles (TheLastDriverLicenseHolder, 2024c).

Other important topics shall be covered, as mentioned within the AVENUE research, for instance, as regards the passive safety of automated minibuses. Their particular (and new) architecture and features require urgently the definition of passive safety requirements. Reflexions and studies are necessary in order to arbitrate on important trade-offs in this field.

### *16.2.4 Recommendations Related to the Regulation of AV Data*

#### 16.2.4.1 Competition

The European Commission has published various legal acts and drafts related to current data issues, such as the Data Act and the Data Governance Act (amongst others). These acts are a positive first step, but further specific acts dedicated to the automotive industry, and automated vehicles, should be enacted quickly. Governments and lawmakers should provide their best efforts in creating legislations that anticipate and guide the technological developments. It should be avoided that, as seen until recently in the field of digitalisation, one or a few players get into a position where they can distort competition, due to their actual power on such market, generated by creating "winner takes it all" and "lock-in" effects within a nearly closed ecosystem (see details in chapter below about the vision of AVENUE). Besides open data (see below), open API and even open map are key issues to keep competition but also to better satisfy the citizen needs and to enable wealth (see vision).

The AVENUE project has demonstrated that vehicle manufacturers and/or tech companies developing automated vehicles and minibuses benefit of a "de facto" control over vehicle data. This means that they have control and can monitor key steps related to the collection of car data. They may, for instance, decide on the collection (for instance, sensors such as video cameras recording the AVs' surroundings), the processing (storage of the data collected on temporary or permanent data storage devices), the transmission of data to third parties, the encryption for data etc.

Thus, vehicle manufacturers can decide, through agreements with other parties, to whom and to which type of data they provide access to. Such a "de facto" position places them in an undesirable gatekeeper position that should be mitigated (Ernst & Young, 2021; Andrasko et al., 2021) by creating a legal framework regulating data-related issues, as contractual arrangements between operators (in a form of self-governance) will not be sufficient.

One of the suggestions of the AVENUE project is to define through dedicated regulations open datasets according to the rules of the Data Governance Act that should be freely shared between the stakeholders, as well as open application programming interfaces to enable the transmission of data and leverage it for the greater public interest (Roedl, n.d.). The trustworthy mobility data-sharing system should accordingly create wealth for the society and provide control to the citizen and trust to the companies. The availability of mobility data also raises the question of privacy (see below) but also of competition as foreign competitor could use AI technology and less trustworthy data to take competitive advantage on European OEM (Original Equipment Manufacturer) competitors. A symmetry in competition should accordingly be guaranteed for European OEMs.

At international level, mutual assistance and questions related to data sovereignty should also be discussed, amongst others, as the data flow might imply that data collected from an automated vehicle located in country A is transmitted, processed and stored in another country B, sometimes outside of the European Union. This will cause problems when personal data gets transmitted to jurisdictions that do not have an adequate level of personal data protection.

#### 16.2.4.2 Data Privacy

The deployment of AVs will raise many privacy concerns, as they are able to collect, store and transmit large amounts of personal data through their various sensors and communication channels.

Here are some of the key privacy issues posed by automated vehicles:

- Definition of personal data: In the context of automated minibuses and mobility-related applications, what data is to be considered personal and non-personal (technical data) is not straightforward in all cases. Further guidance in this area would be necessary to avoid the proliferation of different interpretations of this concept, which has an important impact in practice, for example, on the anonymization of personal data (Campmas et al., 2021; Benyahya et al., 2022).
- Processing of personal data: AVs can collect and store a large amount of personal data that can be used for various purposes, such as research and quality improvement, but also for marketing and establishing data profiles for commercial purposes. This may conflict with important data protection principles such as the data minimisation and the purpose limitation principles (Balboni et al., 2020; Finck & Biega, 2021).
- Transparency: AVs should be transparent about the data they collect, store and share, so that the passengers are aware of the data that is being collected and how it is being used (Horizon 2020 Commission Expert Group, 2020). Achieving transparency towards third parties located outside AVs such as other drivers, cyclists and pedestrians will be harder to achieve, but solutions might be achieved through further research and discussion.

#### 16.2.4.3 Cybersecurity

The storage and transmission of both personal and vehicle data must be protected to prevent unauthorised access and use of the data. This requires strong data security measures to ensure the protection of the data but also to prevent third parties to take control of AVs (see chapters above).

It is important for governments and manufacturers of automated vehicles to address these privacy issues (Vellinga, 2022), to build public trust in the technology and ensure that the personal data of the vehicle's occupants is protected. Adequate privacy protections will help promote the widespread adoption of automated vehicles and ensure that they are used in a responsible and ethical manner.

#### 16.2.4.4 Liability

The general opinion amongst legal scholars (Lohmann & Müller-Chen, 2017) today considers that automated vehicles in their current stage of deployment will not create major gaps in the applicable European liability framework, amongst others, due to the Motor Vehicle Insurance Directive, which requires that all vehicles registered in the EU hold mandatory third-party liability insurance to cover civil liability for the use of motor vehicles. However, various potential issues have been identified.

It should be avoided to place the burden of proof with the victim of an incident (as it might be the case in product liability cases, for instance (Piantoni et al., 2021;

Buiten et al., 2021)). Due to the complexity of AI products and their algorithms, liability claims would become difficult or overly costly to prove for victims, and consequently they may not be adequately compensated. The same applies to (complex) decision-making processes, where the decision-making chain of AI products cannot be explained afterwards and how the (problematic) output resulted (Horizon 2020 Commission Expert Group, 2020).

Information asymmetry is another issue. As mentioned in the chapter dedicated to data, vehicle manufacturers currently benefit of a "de facto" control over the data collected and processed by the sensors of automated vehicles. To understand the liability chain (and causality between a defect and a damage), accessing vehicle data will be of paramount importance. Access to this data, and understanding the data, will be challenging for victims and their representatives.

One solution would lie in the reversal of the burden of proof, or in lowering the burden of proof, as suggested in the current revision of the Product Liability Directive. The creation of a trusted third party like it is existing in the automotive industry could be a solution (see vision chapter below).

In addition, the deployment of automated vehicles and minibuses implies a complex ecosystem with a plurality of actors involved in the services. This could result in increasing difficulties to assess where a potential damage originated and to assign liability (European Commission, 2020b). Forms of shared responsibility, as well as a shift towards liability of vehicle manufacturers and tech companies etc., should be studied, discussed and envisaged.

## 16.3 Accessibility

In the field of automated minibuses, the applicable accessibility requirements could still be improved by vehicle manufacturers, as demonstrated by the existing vehicle constructions, amongst others, in the AVENUE project. Moreover, the mobile applications related to the services provided by automated vehicles and minibuses also lack accessibility.

On the one side, deployers of automated minibuses, public transport operators and public transport authorities should include accessibility as an important criterion in the tender process. In many countries, accessibility is a prerequisite for the funding of vehicles by public authorities.

On the other side, vehicle manufacturers shall apply accessibility requirements defined by European standards in the earliest construction and design phase of automated vehicles (European Commission, 2018) ("accessibility by design" in a similar way than "privacy by design and by default" for data privacy). As a general principle, discriminatory design and service provision should be prevented (Horizon 2020 Commission Expert Group, 2020).

Accessibility of future automated public transport of all types needs to be thoroughly researched, and findings need to be designed into guidelines issued on public transport accessibility levels. Persons with reduced mobility (PRMs) need to become Cooperative Intelligent Transport Systems (C-ITS) nodes for emerging automated vehicles to recognise and protect them, given their potentially lower perceptibility and higher vulnerability. The impact of each new technology on PRM accessibility must be assessed during its design phase, and appropriate adaptability/personalisation means and strategies should be supported.

There is also a need for standardised and enforceable European legislation on accessibility across the mobility chain, as no such dedicated legislation exists for multimodal transport, be it for long-distance or local transport. The European sectorial legislation on air, maritime, railway and road transport does not fulfil this goal, as it (1) allows for exceptions that were largely used by Member States, (2) mainly focuses on long-distance transport and (3) does not support a wide range of the actual requirements of PRMs. The current legislation does not cover the possibilities of spontaneous travel (European Commission, 2018), i.e. there is a duty of prior notification in case of assistance requirements (Regulation on Rail Passengers Rights, n.d.), and provides inadequate support during travel and compensations for the loss or damage to accessibility equipment (for instance, in the cases of denial of travel and limited lost property compensations).

## 16.4 Conclusion

AVs might provide disruptive changes to the mobility of the twenty-first century. The regulatory environment in which they are tested and deployed can be a major incentive (International Association of Public Transport, 2021) or an obstacle to their development. The governance decides also about the citizen centricity, the creation of wealth but also the business viability in particular the creation or not of closed ecosystems. European executive and legislative bodies are now beginning to understand the need for AV regulations and the role they can play. The integration of AV in MaaS/ITS is a new challenge as well (see vision below). Encouraging signals have been emitted recently with the adoption of innovative regulations (Roedl & Partner, 2022). However, much legal work remains ahead, in order to avoid having regulations that are too far behind the technological developments, opening a regulatory "no man's land" that would be beneficial for a few private stakeholders only, and not to society as a whole.

**Acknowledgements** This project has received funding by the European Union's Horizon 2020 research and innovation program under grant agreement No. 769033.

## References

Balboni, P., Botsi, A., Francis, K., & Barata, M. T. (2020). Designing connected and automated vehicles around legal and ethical concerns: Data protection as a corporate social responsibility. Paper presented at Workshop on artificial intelligence, ethics and law, Athens, Greece, 3

September 2020, § 3; see also Finck, A. J., & Biega, M. (2021). Reviving purpose limitation and data minimisation in data-driven systems, technology and regulation, § 4.4.

Campmas, A., Iacob, N., Simonelli, F., Vu, B. (Centre for European Policy Studies—CEPS). (2021, April 20). Big Data and B2B platforms: the next big opportunity for Europe—Report on market deficiencies and regulatory barriers affecting cooperative, connected and automated mobility, Annex III Potential market deficiencies and regulatory barriers, including a common industry-led position in the automotive sector, EASME/COSME/2018/004, p. 198; see also Benyahya, M., Kechagia, S., Collen, A., & Nijdam, N. A. (2022, April 27). The interface of privacy and data security in automated city shuttles: The GDPR analysis. *Applied Sciences, 12*(9), 4413, § 5.

Commission delegated regulation of 20 June 2022 amending Annexes I, II, IV and V to Regulation (EU) 2018/858 of the European Parliament and of the Council as re gards the technical requirements for vehicles produced in unlimited series, vehicles produced in small series, fully automated vehicles produced in small series and special purpose vehicles, and as regards software update, C(2022) 3823 final, and its four Annexes.

Commission implementing regulation of 5 August 2022 laying down rules for the application of Regulation (EU) 2019/2144 of the European Parliament and of the Council as regards uniform procedures and technical specifications for the type-approval of the automated driving system (ADS) of fully automated vehicles, C(2022) 5402 final, and its four Annexes.

Council of Europe, European Committee on Crime Problems (CDPC). (2020). Feasibility Study on a future Council of Europe instrument on Artificial Intelligence and Criminal Law, Document prepared by the Working Group on AI and Criminal Law with Dr. Sabine Gless as General Rapporteur, and in co-operation with the CDPC Secretariat, CDPC(2020)3Rev, Strasbourg, 4 September 2020, § 4.2 p. 13.

Cruise. (n.d.). Cruise Self Driving Cars | Autonomous Vehicles | Driverless Rides & Delivery | Cruise (getcruise.com). Accessed 15 May 2023.

Ernst & Young. (2021, November). Abogados SLP and FIA Region I, Expert study on: "GDPR application in the context of car connectivity", Final Report, p. 103; see also Andraško, J., Hamuľák, O., Mesarčík, M., Kerikmäe, T., & Kajander, A. (2021, September 24). Sustainable data governance for cooperative, connected and automated mobility in the European Union, § 2.3.1.

Ethik-Kommission. (2017, June). (eingesetzt durch den Bundesminister für Verkehr und digitale Infrastruktur), Automatisiertes und Vernetztes Fahren—Bericht.

European Commission (Breemersch et al.). (2018, November 26) Best practices guide on the carriage of persons with reduced mobility, Guide on future measures, policies and strategies aimed at creating a PRM-inclusive transport system in Europe, Final report for: European Commission, DG Mobility and Transport.

European Commission. (2020a, December 9). Sustainable and Smart Mobility Strategy—putting European transport on track for the future, Commission Staff Working Document, SWD(2020) 331 final, N 617, p. 151.

European Commission. (2020b, February 19). Report on the safety and liability implications of Artificial Intelligence, the Internet of Things and robotics. European Commission, COM(2020) 64, p. 14.

Fabolus. (n.d.). https://fabulos.eu/. Accessed 28 Apr 2023.

Foot, P. (1978). The problem of abortion and the doctrine of the double effect. In *Virtues and vices*. Basil Blackwell. (ursprünglich erschienen in Oxford Review, Nummer 5, 1967).

Horizon 2020 Commission Expert Group to advise on specific ethical issues raised by driverless mobility (E03659). (2020, June). Ethics of Connected and Automated Vehicles: recommendations on road safety, privacy, fairness, explainability and responsibility, Luxembourg.

Huonder, S., & Raemy, O. (2016). Autonomes Fahren, Rechtlicher Anpassungsbedarf: ein Werkstattbericht, 4. *Zürcher Tagung, Dike Verlag AG, Strassenverkehr, 1*(2016), 48.

Iacob, N., Campmas, A., & Simonelli, F. (Centre for European Policy Studies—CEPS). (2021, April 20) Big Data and B2B platforms: the next big opportunity for Europe—Annex IV

Recommendations to the EU and national policy makers and an action plan for the creation of shared EU-wide in-vehicle data platforms, EASME/COSME/2018/004, p. 255; see also Nationale Plattform Zukunft der Mobilität (NPM) (2021, October), Ergebnisbericht der Nationalen Plattform Zukunft der Mobilität—Ergebnisse aus drei Jahren NPM (2018–2021), § 3.3 p. 46.

International Association of Public Transport (UITP). (2017, January). Policy brief, Autonomous vehicles: A potential game changer for urban mobility, Brussels, p. 8.

International Association of Public Transport (UITP). (2021, October). GECKO, Project Brief, Governance and regulation models to manage disruptive mobility services, Brussels.

KBA. (n.d.). https://www.kba.de/DE/Presse/Pressemitteilungen/Allgemein/2021/pm49_2021_erste_Genehmigung_automatisiertes_Fahren.html?snn=3662144. Accessed 28 Apr 2023.

Lagrange. (2021, March 21). The importance of UNECE Regulations for the EU legislation on Automated driving, Future Networked Car Symposium, Session 1: Regulatory advances in highly automated driving, slide 8; see also https://unece.org/transport/road-traffic-safety/group-experts-drafting-new-legal-instrument-use-automated-vehicles. Accessed 28 Apr 2023.

Law Commission of England and Wales and the Scottish Law Commission. (2022a, January 26). Automated vehicles: Summary of joint report, HC 1068 SG/2022/15, § 3.30 p. 16.

Law Commission of England and Wales and the Scottish Law Commission. (2022b, January 26). Automated vehicles: Joint report, HC 1068 SG/2022/15, § 6.11 p. 92 et seq.; see also Monti, G., & De Streel, A. (2022, January). Centre on regulation in Europe (CERRE), Improving EU institutional design to better supervise digital platforms, Report.

Lawcom. (n.d.). https://www.lawcom.gov.uk/project/remote-driving/. Accessed 28 Apr 2023.

Lohmann, M. F., & Müller-Chen, M. (2017). Selbstlernende Fahrzeuge—eine Haftungsanalyse. *SZW/RSDA, 1*, 12.

Martin, R., Bruck, E. M., & Soteropoulos, A. (2023). Transformations of European Public Spaces with AVs. In Mitteregger et al. (Hrsg.), *AVENUE21. Politische und planerische Aspekte der automatisierten Mobilität*, § 3 p. 176.

Mobileye. (n.d.). Mobileye | Driver Assist and Autonomous Driving Technologies (last consultation 15 May 2023).

Möltgen-Sicking, K., & Winter, P. (2019). Governance—Eine Einführung in Grundlagen und Politikfelder. https://doi.org/10.1007/978-3-658-24280-0.

Nationale Plattform Zukunft der Mobilität (NPM). (2020, May). Arbeitsgruppe 6, White Paper, Hand-lungsempfehlungen zur Typgenehmigung und Zertifizierung für eine vernetzte und automatisierte Mobilität.

Nationale Plattform Zukunft der Mobilität (NPM). (2021a, December). Arbeitsgruppe 3, Autonomes Fahren als Baustein einer inter- und multimodalen Mobilität, Siebter Zwischenbericht, § 3.1 p. 8.

Nationale Plattform Zukunft der Mobilität (NPM). (2021b, October). Arbeitsgruppe 6, Standards und Normen für die Mobilität der Zukunft, Ergebnisse der Arbeitsgruppe 6 der NPM 2018–2021, Abschlussbericht 10/2021, § 2.3 p. 21.

Piantoni, P., Fiengo, G., Lovaste, G., et al. (Grimaldi studio legale). (2021). Liabilities of Independent Service Providers when providing repair and maintenance under the Secure Onboard Telematics Platform, Legal Study, dated 31 May 2021, published by FIA Region I on 11 June 2021, p. 80; see also European Commission, Report on the safety and liability implications of Artificial Intelligence, the Internet of Things and robotics, COM(2020) 64 final, Brussels, 19 February 2020, § 3; and Buiten, M., De Streel, A., Peitz, M. (2021, March). Centre on regulation in Europe (CERRE), EU Liability rules for the age of artificial intelligence, Report, § 5.2.5.

Pichereau. (2021, July 30). Le déploiement européen du véhicule autonome — Pour un renforcement des projets européens, Assemblée nationale, § 1.2.4 p. 27.

Probst, T. (2021). Das "selbstfahrende" Auto: entfernte Illusion oder nahe Realität?. Strassenverkehr 1/2021, Dike Verlag AG, 2021, pp. 52–58, § H N 1 p. 58.

Punev, A. (2020). Autonomous vehicles: The need for a separate european legal framework. *European View, 19*(I), 95–102.
Regulation on Rail Passengers Rights. (n.d.). Art. 24 litt. a of the Regulation on Rail Passengers' Rights.
Roedl. (n.d.). https://www.roedl.de/themen/kompass-mobilitaet/2023/08/zugang-wettbewerbsrelevante-mobilitaetsdaten-bleibt-weiterhin-umstritten. Accessed 28 Apr 2023.
Roedl & Partner. (2022, July 13). Automatisiertes Fahren (SAE-Level 4) auch auf europäischer Ebene geregelt, Kompass Mobilität Ausgabe 14/2022.
Smith, B. W. (2016, March). How governments can promote automated driving. *New Mexico Law Review, 1*.
Smith, B. W. (2022, May 9). Initial thoughts on "road safety challenges posed by the use of automated vehicles in traffic that an international legal instrument could adequately address". UNECE Informal document No. 2 (GE.3–03-02), N 5 et seq.
Tesladeaths. (n.d.). First fatal collision that lead to the death of Elaine Herzberg in 2018 or the website https://www.tesladeaths.com/. Accessed 28 Apr 2023.
The Verge. (n.d.). https://www.theverge.com/2023/1/27/23572942/mercedes-drive-pilot-level-3-approved-nevada. Accessed 28 Apr 2023.
TheLastDriverLicenseHolder. (2024a). https://thelastdriverlicenseholder.com/2024/02/02/us-driver-unions-protest-self-driving-cars/. Accessed 3 Mar 2024.
TheLastDriverLicenseHolder. (2024b). https://thelastdriverlicenseholder.com/2024/02/11/waymo-vandalized-and-lit-on-fire-in-chinatown/. Accessed 3 Mar 2024.
TheLastDriverLicenseHolder. (2024c). https://thelastdriverlicenseholder.com/2024/03/01/waymo-licensed-to-massively-expand-robotaxi-service-in-california/. Accessed 3 Mar 2024.
TheLastDriverLicenseHolder. (n.d.). https://thelastdriverlicenseholder.com/2022/12/22/waymo-receives-driverless-permit-for-l-a-cruise-starts-first-driverless-rides-in-austin/. Accessed 28 Apr 2023.
UNECE. (2022). https://unece.org/sites/default/files/2022-05/ECE-TRANS-WP.29-2022-59r1e.pdf. Accessed 28 Apr 2023.
Vellinga, N. E. (2022, April). Connected and vulnerable: cybersecurity in vehicles. *International Review of Law, Computers & Technology, 36*, 161.
Waymo. (n.d.). Waymo - Self-Driving Cars - Autonomous Vehicles - Ride-Hail. Accessed 15 May 2023.
World Economic Forum. (2019, April). Filling legislative gaps in automated vehicles. White paper, pp. 8 et seq.

# Chapter 17
# Sustainability Assessment of the Integration of Automated Minibuses in Urban Mobility Systems: Learnings from the AVENUE Project

**Eliane Horschutz Nemoto, Ines Jaroudi, Dorien Duffner-Korbee, Tobias Viere, Gabriele Naderer, Adrian Boos, Nicole van den Boom, and Guy Fournier**

**Abstract** The AVENUE sustainability assessment integrates the environmental, economic and social assessment of the pilot trials of AVENUE. A step further, it adopts an interdisciplinary approach to conduct the analysis and to better understand the complexity of deploying a new form of mobility in urban areas and as part of the transportation system. The goal is to implement new mobility solu-

---

This chapter is written based on the previous publications: Nemoto, E.H.; Jaroudi, I.; Korbee, D.; Viere, T.; Naderer, G.; Boss, A.; Fournier, G. (2022). D8.12 Sustainability assessment. Available at https://cordis.europa.eu/project/id/769033/results.

---

E. H. Nemoto (✉)
CentraleSupélec, LGI, Gif-sur-Yvette, France

Pforzheim University, Pforzheim, Germany

ICLEI Europe, Freiburg im Breisgau, Germany
e-mail: eliane.nemoto@iclei.org

I. Jaroudi
CentraleSupélec, LGI, Gif-sur-Yvette, France

Pforzheim University, Pforzheim, Germany

Fraunhofer ISI, Karlsruhe, Germany

D. Duffner-Korbee
Pforzheim University, Pforzheim, Germany

Fraunhofer ISI, Karlsruhe, Germany

T. Viere · G. Naderer · A. Boos
Pforzheim University, Pforzheim, Germany

N. van den Boom · G. Fournier
CentraleSupélec, LGI, Gif-sur-Yvette, France

Pforzheim University, Pforzheim, Germany



G. Fournier et al. (eds.), *Automated Vehicles as a Game Changer for Sustainable Mobility*, Contributions to Management Science,
https://doi.org/10.1007/978-3-031-61681-5_17

tions that benefit the city and complement public transport. The findings from the social, environmental and economic impact assessments are embedded in a multidimensional set of indicators for the sustainability assessment of the automated minibuses. As a result, mobility radars are built to illustrate the assessment of the AVENUE demonstrator sites. The results reveal the state of the art of the current deployment of the automated minibuses in the AVENUE test sites. In addition, this study highlights other crucial elements for a sustainable deployment that rely on the mode of deployment of the automated minibuses, the level of integration to urban public transport or within MaaS perspective and the alignment with the city's goals, planning and strategies for sustainable mobility. Ultimately, the automated minibuses are an innovative piece within the mobility ecosystem that could support intermodality, MaaS, mobility hubs and the use of soft modes of transport.

## 17.1 Introduction

Automated minibuses for public transport are expected to contribute to sustainable urban mobility. By combining automated, connected, shared and electric technologies, the automated minibuses could improve transport accessibility, efficiency and reduction of greenhouse gases (GHG) (Jones & Leibowicz, 2019). They have the potential to play a role in a shift from vehicle ownership to shared mobility services (Shaheen & Chan, 2016) and to reduce transport externalities (Lim & Taeihagh, 2018). Nonetheless, one cannot take for granted that the deployment of innovation and new technologies per se will contribute to sustainable mobility. It rather depends on certain premises, planning and policies to frame the automated minibuses deployment.

The study by Taiebat et al. (2018) points out the main gaps concerning connected and automated vehicles' impacts; for instance, the net effect of AV technology on energy consumption and emissions in the long term remains uncertain. In addition, the broader society-level impacts and behavioural changes associated with AVs are also unclear. The study highlights that the 'synergetic effects of vehicle automation, electrification, right-sizing, and shared mobility are likely to be more significant than anyone isolated mechanism'.

AVs, especially for private use, could lead to an increase in vehicle kilometres travelled (VKT), reductions of the public transport and slow modes share (Soteropoulos et al., 2019), whereas shared automated vehicles, when considering a high share, could reduce the number of vehicles for the current travel demand, resulting in less parking and more space in the cities (ibid). Yet, it is worth noting that the results of impact assessment for AVs are strongly dependent on model assumptions (Soteropoulos et al., 2019).

The integration of automated minibuses into the public transport of European cities also raises questions regarding their potential benefits and critical points to contribute to the sustainable urban mobility plan (SUMP) and goals towards sustainable mobility of the cities. Therefore, this chapter presents the sustainability assessment of the AVENUE project and demonstrator sites. The goal of the sustainability assessment is to integrate and interrelate the results of the social, environmental and economic impacts and to embed these results by applying the set of indicators for sustainability assessment of the automated minibuses within the AVENUE demonstrator sites. In addition, the concept of SUMP is also a building block for the sustainability assessment.

## 17.2 Research Approach

The AVENUE approach to sustainability assessment comprises the concept of SUMP and indicators for sustainable mobility assessment.

By aiming to achieve sustainable mobility, indicators are used to measure performance and progress towards established goals and objectives (Litman, 2007). Urban sustainability indicators are fundamental to support target setting and performance reviews and to enable communication among the policymakers, experts and general public (Shen et al., 2011; Verbruggen & Kuik, 1991). Therefore, a set of indicators is applied for the sustainability assessment of the deployment of the automated minibuses in AVENUE pilot sites (Nemoto et al., 2021); subsequently, mobility radars illustrate the results.

The disaggregated indicators reveal the strengths and weaknesses of each mobility indicator (World Business Council for Sustainable Development, 2015). As graph representation, the radar (also known as spider chart) enables easy communication and visualisation of the results and comparison among case studies.

## 17.3 Sustainability Assessment of the AVENUE Demonstrator Pilot Sites

The sustainability assessment builds upon the set of indicators (Table 17.1) developed by Nemoto et al. (2021, 2022). The assessment comprehends the mobility multidimensions: social, environmental, economic, governance and system performance. Further, a selection of indicators (Table 17.2) was applied based on data availability from the pilot sites. Table 17.2 describes the indicators, units of measurement and scales of assessment.

Subsequently, the next section presents a brief description of the assessed pilot sites and reports the mobility radars for each pilot site based on the empirical data

**Table 17.1** Set of indicators for sustainable mobility assessment of shared automated electric vehicles from Nemoto et al. (2021)

| Indicators | Unit and methods of measurement | Multidimensions S (x) | En (xx) | Ec (xxx) | G (xxxx) | SP (xxxxx) |
|---|---|---|---|---|---|---|
| Accessibility | • Percentage of the city (area) coverage by the automated minibus service<br>• Percentage of the population that has convenient access (within 0.5 km) to the automated minibus service | x | | | | xxxxx |
| | • Automated minibus digitally accessible (e.g. via apps) | x | | | | xxxxx |
| Accessibility for people with reduced mobility | • External environment facilities, e.g. stops adaption for impaired/disabled people; tactile surfaces information<br>• Internal environment facilities, e.g. audible warning equipment for visually impaired people; facilities for wheelchair users | x | | | | xxxxx |
| | • Usability of the SAEV by people with reduced mobility (PRM)<br>• Rating of users with reduced mobility concerning the automated minibus experience | x | | | | xxxxx |
| Safety | • Risk factor and number of accidents related to the automated minibus (mild injuries, serious injuries, fatalities) considering internal risk (related to passengers) and external risk (related to other road users, pedestrians and cyclists) | x | | | | xxxxx |
| Security | • Number of criminal occurrences; nr/year | x | | | | xxxxx |
| | • Number of cybersecurity threats or attacks; nr/year | x | | | | xxxxx |
| Passenger's affordability | • The price of the ride on the automated minibus | x | | xxx | | |
| Social acceptance | • User's perception about the readiness of the technology<br>• User's willingness to pay<br>• Safety feeling<br>• Security feeling | x | | | | xxxxx |
| User satisfaction | • User rating concerning automated minibus experience (comfort, speed, punctuality, information, frequency, connection to other means of transport) | x | | | | xxxxx |
| Energy efficiency | • Energy consumed for passenger per km (kWh/pkm) | | xx | | | xxxxx |
| Renewable energy | • Use phase: energy source and percentage of renewable energy sources (%) | | xx | | | |

**Table 17.1** (continued)

| | | Multidimensions | | | | |
|---|---|---|---|---|---|---|
| Indicators | Unit and methods of measurement | S (x) | En (xx) | Ec (xxx) | G (xxxx) | SP (xxxxx) |
| Air pollution | • automated minibus emissions of air pollutants: PM levels (ug/m3), NOx, CO emissions | x | xx | | | xxxxx |
| Climate change | • Automated minibus GHG emissions: $CO_2$, $N_2O$, $CH_4$ | x | xx | | | xxxxx |
| Noise pollution | • Automated minibus traffic noise (dB) | x | xx | | | xxxxx |
| Investments on mobility | • Public and private annual average investment on transport concerning automated vehicles (Euro/year), e.g. infrastructure, operational expenditures (cost of personnel, software system, etc.), investments in the vehicle R&D | | | xxx | xxxx | |
| Economic incentives for SAEV and sustainable mobility | • Incentives and subsidies for automated and sustainable mobility, e.g. shared, electric, automated, zero-emission, vehicles (Euro) | | xx | xxx | xxxx | |
| Economic profitability | • TCO (total cost of ownership), TCM (total cost of mobility), cost/km/passenger, revenues (ticketing from passengers, subsidies from authorities and companies) and payback period | | | xxx | | |
| External costs related to the automated minibus | • Automated minibus impacts on congestion avoidance, accidents reduction, noise reduction, air pollution (PM, NOx) reduction, QALY (quality-adjusted life years) reduction, land/parking reduction, vehicle savings | | xx | xxx | | xxxxx |
| Institutional development and innovation | • Existence of policies and regulations concerning automated vehicles<br>• Regulations for open data and/or APIs for transport | | | | xxxx | |
| Technical performance and reliability | • Automated minibus performance:<br>Travel time (speed, frequency of departure or response speed for on-demand, travel-matching, punctuality);<br>On-demand availability;<br>Percentage of operational service; performance on different seasons/weather;<br>Vehicle occupancy (average passenger per km travelled);<br>The average lifetime of the vehicle;<br>Number of disengagements in the urban environment; number of km driven autonomously | | | | | xxxxx |

**Table 17.1** (continued)

| Indicators | Unit and methods of measurement | Multidimensions | | | | |
|---|---|---|---|---|---|---|
| | | S (x) | En (xx) | Ec (xxx) | G (xxxx) | SP (xxxxx) |
| System integration and efficiency | • Automated minibus integration with mobility platform of the operator (planning, reservation, booking, billing, digital ticketing)<br>• System and data interoperability and the existence of open data for the automated minibus (access, static and/or dynamic real-time data, diffusion format, data quality and open APIs for transport)<br>• Intermodality: automated minibus integration with other public or private means of transport or with a multimodal platform for one intermodal trip (planning, reservation, booking, billing, digital ticketing) | | | xxx | xxxx | xxxxx |
| Changes in total kilometres travelled in the transportation system | • Changes in per capita vehicle travelled induced by automated vehicles<br>• Transportation demand management measures introduced congestion pricing, biking lanes, zoning measures, land-use policies | | xx | | xxxx | xxxxx |

from the trials. The four sites assessed comprise Groupama (Lyon), Contern (Luxembourg), Pfaffenthal (Luxembourg) and Nordhavn (Copenhagen).

### *17.3.1 Pilot Sites Assessment and Results*

The indicators were applied for the sustainability assessment of four different demonstrator sites. The description of the sites and respective mobility radar are presented hereinafter. The indicators present a value from 1 to 5—with 1 for the worst performance and 5 for the best performance; therefore, the outside part of the radars represents the optimal results. It is worth noting that the data availability and sample vary from site to site. Table 17.3 summarises the main information on the pilot sites.

Next, Fig. 17.1 presents the sustainability mobility radars from AVENUE pilot sites.

The results from the sustainability assessment (Nemoto et al., 2023) reveal the strong and weak points of the deployment of automated minibuses. Some common results among the sites pointed out that the automated minibuses score poorly on 'energy efficiency' (with the exception of Pfaffenthal) and 'low contribution to climate change' due to the low vehicle occupancy. With the exception of Pfaffenthal

**Table 17.2** Indicators, units and scales of assessment

| Indicators | Parameter | Scale of assessment |
|---|---|---|
| Social acceptance | Average rating reported concerning the (1) willingness to use automated minibus; (2) perception about the readiness of the technology; (3) willingness to pay | City |
| User satisfaction | Average rating satisfaction reported concerning the automated minibuses speed, comfort, punctuality, information, frequency of service, connection to other means of transport and satisfaction with the last ride | Local |
| Passenger's affordability | Costs (euro) passenger-km for passengers | City |
| Climate change | gCO2 eq/passenger-km | Local and global |
| Air pollution | Air pollutant emissions, particular matter, PM2.5 (g/km) and nitrogen oxides, NOx (g/km), from exhaust and non-exhaust | Local |
| Noise pollution | Vehicle noise in decibels (dB) | Local |
| Renewable energy | Percentage of renewable energy in the use phase of the mode of transport | Country |
| Energy efficiency | kWh/passenger-km | City |
| Economic profitability | Costs (euro)/passenger-km for operators | City |
| External costs | €-cent/pkm (with congestion) | City |
| Institutional development and openness to mobility innovations | ROAD index—'the regulation openness for autonomous driving' index.<br>It sets four variables to measure the level of readiness for the implementation of autonomous collective vehicles on open roads:<br>1. National industrial policy<br>2. Local territories autonomy<br>3. National sustainable development policy and declination<br>4. Governance and integration at local level<br>The score for each variable results in the Road Index for a city | City |
| Technical performance and reliability | Assessment of (1) average speed in km/h, (2) frequency or response speed in minutes of waiting time, (3) average occupancy as the average number of passengers on board at any given time and any place within a trip and (4) the percentage of kilometres driven autonomously | Local |
| System integration and efficiency | Five levels of MaaS integration suggested by (Sochor et al., 2018) | City |
| Reduction of risk of induced demand | Percentage of motorised modes of transport—car and buses—that the automated minibuses are replacing based on the reference modal share | City |

**Table 17.3** Description of the demonstrator sites

| City | Pilot | Characteristics of route | Type of passenger | Deployment |
|---|---|---|---|---|
| Lyon | Groupama Stadium | Fixed route with stops 1.3 km. Will become an on-demand, door-to-station service | Regular workers, people with reduced mobility (medical centre nearby) | November 2019 to April 2022 |
| Copenhagen | Nordhavn | Fixed route with stops, 1.2 km, will become an on-demand, door-to-door service | Residents of the area, tourists | September 2020 to April 2022 |
| Luxembourg | Contern | Fixed route with stops, on-demand, 2.2 km | Employees working at Campus Contern | September 2018 to April 2022 |
| | Pfaffenthal | Fixed route with stops, on-demand, 1.2 km | Workers, tourists, residents and visitors of Luxembourg City | September 2018 to April 2022 |

(Luxembourg), all sites presented very low occupancy. This result can be an indication of low demand for the offered mobility services. However, we should be cautious in this conclusion due to the unknown impacts of the Covid-19 restrictions. In addition, the energy efficiency could also be affected negatively in case the automated minibuses were equipped with more hardware and technical features, such as sensors, cameras, Lidars and communications.

As electric vehicles, automated minibuses seem to be a good alternative to tackle 'local air pollution'. However, they are not a significant solution to tackle 'local noise pollution', as their noise level does not differ that much from other motorised modes of transport from 30 km/h speed. It considers that regarding the background noise and traffic density, EV does not differ from ICEV in the usual traffic, except for urban traffic during the night in low-speed areas (Jochem et al., 2016).

As temporary pilot trials, the automated minibuses present low system integration. Nonetheless, they show a high potential in the near future to have information, booking and payment integration within the public transport services, considering that in most cases, they are already deployed by public transport operators. In addition, it is expected that in the future, the automated minibus could be integrated into MaaS systems.

Concerning the technical performance elements (speed, frequency, occupancy rate and km driven autonomously), all sites struggle with low speed and low occupancy rates. The percentage of fully automated driven kilometres is 80–94%. The manual interventions that took place were mainly caused by wrongly parked cars and trucks.

The use of renewable energy for the use phase varies significantly according to the electricity mix of each region or country. In this case, Nordhavn in Copenhagen has the best score (with 62.4% of renewable sources in the electricity mix), and Contern and Pfaffenthal in Luxembourg are the lowest.

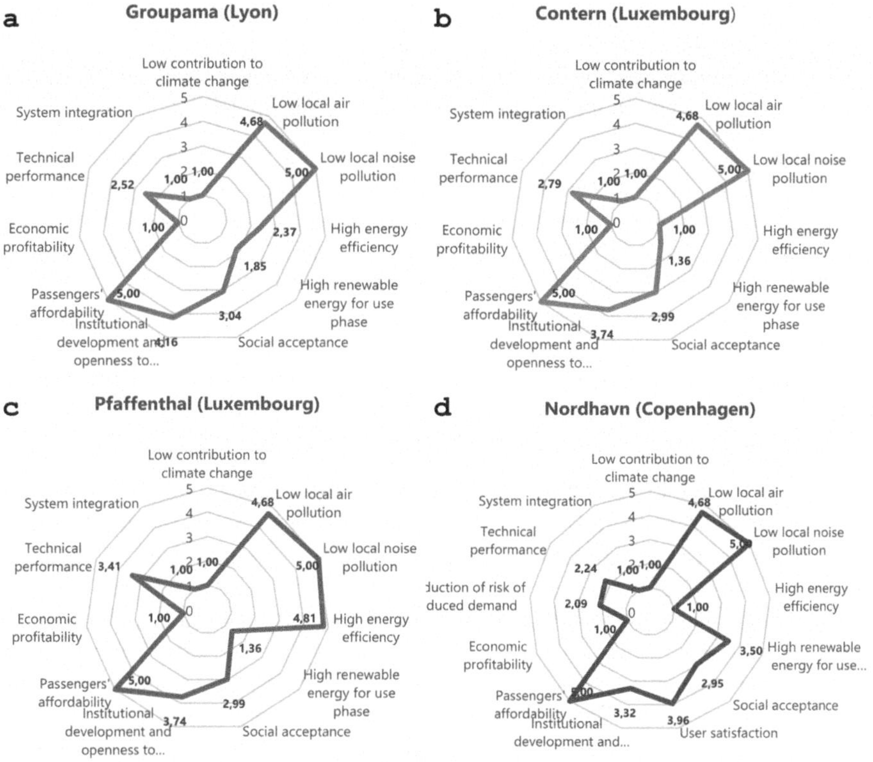


**Fig. 17.1** Sustainability mobility radars from AVENUE pilot sites—with 1 for the worst performance and 5 for the best performance: (**a**) Groupama (Lyon); (**b**) Contern (Luxembourg); (**c**) Pfaffenthal (Luxembourg); (**d**) Nordhavn (Copenhagen)

Currently, the ride on the automated minibus is free of charge. Therefore, passengers' affordability performs high. In the future, it is envisaged to integrate the automated minibus in the ticketing of public transport.

Overall, the economic profitability is still low due to the elevated costs with feasibility studies and legal authorisations, infrastructure works, high annual depreciation and salaries for on-board safety drivers (Antonialli et al., 2021).

Concerning the indicators of social acceptance, respondents show goodwill to use the automated minibus (scoring 3.80 in Lyon and Copenhagen and 3.5 in Luxembourg), while the perception of the readiness of the technology is lower (2.5 in Lyon, 2.4 in Luxembourg and 2.5 in Copenhagen). Based on real experiences, the users in Copenhagen pointed a good satisfaction (3.96) with the ride and the automated minibus services (comfort, information, punctuality, speed).

The indicator on 'reduction of risk of induced demand' scored low in Nordhavn; this is explained by the users' survey, which shows that the automated minibuses have been replacing walking and cycling (17% and 45%, respectively). In parts, this can be explained due to the vehicles' low speed.

All in all, the indicators reflect an incipient phase of deployment and development of the technology. In the short term, key factors for improvement are:

1. The minibuses' occupancy, a key factor in fostering environmentally friendly mobility. The automated minibuses should be deployed to cover real mobility gaps and to provide rides with great potential to replace private cars. These factors are crucial to guarantee higher occupancy and reduction of the risks of induced demand and increase in vehicle kilometres travelled.
2. Better mobility services integration, as the integration of information, booking and payment.
3. Offer of permanent lines/services, on-demand services and higher speed as a factor to improve flexibility and reduce travel time.
4. Monitoring and planning the deployment in order to replace car trips.

In the medium and long term, the economic profitability of deploying the automated minibuses should become more attractive with the development of a legal framework and lower costs with feasibility studies, authorisations and exemption of safety drivers.

Concerning the SUMP concept, it is worth noting that automated vehicles and minibuses will not be sustainable per se, but rather their mode of deployment is very important, and factors such as shared mobility, ride-matching capacity and efficiency, system integration and means of transport it will replace proper policies and regulations. The automated minibuses should be integrated into urban public transport or within MaaS perspective and fundamentally aligned with the city's goals, planning and strategies for sustainable mobility. Also significant is to keep an integrated vision of the mobility system and, as highlighted by the SUMP approach, to develop all modes of transport in an integrated manner. Thus, the automated minibuses are a piece within the mobility ecosystem that could support intermodality, MaaS, mobility hubs and the use of soft modes of transport.

Concerning SUMP principles, the deployment of this new mode of transport and new mobility technologies requires more than ever long-term vision and planning, development of all transport modes in an integrated manner, cooperation across institutions, stakeholders and citizens' participation, performance assessment and monitoring towards established sustainability goals.

Therefore, the SUMP principles and the four steps/guidelines are a valuable tool for planning and implementing automated minibuses aiming at people's mobility needs and better quality of life. In this regard, the indicators are a tool to measure and monitor the progress and achievement of sustainable mobility planning and goals.

## 17.4 The Mobility Radar to Assess Two Potential Scenarios

This section explores the application of the indicators for two potential scenarios: (i) automated minibus integrated in MaaS and (ii) AV deployed as robotaxis. According to these two scenarios (detailed in Chap. 14), the level of integration of the automated minibus varies significantly and, as consequence, the potential impacts on mobility systems as well.

Figure 17.2 presents the mobility radar for assessment of the two scenarios (Nemoto et al., 2023). This is an explorative approach, these future perspectives need further research, and the enclosed developed hypothesis will be deepened on sites in the EU-funded ULTIMO project (2022–2026).

The assessment suggests that, overall, automated minibuses in MaaS tend to present a better performance than AVs deployed as robotaxis. As electric vehicles, the external costs for climate change and air pollution of both scenarios score high, whereas automated minibuses in MaaS score much better in terms of external costs for congestion and energy efficiency, considering that the average occupancy and sharing rate for automated minibuses in MaaS are higher.

The calculation of the external costs is detailed in Chap. 14, and it comprises air pollution, climate change, well-to-tank, noise, accidents and congestion (Jaroudi, 2021).

The social acceptance for robotaxis could be higher than for the automated minibus in MaaS, since they are akin to individual mobility; they could be a faster option (shorter waiting time), cheaper and not intermodal. Therefore, the social acceptance could be higher; although this form of deployment would be less sustainable. The robotaxis are very attractive for users and compete with public transport. The passenger traffic would thus be displaced from public transport to robotaxi and increase congestion (WEforum, 2020).

The economic profitability is assessed according to the costs (Euro) per vehicle-km; therefore, in VKM the operation of robotaxis is cheaper than automated

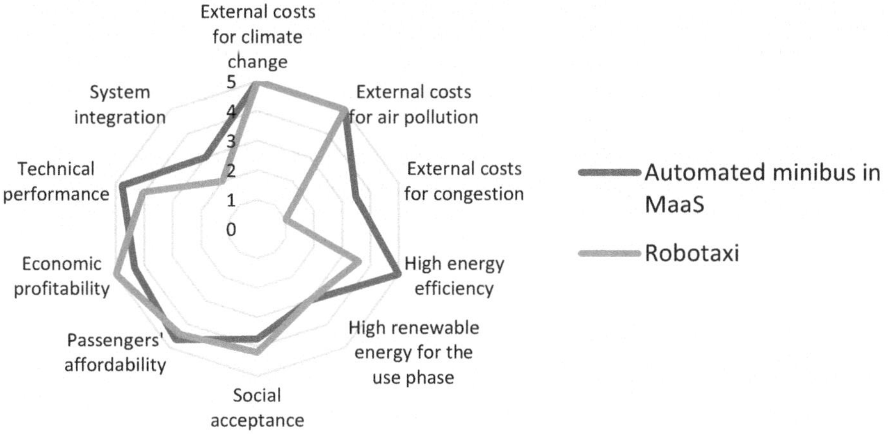


**Fig. 17.2** Sustainability mobility radars for scenarios deploying automated minibus in MaaS and robotaxis

minibuses. However, when it comes to costs (Euro) passengers-km, the price for the automated minibus services is more attractive and affordable (Bösch et al., 2018).

Regarding the technical performance, robotaxis are expected to have a higher speed, with less waiting time; however, the automated minibus would present higher occupancy. And in terms of system integration, the automated minibus would be integrated in MaaS for data, information, booking, ticketing, billing for different mobility services etc. On the contrary, robotaxis would present a lower level of integration. This means that the integration of data, information and the related partners within the transport system is low, and no synergies or positive externalities can be enabled.

The sustainability assessment of the scenario is in line with MaaS and SUMP approaches, reinforcing that mobility system integration is crucial to fostering intermodality and sustainable mobility. Additionally, the development of policy instruments and push-and-pull measures is a lever for a mobility shift from private and individual mobility towards public transport-centred MaaS.

## 17.5 Limitations of the Assessment

As this study analyses pilot trials, the sustainability assessment in this study is limited to a local and small-scale deployment, in addition to the technological limitations due to the development of the automated minibuses (Levels 3 and 4 of driving automation) and software. The assessment is also limited to just one type of vehicle provided by the same vehicle manufacturer.

Further, the application of the indicators is limited to data availability and data asymmetry from the pilot sites. For instance, the AVENUE representative and user surveys among the four demonstrator cities have different samples (n), hence, varying their representativeness.

It is also important to note that all the pilot tests within AVENUE project have been directly affected by the Covid-19 pandemic. The trials had interruptions; the number of passengers dropped, as had happened to public transport in general; and in some trials, the maximum number of passengers was limited to four in order to keep the social distance. For these reasons, the data collection and data availability for assessment were affected.

For some indicators, the assessment was simplified considering standard units of measurement available in the literature and commonly applied to other modes of transport.

## 17.6 Concluding Remarks

The sustainability assessment embedded results from the AVENUE project, targeting the social, environmental and economic impact assessment.

The assessment of the pilot trials points out that at the current stage, the automated minibus does not fulfil all the premises for sustainable mobility. However, the automated minibuses prove to be feasible as new alternative mobility and with the potential to support cities to achieve sustainable mobility under certain premises (e.g. technological improvements, vehicle usability and occupancy, integration into the mobility systems and intermodality, policies and strategies for sustainable mobility).

Further variables and questions will influence the performance and assessment of the automated minibuses in urban mobility, such as which modes of transport it will replace, what will be the occupancy rate, how fast the technology and policy development will occur, at to which extent automated minibuses will be integrated into the mobility system and under which policies and incentives. Additionally, the development of policy instruments is a lever for a mobility shift from private and individual mobility towards public transport-centred MaaS. Such elements are key for system innovation, and it comprises changes in governance, from a laissez-faire approach to a 'governing by enabling' approach.

Finally, the perspective is that automated minibuses could be integrated into urban mobility to improve the transport network, cover mobility gaps and foster intermodality by substituting motorised vehicles and offering on-demand and door-to-door services. The automated minibuses can be seen as a game-changer by improving mobility services and offering attractive private mobility, being part of the mobility innovations that target a system innovation and a shift from private to a mobility that serves the general interest. Indeed, automated minibuses could support MaaS approach, electrification and shared mobility, and accordingly to the recommendations in our study, they can foster SUMP and the sustainable agenda of cities.

**Acknowledgements** This project has received funding by the European Union's Horizon 2020 research and innovation program under grant agreement No. 769033.

## References

Antonialli, F., Fournier, G., Jaroudi, I., Mira-Bonnardel, S, & Thalhofer, M. (2021). D8.2 Second Iteration Economic impact. Available online at https://h2020-avenue.eu/public-delivrables/. Accessed 20 Apr 2021.

Bösch, P. M., Becker, F., Becker, H., & Axhausen, K. W. (2018). Cost-based analysis of autonomous mobility services. *Transport Policy, 64*, 76–91. https://doi.org/10.1016/j.tranpol.2017.09.005

Jaroudi, I. (2021). *What is the potential impact of the transition from traditional transport to new mobility (electric automated minibuses) in European cities?* Gerpisa Colloquium. Available online at https://gerpisa.org/node/6370

Jochem, P., Doll, C., & Fichtner, W. (2016). External costs of electric vehicles. *Transportation Research Part D: Transport and Environment, 42*, 60–76. https://doi.org/10.1016/j.trd.2015.09.022

Jones, E. C., & Leibowicz, B. D. (2019). Contributions of shared autonomous vehicles to climate change mitigation. *Transportation Research Part D: Transport and Environment, 72*, 279–298. https://doi.org/10.1016/j.trd.2019.05.005

Lim, H., & Taeihagh, A. (2018). Autonomous vehicles for smart and sustainable cities: An in-depth exploration of privacy and cybersecurity implications. *Energies, 11*(5), 1062.

Litman, T. (2007). Developing indicators for comprehensive and sustainable transport planning. *Transportation Research Record: Journal of the Transportation Research Board, 2017*(1), 10–15. https://doi.org/10.3141/2017-02

Nemoto, E. H., Issaoui, R., Korbee, D., Jaroudi, I., & Fournier, G. (2021). How to measure the impacts of shared automated electric vehicles on urban mobility. *Transportation Research Part D: Transport and Environment, 93*, 102766. https://doi.org/10.1016/j.trd.2021.102766

Nemoto, E. H., Korbee, D., Jaroudi, I., Viere, T., Naderer, G., & Fournier, G. (Eds.). (2022). Sustainability assessment of the deployment of automated minibuses in urban mobility of European cities. In *Transport Research Arena (TRA) Conference*. Lisbon, 14–17 November.

Nemoto, E. H., Korbee, D., Jaroudi, I., Viere, T., Naderer, G., & Fournier, G. (2023). Sustainability assessment of the deployment of automated minibuses in urban mobility of European cities. *Transportation Research Procedia, 72*, 1364–1371. https://doi.org/10.1016/j.trpro.2023.11.599

Shaheen, S., & Chan, N. (2016). Mobility and the sharing economy: Potential to facilitate the first- and last-mile public transit connections. *Built Environment, 42*(4), 573–588. https://doi.org/10.2148/benv.42.4.573

Shen, L.-Y., Jorge Ochoa, J., Shah, M. N., & Zhang, X. (2011). The application of urban sustainability indicators—A comparison between various practices. *Habitat International, 35*(1), 17–29. https://doi.org/10.1016/j.habitatint.2010.03.006

Sochor, J., Arby, H., Karlsson, I. M., & Sarasini, S. (2018). A topological approach to mobility as a Service: A proposed tool for understanding requirements and effects, and for aiding the integration of societal goals. *Research in Transportation Business & Management, 27*, 3–14. https://doi.org/10.1016/j.rtbm.2018.12.003

Soteropoulos, A., Berger, M., & Ciari, F. (2019). Impacts of automated vehicles on travel behaviour and land use: An international review of modelling studies. *Transport Reviews, 39*(1), 29–49. https://doi.org/10.1080/01441647.2018.1523253

Taiebat, M., Brown, A. L., Safford, H. R., Qu, S., & Xu, M. (2018). A review on energy, environmental, and sustainability implications of connected and automated vehicles. *Environmental Science & Technology, 52*(20), 11449–11465. https://doi.org/10.1021/acs.est.8b00127

Verbruggen, H., & Kuik, O. (1991). Indicators of sustainable development: An overview. In O. Kuik & H. Verbruggen (Eds.), *Search of indicators of sustainable development* (Vol. 1). Springer.

WEforum. (2020). Will robo-taxis cause radical disruption - or gridlock? Available online at https://www.weforum.org/agenda/2020/01/will-robo-taxis-bring-radical-disruption-to-our-streets-or-gridlock/. Accessed 10 May 2022.

World Business Council for Sustainable Development. (2015, March). Methodology and indicators for sustainable urban mobility. Sustainable Mobility Project 2.0 (SMP2.0).

# Part III
# Future Vision of AVENUE

# Chapter 18
# System Innovation in Passenger Transportation with Automated Minibuses in ITS: The Citizen-Centric Approach of AVENUE

**Guy Fournier, Michael Thalhofer, Johannes Klarmann, Philippe Chrétien, Dorien Duffner-Korbee, Adrian Boos, Ines Jaroudi, Eliane Horschutz Nemoto, Lionel Binz, Gabriele Naderer, Dimitri Konstantas, and Tobias Viere**

G. Fournier (✉)
Pforzheim University, Pforzheim, Germany

CentraleSupélec, LGI, Gif-sur-Yvette, France
e-mail: guy.fournier@hs-pforzheim.de

M. Thalhofer
Software-driven Ecosystems and Market Places, Siemens Technology, Munich, Germany

J. Klarmann · A. Boos · G. Naderer · T. Viere
Pforzheim University, Pforzheim, Germany

P. Chrétien
Centre Européen d'Etudes de Sécurité et d'Analyse des Risques, CEESAR, Nanterre, France

D. Duffner-Korbee
Pforzheim University, Pforzheim, Germany

Fraunhofer ISI, Karlsruhe, Germany

I. Jaroudi
Pforzheim University, Pforzheim, Germany

CentraleSupélec, LGI, Gif-sur-Yvette, France

Fraunhofer ISI, Karlsruhe, Germany

E. H. Nemoto
Pforzheim University, Pforzheim, Germany

CentraleSupélec, LGI, Gif-sur-Yvette, France

ICLEI Europe, Freiburg im Breisgau, Germany

L. Binz
Cantonal Transports Office, Geneva, Switzerland

D. Konstantas
Geneva School of Economics and Management, University of Geneva, Geneva, Switzerland



G. Fournier et al. (eds.), *Automated Vehicles as a Game Changer for Sustainable Mobility*, Contributions to Management Science,
https://doi.org/10.1007/978-3-031-61681-5_18

**Abstract** The first chapter explores three pathways for incorporating automated vehicles (AVs) into future mobility ecosystems: privately owned AVs, robotaxis and automated minibuses in a Mobility-as-a-Service (MaaS) and (later) in an Intelligent Transport System (ITS). The chapter emphasises that automated minibuses, when seamlessly integrated into a MaaS, could emerge as pivotal "game changer", complementing and fostering other means of transport, in particular mass transport. Integrating in a next step AVs within an ITS could further make it possible to use mobility data and artificial intelligence (AI) to improve the transport system to a higher level of mobility evolution. By employing both fast and slow closed AI loops, it envisions a transport ecosystem that not only operates more efficiently and flexibly but is also consequently citizen-centric and addresses sustainability challenges. The successful implementation of this concept hinges on open data, Application Protocol Interface (APIs) and the potential of AI to create a self-learning transport system to serve the general interest. A sustainable and citizen-centric mobility is thus possible without a coercive (technology push) transport policy. Instead, it champions an approach that increases the transport options and enhances the provided mobility services (demand & attractiveness pull). Depending on the local governance, even privately owned or privately shared cars can be part of the model, justified, e.g. in areas where mobility infrastructure deficits loom large or when travel time is too high. This chapter therefore forms the basis for a vision to redesign our mobility ecosystem and marks the beginning of a disrupting system innovation, where integrated sustainability and citizen centricity reshape the nature of mobility.

## 18.1 Introduction

In this chapter, we will discuss and present the vision of AVENUE for a medium-term horizon (2030) of future public transportation. Our aim is to integrate automated minibuses into transportation systems available in a city. A central issue in this vision is the mobility needs of citizens which have to be satisfied in an optimal way: an abundant service offer portfolio with a high variety of private and public mobility modalities combined to one individualised intermodal trip. Automated minibuses play a central and critical role in this model: (1) as a feeder for the other means of transport, in particular mass transit, and (2) as a mobility complementor for the entire transport system in case of weak or not existing transport offerings (e.g. rural or remote areas), incomplete transport chains (e.g. for tangential connection) and disturbances of the transport system. Public as well as private transport operators (PTOs) are forming an enhanced public-private partnership (PPP) to utilise and synergise their multifaceted complementarity in this MaaS (Mobility-as-a-Service) vision. The application of advanced self-learning systems (e.g. based on human and/or artificial intelligence) could in a further stage let this visionary MaaS concept become a self-learning ITS (intelligent transport system). This disruptive transport system innovation could create best citizen-centric transportation with optimised and balanced private and public value. Further, it improves economic performance, supports environmental benefits and also increases acceptance by citizens, as this intelligent transport system is enabling smooth transition towards efficient, safer, nearly carbon-free, inclusive and sustainable transport in cities.

The AVENUE vision of the future mobility system builds upon insights out of the AVENUE sustainability assessment and the conceptual transition studies and recommendations (see Chap. 19). The following research questions have been guiding the analysis:

How can citizens' mobility needs, stakeholders' interests and general interest be optimally satisfied with AVs?

How can AVs and automated minibuses be integrated in a future transport system in the city in a meaningful and optimal way?

How should EU regulations evolve to support a reliable transport system that balances economic values of all stakeholders, regards privacy requirements and ensures safety and security issues and governance; general interest on social, economic and environmental level; freedom; and sovereignty (mobility data, technological sovereignty)?

These questions will be answered in the next sections of this chapter. The integration of automated minibuses into an intermodal and citizen-centric perspective is discussed in Sect. 18.2; the integration of the automated minibus within a MaaS system is designed in Sect. 18.3; and a deployment of automated minibuses integrated with artificial intelligence (AI) and intelligent transport systems (ITS) is proposed in Sect. 18.4. After a discussion in Sect. 18.5 about limitations of the vision, conclusions are presented in Sect. 18.6.

In terms of methodology, numerous interdisciplinary workshops have been organised with scientists, public transport operators, original equipment manufacturers (OEMs), industrial companies, public transport authorities and other stakeholders, both within and outside the project (see Chaps. 9 and 10). Various alternatives were developed and evaluated using a holistic approach. Experts from a wide range of fields, including automation engineers, IT specialists, economists, social scientists, psychologists, environmental scientists and lawyers, contributed to the discussions. The aim of this multidisciplinary dialogue was to ensure a comprehensive understanding of the interrelationships between different impact areas, potential rebound effects and conceivable futures. The results have been accepted by all project partners and the advisory board of AVENUE. A more comprehensive summary of this chapter has been published in 2023 by Fournier et al. (2023).

## 18.2 Integration of Automated Vehicles in Future Urban Transport Systems: From Product Innovation to Citizen- and Purpose-Centric Transport System Innovation

Automated vehicles can be integrated into future transport systems in different ways. Among others, three use cases are possible (based on Grisoni & Madelenat, 2021; Heineke et al., 2019; UITP, 2017):

Private automated vehicles (AVs): Like the currently dominant choice of privately and individually owned cars, AVs can be integrated into an urban mobility system. An integration of AVs in such way would lead to an incremental improvement of the individually used vehicle technology from today's driver-assisting systems to fully automated driving systems (Fraedrich et al., 2015; Deloitte, 2022).

Robotaxis: The definition for robotaxis inhere follows the description as shared automated vehicles (Fagnant & Kockelman, 2018). This definition is quite general and does not distinguish between vehicle sharing (car-sharing) and service sharing (ride-sharing, ridepooling, ride-hailing etc.; see MOQO, 2020). Within the understanding of the AVENUE project, robotaxis mostly appear as driven by private operators with commercial interest. They are meant to be not or just poorly integrated with the public transportation and therefore stand consequently mostly in direct competition to it. They are further mainly used for single ridership although those capabilities would support ridepooling services. Examples for robotaxis are Waymo, Uber Pool or Moia in the German city Hanover (see, e.g. WEForum, 2020; May et al., 2020; Clewlow & Gouri, 2017; Merlin, 2017; Jaroudi et al., 2021; https://www.moia.io/de-DE/stadt).

Automated minibuses in a MaaS system: Another pathway of integrating AVs into an urban transport system is the usage of automated minibuses as one optional transportation mode in a Mobility-as-a-Service (MaaS) system (see Fig. 18.1). The automated minibuses within the scope of this AVENUE project are defined as fully electric, automated vehicles without a driver [SAE Level 4 of automation; see SAE International (06-May-23)] and with a transportation capacity of up to 15 passengers. This vehicle can be combined with a means of either public (e.g. train, tramway, traditional or even automated fixed-scheduled bus line etc.) or private (e.g. bike-sharing, car-sharing etc.) transportation to a seamless journey.

The integration into a MaaS is meant to pool multiple trips of users that request an on-demand, door-to-door service. Furthermore, the transportation service can cover mobility gaps in the public transportation network (Sochor et al., 2016; Shen et al., 2018). Aside from their capability to fill the gaps of public transportation networks, automated minibuses via MaaS enable the collaboration of private and public transport operators (PTOs) to provide one individual intermodal trip to citizens that optimises travel time, costs and/or environmental impacts (Sochor et al., 2018; Kamargianni et al., 2015; Vleugel & Bal, 2017). MaaS in this case provides citizens the freedom to select the optimised transportation modes for its travel depending on its user profile, persona and even external factors (such as weather, etc.) (see Fig. 18.2). Technically, MaaS systems rely on the commitment to a standardised application programming interface (API) for all participating transport operators to offer seamless intermodal trips to its users (Sochor et al., 2018; Kamargianni et al., 2015).

These described three pathways each have advantages and disadvantages. In the development of privately owned AVs (1), the service can be seen as very convenient due to exclusive usage, achieving probably the highest degree of individuality and

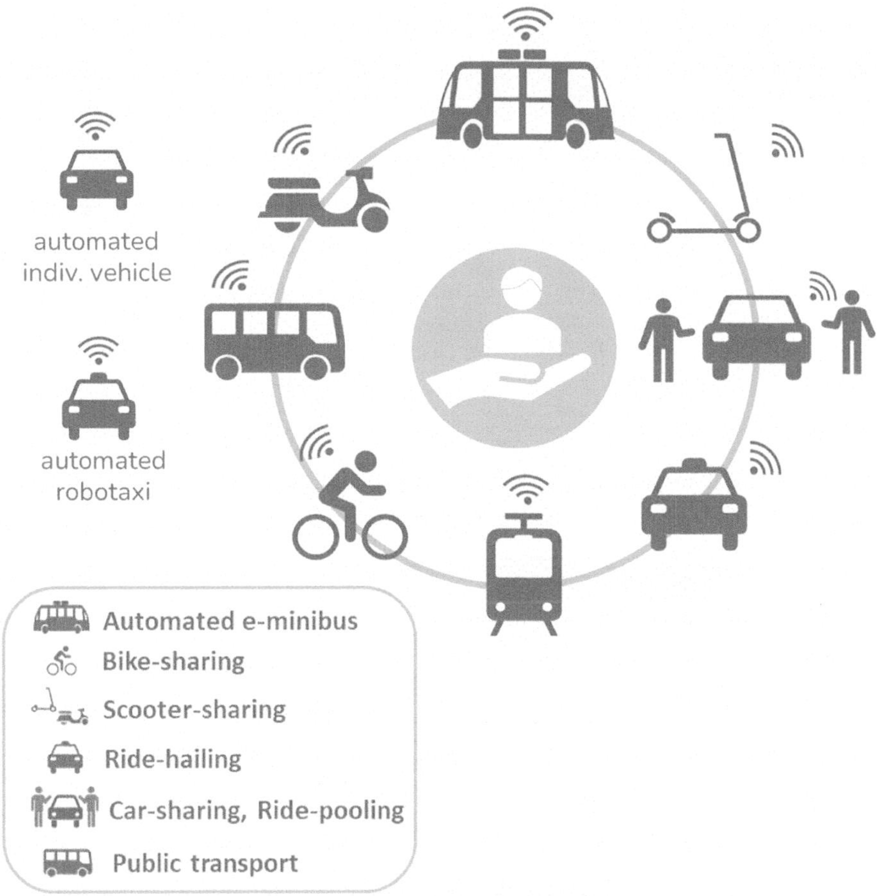


**Fig. 18.1** The integration of the automated minibus into a MaaS system

freedom. On the other hand, the costs of individual mobility would be very high for the users and not affordable for most citizens. A mass diffusion of this solution is accordingly not expected yet. It is also assumed to be one of the reasons why the automotive industry currently focuses on SAE Level 2+.[1] Nevertheless, in case of a wider diffusion of the privately owned fully automated vehicles, the convenience of these vehicles would encourage more vehicles on the road and lead to higher traffic

[1] The term Level 2+ in automated driving context reflects a combination of SAE Level 2 automated driving with technologies to monitor the driver. This technology involves a powerful form of sensing and automated control while keeping the driver engaged and able to monitor. It is important to note that liability remains on the driver's side of the equation, which still makes a driver mandatory in the car (https://www.spglobal.com/mobility/en/research-analysis/fuel-for-thought-waiting-for--autonomy.html).

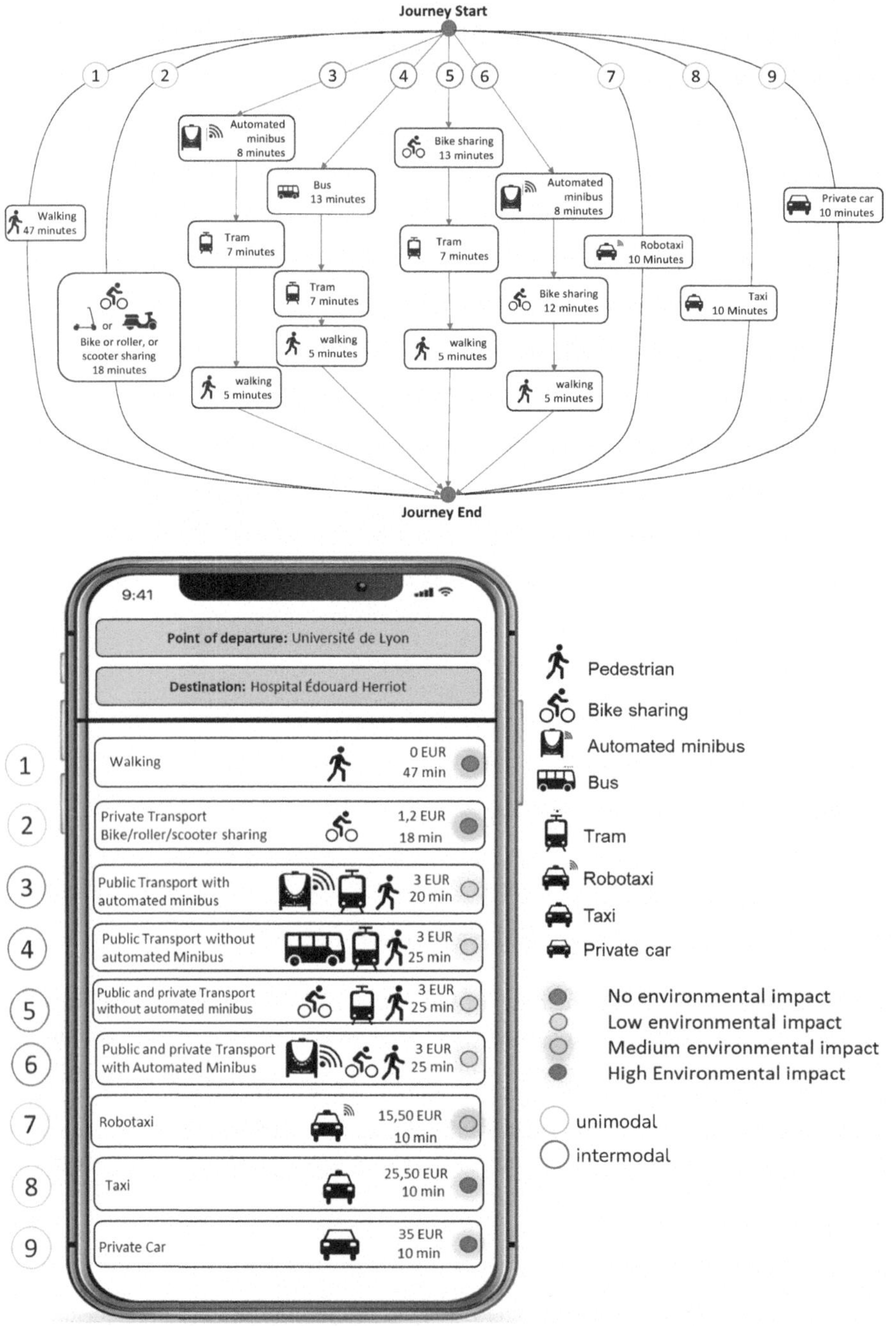


**Fig. 18.2** Illustration of various citizen journeys as combinations of all means of transport offered to a citizen by the MaaS concept

volume and congestion costs (externalities). A weakening of public transport would be likely as well with growing market share (Fournier et al., 2020; Milakis et al., 2017).

It is also expected that robotaxis (2) will be widely accepted by citizen due to their affordability, availability and comfort, but they may not complement other means of transport in a synergistic way (Korbee et al., 2023). Robotaxis could in particular be in competition with and may substitute other means of transport, including cars, walking and public transport (WEF Forum, 2020). Individual unimodal mobility could be more convenient and strengthened. Simultaneously, the transportation system as a whole would become less efficient and sustainable, also resulting in increased external costs due to additional traffic and displaced public transport (Meyer et al., 2017; Niles, 2019; May et al., 2020; Clewlow & Mishra, 2017; Rayle et al., 2016; Merlin, 2017; Childress et al., 2015; WEForum, 2020). Market fragmentation and asymmetric competition between this market leader and other private and public mobility providers could also result in the emergence of a dominant robotaxi provider (Hassani, 2018; Niles, 2019). For example, if private transport platforms are not providing open mobility data in the same way as public transport providers, data asymmetries will arise and lock in the customer in the robotaxi ecosystem. As a result, a dominant market position for private transport providers and the associated "winner-takes-it-all" effect could appear and bear the risk of rising monopolies within the transport system (Hassani, 2018; Hoffmann, 2021; Cabral et al., 2021). Similar situations can be observed for other online services. There is currently an oligopoly dominated by a group of companies known as GAFAM (Google (Alphabet), Apple, Facebook (Meta), Amazon and Microsoft) or now MAAMA (Meta, Alphabet, Amazon, Microsoft and Apple) (The Economist, 2022; European Commission, 2021; Cabral et al., 2021; Toledano, 2020).

Integrating automated minibuses into a Mobility-as-a-Service system (3) could finally be the game changer: public and private transport could become individualised, by offering, e.g. first and last mile trips, as well as a critical mobility gap-filler for the feasibility of seamless citizen journeys. The concept of providing better transport services, improved efficiency and flexibility in the transport system and positive externalities in this system puts citizens at the centre (Coyle et al., 2020; Coyle & Diepeveen, 2021). Automated minibuses in MaaS create multifaceted value for the stakeholders involved, resulting in a win-win situation for all.

Automated minibuses in a MaaS thus seem to offer more benefits for the stakeholders (travellers, TPOs, cities) and for the general interest (Becker et al., 2020). The integration of automated minibuses into public transport in urban, suburban and rural areas and into a MaaS system (our vision) is highly beneficial and advantageous for its stakeholders, in particular for citizens and private and public mobility providers as well as cities and regions (administrations/government) in general. This integration is understood as follows:

A customer-/citizen-centred approach allows the user to orchestrate an individually selected combination of all different types of transport into a gap-less and reliable trip that meets the different preferences of all population groups. As such, automated minibuses in MaaS are seen as a transformative force in urban mobility, capable of providing personalised public transport that is diverse, accessible (including for PRM), inclusive and affordable. This could be an attractive and beneficial alternative to the private car transport. "The results of a representative survey among 1816 citizens (of which 1526 have privately-owned vehicles) in Lyon, Copenhagen, Luxembourg and Geneva confirm that 45% of car drivers are 'willing' (22%) or even 'very willing' (23%) to give-up using their own car to use automated minibuses to bridge the first and the last mile if this were available. If the service is on-demand and door-to-door, the acceptance could be even 52% (respectively 28% and 24%) in total" (Fournier et al., 2023, p. 3; Korbee et al., 2023).

For the transport operators, the "automated minibus in a MaaS concept provides a better exploitation of existing capacities and resources and a positive experience of fair competition" (Fournier et al., 2023, p. 3). In particular, automated minibuses are an important factor and a missing complementarity modality in making large mass transit systems such as rail, metro, bus etc. more attractive, more efficient and thus more profitable. By taking this perspective, all offered means or modalities of transport are potentially available for an individual trip design. The mobility choice and subsequently the modality choice of the citizen (the end user/customer as central "king/queen of the system") are considered as right and definitive and not limited by the transportation portfolio of a single (private or public) transport provider (Fig. 18.2). The automated minibus integration into the urban mobility system contributes to avoid thereby the "winner-takes-it-all phenomenon". On contrary, a win-win situation for all stakeholders in terms of value creation, fair value sharing, sustainable mobility and long-term purpose is achieved, as stands the Purpose Economy (Hurst, 2016; Business Roundtable, 2019).

For the cities (administrations), the designed concept would offer a significant optimisation of mobility excellence (efficiency, effectiveness, safety, quality of transport, user-friendliness etc.), serving the wider public interest and providing financially, environmentally and socially sustainable transport. The complementary use of public and private transport together, with automated minibuses, would therefore be able to make travel less costly, more competitive and more predictable and would support a higher level of resilience in the overall mobility system, decreasing external impacts of mobility (Alazzawi et al., 2018; Vleugel & Bal, 2017), and this will contribute to a new paradigm of sustainable mobility and to the Sustainable Urban Mobility Plan (SUMPs) that involve citizens and stakeholders in a goal-oriented approach that serves public interest. Moreover, it will increase the public acceptance through higher effectivity and flexibility in the whole system (Hurst, 2014; Nemoto et al. 2021; Korbee et al., 2023). Based on the results of the aforementioned survey, a transformation of the mobility paradigm could be realised without a coercive policy (push strategy). An improvement of the offer would change the behaviour (pull strategy).

Open data and interaction platforms and open APIs are prerequisites to enable access to all the means of transport, their interoperability and intermodality, to ensure balanced win-win situations among all stakeholders and thus a kind of democratisation of the automated minibus in a MaaS ecosystem (ERTICO – ITS Europe, 2019; Coyle et al., 2020). Open data and open APIs are further a condition to enable innovation and evolution of mobility concepts and develop the MaaS towards an intelligent transport system (ITS). Using data makes it easier to generate information and knowledge and provide the basis for decisions to improve the transport system and to generate positive externalities. The transport system could become a self-learning system with the support of artificial intelligence and humans (see details in chapter below). A "circulus virtuosis" (virtuous cycle/loop) and positive externalities could be created, to guarantee safety and to promote global improvement. To achieve this ITS vision, several improving loops have been identified which could be implemented and which make the transport system more resilient and future oriented (see Sect. 18.4).

The third path, "automated minibus in MaaS", can thus provide the best public value and therefore represents the long-term vision of AVENUE. It is the main enabler of a holistic system innovation and is not only regarded as a classic product (e.g. automated minibuses) innovation. However, this is not to underestimate the importance of automated minibuses as a product innovation: combining automated minibuses and MaaS represents a disruptive game changer to take the entire transport system to a higher level of mobility evolution. This evolution will be described in the sections with AV in MaaS (Sect. 18.3) and the next progress with AV in ITS (Sect. 18.4).

## 18.3 Integration of Automated Minibuses in a MaaS System: A Citizen- and Purpose-Centric Approach

The concept of purpose economy refers to a new way in which people and organisations create value and define the principles for innovation and growth (Hurst, 2016). The value lies in establishing purpose and creating meaning value for employees and customers beyond their own benefits, but aiming personal and community development (Hurst, 2016).

> The Purpose Economy creates purpose for people. It serves the critical need for people to develop themselves, be part of a community, and affect something greater than themselves. (Hurst, 2016)

The "citizen-centric automated minibus in MaaS" approach is in line with the purpose economy and targets a mobility system built by all and for all. The concept is focusing on the compilation of various transportation service modalities to one single seamless trip on demand and, according to the preferences of the citizen, including related services like trip planning, ticketing and others (European Commission, 2016).[2] Citizens have the choice to choose from various transportation

[2] The citizen-centric approach naturally also includes older people and people with disabilities, as these groups are dependent on accessible public transport and expected to change fast future mobility demand. This is also required by law (European Accessibility Act). The ageing society and 80 million people with disabilities in the EU create a new demand for products and services and will have a major impact on mobility in the future, especially regarding the fact that public transport can influence how and where people live and whether they can maintain their independence. Future mobility services must take this into account. For example, booking services need to include special needs in their service portfolio. Aspects such as additional space (e.g. for guide

options for their journey within or outside the city or even the country. To support them in their decision-making, the MaaS platform provides various functionalities such as simulations of time, distance, cost and CO2 footprint, as well as options for comparison. Once the citizen has made their decision, they are requested by the mobility integrator app to make the booking and place a binding order. The data from the selected trip option is then sent back to the MaaS platform and onwards to the private and/or public transport platform and transportation provider or possibly to other MaaS system platforms. "The purpose is to better satisfy citizen mobility needs and evolve towards a sustainable mobility system, create value for all stakeholders, contribute to sustainable societal transformations, and provide an answer to societal challenges (the so-called 'purpose')" (Fournier et al., 2023).

To achieve this, it is necessary to consider factors such as the availability and interoperability of both hardware and software devices, which are provided through standardised interfaces (APIs), as well as management and coordination software and services offered by an open ecosystem of service aggregators and other intermediaries. By combining digitalisation with a citizen-centric approach, the distinction between private and public transport operators will be blurred, leading to the introduction of coopetition (Haan et al., 2020). Coopetition is a new perspective that involves both competition and cooperation existing together, utilising complementary resources cooperatively (Liu et al., 2015). Additionally, democratised and federated governance of ecosystem stakeholders could be established, where beneficiaries are determined by their value-creating contributions to the defined purpose (Gassmann & Ferrandina, 2021; Schmück, 2022).[3]

Figure 18.3 depicts the integration of automated minibuses in a citizen-centric MaaS, with the citizens at the core. A variety of transport means is available for the citizens to choose from, as depicted in the middle circle in Fig. 18.3. The transport options are provided by various stakeholders (the actors named in the yellow blocks, at the outer circle of Fig. 18.3). Other governance scenarios to coordinate the different stakeholders are of course possible depending on market (private, public and coopetition) and data schemes (private or open data, open interfaces and protocols). These scenarios have been developed by UITP 2019a (UITP, 2017, 2019; Ertico ITS-Europe 2019; Capgemini, 2020). The scenario which fulfils best the needs of the customer, enables fair competition through avoiding the "winner-takes-it-all" phenomenon, enables positive externalities and fulfils sustainable and societal goals is the citizen-centric MaaS approach chosen by AVENUE. It defines a purpose, balancing the interest of the numerous stakeholders to serve the general interest.

Important for the success of the AVENUE vision is the full integration of automated minibuses into the described citizen centric MaaS system. To ensure a fair understanding of how automated minibuses could be integrated in the MaaS system,

---

dogs or walkers) and booking assistance for entering/changing means of transport, as well as barrier-free design of the apps used to book these services, need to be considered.

[3] Private transport operators are the owners of a private transport operation platform providing and managing private transportation services from various modalities by subcontracting private suppliers.

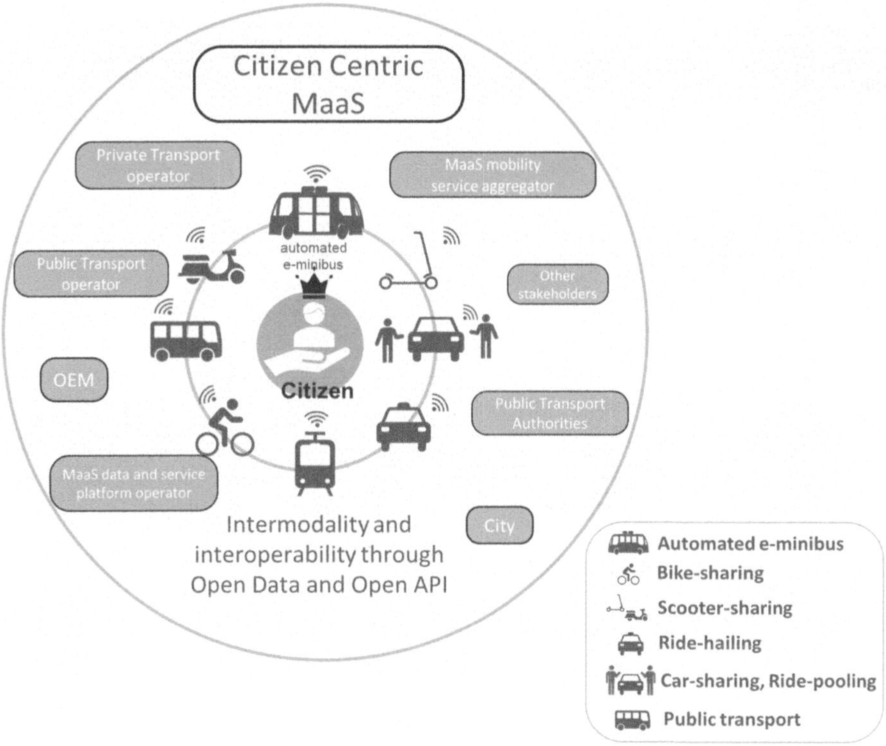


**Fig. 18.3** Automated minibuses integrated into a citizen-centric MaaS

a typical booking process which integrates automated minibuses is set up with the following five steps:

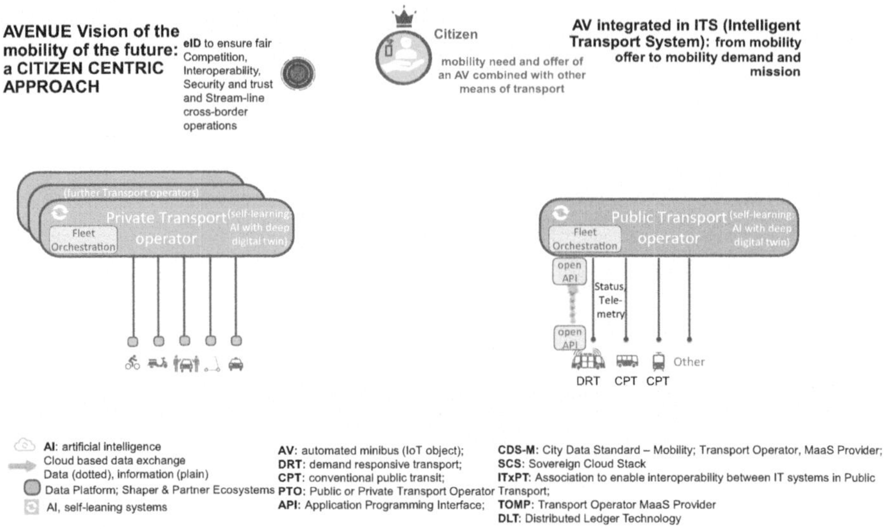


**Fig. 18.4** Vision of the mobility of the future: Step 1

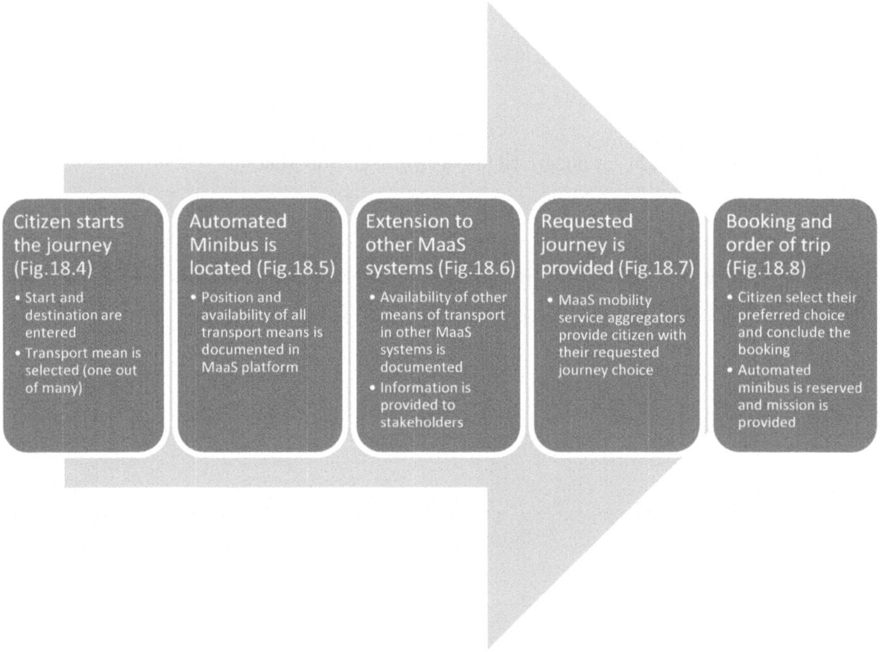


These steps are detailed and explained below; each of the steps is visualised by a figure (Figs. 18.4–18.8). The step-by-step description starts with a current, conventional outline of the system, in which the citizen has to enquire about options and

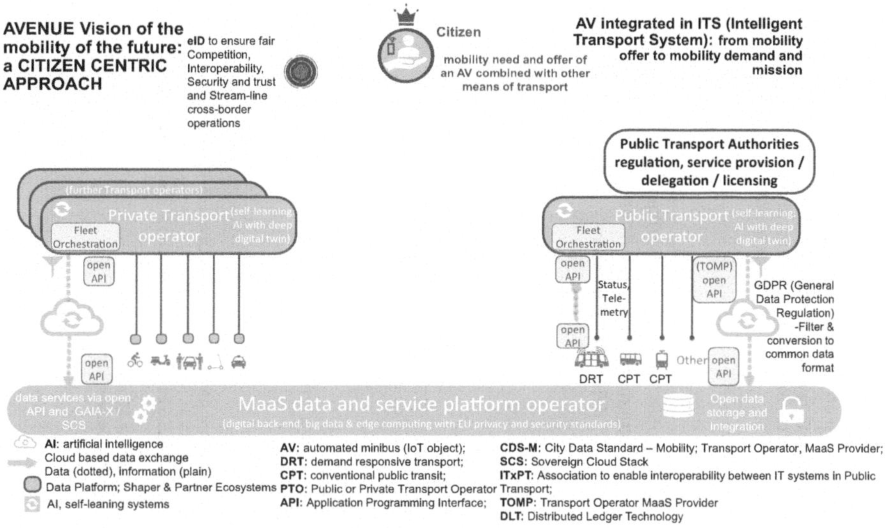


**Fig. 18.5** Vision of the mobility of the future: Step 2

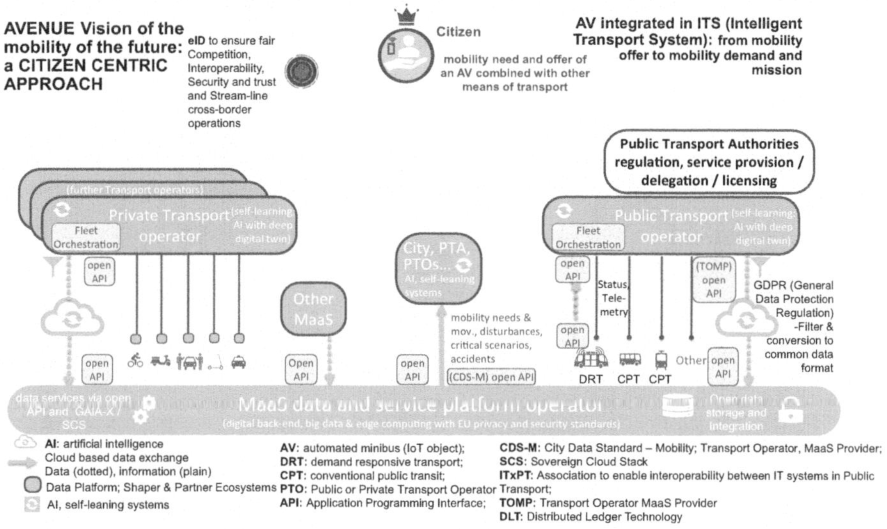


**Fig. 18.6** Vision of the mobility of the future: Step 3

has to make the decisions themselves, building up to a fully integrated system. With each of the steps, more complexity (through the integration of stakeholders and additional platforms) is added to the MaaS system.

Step 1: The first step in a MaaS booking process consists of the citizen choosing the destination of its trip. All combinations of several means of transport, offered by both public and private operators, can be chosen. Public transport operators provide

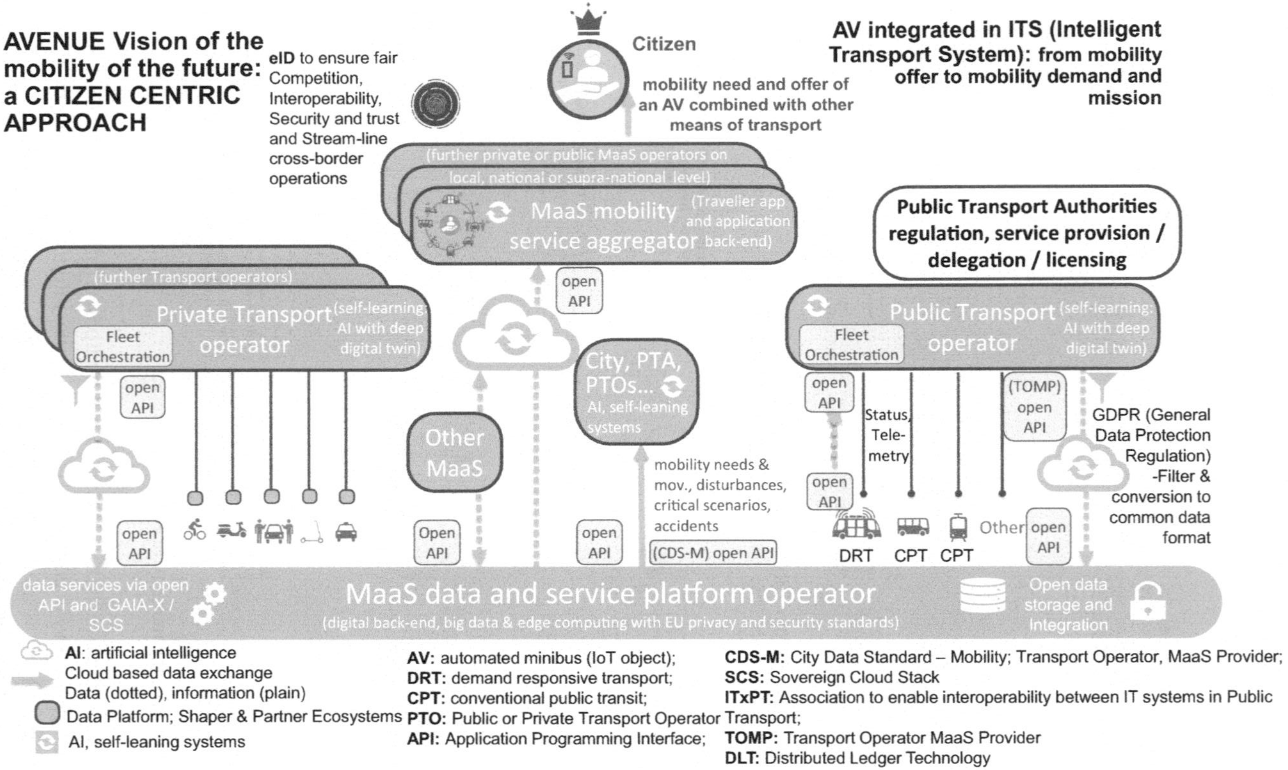


**Fig. 18.7** Vision of the mobility of the future: Step 4

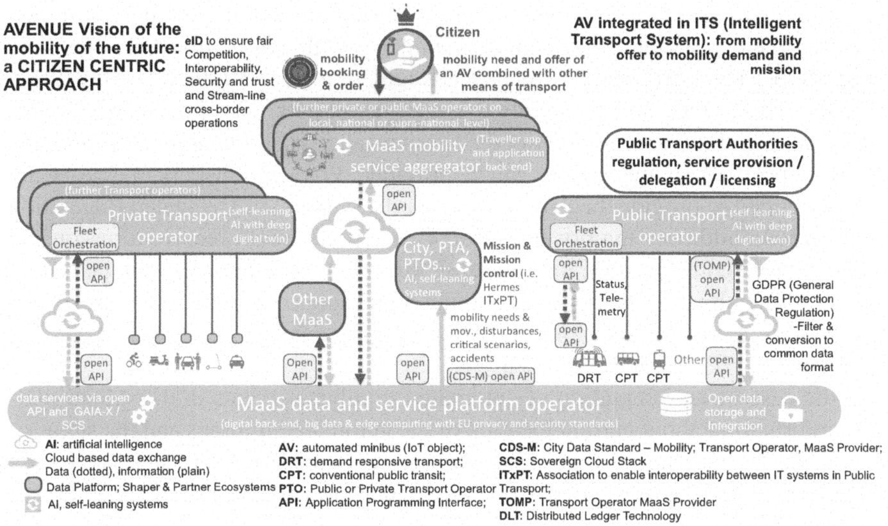


**Fig. 18.8** Vision of the mobility of the future: Step 5

demand-responsive transport (DRT), which aims to optimise the match of transportation supply with transportation demand efficiently and uses automated minibuses, for the first and last mile and as mobility complementor, and conventional public transport (CPT) services, like bus, metro and train. Private transport operators provide transportation modes like taxi, ride-hailing like Uber, car-sharing, bike-sharing and other micro mobility devices like e-scooter.

Traditionally, the citizen is forced to select and schedule separately every modality from private or public transport operators or a combination of it on his own and singularly according to his trip planning from A to B and related modality compilation, as is visualised in Fig. 18.4. This also includes a multi-factor optimisation of the modality-portfolio sample regarding time, cost, travel requirements (e.g. luggage, wheelchairs), personal preferences and others, including inherent error-proneness.

Step 2: In the next step, the position and availability of an automated minibus is documented together with other public and private means of transport on the MaaS platform. This connective platform is visualised as the yellow bar in Fig. 18.5. To optimise the decision and selection process, these tasks are transferred from citizens to specialised intermediaries: a MaaS data service platform is installed. This data service platform is designed, managed and controlled by a MaaS data and service platform operator and governed by Public Transport Authorities (PTA) as regulation bodies, for instance, by extending the existing National Access Points (Art. 3 Delegated Regulation 2017/1926 of 31 May 2017). In this platform all necessary data about modality and transportation provider offerings and related metadata from private and public transport operators are integrated, managed and exchanged. In particular private transport operators should provide all of their journey-relevant data which are necessary to offer a seamless intermodal trip to the customer. Providing data avoids an information asymmetry with public transport operators which are per law (Art. 4 and 5 Delegated Regulation 2017/1926 of 31 May 2017) obliged to share static and eventually dynamic travel data. This information asymmetry could lead to the aforementioned "winner-takes-it-all phenomenon" and to a dominant position of the private MaaS provider. Private transport operators are already obliged to provide the data foreseen under Annex I of Delegated Regulation 2017/1926. In addition, the private transport operators might receive a reasonable and cost-based compensation for providing this data: "Any financial compensation shall be reasonable and proportionate to the legitimate costs incurred of providing and disseminating the relevant travel and traffic data" (Art. 8 (4) Delegated Regulation 2017/1926 in fine).

In the case of an integration of automated minibus services in this MaaS platform, information about current location of the automated minibus and the status of occupation by passengers should be made accessible. However, this should not include personal data or only depersonalised data in accordance to privacy and security regulations.[4]

For this reason, only the status and telemetry of the data are provided to the private or public transport operators. To make the data available to all users in a standardised and quality assured form,[5] an open application programming interface (API) is required, as well as commonly agreed standards. These APIs[6] enable authorised applications to communicate, use one another's functions and exploit data sets provided by other applications or databases (Matthes & Bondel, n.d.). They make it in fact possible for the stakeholder to communicate with other stakeholders in order to provide a uniformed offer to the MaaS user (citizen). APIs are considered to be the "connective tissue of the cloud"; they are essential to integrate the transport system. However, the urban mobility ecosystem is still very fragmented (Bestmile, 2020). To overcome this barrier, initiatives envision setting standards of open APIs to enable mobility providers to integrate services. Examples of such an initiative are the projects MyCorridor, MaaS4EU and IMOVE. These projects foresee the use of a "common language" to designing a transport service API, comprehending "the use of communication protocol and data format to security standards, basic methods and service calls, responses and general behaviour of an API" (MaaS Aliance, 2019). Another initiative, the Information Technology for Public Transport (ITxPT),

[4]This has of course to be deepened in further research: according to some studies, "Merely technically collected data could, due to the transformational impact of big data analytics, become personal data or even sensitive data and thus trigger the application of privacy and data protection laws" (LeMO D4.1 p. 16 and quoted references link).

[5]This should be documented as required by the GDPR. The data users are considered as data controllers who have obligations regarding the processing of data. This obligation could be placed in another (existing or upcoming transport dedicated) legal basis than in the "generic" GDPR.

[6]Applications need policies to communicate with each other. In the digitised world, access to systems through open interfaces is called an open API (Termer, 2017). API's are blocks of code that allow developers to access smartphone functionality, such as using GPS data or turning the phone camera light on and off.

focuses on open standards and procedures for integrated Information and Technology Systems for public transport (Rogg, 2021). Therefore, open interfaces, protocols and standards are key factors for MaaS ecosystems.[7]

Consequently, an open standardised API[8] concept is initially required for a functioning MaaS platform. It has to connect the vehicle modality and its providers, the transport operators and the exchange platform and its platform operator. Additionally, they are utilised by further relevant parties of the ecosystem, like other MaaS systems (e.g. when trips are overreaching MaaS systems of other cities), or regulatory tasks of PTAs and city infrastructure management. For the automated minibus, its MaaS relevant information is transferred to the open data MaaS platform to be exchanged between the public transport operator (PTO) on the one side and the "MaaS data and service platform operator" on the other side. In addition to the exchange of data, open APIs ease the management of automated minibuses from different OEMs. This allows a flexible use of the individual vehicles as well as the entire fleet. A functioning system results in lower costs of operation as less vehicles are needed.

An important reason for an open API for the "MaaS data and service platform operator" is that in case a Public Transport Authority (PTA) desires to redesign the "MaaS data and service platform" itself, there is no barrier due to dependencies with the data model of the original equipment manufacturer (OEM).

In Step 3, the availability of other means of transport of other MaaS in a particular country or other countries of the EU is documented in the MaaS platform. The MaaS platform provides further data to stakeholders (city, PTA, PTO etc.) which can be used to improve the transport system (city planning, traffic, incidents, accidents, fleet management, vehicles safety). This is visualised in Fig. 18.6 by adding yellow arrows connecting the MaaS platform to the private operators, other MaaS systems and other stakeholders.

Step 4 integrates MaaS mobility service aggregators, as intermediary organisations between public and private transport operators and the citizen to provide the existing choice for the requested passenger journey. This new layer is introduced by local, national or supra-national private or public mobility aggregators and even by PTAs themselves. PTAs regulate, delegate and supervise this market. The mobility service aggregators are designed to analyse and process citizen profiles (in an

[7] A study conducted in nine cities divided data formats into the following three categories and found that all of these cities already have developer-friendly data formats. This can facilitate both application development and the handling of this data in the future. Therefore, a standard for data sharing should be created so that both small and large transit agencies can connect to the MaaS solution (UITP, 2019).

[8] An open API allows universal access and allows application programmes to interact with each other and share data; those elements are crucial for the mobility system interoperability and intermodality. Data interoperability is essential in this case, as the quality and consistency of shared open data and the data format are crucial for MaaS (UITP, 2019). For data mobility, the free flow of information must be enabled, interoperability must be ensured, and the values and interests of the broader society must be safeguarded, as private actors will play a central role in exploiting the value chain of data (Passau Declaration 2020).

anonymous manner, under Art. 4 (4) GDPR) and trip requests as well as general data about transport providers, vehicles and services and specific data about current positions and status of vehicle usage from public and private transport operators. Hence, this additional layer integrates information from both sides: trip request or transportation demand, and transportation offering services or vehicles, in combination with further information from other MaaS ecosystems and city infrastructure.

All relevant data are utilised for compilation of alternative trip options, as displayed in Figs. 18.2 and 18.7. Alternative trip options are selected based on the trip request, including individual preferences, but have to be protected by adequate privacy and security concepts and measures. These concepts must be completed/developed and deployed. Especially in the case of disruptions, critical scenarios, and accidents, these incidents will be analysed and evaluated using concepts and tools of self-learning systems and artificial intelligence (AI).

By applying algorithms of artificial intelligence (AI) to a MaaS ecosystem, the infrastructure and the value chain of transport companies can be significantly improved according to the needs of the citizens. Furthermore, with the support of AI all kinds of incidents, risks or problems within the MaaS ecosystem can be identified, tracked, traced and analysed in order to define and conduct strategies and measures for detecting, preventing and solving them.

AI applications comprise multiple functionalities for supporting success critical tasks, such as finding patterns and new insights, making predictions, interpreting unstructured data and interacting with the physical environment, their machines and humans (Scherk et al., 2017). One of the central topics of AI are self-learning algorithms of systems (e.g. based on neuronal networks) with the goal of finding independent solutions to new and unknown problems (Scherk et al., 2017).

To summarise, all potential trip-relevant information from the MaaS stakeholders has to be available, and the quality of related data has to be assured (i.e. static and dynamic real-time data, data validity, data privacy and security). This information should be provided to the MaaS mobility service aggregators, who are able to aggregate all transportation needs or requests from citizens on the one hand and all transportation modality offerings (including automated minibuses) on the other hand and match them in an intelligent way (supported by AI algorithms) according to citizen preferences (first priority, e.g. time efficiency, cost-efficiency, carbon footprint) and offering goals (second priority, e.g. route efficiency, vehicle utilisation) in order to subsequently provide the information of alternative transportation options to each citizen.

In Step 5, individual citizens can select their preferred transportation option for his trip and can place a booking and ordering with payment. The automated minibus is planned and reserved for the citizen. The mission is provided to the vehicle, which requires a return flow of the data (indicated by the blue dotted arrows in Fig. 18.8). The citizen has the choice to select between different transportation options for his trip from A to B in or outside of the city or even outside of the country. Hereby various valuable functionalities (like simulations about time or distance or cost or CO2 footprint, together with option comparisons) are provided to the citizen. After making this decision, the citizen is requested by the ticketing app to make the booking and place the binding order. After ordering, the data from the selected trip option is sent back to the MaaS data and service platform and hereafter to the private and/or public transport platform and transportation provider or possibly to other Maas system platforms.

As mentioned above, automated minibuses in MaaS could be used to generate information and knowledge and provide in this way the basis for human or AI-based decisions to enable a self-learning resilient transport system. This will be the endeavour of Sect. 18.4.

## 18.4 Automated Minibuses Integrated in Intelligent Transport Systems: A System Innovation Approach for a Resilient Self-Learning Citizen- and Purpose-Centric City Transport

The concept of "citizen-centric intelligent transport system" combines the intelligent transportation system with the citizen-centric approach and sustainability. Intelligence in this context is provided by artificial intelligence (AI) technologies. An ITS aims to provide services relating to different modes of transport and traffic management, enabling users to be better informed and make safer, more coordinated and "smarter" use of transport networks. They include advanced telematics and hybrid communications including IP-based communications as well as ad hoc direct communication between vehicles (V2V) and between vehicles and public and private infrastructure (V2X) (ETSI, 2021). Another keystone refers to a self-learning system with data-driven approach and AI for the development of intelligent transport system for smart cities and sustainable mobility (Iyer, 2021; UNESCO, 2021). AI and digital transformation could support to identify common patterns of mobility and quantify the crucial factors affecting the efficiency of the whole system (Lucca, 2022).

Figure 18.11 depicts the citizen-centric intelligent transport system. At the core are the citizens as it follows a citizen-centric approach. A variety of transport means is available for the citizens to choose from; these are depicted in the middle circle in Fig. 18.11 in a similar way to Fig. 18.5. The transport options are provided by various stakeholders (the actors named in the yellow blocks, at the middle circle of Fig. 18.11). In addition to Fig. 18.5, data are used to generate information and knowledge about and within the transport system. An intelligent transport system can learn and improve through experience gained by diverse data; this procedure is visualised by seven AI loops in the outer circle. Highly relevant is that those positive externalities only work in closed loops. This means that collected data are not enough since there are still decisions that need to be taken to close the loop (e.g. to improve the transport system). This is the case when automated vehicles have "learned" from data how to avoid an obstacle or a critical scenario in the city and the improved drive algorithm and eventual related hardware have been uploaded (fast

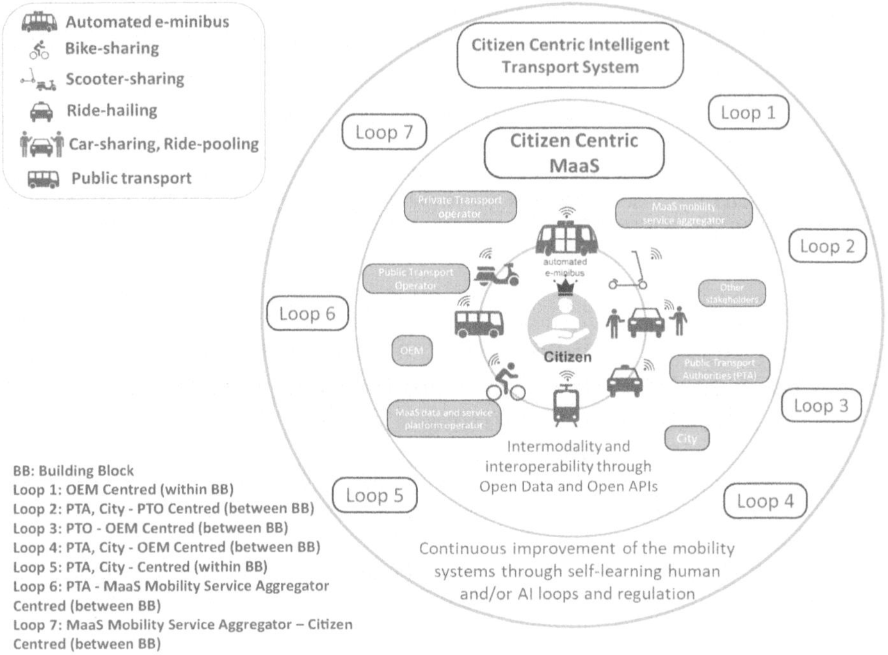


**Fig. 18.9** Citizen-centric intelligent transport system (ITS)

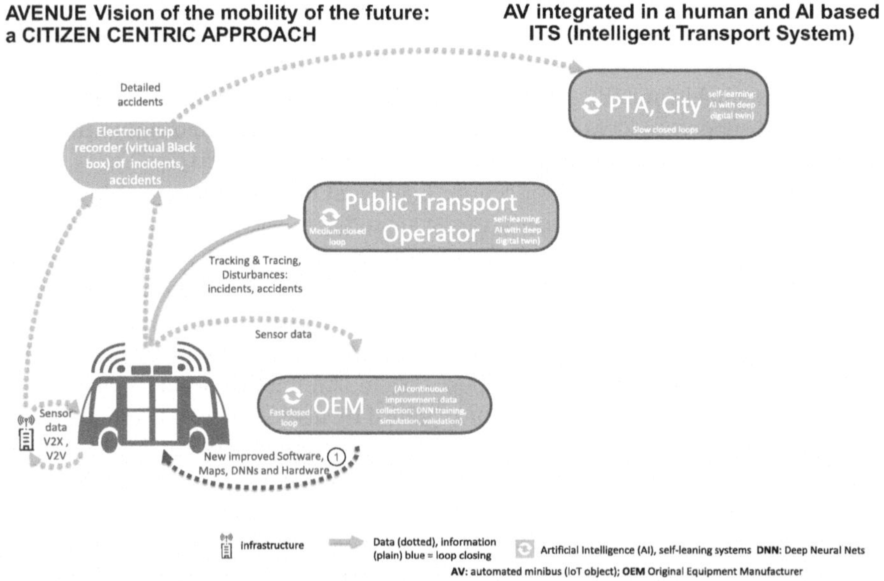


**Fig. 18.10** Automated minibus integrated in ITS: Loop 1

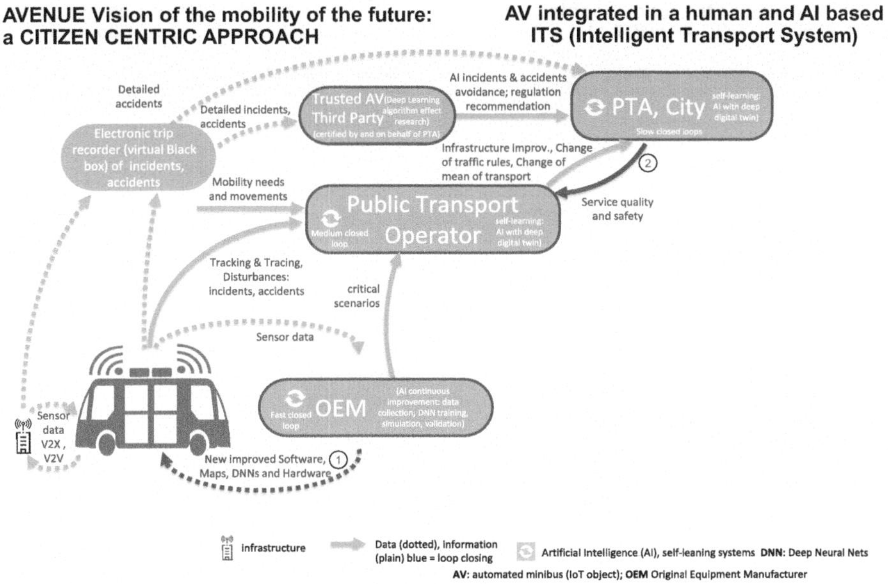


**Fig. 18.11** Automated minibus integrated in ITS: Loop 2

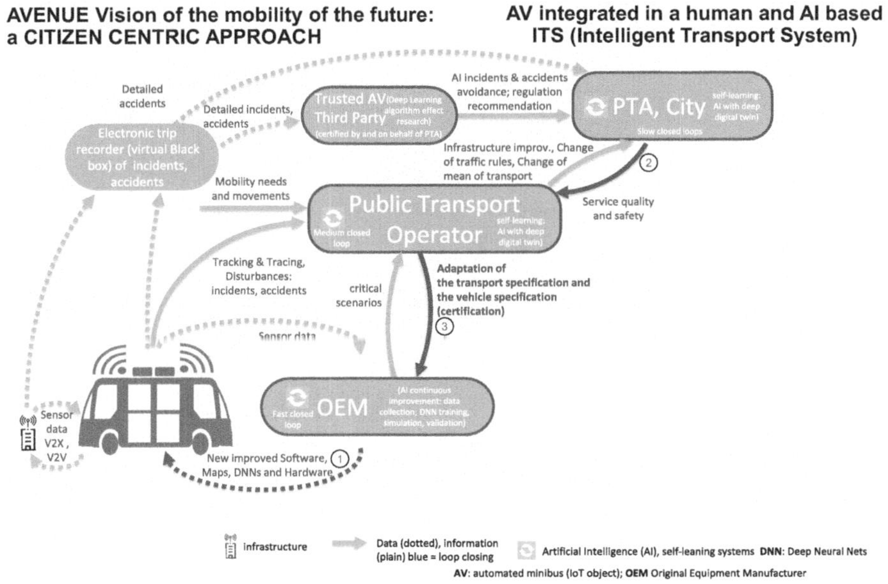


**Fig. 18.12** Automated minibus integrated in ITS: Loop 3

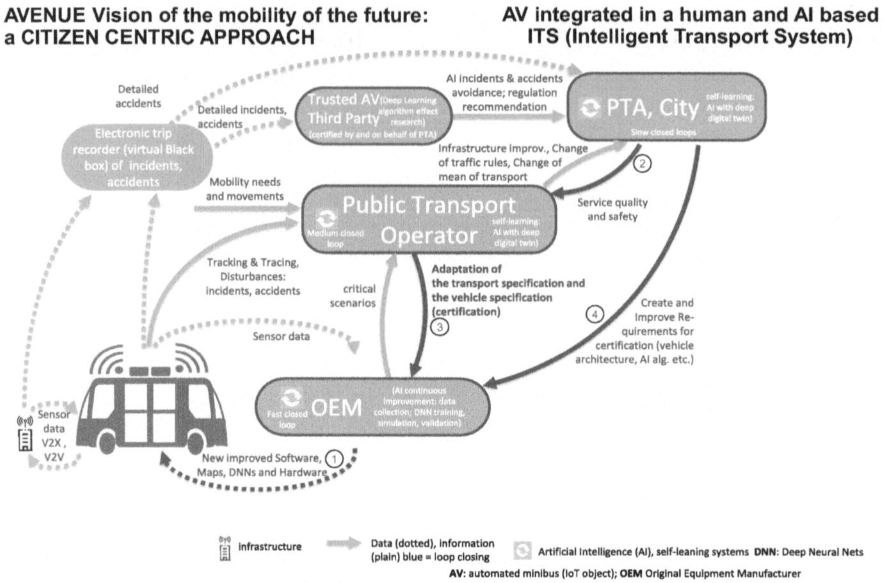


**Fig. 18.13** Automated minibus integrated in ITS: Loop 4

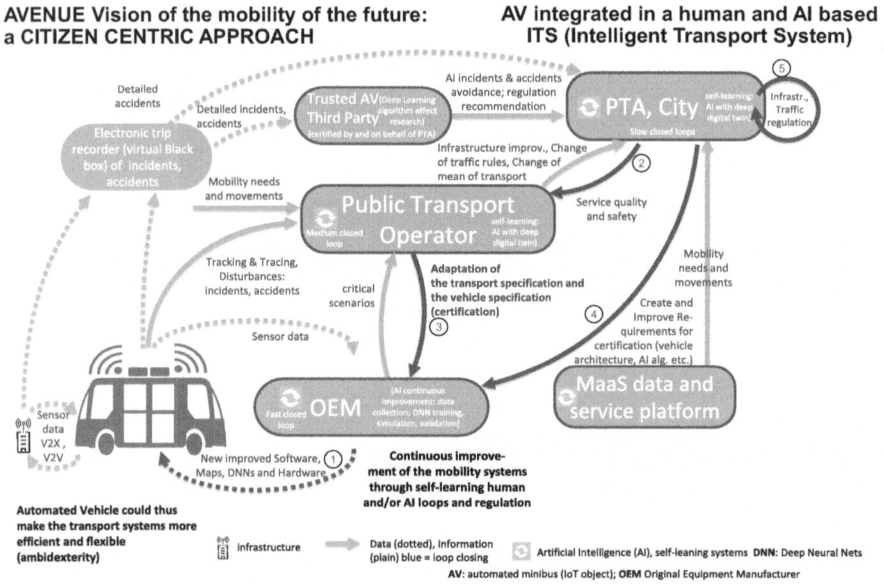


**Fig. 18.14** Automated minibus integrated in ITS: Loop 5

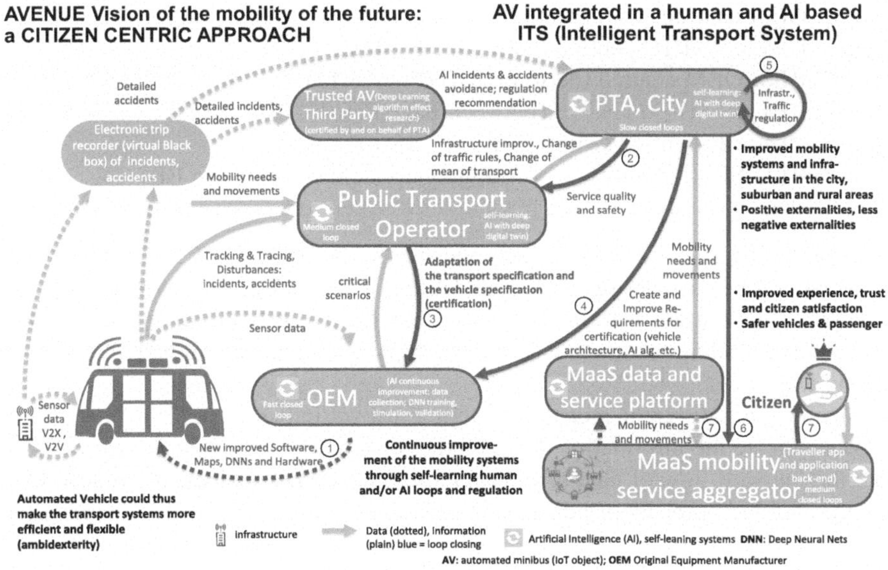


**Fig. 18.15** Automated minibus integrated in ITS: Loops 6 and 7

closed loop). This can happen as well when a city chooses to change the traffic rules on places with a high level of safety problems (slow closed loop).

On-demand and door-to-door data, compliant with the General Data Protection Regulation (GDPR), could be delivered with unprecedented precision in quality and quantity. In fact, this data could be used to generate information, including decisions and knowledge about the transport system, enabling continuous improvement of the urban mobility system through state of the art, self-learning human and/or AI loops and controls. This is visualised by the loops in the outer circle of Fig. 18.11, which show the loops in detail. The loops in the outer circle of Fig. 18.11 illustrate this as they are designed around stakeholders to implement lessons learned to improve locally and globally. Data and information are generated by all components of the transport system such as the automated minibus, infrastructure, platform, applications and especially the user. These generated data add in return value to the other stakeholders which closes the loop and results in an optimisation of the overall transport system (ERTICO – ITS Europe, 2019).

The functional context of an intelligent transport system is systematically described by the following step-by-step illustrations.

Depicting the seven relevant loops (as indicated in Fig. 18.9):

Loop 1: OEM centred (within building blocks (BB) (Fig. 18.10)
Loop 2: PTA, city–PTO centred (between BB) (Fig. 18.11)
Loop 3: PTO–OEM centred (between BB) (Fig. 18.12)
Loop 4: PTA, city OEM centred (between BB) (Fig. 18.13)
Loop 5: PTA, city centred (within BB) (Fig. 18.14)
Loop 6: PTA–MaaS mobility service aggregator centred (between BB) (Fig. 18.15)
Loop 7: MaaS mobility service aggregator–citizen centred (between BB) (Fig. 18.15)

Each of the loops will be described and explained below. The descriptions are accompanied by visualisations (0, Fig. 18.15) of the seven self-learning loops to achieve a complete self-learning ITS ecosystem. Each loop builds upon the previous system. The coding within the graphical representations is defined in the following way:

Arrows have different meanings: Dotted arrows indicate the data flow; plain arrows indicate the information flow and the related decisions. Yellow arrows indicate the outward flow of data or information, blue arrows indicate the return flow and the clothing of the loop; a data exchange loop is represented by the combination of outward and return flow. The determined loops can be fast (because, e.g. just one stakeholder or one building block like an OEM is involved), medium or slow (when a consent between stakeholder (building blocks) or laws/vehicle specification/vehicle certification must be developed and published).

Loop 1, as depicted in Fig. 18.10, is an OEM-centred fast loop. Three data flows and one information flow are emitted by the automated minibus to the surrounding systems. The closed data loop on the left in Fig. 18.10 illustrates the data flow between the automated minibus and the road infrastructure and environment, generating both sensor data from the automated minibus' on-board and roadside systems. Characteristic for the second data stream is the long-time data storage in a virtual black box similar to those used in air transport. The automated minibus uses the backend systems to store permanently real-time data about recorded incidents and accidents. Access to this real-time data is also available for the PTAs (Public Transport Authorities), who then can optimise service quality and safety (Loop 2) for cities, PTOs (Public Transport Operator) and OEMs (Original Equipment Manufacturer). A further beneficial data flow is the direct connection to the PTOs. This linkage allows the PTO to derive route and service optimisations based on the real-time data about incidents or accidents. Further recommendations to optimise the road infrastructure, traffic rules or change of transport means become possible based on this data flow. Data of the vehicle can finally be used to monitor and optimise the charging of the vehicle. Maintenance costs of the vehicle can be improved as well to move from a curative (in case of breakdown) to a preventive (plan is better than cure) and later to a predictive (forecast breakdowns and predict a malfunction) maintenance.

Loop 2—PTA, city–PTO centred: The second loop, as depicted in Fig. 18.11, is PTA, city–PTO centred.

To ensure a trustworthy accessibility of the generated data for PTA, an intermediary scientific service provider (trusted AV third party) shall be established. This trusted third party gets certified and operates on behalf of the PTA as non-biased, neutral and objectively acting institution. As neutral institution it could analyse, process and forward all data about any incidents and accidents to the PTA and propose improvements but also support for juridical treatment when necessary. They could also proof the AI used by the OEM for their vehicles and transmit this information to the PTA and provide recommendation for improvement measures for regulation.

Based on the PTO-generated data, the PTO is as well able to provide learnings and suggestions for technology or regulatory improvements to the PTA. Subsequently, the data cycle of permanent information between the PTO and PTA to drive optimisations on the regulations for service quality and safety is captured in Loop 2 and also with the connection to OEM (Loop 4) (see in Fig. 18.13).

Loop 3—PTO–OEM centred: The third loop, (see Fig. 18.12) is PTO–OEM centred, and further completes the intelligent transport system.

Critical automated minibus scenarios are transferred from the OEM to the PTO empowered by AI algorithms. Subsequently, the data is then returned back from the PTO to the OEM. This provides the OEM with information on adjustments to the transport specification such as capacity, range, availability and persona profiles, adjustments to the vehicle specification and new and passive safety requirements. This enables the OEM to analyse the data and provide an optimised utilisations of automated minibuses by taking into account the new road safety requirements.

Loop 4—PTA, city–OEM centred: The fourth loop, depicted in Fig. 18.13, is PTA–OEM centred. Based on mobility data from the trusted AV third party, PTO and the PTA, PTAs can create and improve OEM certification requirements and (type) approval requirements and define automated driving regulations (e.g. vehicle architecture, AI algorithms etc.). For example, new active and passive safety requirements can be introduced from accident analysis and lessons learned coming from the third party.

Loop 5—PTA, city centred: The fifth loop, depicted in Fig. 18.14, PTA, City centred. This data cycle shows a slow closed loop of PTAs receiving information and data provided by all stakeholders and using them as a basis for improving urban infrastructure and traffic management, updating, monitoring and mobility (re) planning.

Loops 6 and 7 complete the ITS by adding both the citizens and the MaaS data and service platform. The two-way information sharing between the PTAs and the MaaS mobility service aggregator is key:

The data provided by the mobility aggregator and the MaaS data and service platform provider on one hand informs the PTAs anonymously about the mobility demands and patterns of the citizen (attitudes and behaviours).

In return, the PTAs are in a better position to enhance the mobility system and the mobility infrastructure (Loop 5) aimed at inducing positive and reducing negative externalities of mobility. Eventually, the data collected by the mobility aggregator will allow the provision of customised unimodal or intermodal mobility alternatives to optimise the customer's choice according to their profile, persona or even the weather. Trip duration, expense and environmental effects provide a foundation for decision-making. To ensure a positive mobility experience, trust and satisfaction, safe travel and acceptance, these last loops are crucial.

Through the seven loops, the integration of automated minibuses in the future transport system could thus be beneficial for all the stakeholders, lower negative externalities, provide positive externalities and satisfy better the citizen needs and their acceptance. All the means of transport can be offered and combined to a seamless offer which is nearly as individual, safe and attractive as an individual car. With the integration of automated minibuses into an intelligent transport system, the whole transportation system could further be improved continuously. The transportation system would become more efficient but at the same time more resilient and flexible as failures could be determine in a predictive way and in case of failure in the transport system, automated minibuses could bridge the mobility gaps. Achieving these antinomic economic goals (efficiency and flexibility) into one transport system could make the transport system ambidextrous (Raisch & Birkinshaw, 2008).

Various upcoming digitalisation technologies like digital twins of automated (smart) vehicles as transportation objects, digital twins of (smart) citizens, digital twins of (smart) infrastructures and even digital twins of all kinds of stakeholders (PTOs, intermediaries, PTAs etc.) and mobility subsystems are enabling the governance, simulation, monitoring, tracing, analysis and optimisation of the entire mobility ITS/MaaS ecosystem. The AI-based self-learning loops of the mobility ecosystem could be ready for design and implementation into a future citizen-centric "Mobility Metaverse" aiming to optimise the physical/real "Mobility Universe" as a digital twin.

As the herein described concept is combining two disruptive and technologically complex innovations, there are also limitations that need to be reflected: on the one hand, there is the automated minibus itself as a product innovation and on the other hand MaaS/ITS as a mobility system innovation. The automated minibus depends on technical capabilities (perception and ability to determine, speed, level of automation etc.), the automated minibus environment (mapping and modification of routes, e.g.) and human involvement (for disabled access or the necessity of on-board safety driver, e.g.) that can be questioned even if fast improvements are expected via AI deep learning. These include the highly sensitive issues of user privacy and data and personal security, as well as the ethical issues that will always accompany automated driving (such as the trolley problem) (Fagnant & Kockelman, 2015; Geisslinger et al., 2021). The MaaS innovation is further driving a wider social, economic and environmental "ecosystem" innovation and will be impacted by an appropriate governance approach focusing on the general interest.

The main risks and limitations of the automated minibus in MaaS concept are addressed in the following section.

## 18.5 Discussion and Limitations

The aforementioned advantages of the AVENUE vision 2030 can only be achieved through the described disruptive product and system innovation. A prerequisite for achieving the AVENUE vision 2030 is an alteration of the current mobility paradigm. This paradigm is based on cheap fossil fuel energy, high CO2 exhausts, individual mobility (product orientation) and a linear economy (Fournier, et al., 2018).[9] Altering this mobility paradigm requires social transformations in addition to technical innovations (such as automated minibuses) to accomplish a socio-technical transformation of the mobility system (Geels, 2011). A socio-technical transformation requires alterations in society, in business (ecosystem collaboration & market places, governance & management, vehicle/fleet demand & offering side) and in particular for stakeholders such as passengers, transport operators and related companies, technology provider, governance etc. For one, this entails an integration of stakeholders in designing, implementing and evaluating sustainable solutions for the mobility sector. As an optimum solution, all involved stakeholders should cooperate and agree upon a common purpose-centric strategy. This strategy should evolve towards a sustainable mobility system and, like mentioned above, create value to all stakeholders including citizen, public and private PTOs, OEM, city etc. and contribute to the transformation of our society in providing a no coercive answer to societal challenges.[10]

To fulfil this overarching purpose of serving general safety and sustainability interest, there is a strong need for stakeholders to harmonise different interests and cooperate with each other in an innovative coopetition mode. Known difficulties are that stakeholders will make choices based on knowledge available to them with the aim of maximising their individual gain. The combined outcome of the individual strategies could result in a situation that was not intended and not desired by any of the involved stakeholders [see, e.g. the prisoner's dilemma from game theory by Tucker (1950/1983) in Morrow].

To establish a mobility system as shown in the AVENUE vision 2030, some stakeholders could still try to promote individual mobility with private cars or take advantage of the "winner-takes-it-all" phenomenon to capture the value creation to the detriment of other stakeholders or even take control of the entire mobility system by taking control over a significant mass of captured system relevant data (private MaaS) of the mobility ecosystem and selling it to the system's customers (such as city governments) and public transport authorities (PTAs). It is therefore crucial

[9] A linear economy is based on the "take-make-dispose" strategy. For a transition to a more sustainable society, one standard assumption is that the economy should transform into a circular economy. A circular economy is an economic system of closed loops in which raw materials, components and products lose their value as little as possible and systems thinking is at the core.

[10] Discussions about stakeholders have existed since the 1930s and were raised in the World Economic Forum (WEF) in 1973 by Klaus Schwab. Freeman saw in his comprehensive approach stakeholders as essential to every management process. The stakeholder approach became afterwards so popular that in 2018, even BlackRock—the largest asset manager in the world—preferred the stakeholder instead of the shareholder approach to create sustainable value for all stakeholders.

to coordinate the actors involved and to synchronise operations both vertically (between the actors along the different modes of transport of the journey) and horizontally (among actors operating the same modality, such as different automated minibuses) to ensure seamless intermodality and interoperability of the transport system, similar to a supply chain (Giusti et al., 2019).

For a successful socio-technical transformation of the mobility sector, this process needs to be managed, to avoid and overcome divergent interests and resistance. Research has shown that 70% of transformations fail, and part of the solution for the path to transformation includes moving fast, "shared vision and stretching aspirations" by leaders and a "more holistic and expansive approach to transformation" (McKinsey, 2020). In addition, cornerstones of public sector transformation relate to performance and organisational health (McKinsey, 2021). In the following, we have identified transformation challenges at the technical, environmental, social, economic and governance dimensions. These dimensions are part of a holistic and expansive approach to transformation, and they can influence each other through the interdependence of success factors and barriers.

Next, we present in more detail the success factors and barriers that need to be addressed in order to successfully implement the AVENUE vision 2030. For each of the dimensions, we subdivide into success factors and obstacles for the automated driving system and separately for the automated minibus integration in an ITS/MaaS system.

### *18.5.1 Technical Success Factors and Obstacles*

In order to achieve the desired vision, there are several technical obstacles and success factors that need to be addressed and solved. The deployment of automated minibuses in all environments, both in urban and rural, requires the technology to be able to handle all safety-related scenarios and to be effective, intelligent, connected and fully implemented within the ITS infrastructure. The technical obstacles and success factors related to the automated driving ecosystem concern automated technologies, such as integral sensor systems and technologies. The automated driving ecosystem can be defined in the following categories, with reference to Table 10.3 (Chap. 10), which is specific to the on-demand and door-to-door solution developed by AVENUE.

Automated shuttle capabilities: In terms of active safety, LIDAR and camera technologies that detect and avoid obstacles as an integrated part of the decision-making engine will be key. Machine learning algorithms and AI technologies will enable continuous development of driving capabilities and vehicle positioning in traffic patterns and infrastructure. Perception and judgement, which are also related to AI technologies, will allow the vehicle to build up and constantly update a library of objects and obstacles, so that it knows the difference between a car, a bicycle, a truck, a pedestrian and so on. Radar technologies are needed to perform emergency braking and detect large objects. The processing power of automated minibuses s and AVs is developing rapidly around the world, and technologies need to be shared to achieve the highest possible processing power—and therefore both speeds and braking algorithms and other safety features—ultimately making AVs safer and more reliant than human drivers. Accident analysis will also lead to passive safety requirements, which have to be anticipated, as they are structural to the vehicle platform. Conventional crash tests are not applicable to existing automated minibuses, and a compromise will need to be found quickly.

Automated shuttle environment: The main factors here are mapping and commissioning of new routes and updates of already existing routes. Current automated minibuses can yet only drive on maps recorded by the vehicle manufacturer, which is time-consuming and costly to record. These recorded maps are not shared with other manufacturers. In the future, these "map routes" will need to be shared and constantly updated across Europe to allow automated minibuses to drive anywhere. Similar strategies are followed in Japan and South Korea. The processing power of automated minibuses and AVs needs to be developed at a rapid pace in order to achieve speeds, braking algorithms and other safety-related features. Such elements are important to ensure that AVs are safer and more reliable than regular drivers (as already described in the prior paragraph). All weather-sensing systems should be able to operate in all conditions: this robustness is fundamental to vehicle reliability, a key requirement for service quality and safety. Currently, heavy rain and snow still cause sensors to malfunction and lead to many operational problems. In these cases GNSS makes it possible to position the vehicle precisely. The mobility infrastructure of cities and other application areas (e.g. roads, transport facilities and services, roadside sensors, charging facilities, communication technologies, future: AI and digital twin-based traffic management) also plays a crucial role in the implementation of automated minibuses and should be considered in local development planning. Finally, cybersecurity is critical as laws and standards need to be further developed to include automated minibuses.

Human involvement: The key factor here is the need for a safety operator on board of automated minibuses. Human interaction with automated minibuses will not be eliminated, but the transition from on-board safety drivers to remote supervisors and local technicians need to be developed and tested. Currently, automated minibuses can only operate with a safety driver on board and, in many scenarios, with a remote supervisor at the same time. This is making the current implementations very costly. Finally, access for disabled people is critical, as human intervention is still required. Human-machine interaction and interface technologies play an increasingly important and critical role for passengers, supervisors and other stakeholders in operations and service applications.

With regard to automated minibuses in a MaaS/ITS ecosystem, as previously mentioned in Sect. 18.2 of this document, open data, open platforms and open APIs are key factors to ensure a consistent citizen-centric approach and a single-point-of-contact applications for travellers in ITS. This openness is a prerequisite for multi-sided access to stakeholders, in particular passengers, and for integrating all modes of transport into a seamless journey. Interoperability and intermodality can improve performance and flexibility of the system, enable positive mobility externalities and reduce negative externalities. Open data, open platforms and open APIs are therefore a prerequisite for developing the benefits of automated minibuses in MaaS. The described AI loops are the precondition to enable an ITS with its benefits, but they require a collective effort for the common good. This process is challenging to manage as it involves several private economic interests of the related ecosystem.

These advantages however raise the risk of unauthorised tracking and traceability of passengers (privacy and safety of passenger data) and thus the risk of multiple undesired monitoring possibilities of mobility. In the public discussion on automated minibuses in MaaS, open data could lead to a major limitation and therefore needs to be technically secured (Meijer et al., 2014). With the definition of the open standards and regulations for data (e.g. GDPR) and APIs, as well as specific software-based technologies (ledger technologies such as blockchain), strong economic interests are further impacted as an ecosystem is open to all. A closed ecosystem could enable companies to capture and privatise the value of data in particular the positive externalities associated with it. The technical definition of the communication protocol, data format and interfaces has thus to be managed by the PTA and probably at European level to avoid security issues and "the winner-takes-it-all phenomenon" (see Chap. 2). The last point is important, as in an extreme case, the creation of private MaaS based on automated vehicles could limit the profitability of running mobility businesses but even severely limit the influence of the PTA in creating own rules for the urban mobility ecosystems.

### 18.5.2 Economic Success Factors and Obstacles

The results of AVENUE indicate that the business with automated minibuses is not yet profitable within the current automated driving ecosystem due to high driving-related personnel costs. This is mainly because safety drivers are still required to be inside the vehicle, and remote supervisors are currently legally forbidden or restricted in most European countries. The profitability of using automated minibuses in the fleet of PTOs is heavily impacted, making it a main restriction for usage in urban, suburban or rural areas. Furthermore, the high acquisition costs of the vehicles, elevated costs of feasibility studies and legal authorisations, infrastructure works and high annual depreciation are also seen as constraints (see Chap. 12; Antonialli et al., 2021).

The non-integration of automated minibuses with other means of transport within a MaaS reduces further the attractivity for passengers in terms of time and usability as several apps have to be used. Currently, there is no customer-centric approach that integrates automated minibuses with other means of transportation.

In both cases the technology has to be improved (see above), and governance (see below) has to be adapted. Regulatory sandboxes for self-driving vehicles could facilitate the testing and implementation of new technologies, leading to economies of scale and accordingly lower prices for wider adoption (BMWi, 2019). The limited acceptance and integration of automated minibuses' into mobility networks by private and public PTOs, fleet owners and city/regional governments as B2B customers is due to technology readiness and operational business risks, as well as low subsidies for this innovative technology.

Integrating automated minibuses in MaaS/ITS is in fact one of the several possible future scenarios where the expected benefits are the highest due to the integration of societal goals. The governance discussed later will determine the benefits for the stakeholders and the general interest. Accordingly, it is expected that the stakeholders which will be negatively affected by serving the general interest could be inhibitors of the required mobility transformation. The leading stakeholders of the current mobility paradigm, such as the automotive industry and its employees, may be inhibitors and unwilling to adapt to the new constraints like it could be observed with environmental constraints (particulate matter) or electric vehicles. This is understandable as a real alternative to the spatial and temporal flexibility of cars could be provided and could satisfy the needs of the citizen (see Chap. 15): individual mobility with private cars could be more easily substituted by automated minibuses in MaaS/ITS, with the consequence of losing the shaping of the mobility value chain. The PTOs and PTAs will have to further adapt to the new digital ecosystem and customer-centric approach. Cities and regional governments, which would have to transform not only their operational mobility concepts but also their strategic business ecosystem (including open data, platforms, APIs, towards a new IT concept) from a "traditional" type of vehicle and (government) business ecosystem to a balanced, federated or democratised governance and operational IT concept with automated minibuses, could also face multilateral obstacles regarding the cost and effort of transforming a "running system" and the often notorious lack of budget as well as the difficulty of a necessary paradigm (mindset) change of public authorities that are often risk-averse (non-entrepreneurial). However, even this kind of disruptive (game changing) transformation of the business ecosystem can be guided by individually adapted and designed approaches (e.g. incremental transformation steps, "think big, but start small and scale fast").

Finally, all stakeholders tend to maintain their current habits: passengers, members of organisations or the organisation itself try to reduce the psychological costs of change.

Automated minibuses in ITS will last but not least require investments in digitalisation, hardware, software, training etc. In particular, as cited in the Digital Economy and Society Index (DESI) 2021 (European Commission, 2022), the availability of employees with digital skills could in particular slow down the digitalisation and dematerialisation of mobility. The comparatively slow digital transformation of companies in many EU countries is linked to "a lack of employees with advanced digital skills" (European Commission, 2022). For example, by 2020, 55% of companies will report difficulties in recruiting IT specialists (European Commission, 2022).

### 18.5.3 Environmental Success Factors and Obstacles

The automated driving ecosystem of AVENUE shows a low environmental impact (see Chap. 13). A Life Cycle Assessment (LCA) of the automated minibus deployed within AVENUE indicates that its automated technologies account for less than 5% of the total energy used. In a near-future use case, 59% of the automated minibus impact comes from the use phase, while 39% comes from the component manufacturing phase (Huber et al., 2022). In addition, the automated minibus' environmental credentials depend on many factors such as occupancy, vehicle speed, mileage and lifetime.

The analysis of the energy demand of automated driving technology and the connectivity-related demand shows that, on the one hand, predictive, adaptive and information sharing through vehicle communication with infrastructure and other vehicles improves driving performance (e.g. braking performance) and, consequently, energy consumption. On the other hand, a highly connected vehicle means more data processing inside and outside the vehicle, which may outweigh the sustainability of V2X. Overall, the energy-saving potential of predictive driving features is likely to exceed the energy used for data transmission.

The AVENUE environmental impact assessment reveals that if automated minibuses are well utilised in terms of mileage and are used regularly by passengers, they will have major environmental advantages over single-occupancy vehicles (Huber et al., 2022). The automated minibuses are seen as complementary vehicle in public transport and could increase the availability, flexibility, efficiency, effectiveness and reliability of regional public transport (Huber et al., 2022).

More relevant is whether and how automated minibuses could be integrated into a MaaS/ITS ecosystem. Depending on the governance scenario chosen, automated minibuses can be used in a separate ecosystem competing with current transport modes (laissez-faire strategy). In this strategy, for example, the convenience of the robotaxi and the customer centricity of the robotaxi provider could substitute public transport, private cars, bicycles and even walking. This is expected to generate additional traffic in the city (rebound effects), with associated external costs such as congestion, CO2 emissions, additional space etc. The proposed automated minibus in a MaaS/ITS, on the contrary, aims to satisfy the best needs of the citizens and the general interest in the leveraging of positive externalities and the reduction of negative externalities. Positive externalities like cost-efficiencies gained from increased usage (Mohring effect), strengthening network effects, enhanced accessibility etc. can be generated additionally through the use of data, intermodality and interoperability.

### 18.5.4 Social Success Factors and Obstacles

An important finding of the AVENUE social impact assessment (see Chap. 15) is that users and potential users demonstrate a positive attitude and a receptive (goodwill) attitude towards the automated minibus. There is therefore potential to convince those who are not yet refusing, but who are open-minded, by means of targeted communication campaigns, particularly on social media. We defined five target groups of potential users (unreserved goodwill; sceptical goodwill; undecided; critical reservation; unconvinced refusal), which differ not only in their perception of perceived benefits and concerns but also in their level of knowledge, preferred transport system, willingness to use and pay for travel by automated minibus and sources of information used to form their attitudes. The results show that high levels of goodwill among potential users (e.g. car drivers) and high levels of satisfaction among users of public transport translate into high levels of willingness to use (again). In general, users were very satisfied with their experience, and most were willing to use the automated minibus again. In Copenhagen 76% of users are very willing to use the automated minibus again, and in Sion 55% of respondents are willing or very willing to use the automated minibus again.

The most important factors for social acceptance are the (perceived) need for improvement of the current situation and whether the proposed alternative service meets this need for improvement. Fears of a lack of safety or security are currently less important for social acceptance. In addition, real experience with the automated minibus has a generally positive effect on trust in the system. Therefore, an important success factor for the social acceptance of automated minibuses is to allow citizens to use and experience the advantages of automated minibuses.

Some of the critical points encompass the highly sensitive issues of user data handling and data security, as well as the ethical questions that constantly accompany automated driving (such as the trolley problem, which addresses the ethical dilemmas of the trolley's role in the car's life, and the question of whether or not the trolley's role is to act as a safety device for the driver) (Fagnant & Kockelman, 2015; Geisslinger et al., 2021). In addition, the results of a representative survey of citizens (n = 1816) indicated perceived concerns about the use of automated minibuses. The main concerns are related to the interaction of the automated minibus in unforeseen situations and with other road users (motorised and non-motorised), liability in the event of an accident, security issues and the risks of hacking and misuse of the software (Korbee et al., 2023).

In addition to this, the safety operators also mention that they are observing a high level of goodwill towards the innovative service. In the opinion of the safety operators, the users are highly satisfied, especially when subjective aspects such as the good atmosphere are taken into account, not least because of the lower number of passengers.

Based on their observations, safety and accessibility are qualities that are also evaluated as being satisfactory. In the viewpoint of the safety perspective, the automated minibuses meet the needs and requirements of users and contribute to a positive user experience.

A risk for the target groups of enthusiasts, uncritical goodwill and sceptical goodwill is that they may be disappointed if they recognise actual performance in terms of speed and flexibility. It is very important to increase both the speed and the flexibility of use through an on-demand service, or at least improved temporal and area flexibility (nearly as flexible as private cars) compared to existing public transport services, in order to ensure that the high level of goodwill actually leads to a high level of acceptance of the new systems. This positive perception is a good point, but it cannot be taken for granted: the perception can change in the event of an accident, for example.

Beyond the current AVENUE approach, integrating automated minibuses into a MaaS/ITS system aims to best satisfy passenger needs and increase social acceptance. We could therefore expect automated minibuses to be a real "game changer" in the future, making public and private transport more personalised and providing a real alternative to individual privately owned vehicles as it has the potential to increase flexibility for users and choice for passengers while serving the public interest. The results of a representative survey of 1816 citizens (of whom 1526 have privately owned vehicles) in Lyon, Copenhagen, Luxembourg and Geneva confirm that 45% of car drivers would be "willing" (22%) or even "very willing" (23%) to give up using their own car in order to use automated minibuses for the first and last mile, if it were available. If the service is on demand and door to door, acceptance could be even higher (Korbee et al., 2023). Of course, this needs to be explored in future research if these measured attitudes can be changed in behaviour.

Another alternative could be the robotaxi which could satisfy best individual mobility needs without a change in transport mode but at the same time would reduce and privatise positive externalities and increase negative externalities through additional traffic and space. This choice could be better accepted than automated minibuses in MaaS in ITS but without serving the general interest. The choice between a non-integrated robotaxi in a MaaS and automated minibuses in a MaaS/ ITS therefore needs to be discussed, organised and formalised in a future governance. Automated minibuses in an ITS/MaaS is of course the better alternative for the city and its citizens, but this requires a transformation towards a new social contract for mobility to avoid social frustration and crisis (Shafik, 2021; Rousseau, 1762). Special attention should be paid to digital illiteracy to avoid excluding parts of the population from the transformation process.

### *18.5.5 Governance Success Factors and Obstacles*

The above limitations are all influenced by the chosen governance. Thus the governance is a key issue to enable the technical ecosystem and the integration of automated minibuses in MaaS or ITS.

For the automated minibus and its vehicle-related ecosystem, the development of regulations for the automated driving ecosystem encompasses amendments of the Vienna Convention on Road traffic and the adoption of a new legal instrument on the use of automated vehicles in traffic at the international level.[11] At regional and

[11] See among others the work of UNECE's Group of Experts on drafting a new legal instrument on the use of automated vehicles in traffic (GoE LIAV).

national level, it should be possible to approve the operation of automated minibuses without strict requirements for the safety operator (in a first stage) and with a remote supervisor only (in a second stage),[12] as the costs and time for type approval are a key issue for the diffusion of automated minibuses. In Member States that have not yet adopted AV-related legislation, regulatory sandboxes[13] for self-driving vehicles, as mentioned above, could ease the disruption process and provide EU stakeholders with similar competitive conditions to other more liberal parts of the world to test and trial new technologies and facilitate their diffusion. Moreover, a legislative database at European level which brings up to date and real-time information about the fast-growing automated vehicles legislation (e.g. type approval, commercial, cybersecurity of vehicle, definitions, infrastructure and connected vehicles, insurance and liability, licensing and registration, operation on public roads, safety operator requirements etc.) on local, state and regional regulations is missing. This could simplify the planning of all the stakeholders like vehicle manufacturers, deployers of automated minibuses, transport operators, states, municipalities etc. and encourage the diffusion of the automated minibus technologies and the improvement of the transport system. The USA is a good example of how such a database could be managed (NCSL, 2022).

At the city or local authority level, it will be necessary to design, implement and monitor individual technical (vehicle, IT) certification and licensing schemes for automated minibuses and their integration into MaaS according to relevant regulations or standards (similar to TÜV in Germany) as well as related training courses for operational personnel (experts).

City or local authorities are also in the best position to understand the local mobility needs and where automated minibuses could provide new services and/or improve the existing services (as feeders).

Governance is therefore crucial for balancing these interests, although it is possible that the creation of closed ecosystems supported by the users who are individually highly satisfied by customer-centric affordable robotaxi could weaken the power of the PTA and the ability to balance interests.

Implementing AV for mobility is also a technical and innovative approach that believes in progress and science, using dematerialised consumption to save resources for a sustainable development (mobility) path. It will have to overcome the growing scepticism generated by debates on concepts such as "degrowth" or "frugality", which promote a change in consumer behaviour through sacrifice in order to save energy and resources (sufficiency strategy).

[12] See, for instance, the French and German legal frameworks adopted in 2021, as well as the EU Commission's Commission Delegated Regulation of 20 June 2022 as regards the technical requirements for [...] fully automated vehicles produced in small series [...], C(2022) 3823 final, and Implementing Regulation of 5 August 2022 [...] as regards [...] the type approval of the automated driving system, C(2022) 5402 final, described in Chap. 16.

[13] As suggested by Art. 53 and 54 proposal for a regulation laying down harmonised rules on artificial intelligence (COM (2021) 206 final).

For automated minibuses integrated into a MaaS or ITS ecosystem, public institutions have a crucial role to play in developing an interoperable, standardised and connected data landscape across different sectors, including the mobility sector in particular, through specific regulatory frameworks. The benefits of interoperable, standardised and connected data include improvements and new offerings in public services, an increased government efficiency, data-driven policymaking and the value of open data (Domeyer et al., 2021). The regional and national data strategies are also important: these address data protection and privacy requirements. Some of the challenges to interoperable and connected data stem from the fact that public and private data remain dispersed, not digitally accessible and not interoperable (referring to obstacles that prevent the combination and joint processing of data) (Domeyer et al., 2021). Currently, vehicle manufacturers can control who has access to which vehicle data, creating an undesirable "gatekeeper" position that will probably only be resolved partially by the cross-sector Data Act (European Commission, 2022).

To address these issues, European and national strategies for integrated data management are needed, as well as infrastructure and the setting of technical standards, as "fast and automated data exchange is only possible through harmonized data formats and standards" (Domeyer et al., 2021). The "European Data Strategy" aims to create a single market for data, and one of the main goals is to create safer and cleaner transport systems (European Commission, 2020c). For automated minibuses in MaaS or ITS, open APIs and open data are accordingly a prerequisite and a key to achieving interoperability and coordination of all stakeholders in the mobility system. As a result, seamless trips to meet the needs and acceptance of citizens become possible.

At the same time, this raises the difficult question and debate for regulation: who benefits from the use of data? This is clear for data collected in the event of an accident, as it will help to understand the causes and build anonymised lessons learned, to the benefit of all mobility stakeholders (see Figure Loop 1, Fig. 18.10), as feedback is shared. For other data, while individuals may have an interest in not sharing their personal data for privacy, safety or security reasons, private companies are interested in collecting these data to create new profitable business models ("winner takes it all", advertising or other forms of value capture) but also to better meet customer needs. The use of data is further necessary to allow OEMs to optimise automated minibuses and make the drive more reliable and safer or to realise the loops described in Sect. 18.4. PTOs could use the data to optimise the use of the fleet and the PTA to serve the general interest. A balance therefore needs to be found and managed between the interests of the individual data provider (the mobility user and its stakeholders, GDPR) and the legitimate interests of the community in using the data to serve the general interest. These discussions should take place at European, national and local levels. This will require a less administrative corporate culture on the part of PTAs in order to manage the transformation and, in particular, to find the right balance between the numerous stakeholders towards the unity of society for sustainable mobility (purpose). A decentralised governance model which unifies the mobility network and supports a federated, democratic, decentralised mobility platform using distributed ledger technologies (DLT, so-called Web3) to "integrate stakeholders and different technologies in a non-discriminatory and equal way" (Schmück et al., 2021) could accordingly be chosen. As a result, the beneficiary of the system would be directly the data subject rather than an intermediary party that owns the mobility platform and the customer touchpoint (Schmück, et al., 2021). As an alternative, other federation-based governance strategies could be used, where so-called substantive platforms are used to serve the general interest between stakeholders. The aim is to integrate economic dimensions to satisfy (mobility) needs on the one hand and political issues (through collective deliberation) to preserve the commons and enable purpose within a community on the other. Linking all stakeholders allows defining the needs, rules and structures of the ecosystem without intermediaries. The substantive platforms can vary and depend on the valuation chosen, which can be based on (classical) markets, on off-markets (based on donation, reciprocity and/or redistribution) or on hybrid forms combining market and off-market valuation. These new models are still exploratory. These respect GDPR by design (ethics by design).[14]

In terms of data protection and data privacy, the GDPR is an asset for personal data protection, safety and security, but it brings complex and challenging implementation in the field of connected and automated vehicles, as well as obstacle for AI (see the aforementioned loops within this chapter).

From a legal perspective, there are many challenges ahead, including the following. With regard to privacy and personal data protection, certain key concepts (e.g. "personal data", "identifiable natural person", "anonymous information" and "processing") are defined and could be interpreted in such a way that the rules of the GDPR would apply, to the extent that this would require additional investment or hinder or slow down the development of the processing activities described in previous sections above (Sects. 18.3 and 18.4).

With respect to the data flows between the different stakeholders described in Sect. 18.4, the distinction between "controller" and "processor" is complex, taking into account the concepts of joint controllers, controllers in common and sub-processors.[15]

The "controller" is any person or entity that determines the purposes and means of the processing of personal data, and the "processor" only processes personal data on behalf of the controller. Hence, it may be difficult to identify the role of each actor intervening in this context their underlying obligations.

Only technically collected data (i.e. non-personal data) or personal data anonymised by the "GDPR filter" (described in Fig. 18.8 above) or by another

---

[14] Vercher-Chaptal, C., Acosta Alvarado,A. S., Aufrère, L., Brabet, B., Broca, S., et al. There Are Platforms as AlternativeS. Entreprises plateformes, plateformes collaboratives et communs numeriques. [Rapport de recherche] DARES - Ministère du Travail, de l'Emploi et du Dialogue social; DREES. 2021. https://hal.archives-ouvertes.fr/hal-03413930/document

[15] See LEVERAGING BIG DATA FOR MANAGING TRANSPORT OPERATIONS (LeMO), Deliverable 4.1 Report on the characterization of the barriers and limitations, September 2019, § 4.2.1 p. 17.

**Table 18.1** Expected actions from public authorities in the development of MaaS and the different governing approaches by Audouin and Finger (2019)

| Governing approaches | Corresponding actions expected in the development of MaaS schemes |
|---|---|
| Governing by authority | Develop specific legislation/regulation enforcing the development of MaaS in a top-down fashion<br>Procure MaaS to a third party through traditional tender mechanisms |
| Governing by enabling | Initiate public-private interactions<br>Define vision with strong quantitative objectives<br>Provide funding<br>Influence negotiations in favour of MaaS, and leverage MaaS opponents using horizontal network governance |
| Governing by doing | Develop a MaaS solution in-house in a closed manner<br>Minimise collaboration with third parties |
| Self-governing | Provide all government employees with a MaaS solution to show the example for citizens to follow |
| Governing by laissez-faire | Refuse to get involved in the development of MaaS<br>Adopt a wait-and-see approach |

Source: Audouin and Finger (2019)

technique will be provided to the MaaS data and service platform operator. This data may become personal data or even sensitive data due to the transformative impact of big data analytics and thus trigger the application of the GDPR and further privacy and data protection laws.[16]

The potentially high volume of data (personal and non-personal) collected by automated minibuses can be assimilated to big data analytics, challenging certain core assumptions of EU data protection law, such as data minimisation and purpose limitation.[17]

The scope of the e-Privacy Directive (lex specialis to the GDPR) is much broader and applies not only to personal data but also to all information regardless of its nature. Consequently, this framework requires an assessment of which processing activities may fall within the scope of the e-Privacy Directive and, subsequently, what specific requirements the e-Privacy Directive imposes on stakeholders in relation to these processing activities. Furthermore, where personal data are involved, the interaction between the GDPR and the e-Privacy Directive needs to be closely examined.[18]

[16] See LEVERAGING BIG DATA FOR MANAGING TRANSPORT OPERATIONS (LeMO), Deliverable 4.1 Report on the characterization of the barriers and limitations, September 2019, § 4.2.1 p. 16.

[17] See LEVERAGING BIG DATA FOR MANAGING TRANSPORT OPERATIONS (LeMO), Deliverable 4.1 Report on the characterization of the barriers and limitations, September 2019, § 4.2.1 p. 17.

[18] European forum and oBsErvatory for OPEN science in transport (BE OPEN), Open Science in transport research: legal issues and fundamental principles, 31 August 2020, p. 8.

The (current) lack of standards for open APIs could lead to coordination issues, and it should be ensured that Member States apply the same standards to avoid the creation of technical barriers.

Finally, it is important to note that there are different governing approaches from public authorities for MaaS development (see Table 18.1). The aforementioned economic and environmental scenarios show that the governance determines if the rules serve the general interest and promote sustainable mobility and accordingly the technical, social, economic and environmental impacts. A governing by authority on one side which puts forward the general interest can be imagined as well as a governing by "laissez-faire" (e.g. voluntary lack of regulation or minimalistic regulation) on the other side where few private companies could take advantage from the "winner-takes-it-all phenomenon" of the mobility ecosystem and capture the expected benefits for private purposes. In the case of "laissez-faire", potential competitors would be excluded (vendor lock-out strategy) and could not find a viable business model anymore. Vendor could also be complementary to the private ecosystem (vendor lock-in strategy) when an alternative market is not viable anymore and the competitor has to become a complement of the vendor (proprietary ecosystem owner). Even a threatening of Public Transport Authorities might be imagined when the private terms and conditions of the ecosystem of a private MaaS substitute to the public laws of the PTA. This has already been observed in other European markets. The lack of speed of adapting the rules of the PTA to the disruptive changes in mobility could hence generate the same problem that the EU had to experience before introducing the DMA (Digital Market Act). The aim of the DMA is to ensure the proper functioning of the internal market, by promoting effective competition in digital markets, in particular a contestable and fair online platform environment. More specifically, the DMA's objectives are (1) to address market failures to ensure contestable and competitive digital markets for increased innovation and consumer choice, (2) to address gatekeepers' unfair conduct and (3) to enhance coherence and legal certainty to preserve the internal market (European Commission, 2020d). As the DMA is not likely to apply to the mobility sector yet, an anticipatory approach for the local, national and European mobility markets which takes the specificities of the transport sector into account through a dedicated "mobility data act" is most likely needed and would accordingly be recommendable as well.

## 18.6 Conclusion

For all initial questions about "how to integrate AV in the city transport system to serve general interests", the AVENUE vision could provide a convincing conceptional solution approach and a pragmatic transformation concept based on the hypothesis that automated minibuses integrated into intermodal transport and MaaS or better ITS can be a promising game changer in urban mobility. The automated minibus, deployed on demand and door to door, will provide more mobility choices and flexibility for all citizens, including better accessibility for people with reduced mobility (PRM) and potentially better acceptance of AVs based on positive experiences.

As argued above (Sect. 18.3), a coopetition governance scenario and open data schemes (open data, open platforms, open interfaces and protocols) are further key factors to guarantee fair competition between public and private mobility providers, avoiding dominant position and "the-winner-takes-it-all" effects. In addition, AVs coupled to ITS and AI are expected to make the transport systems more reliable, safe, efficient and flexible (concept called ambidexterity), and thus the antinomic goals—incremental and disruptive innovation—are combined. As a result, the transport system becomes, through ITS and automated minibuses, citizen-centric, inclusive and sustainable, enabling positive externalities (Mohring effect, network effects, enhanced accessibility etc.) and lowering negative externalities (e.g. congestion costs, space, use of energy and materials). The citizen-centric approach could thus become purpose-centric, serving the general interest to the best for all stakeholders. This vision is a key for acceptance and thus coherent with the SUMP concept of a holistic approach and the requirement of the EU Green Deal (European Commission, 2020a), EU Sustainable and Smart Mobility Strategy (European Commission, 2020b, 2021) and the European Data Strategy (European Commission, 2020c).

Through the vision concept, the EU could gain on sovereignty with automated minibuses in ITS and become the worldwide mobility system leader in purpose-centric—this means citizen-centric, responsible, independent and sustainable for all the stakeholders—mobility which respects human and individual rights (data privacy and security, GDPR), by aligning and utilising future (technical and business) mobility product and system innovations for this purpose. This could be an alternative to the path of development chosen in other continents.

Of course, this approach is a socio-technical and innovative approach which believes in progress and science and uses dematerialised consumption to save resources for a sustainable development (mobility) path (McAfee, 2019). It will have to convince the growing scepticism which arouses in discussions around concepts such as "degrowth" or "frugality" which promotes a paradigmatic change in consumer behaviour through renunciation to save resources. This so called sufficiency strategy asks people to consume consciously and since around 13% of consumer spending by European citizens is spent on mobility, this issue is an important factor (Statista, 2020). Our vision shows a feasible and promising path of how dematerialisation of mobility through automated minibuses and ITS could save energy and resources without sacrificing individual mobility needs and comfort (Linz, 2004; Mauch et al., 2001).

This vision will be the basis to structure and deduct our transition goals, recommendation and transition roadmap to design the future (public) transportation service. The Horizon Europe project ULTIMO (Safe, Resilient Transport and Smart Mobility services for passengers and goods) implements parts of the concept from 2022. The concept has been further presented at Transportation Research Area Conference 2022 (TRA, Lisbon) (Fournier et al., 2023) and at the High-Level

Dialogue on Connected and Automated Driving organized in June 2024 in Ghent by the Belgian Presidency of the European Council (Flemish Ministry of Mobility and Public Works 2024).

**Acknowledgements** This project has received funding by the European Union's Horizon 2020 research and innovation program under grant agreement No. 769033.

## References

Alazzawi, S., Hummel, M., Kordt, P., Sickenberger, T., Wieseotte, C., & Wohak, O. (2018). Simulating the impact of shared, autonomous vehicles on urban mobility – a case study of Milan. *SUMO 2018 – Simulating Autonomous and Intermodal Transport Systems, 2*, 94–76. https://doi.org/10.29007/2n4h

Antonialli, F., Fournier, G., Jaroudi, I., Mira-Bonnardel, S., & Thalhofer, M. (2021). *D8.4 Second Iteration Economic impact*. https://cordis.europa.eu/project/id/769033/results

Araghi, Y., Larco, N., Doll, C., Vonk Noordegraaf, D. M., Bouma, G. M., & Krauss, K. (2020). *Drivers and barriers of Mobility-as-a-Service in urban areas*. Accessed September, 7, 2021, from https://www.isi.fraunhofer.de/content/dam/isi/dokumente/ccn/2020/tra2020_drivers_and_barriers_final-after-reviews-19-10-2019.pdf

Audouin, M., & Finger, M. (2019). Empower or thwart? Insights from Vienna and Helsinki regarding the role of public authorities in the development of MaaS schemes. *Transportation Research Procedia, 41*, 6–16. https://doi.org/10.1016/j.trpro.2019.09.003

Becker, H., Balac, M., Ciari, F., & Axhausen, K. W. (2020). Assessing the welfare impacts of Shared Mobility and Mobility as a Service (MaaS). *Transportation Research Part A: Policy and Practice, 131*, 228–243. https://doi.org/10.1016/j.tra.2019.09.027

Bestmile. (2020). *The critical role of APIs in mobility services*. Accessed July 13, 2021, from https://bestmile.com/company-blog/the-critical-role-of-apis-in-mobility-services/

BMWi. (2019). *Making space for innovation – The handbook for regulatory sandboxes*. Accessed October 11, 2022, from https://www.bmwk.de/Redaktion/EN/Publikationen/Digitale-Welt/handbook-regulatory-sandboxes.pdf?__blob=publicationFile&v=2

Business Roundtable. (2019). *Business roundtable redefines the purpose of a corporation to promote 'An Economy That Serves All Americans'*. Accessed October 11, 2022, from https://www.businessroundtable.org/business-roundtable-redefines-the-purpose-of-a-corporation-to-promote-an-economy-that-serves-all-americans

Cabral, L., Haucap, J., Parker, G., Petropoulos, G., Valletti, T., & van Alstyne, M. (2021). *The EU Digital Markets Act. JRC122910*. Publications Office of the European Union. Luxembourg (JRC). https://publications.jrc.ec.europa.eu/repository/handle/JRC122910

Capgemini. (2020). *The future of Mobility as a Service (MaaS). Which model of MaaS will win through?* https://www.capgemini.com/wp-content/uploads/2020/12/Capgemini-Invent-POV-Maas.pdf

Childress, S., Nichols, B., Charlton, B., & Coe, S. (2015). Using an activity-based model to explore the potential impacts of automated vehicles. *Transportation Research Record, 2493*(1), 99–106. https://doi.org/10.3141/2493-11

Clewlow, R. R., & Mishra, G. S. (2017). *Disruptive transportation: The adoption, utilization, and impacts of ride-hailing in the United States*. Institute of Transportation Studies, University of California, Davis, Research Report UCD-ITS-RR-17-07. https://escholarship.org/uc/item/82w2z91j

Coyle, D., Diepeveen, S., Kay, L., & Tennison, J. (2020). *The value of data*. Bennett Institute, ODI.

Coyle, D., & Diepeveen, S. (2021). Creating and governing social value from data. *SSRN Journal*. https://doi.org/10.2139/ssrn.3973034

Deloitte. (2022). What is digital economy? *Unicorns, transformation and the internet of things*. Available online at https://www2.deloitte.com/mt/en/pages/technology/articles/mt-what-is-digital-economy.html

Domeyer, A., Hieronimus, S., Klier, J., & Weber, T. (2021). *Government data management for the digital age*. McKinsey. Accessed August 11, 2021, from https://www.mckinsey.com/industries/public-and-social-sector/our-insights/government-data-management-for-the-digital-age?cid=other-eml-alt-mip-mck&hdpid=faf57223-411c-4522-a595-23ac64af0812&hctky=2808780&hlkid=6bb6d6343b2340ab873cc6668394c8b0#0&cid=app

ERTICO – ITS Europe. (2019). *MOBILITY AS A SERVICE (MAAS) AND SUSTAINABLE URBAN MOBILITY PLANNING*.

ETSI. (2021). *Automotive Intelligent Transport Systems (ITS)*. Accessed July 20, 2021, from https://www.etsi.org/technologies/automotive-intelligent-transport

European Commission. (2016). *European accessibility act - Related documents*. Accessed October 11, 2022, from https://ec.europa.eu/social/main.jsp?catId=1202&langId=en&moreDocuments=yes

European Commission. (2020a). *A European Green Deal*. https://ec.europa.eu/info/strategy/priorities-2019-2024/european-green-deal_en#actions

European Commission. (2020b). *Sustainable and smart mobility strategy*. https://ec.europa.eu/info/law/better-regulation/have-your-say/initiatives/12438-Sustainable-and-Smart-Mobility-Strategy

European Commission. (2020c). *The European data strategy*. https://ec.europa.eu/commission/presscorner/detail/en/fs_20_283

European Commission. (2020d). *dafür gibt es seeehr viel in Citavi, ich weiss leider nicht welche Quelle gemeint ist*

European Commission. (2021). *Mobility strategy. A fundamental transport transformation: Commission presents its plan for green, smart and affordable mobility*. https://ec.europa.eu/transport/sites/default/files/2021-mobility-strategy-and-action-plan.pdf

European Commission. (2022). *(EU Comm) Implementing regulation C(2022) 5402 final*.

European forum and oBsErvatory for OPEN science in transport. (2020). *Open Science in transport research: Legal issues and fundamental principles*. Accessed October 11, 2022, from https://beopen-project.eu/storage/files/beopen-d41-open-science-in-transport-research-legal-issues-and-fundamental-principles.pdf

Fagnant, D. J., & Kockelman, K. (2015). Preparing a nation for autonomous vehicles: Opportunities, barriers and policy recommendations. *Transportation Research Part A: Policy and Practice, 77*, 167–181. https://doi.org/10.1016/j.tra.2015.04.003

Fagnant, D. J., & Kockelman, K. M. (2018). Dynamic ride-sharing and fleet sizing for a system of shared autonomous vehicles in Austin, Texas. *Transportation, 45*(1), 143–158. https://doi.org/10.1007/s11116-016-9729-z

Flemish Ministry of Mobility and Public Works (MOW): High-Level Dialogue on Connected and Automated Driving, 18 and 19 June 2024 in Ghent, Belgium, https://www.vlaanderen.be/en/mobility/flanders-together-in-motion/high-level-dialogue-on-connected-and-automated-driving#about-the-break-out-sessions

Fournier, G., Baumann, M., Gasde, J., & Yasin, K. K. (2018). Innovative mobility in rural areas—The case of the Black Forest. *IJATM, 18*(3), 247. https://doi.org/10.1504/IJATM.2018.093420

Fournier, G., Boos, A., Wörner, R., Jaroudi, I., Morozova, I., & Nemoto, E. H. (2020). Substituting individual mobility by mobility on demand using autonomous vehicles. A sustainable assessment simulation of Berlin and Stuttgart. *International Journal of Automotive Technology and Management, 20*, 369–407. Article 4.

Fournier, G., Thalhofer, M., Chrétien, P., Boos, A., Korbee, D., Jaroudi, I., et al. (2023). System innovation in transport with automated minibuses and ITS: The citizen centric approach of AVENUE. Transport Research Arena (TRA) Conference. Lisbon.

Fraedrich, E., Beiker, S., & Lenz, B. (2015). Transition pathways to fully automated driving and its implications for the sociotechnical system of automobility. *European Journal of Futures Research, 3*(1), 91. https://doi.org/10.1007/s40309-015-0067-8

Gassmann, O., & Ferrandina, F. (Eds.). (2021). *Connected business*. Springer International Publishing.

Geels, F. W. (2011). The multi-level perspective on sustainability transitions: Responses to seven criticisms. *Environmental Innovation and Societal Transitions, 1*(1), 24–40. https://doi.org/10.1016/j.eist.2011.02.002

Geisslinger, M., Poszler, F., Betz, J., Lütge, C., & Lienkamp, M. (2021). Autonomous driving ethics: From trolley problem to ethics of risk. *Philosophy & Technology, 34*(4), 1033–1055. https://doi.org/10.1007/s13347-021-00449-4

Giusti, R., Manerba, D., Bruno, G., & Tadei, R. (2019). Synchromodal logistics: An overview of critical success factors, enabling technologies, and open research issues. *Transportation Research Part E: Logistics and Transportation Review, 129*, 92–110. https://doi.org/10.1016/j.tre.2019.07.009

Grisoni, A., & Madelenat, J. (2021). *Le véhicule autonome : quel rôle dans la transition écologique des mobilités ?* With assistance of FORUM VIES MOBILES. LA FABRIQUE ECOLOGIQUE. https://www.lafabriqueecologique.fr/app/uploads/2020/02/Rapport-Complet_Ve%CC%81hicule-autonome-et-Transition-e%CC%81cologique_La-Fabrique-Ecologique-Forum-Vies-Mobiles-1.pdf

de Haan, P., Straumann, R., Bianchetti, R., Stetter, A., Oehry, B., & Jermann, J. (2020). *Verkehr der Zukunft 2060: Tech-nologischer Wandel und seine Folgen für Mobilität und Verkehr. Forschungsprojekt SVI 2017/003 auf Antrag der Schweizerischen Vereini-gung der Verkehrsingenieure und Verkehrsexperten (SVI)*. Bundesamt für Strassen (ASTRA). http://www.mobilityplatform.ch

Hassani, J. E. (2018). Les entreprises de mobilités bientôt dépossédées de leurs données ? Journal du Net. https://www.journaldunet.com/economie/transport/1417359-loi-mobilites-ouverture-donnees-transport-open-data/

Heineke, K., Kampshoff, P., Kellner, M., & Kloss, B. (2019). *Change vehicles: How robo-taxis and shuttles will reinvent mobility*. McKinsey.

Hoffmann, I., Jensen, N., Cristescu, A. (Eds.) (2021). Decentralized governance for digital platforms—Architecture proposal for the mobility market to enhance data privacy and market diversity. *IEEE 18th Annual Consumer Communications & Networking Conference (CCNC), 9–12 Jan 2021*. IEEE.

Huber, D., Viere, T., Nemoto, E. H., Jaroudi, I., Korbee, D., & Fournier, G. (2022). Climate and environmental impacts of automated minibuses in future public transportation. *Transportation Research Part D: Transport and Environment, 102*, 103160. https://doi.org/10.1016/j.trd.2021.103160

Hurst, A. (2014). *The purpose economy. How your desire for impact, personal growth and community is changing the world*. Elevate.

Hurst, A. (2016). *The purpose economy. Expanded and updated how your desire for impact, personal growth and community is changing the world. Second paperback printing; expanded and updated*. Elevate.

Iyer, L. S. (2021). AI enabled applications towards intelligent transportation. *Transportation Engineering, 5*, 100083. https://doi.org/10.1016/j.treng.2021.100083

Jaroudi, I., Nemoto, E. H., Korbee, D., Bulteau, J., Viere, T., Fournier, G. (Eds.) (2021). What is the potential impact of the transition from traditional transport to new mobility models (electric automated minibuses) in European cities? *The transformations of the global automotive industry: digitalisation, ecological transition and the impact of the Covid 19 crisis. Paris, 8 June. Gerpisa colloquium*. Available online at https://gerpisa.org/node/6370

Kamargianni, M., Matyas, M., Li, W., & Schäfer, A. (2015). *Feasibility study for "Mobility as a Service" concept in London*.

Korbee, D., Naderer, G., & Fournier, G. (2023). *The potential of automated minibuses in the socio-technical transformation of the transport system*. Transport Research Arena (TRA) Conference. Lisbon.

Leveraging Big Data for Managing Transport Operations. (2019). *Deliverable 4.1 Report on the characterization of the barriers and limitations*.

Linz, M. (2004). Weder Mangel noch Übermaß: Über Suffizienz und Suffizienzforschung. *Wuppertal Papers, 145*.

Liu, J., Li, Y., & Li, J. (2015). Coopetition in intermodal freight transport services. *Discrete Dynamics in Nature and Society, 2015*, 1–11. https://doi.org/10.1155/2015/680685

Lucca, N. (2022). *AI-research project started for sustainable mobility - WirelessCar*. Accessed October 11, 2022, from https://www.wirelesscar.com/ai-research-project-started-for-sustainable-mobility/, updated on 17-Mar-22.

MaaS Aliance. (2019). *EU projects explore common standard for digital mobility services*. Accessed July 14, 2021, from https://maas-alliance.eu/2019/01/24/eu-projects-explore-common-standard-for-digital-mobility-services/

Mauch, U., North, N., & Pulli, R. (2001). Between efficiency and sufficiency. The optimal combination of policy instruments in the mobility sector towards sustainable development. In R. Kaufmann-Hayoz & H. Gutscher (Eds.), *Changing things — Moving people* (pp. 133–150). Birkhäuser Basel.

May, A. D., Shepherd, S., Pfaffenbichler, P., & Emberger, G. (2020). The potential impacts of automated cars on urban transport: An exploratory analysis. *Transport Policy, 98*(2), 127–138. https://doi.org/10.1016/j.tranpol.2020.05.007

McAfee, A. (2019). *More from less. The surprising story of how we learned to prosper using fewer resources–and what happens next* (First Scribner hardcover edition). Scribner.

McKinsey & Company. (2020). *The path to true transformation*. Accessed August 15, 2022, from https://www.mckinsey.de/business-functions/transformation/our-insights/the-path-to-true-transformation

McKinsey & Company. (2021). *Government transformations in times of extraordinary change: Key considerations for public-sector leaders*. Accessed August 22, 2022, from https://www.mckinsey.com/industries/public-and-social-sector/our-insights/government-transformations-in-times-of-extraordinary-change-key-considerations-for-public-sector-leaders

Meijer, R., Conradie, P., & Choenni, S. (2014). Reconciling contradictions of open data regarding transparency, privacy, security and trust. *Journal of Theoretical and Applied Electronic Commerce Research, 9*(3), 32–44. https://doi.org/10.4067/S0718-18762014000300004

Merlin, L. (2017). Comparing automated shared taxis and conventional bus transit for a small city. *JPT, 20*(2), 19–39. https://doi.org/10.5038/2375-0901.20.2.2

Meyer, J., Becker, H., Bösch, P. M., & Axhausen, K. W. (2017). Autonomous vehicles: The next jump in accessibilities? *Research in Transportation Economics, 62*(1), 80–91. https://doi.org/10.1016/j.retrec.2017.03.005

Milakis, D., Snelder, M., van Arem, B., van Wee, B., De Almeida, H., & Correia, G. (2017). Development and transport implications of automated vehicles in the Netherlands: Scenarios for 2030 and 2050. *European Journal of Transport and Infrastructure Research, 17*(1), 10.18757/ejtir.2017.17.1.3180.

MOQO. (2020). *Shared mobility status report*. https://moqo.de/shared-mobility-status-report-2020

NCSL. (2022). *Autonomous vehicles state bill tracking database*. Accessed August 23, 2022, from https://www.ncsl.org/research/transportation/autonomous-vehicles-legislative-database.aspx

Nemoto, E. H. (2022). *Towards a new sustainable mobility paradigm. The role and impacts of automated minibuses in the mobility transition*. PhD Thesis. CentraleSupelec, Paris. Université Paris-Saclay.

Nemoto, E. H., Jaroudi, I., & Fournier, G. (2021). Introducing automated shuttles in the public transport of European cities: The case of the AVENUE project. In E. G. Nathanail, G. Adamos, & I. Karakikes (Eds.), *Advances in Mobility-as-a-Service Systems. CSUM 2020* (Advances in Intelligent Systems and Computing) (Vol. 1278). Springer.

Niles, J. (2019). *Full potential of future robotaxis achievable with trip-based subsidies and fees applied to the for-hire vehicles of today*. Mineta Transportation Institute Publications.

Raisch, S., & Birkinshaw, J. (2008). Organizational ambidexterity: Antecedents, outcomes, and moderators. *Journal of Management, 34*, 375–409. https://doi.org/10.1177/0149206308316058

Rayle, L., Dai, D., Chan, N., Cervero, R., & Shaheen, S. (2016). Just a better taxi? A survey-based comparison of taxis, transit, and ridesourcing services in San Francisco. *Transport Policy, 45*, 168–178. https://doi.org/10.1016/j.tranpol.2015.10.004

Rogg, J. (2021). *White Paper: ITxPT and Information Technology Standards for Public Transport. Delivering flexible, advanced Intelligent Transport Systems through Open Standards*. Trapeze Group. Accessed July 13, 2021, from https://trapezegroup.co.uk/itxpt-technical-white-paper/?utm_source=Email+marketing&utm_medium=email&utm_campaign=IT+-+industry+Insight+-+E-mobility+-+Trapeze+Group+-+09+July+2021&utm_term=Read+our+latest+Transport+Tech+Industry+Insight&utm_content=https%3a%2f%2femails.intelligent-transport.com%2frussellpublishinglz%2f&gator_td=MYOPaxduAAShzmyuCrZFtBL0rc3UKIX7gIrZeKyipk9sAOGS7ajoOkC1Ii7wzNnICooBw1Cf7rYkn1CPBkxjdwZuFxAt0kbvcukaWd%2fDPIePpGLZRb5lEnM1Mx%2bBO7yoldDwU7d5B095%2byCInpINxGnari%2bfbDc14U0gPtFDfQjs35%2bY0Ef9qM9k2lJzjCZ9TTpXdEoytTa%2f%2bIFgYFQhcfLgplKZGBMsQAZxtNDwmo8rdzDStOuo2DLtsi6JmpwX

Rousseau, J. J. (1762). *Du contrat social*. Amsterdam.

Scherk, J., Pöchhacker-Tröscher, G., & Wagner, K. (2017). *Künstliche Intelligenz. Artificial Intelligence*. BMVIT Bereich Innovation.

Schmück. (2022). *Less trust, more truth: Affordances of distributed ledger technologies for decentralized platform ecosystems*. https://www.alexandria.unisg.ch/265924/1/Dis5197.pdf

Schmück, K., Sturm, M., & Gassmann, O. (2021). Decentralized platform ecosystems for data-sharing and digital trust in industrial environments. In O. Gassmann & F. Ferrandina (Eds.), *Connected business* (pp. 127–136). Springer International Publishing.

Shafik, M. (2021). *What we owe each other. A new social contract / Minouche Shafik*. Vintage Digital.

Shen, Y., Zhang, H., & Zhao, J. (2018). Integrating shared autonomous vehicle in public transportation system: A supply-side simulation of the first-mile service in Singapore. *Transportation Research Part A: Policy and Practice, 113*, 125–136. https://doi.org/10.1016/j.tra.2018.04.004

Sochor, J., Arby, H., Karlsson, I. C. M., & Sarasini, S. (2018). A topological approach to Mobility as a Service: A proposed tool for understanding requirements and effects, and for aiding the integration of societal goals. *Research in Transportation Business & Management, 27*, 3–14. https://doi.org/10.1016/j.rtbm.2018.12.003

Sochor, J., Karlsson, I. C. M., & Strömberg, H. (2016). Trying out Mobility as a Service: Experiences from a field trial and implications for understanding demand. *Transportation Research Record, 2542*(1), 57–64. https://doi.org/10.3141/2542-07

Statista. (2020). *Europäische Union: Anteil von Verkehr an den Konsumausgaben der privaten Haushalte, aufgeschlüsselt nach Mitgliedstaaten im Jahr 2020*. Accessed October 11, 2022, from https://de.statista.com/statistik/daten/studie/1219892/umfrage/anteil-der-konsumausgaben-fuer-verkehr-an-den-gesamtausgaben-in-den-laendern-der-europaeischen-union/, updated on 11-Oct-22.

Termer, F. (2017). Open Data—Neue Konzepte erfolgreich umsetzen, herausgegeben von bitkom e.V.

The Economist. (2022). *The era of big-tech exceptionalism may be over*. Accessed October 11, 2022, from https://www.economist.com/leaders/2022/07/27/the-era-of-big-tech-exceptionalism-may-be-over, updated on 11-Oct-22.

Toledano, J. (2020). *GAFA. Reprenons le pouvoir!* O. Jacob.

UITP. (2017). *Autonomous vehicles: A potential game changer for urban mobility*. Accessed September 2, 2020, from https://www..org/sites/default/files/cck-focus-papers-files/PolicyBrief_Autonomous_Vehicles_LQ_20160116.pdf

UITP. (2019). *Ready for MaaS—Easier mobility for citizens and better data for cities*. Policy brief. Available online at https://www.uitp.org/publications/ready-for-maas-easier-mobility-for-citizens-and-better-data-for-cities/

UNESCO. (2021). *AI for the Planet: Highlighting AI innovations for sustainable mobility and smart cities, 2021*. Accessed October 11, 2022, from https://www.unesco.org/en/articles/ai-planet-highlighting-ai-innovations-sustainable-mobility-and-smart-cities

Vercher-Chaptal, C., Acosta Alvarado, A. S., Aufrère, L., Brabet, J., Broca, S., Smichowski, B. C., et al. (2021). *There Are Platforms as AlternativeS. Entreprises plateformes, plateformes collaboratives et communs numeriques*. Accessed October 11, 2022, from https://hal.archives--ouvertes.fr/hal-03413930/document

Vleugel, J. M., & Bal, F. (2017). More space and improved living conditions in cities with autonomous vehicles. *International Journal of Design & Nature and Ecodynamics, 12*(4), 505–515. https://doi.org/10.2495/DNE-V12-N4-505-515

WEforum. (2020). *Will robo-taxis cause radical disruption - or gridlock?* Accessed October 5, 2022, from https://www.weforum.org/agenda/2020/01/will-robo-taxis-bring-radical-disruption-to-our-streets-or-gridlock/

# Chapter 19
# Transition Planning Towards a Sustainable Urban Mobility Ecosystem

**Michael Thalhofer, Adrian Boos, Eliane Horschutz Nemoto, Dorien Duffner-Korbee, Markus Dubielzig, Christian Zinckernagel, Inès Jaroudi, Dimitri Konstantas, and Guy Fournier**

**Abstract** In this chapter we aim to provide pragmatic recommendations and exemplary concrete transition measures for consultants and ecosystem stakeholders to get valuable practice proven cornerstones and hints for successfully achieving their individually designed mobility vision. Hereby we describe the transition

---

M. Thalhofer (✉)
Software-driven Ecosystems and Market Places, Siemens Technology, Munich, Germany
e-mail: michael.thalhofer@siemens.com

A. Boos
Pforzheim University, Pforzheim, Germany

E. H. Nemoto
Pforzheim University, Pforzheim, Germany

CentraleSupélec, LGI, Gif-sur-Yvette, France

ICLEI Europe, Freiburg im Breisgau, Germany

D. Duffner-Korbee
Pforzheim University, Pforzheim, Germany

Fraunhofer ISI, Karlsruhe, Germany

M. Dubielzig
Siemens Technology, Accessibility Competence Center, Paderborn, Germany

C. Zinckernagel
Holo, København N, Denmark

I. Jaroudi
Pforzheim University, Pforzheim, Germany

CentraleSupélec, LGI, Gif-sur-Yvette, France

Fraunhofer ISI, Karlsruhe, Germany

D. Konstantas
Geneva School of Economics and Management, University of Geneva, Geneva, Switzerland

G. Fournier
Pforzheim University, Pforzheim, Germany

CentraleSupélec, LGI, Gif-sur-Yvette, France

 
G. Fournier et al. (eds.), *Automated Vehicles as a Game Changer for Sustainable Mobility*, Contributions to Management Science,
https://doi.org/10.1007/978-3-031-61681-5_19

planning as a systemic and systematic approach designed as a pragmatic process with essential key methods from the status quo (2020) of all existing states towards a desired common future vision of AV in ITS to be achieved in 2030. For this purpose we visualised a 'Big Picture' of transition management work packages and a roadmap as a 'common thread', conceptional guide and orientation for all stakeholders of the transition. This roadmap of multifaceted recommendations is structured into phases which are systematically building on one another, outlining the transition from the status quo with no automated minibuses (AM) via the milestones: AM, AM System, AM in MaaS, AM in MaaS/ITS to the future mobility Vision 2030 (Chap. 18). These recommendations are the basis for individual operational measures which have to be documented in an implementation plan. Finally, a general transition plan is suggested consisting of all scenarios, steps and planning results of the transition concept. This should serve all stakeholders as an orientation guide as well as a strategic and operational basis for the entire mobility transition.

## 19.1 Introduction

It is of course not sufficient to describe a desirable mobility future with AVs as mobility game changer (AV in MaaS/ITS) (see Chap. 18). A status quo state as a starting basis has to be considered as well and transition planning to understand how to realise and shape a future sustainable mobility ecosystem. This transformation has also to consider a high complexity of multifaceted challenges in a volatile, uncertain, complex ambiguous (VUCA) mobility environment.

To meet all these goals and framework conditions, a systemic and systematic approach to provide a clear structure and process for the stakeholders of the transition, especially providing strategic guidance to be manageable for the transition managers, has accordingly to be followed. Furthermore, the transition planning must be operationally feasible and deployable for the operational actors in a way that detailed goals and concrete pragmatic recommendations and roadmap-guided measures must be provided. Finally, all these transition deliverables must be compiled into an agile (incremental-iterative) comprehensive transition plan that ensures learning loops on strategic as well as on operational transition management level. A special focus must be set on the mindset and cultural issues of the transition in form of an explicit 'soft fact change management' since the transition of the mobility ecosystem requires a complex and disruptive paradigmatic change for all stakeholders of the transition together with accompanying and enabling management measures and roles like enablers, promoters, accelerators, multiplicators, change management board, etc.

Considering the goals, the giving framework and the mentioned guiding principles, the transition approach in this chapter will thus first explain the purpose and goals of the transition (Sect. 19.2). In the third section, the design (Big Picture) of the transition management concept will be developed. After a description of the status quo states and the future vision (Sect. 19.4), the transition steps from the status quo to the future vision will finally be explained in Sect. 19.6.

The methodology for this transition concept has been developed and elaborated by experts from academia (e.g. Pforzheim University, CentraleSupélec), industry (e.g. Siemens AG) and transport organisations (PTOs) of the involved pilot sites (Geneva, Luxembourg, Copenhagen, Lyon).

## 19.2 Purpose and Goals of the Transition

The transition planning towards a sustainable urban mobility ecosystem is following a generic process in three phases: (1) analysis of the status quo of all relevant scenarios, (2) definition of a future vision to be achieved and (3) design of the transition from status quo states to achieve the defined future vision (see Fig. 19.1). This procedure ensures that the transition design is based on the balanced integration of both transitional forces: the 'status quo push' effect and the 'future vision pull' effect.

Purpose of the transition is finally to generate and suggest pragmatic recommendations and concrete transition measures for consultants and mobility B2B customers or ecosystem owners to get proven, structured and valuable cornerstones and hints for successfully achieving their individually designed vision starting from their individual status quo position.

Goal of this transition approach is to provide a pragmatic process and methods for achieving the desired common future vision of AV in ITS. This is based on a validated, detailed and consolidated understanding of the status quo and conducted in systematically designed and efficiently manageable process steps.

The following illustrations and descriptions are partially excerpts from the documentation of the work results from chapter D9.3 of the AVENUE project (roadmap for cost viable AV in MaaS business (Fournier, 2022)).

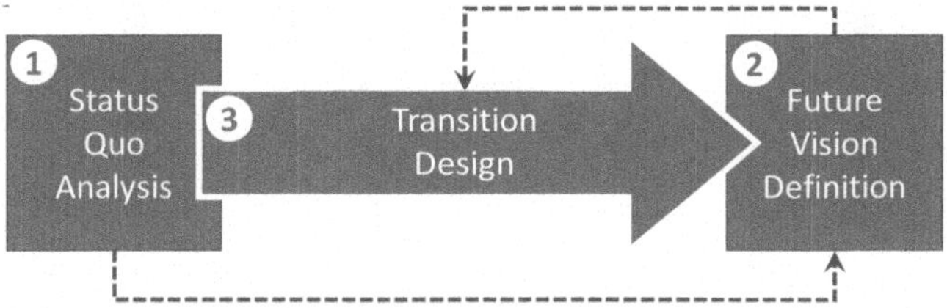


**Fig. 19.1** Generic phases of transition planning

## 19.3 Design (Big Picture) of the Transition Management Concept

The following overview represents a 'Big Picture' of all identified possible scenarios as levels (vertical dimension) and development steps (horizontal dimension) of the transition management concept from the status quo to the vision of AVs in an ITS. This is regarded as a systemic 'work package map' and a 'common thread' and thus as the systemic and orientational task basis for all partners to collaboratively contribute to this concept with defined deliverables (see Fig. 19.2). The so defined and structured work packages (in total regarded as a 'work breakdown structure') can be put in a sequence (as prioritised actions) for systematic elaborations.

Regarding the status quo analysis in *generic phase 1* (see Figs. 19.1 and 19.2: *step 1*), the status quo considered has been identified, described, analysed, summarised and compared each, and cases are built for why a scenario of automated vehicles (AVs) in a Mobility-as-a-Service (MaaS)/Intelligent Transport Service (ITS) environment is preferable to a pure laissez-faire strategy resulting in the development of robotaxis for individual transport. A laissez-faire strategy means no changes in governance are planned and new players enjoy a high level of autonomy, e.g. for creating an own proprietary ecosystem which could compete with public transport or enable congestion costs. In an evolutionary logic of scenarios, both—an existing scenario 1 and scenario 2—could be a preliminary stage of scenario 3, which would be a preliminary stage of a future vision itself. These activities are collectively represented in step 1 of the detailed overview of the transition management concept (see Fig. 19.2).

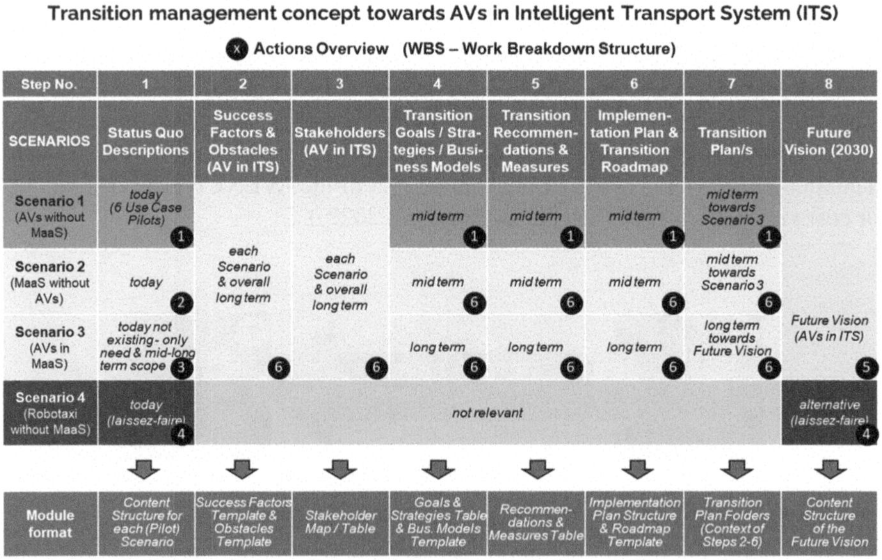


**Fig. 19.2** Transition from status quo to AVs in ITS

Focusing *generic phase 2* (see Figs. 19.1 and 19.2: *step 8*), the vision of AVENUE for a medium-term horizon (2030) of future public transportation has been discussed and defined (see details in Chap. 18). The aim hereby is to integrate automated minibuses into all citizen transportation systems available in a city. A central issue in this vision is the mobility needs of citizens which must be satisfied in an optimal way: an abundant service offer portfolio with a high variety of private and public mobility modalities should be provided and combined to one individualised intermodal trip. Automated minibuses play a central and critical role in this model (a) as a feeder for the other means of transport and (b) as a mobility gap filler for the entire transport system. Public as well as private transport operators (PTOs) are forming an enhanced public-private partnership (PPP) to utilise and synergise their multifaceted complementarity in this MaaS vision. The application of advanced self-learning systems (e.g. based on human and/or artificial intelligence) could in a further stage let this visionary MaaS concept become a self-learning ITS. This disruptive transport system innovation could create best human centric transportation, optimised and balanced private and public value and raised acceptance by citizens. This task of future vision generation is represented in step 8 of the detailed overview of the transition management concept (see Fig. 19.2).

In *generic phase 3* (see Figs. 19.1 and 19.2: *steps 2–7*), the essential building blocks for the systematic development of a comprehensive transition plan are represented. The elaboration of each building block has been conducted on every scenario level of the three identified scenarios (scenarios 1–3) for AV to ITS which avoids a not desired laissez-faire situation (scenario 4).

- The focus of *step 2* is to identify the main critical success factors as well as the main obstacles of the transition of AV in ITS for each scenario (mid- or long-term relevance), from their status quos to the common future vision. This knowledge is of high importance for putting special attention on these factors at further elaborations of the transition concept. These factors include all facets of the comprehensive (business, technical, social, environmental and governance related) transition concept and thus also the main aspects of a stakeholder-focused change management.
- *Step 3* aims to identify and characterise all stakeholders relevant for the transition towards the future vision of AV in ITS represented by a variety of structural designs of stakeholder maps from different perspectives (e.g. promoters, opponents, catalysts, multipliers) as well as characterisation tables.
- In *step 4* the goals and strategies as well as business models of AV in ITS are defined for the transition of each scenario (mid- or long-term relevance), aiming at the achievement of the common future vision. These are based on the success factors and obstacles previously identified in step 2.
- *Step 5* focuses on the identification or derivation of strategic and operational recommendations and their translation or refinement into concrete measures for action of AV in ITS for the transition of each scenario (mid- or long-term relevance).

- The task of *step 6* is to design an implementation plan, integrating the previously elaborated operational strategies and measures in steps 4 and 5, as well as their prioritisation and timely assignment as a transition roadmap for each scenario (mid- or long-term relevance).
- *Step 7* is regarding the compilation of all results of steps 2–6 towards a comprehensive, systemic concept of a transition plan(s) of each scenario (mid- or long-term relevance). As already mentioned in the evolutionary scenario logic in phase 1, the transition plans of scenario1 as well as of scenario 2 are oriented to the future vision of AV in ITS, but pragmatically aiming to prepare and achieve scenario 3 as mid-term transition plans. Scenario 3 as preliminary stage of AV in ITS as the future vision is adapting these previous stages focusing on the achievement of AV in ITS as a long-term transition plan towards the future vision.

## 19.4 Description of the Scenarios, the Status Quo (Step 1) and the Future Vision (Step 8)

By 2050, almost 68% of the world's population is expected to live in cities, indicating that urban areas will continue to experience rapid growth (Khor et al., 2022). Therefore, it is essential for metropolitan areas in particular, but also for connecting these areas with rural regions, to develop and use new forms of mobility as a complement to existing transport systems. One recent example for this is the rentable e-scooters that can now be used in most cities.

Conventional transport systems are made up of a mix of public and private transport. Local public transport is a vital economic and location factor, particularly in conurbations, but also in numerous medium-sized and small towns. From an environmental perspective, there is a strong interest in expanding and enhancing local public transport by decreasing the traffic volume from private vehicles while at the same time lowering environmental pollution through improved journey timing and further expansion of local public transport. It also offers a wide range of transportation options in the urban environment. Except for some local special forms such as cable cars, the commonly used forms of public transport throughout the world are buses, trams, metros and trains.

However, with a share of up to 70%, private cars are used far more frequently. In Germany, for instance, the share of total mobility in 2017 amounted to 43% as a driver and 14% as a passenger, which shows that 57% of approximately 3.2 billion total passenger kilometres per day were travelled by car (Federal Ministry for Digital and Transport 2019). Additionally, there exist private-sector mobility systems, like taxi and ride-hailing, regional and suburban rail, car and bicycle sharing, or the previously mentioned e-scooters.

Within the AVENUE project, the boundaries of automated driving representing different nested levels of mobility action fields are divided as shown in the diagram below (see Fig. 19.3). The first and most condensed level includes only the vehicle and the related minibus ecosystem (electric, automated, connected and

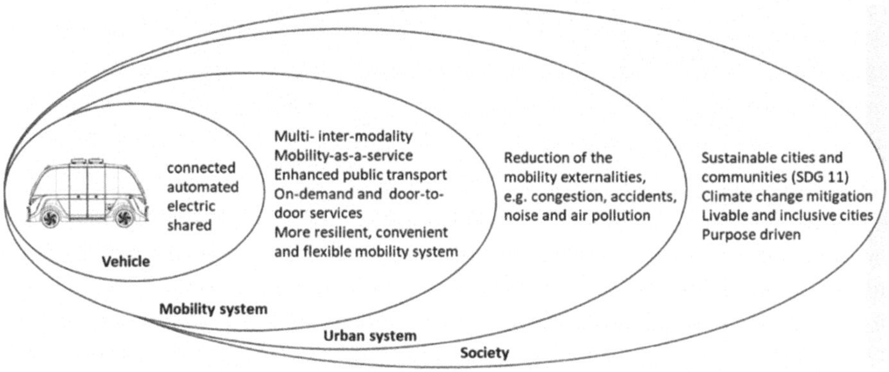


**Fig. 19.3** System boundaries (Nemoto, 2022)

rideshared). The mobility system level includes additionally the integration in the MaaS system to provide a seamless journey and a more resilient, convenient and flexible on-demand and door-to-door public transport system. The urban mobility system level builds up on an even higher level, including the beforementioned vehicle and mobility system levels. Here the effects of AV in MaaS are shown by a reduction of negative externalities as indicated in Chap. 14. The last and most inclusive level is the society. On this level, SDG 11, 'Sustainable cities and communities', is finally measurable. This shows the impact of AV in MaaS on cities, making them livable and more inclusive and actively working towards climate change mitigation.

### *19.4.1 Scenario 1 (at Generic Phase 1 or Step 1): AVs Without MaaS*

The status quo of AVs, especially automated minibuses within the project, is explained in detail in previous chapters of this book (Chaps. 2, 3, and 11), regarding the trial use cases, international pilots, in-vehicle services (Chap. 5) and out-of-vehicle services (application or panel for information services, ticketing etc.), integration in public transport and much more. The main objective of the whole AVENUE project was to test and present this status quo and to identify what can be possible in future mobility scenarios. That's why the description of different topics and details continues naturally throughout all deliverables of the project, e.g. 'vehicle-to-platform interfaces' or the reports from the pilot sites. The main part, however, has been regarded in the assessments for social impact, environmental impact and especially economic impact. The economic impact describes how the direct and indirect costs of the status quo are distributed. However, a meaningful summary is not feasible in a short form that makes sense, which is why the reader is referred to the economic impact chapters.

### *19.4.2 Scenario 2 (at Generic Phase 1 or Step 1): MaaS Without AVs*

Considering the current state of transport options, it can be observed that mobility in cities has been transformed by mobility services such as carsharing, ridesharing and on-demand services. These provide flexibility, convenience and customisation while helping to reduce the number of private vehicles and increase the use of mobility service providers. Hence, a multimodal transport infrastructure, which refers to the parallel use of different means of transport, is in place already. The emphasis is on making the serial use of different modes of transport on one route efficient and attractive. It is already a possibility right now to use different means of transport in every city, but it is in many cases not very intuitive to change and switch between them. Because of changes in travel behaviour and customer demand, new players are entering the market besides the traditional transport providers. This presents new challenges for public infrastructure and public transport services regarding planning, design, operation, regulation and financing. To enable citizens to use these different modes of transport and mobility offers in combination, and thus providing them with optimised route planning tailored to their needs, the MaaS concept is being adopted in a growing number of cities. This concept provides users with access to various means of transport with different mobility offers, which can be chosen, booked and paid for using an app. MaaS marks a shift away from a transport model where individuals travel by private vehicle to a model where individuals can choose between different travel service providers to create the optimal seamless itinerary.

In consequence, mobility is no longer viewed as a 'commodity', since owning a vehicle is no longer required to move around, but as a service where the journey is purchased. It is believed that the travel behaviour of citizens is shifting, and as a result the use of private vehicles could decrease. MaaS is made up of two components:

1. The mobility modes and mobility service providers like bike-sharing, scooter-sharing, ride-hailing, micro-transit and carsharing in addition to the existing transport modes (tramway, bus, train etc.).
2. A mobility platform where all mobility forms and mobility service providers are integrated, and the user can map out his routes and obtain the travel costs to then book the route and pay the mobility providers. This mobility platform will be provided to the user as an app.

Given the global trend of urbanisation, MaaS is projected to grow at a gigantic rate (Araghi et al., 2020). In certain cities, the first MaaS are already in operation, including companies and start-ups with promising concepts, such as the 'Whim' app (Whim online) or 'Trafi' platform (Trafi online), both with millions of users worldwide. In addition to examples like these, there are now also normal public transport providers such as Ruter AS in Norway, who already offer a functioning MaaS in Oslo and beyond, although without AVs so far (Ruter, 2019).

### 19.4.3 Scenario 3 (at Generic Phase 1 or Step 1): AVs in MaaS

To our knowledge, there is currently no MaaS that includes AVs. Even in AVENUE, this is only a vision for the year 2030, but it is far from being a reality. As mentioned above, a variety of MaaS environments exist worldwide in which AVs can be included. For this to happen, they must first be recognised and accepted as one of the reasonable building blocks of modern mobility. Therefore, the AVENUE vision for a future mobility of a MaaS with AVs is presented in detail in Chap. 18.

### 19.4.4 Scenario 4 (at Generic Phase 1 or Step 1): Robotaxis Without MaaS

This section details the status quo of development and testing phases of existing robotaxi players. A robotaxi is an automated taxi that can be hired by up to two passengers for individual trips.

As the population will continue to move to urban areas in the future, mobility in these areas will also undergo changes. The development and use of automated vehicles through robotaxis appears to be a realistic future scenario. A study on 'Urban mobility and autonomous driving in 2035' by Deloitte identifies five possible developments resulting from the use of automated driving services. Firstly, automated driving services could become the primary mode of transportation, with one in three journeys made by citizens in urban regions being made by automated driving services. In addition, the use of automated vehicles could result in a price war since they could be up to 25% cheaper compared to current public transport and private vehicles (Deloitte, 2019). Furthermore, the market potential of automated vehicles is very considerable. The Deloitte study estimates a sales volume of up to 16.7 billion euros per year. Nonetheless, this is also dependent on the business model of the providers and on future regulations. Still, it is expected that the use of automated vehicles will decrease the number of private vehicles in cities, but that the use of driving services will increase traffic volumes in total. In consequence, up to 40% more vehicles can be on the road simultaneously during peak times. With higher traffic volume, the risk of congestion rises, and the average speed in cities declines (Deloitte, 2019).

The forecasts of this and other studies reveal a large market potential for established mobility companies as well as for newly created start-ups globally. The most important, or at least most recognised, companies with respect to the development of automated vehicles are Cruise, Waymo, EasyMile and Mobileye.

### 19.4.5 The Future Vision 2030 (at Generic Phase 2 or Step 8): Automated Vehicles (AV) Within MaaS and ITS

The vision of the AVENUE project for future public transportation in an urban environment integrates both personal transportation and public transportation in mass transit. This vision of future mobility is outlined using the citizen-centred approach.

In this vision, the citizens' mobility needs are depicted with the possibility of using an automated minibus combined with other modes of transport, depending on the citizens' preferences. In this regard, the AVs enable the first and last mile travel serving as mobility gap fillers for seamless transport. Furthermore, AVs are intended to compensate in the event of private and public transport failures (disruptions, accidents, etc.). This is expected to reduce the negative externalities and thereby make public transport more attractive for citizens (see details in Chap. 18 and Fournier et al., 2023). This includes automated vehicles in addition to the established means of transportation like bus, train, cab, carsharing, bicycle and others. This necessitates factors such as the interoperability of hardware and software devices providing standardised interfaces (APIs), coordination and management software, as well as management services provided by service aggregators and other intermediaries (see details in Chap. 18 and Fournier et al., 2023).

## 19.5 Description of the Transition Steps from the Status Quo to the Future Vision

Following the logic of the Big Picture of the transition management concept (see Fig. 19.2) representing the systematic development of a comprehensive transition plan conducted for each of the three identified scenarios (scenarios 1–3) for AV to ITS, the essential building blocks for a transition (*generic phase 3 or steps 2–7*) are represented in this section and illustrated by typical examples.

### *19.5.1 Identification of Success Factors and Obstacles (Step 2)*

Deploying automated minibuses in all environments, both in city and rural areas, requires the technology to be able to handle all safety-related scenarios as well as to be effective, smart, connected and fully implemented within an ITS infrastructure. The *technical obstacles and success factors* are, e.g. related to the automated driving ecosystem, automated technologies, open data, open platforms and open API. More detailed elaborations on technical success factors and obstacles can be found in Chaps. 11 and 19 (limitations) of this book.

One key issue for AVs is the level 4 of automation (see Chap. 11) where safety drivers are not necessary any more in the automated vehicle to enable savings in salaries. This is relevant for *the economic performance* of the PTO. Business with automated minibuses is not profitable yet due to high driving-related personnel costs (see Chap. 12). The non-integration of automated minibuses with other means of transport within a MaaS would further reduce the attractivity for passengers. Another obstacle is the comparatively slow digital transformation of companies in many EU countries. Further details and issues on economic success factors are elaborated in Chaps. 12 and 19 (limitations) of this book.

Regarding the automated driving technology and connectivity-related demand, the predictive, adaptive and information sharing through vehicle communication with infrastructure and other vehicles improves driving performance and energy consumption. On the other hand, a highly connected vehicle means more data processing within and outside the vehicle, which may outweigh the sustainability of interconnected vehicles. If the automated minibuses are well utilised in terms of mileage and regularly used by passengers, they will present great advantages over individual vehicles from an environmental point of view. The suggested automated minibuses in a MaaS/ITS on contrary wants to satisfy the best citizen needs and the general interest in *leveraging positive externalities and lowering negative externalities*. Enabling positive externalities are possible due to the use of data, intermodality and interoperability. Further environmental issues are discussed in Chaps. 13, 14 and 19 (limitations) of this book.

An important *social success factor* is the high level of goodwill among potential users and a high level of satisfaction among users that can be translated into a high level of willingness to use the automated minibus (again). Most important factors for the social acceptance are the (perceived) need for improvement of the current situation and whether the proposed alternative service fulfils this need for improvement. Some critical points however encompass the highly sensitive issues of handling user data and data security and the ethical questions that constantly accompany automated driving. If automated minibuses could be a real 'game changer' making public and private transports individual and providing a real alternative to individual privately owned vehicles, it has the potential to increase effectiveness and flexibility for the users and increase the choice for the passenger, following general interest at the same time. Another alternative in the sense of an obstacle for automated minibuses in MaaS/ITS could be robotaxi which could satisfy best individual mobility needs without changing the means of transport but at the same time lower and privatise positive externalities and increase negative externalities through additional traffic and space. This is not serving general interest. A more detailed discussion on social issues is conducted in Chaps. 15 and 19 (limitations) of this book.

Like raised above, the beforementioned limitations are all influenced by the chosen governance. The governance for automated minibus and its ecosystem (especially regulations and standardisation for data, data security and privacy, technical interoperability, coopetition [cooperate to compete] of stakeholders, licensing on regional and national levels) is thus a central issue to enable automated minibuses' technical ecosystem and the integration of automated minibuses in MaaS or ITS. Coopetition means when competition coexists with cooperation to use complementary resources cooperatively. In this sense the governance is crucial to balance the interests of stakeholders of an ecosystem (Fournier et al., 2023). However, there's currently a lack of standards for open APIs and unaligned technical standards existing that might trigger coordination issues. More issues and details on this governance topic are discussed in Chaps. 16 and 19 (limitations) of this book.

### *19.5.2 Identification and Characterisation of Stakeholders (Step 3)*

For the transition of the scenarios from status quo to the future vision, it is precondition to identify, characterise and analyse the stakeholders and especially the key actors involved in testing and deploying automated minibuses for public transport within MaaS and ITS in European cities. The task hereby is to construct a strategic overview of the expectations, needs and impacts of the stakeholders and the connections between them. For a general overview of the stakeholders for the transition from status quo states to achieve the defined future vision, their characterisation and analysis, we refer to the general elaborations in the specific chapter (see Fig. 19.1 and Chap. 9: Final stakeholder Analysis and Stakeholder Strategies) within this book, comprising not only stakeholders of the status quo states and the future vision but also stakeholders which are relevant for the focused transition process.

The following discussion additionally focuses the characterisations of the main selected stakeholders of scenario 3 (AV in MaaS, see Fig. 19.2) due to the special relevance for the transition towards the future vision (AV in MaaS/ITS):

The *stakeholder group of manufacturers and software developers* is focused on developing and advancing the required technologies for the automated minibuses following the detailed work from Chap. 9 (Nemoto et al., 2021).

This is because *software providers* are directly involved in both the MaaS system and AVs, providing platforms that permit the operation and optimisation of automated mobility services, overseeing both scheduled and on-demand services. These cloud-based platforms act as an intermediary between the mobility providers and customers; therefore, they have a major influence on the overall project. In addition, it is very important to develop a trustworthy system where users take advantage of the rides in automated vehicles to pursue other activities without having to focus on the road. This generates added value for the user, which is what the project seeks to accomplish. Nevertheless, a close collaboration with other stakeholders, the public transport operators (*PTOs*), is also needed. Since the PTOs must operate these new services already during the transitional phase, a high level of involvement is required from them, particularly at the beginning. The PTOs are commissioned by the municipality, making them one of the stakeholders with a great amount of influence on the project. This is the case because the municipality is responsible not only for commissioning the PTOs but also for the requirements for the public transport system and for the requirements for the registration of the automated minibuses. They are equally charged with urban planning. Urban planners concentrate on decreasing urban space for private transport and preparing infrastructure for transformation by assisting the PTOs. In this way, these stakeholders define the rules of the game for the project launch (Nemoto et al., 2021). However, since the *city administration* is also subject to the *national government* and the *European Commission*, it is mainly the European Commission and the national government that shape the scope of action for the city government and consequently also for the PTOs (Nemoto et al., 2021). The European Commission can impose regulations and thereby support or obstruct the development of the technology. National governments can step in and impact the project through regulations, incentives and rules (Nemoto et al., 2021).

Besides the actors of software development, the government and the city administration, the *manufacturers* of the automated minibuses are crucial stakeholders which have a great effect on the implementation of the project. As they are in charge of developing and offering mobility solutions, they have to build trust and acceptance for their vehicles among consumers, and they must secure a good market position to establish their own company in the market. To achieve these three key issues, they put their focus on the quick introduction of products into the public system, where well-designed standards can reduce the total cost per vehicle while increasing the speed of development (Nemoto et al., 2021). The *drivers' union* will be involved in the implementation of the project by engaging in driver retraining and education, optimised road conditions, safety and arrangements with governments and employers' organisations (Nemoto et al., 2021). However, drivers' unions and environmental *non-governmental organisations* (NGOs) have up to now taken a negative position towards automated minibuses, and to lessen the negative impacts on the environment and society, these stakeholders should be implicated in decision-making and discussions in a more purposeful way (Nemoto et al., 2021). The final critical stakeholders are the *end users* of the system. The project will only succeed if it is accepted by the users. To ensure that users alter their mobility behaviour in the future, the automated vehicles must afford the flexibility and convenience of customised mobility solutions (Nemoto et al., 2021).

Given that there are several more stakeholders, the following stakeholder groups with specific relevance in the context of transition are still enumerated for completeness: insurance companies, electricity charging infrastructure providers, energy providers, research institutes, financial services, recycling industry, emergency aids, industry lobbies, consulting companies and others.

### 19.5.3 Definition of Transition Goals, Strategies and Business Models (Step 4)

#### 19.5.3.1 Transition Goals

The definition of transition goals depends on the individual status quo of the mobility ecosystem which determines the setting of the goals from this starting point to the target (next scenario level or future vision). As shown in an evolutionary logic of scenarios (see Fig. 19.2 of this chapter), an existing scenario 1 could be a preliminary stage of scenario 3, and an existing scenario 2 could be also a preliminary stage of scenario 3 which would be a preliminary stage of a future vision itself. Against the background of such a viewpoint, the identification of transition goals to the next scenario level or even future vision level requires a very individual analysis. For this reason, only a few broad categories of transition goals with some typical examples will be mentioned here:

**Technological Implementation of Vehicles and Infrastructure**
The possibilities for the means of transport and the AVs as well as for the AVs and the infrastructure to interact should be adjusted and increased. The achievement of level 4 vehicles and interoperability (open API) of AVs with other means of transport within a MaaS is important to leverage business benefits for the PTOs and better services for citizens.

**Software and Hardware Implementation**
The software-technical and hardware-technical point of view comprises a fully automatic order processing of the booking, considering the personal preference for one of the transition goals. A further goal is to inform the user about the next mobility alternative entirely automatically in the case of a change in scheduling. The data should be exchanged between the mobility providers and the means of transport as well as the administration between different AV manufacturers without any problems.

**Customer-Oriented Needs**
This should ensure a customer-optimised provision of seamless transport options without any waiting times. Furthermore, attention should be paid to take into account the transport needs of customers and to automatically offer a satisfactory solution in case of unforeseen events, like accidents or breakdown of transport means. The optimised transport system should reduce negative externalities.

#### 19.5.3.2 Transition Strategies

Regarding the three scenarios of the transition management concept (see Fig. 19.2), the largest and most important innovation leap within the transition phase is made by the implementation of scenario 3 (AVs in MaaS—business ecosystem) especially in urban areas, where transition strategies play a central and crucial role. This is the mandatory precondition for creating the final 'intelligence'-based innovation leap towards the future vision (AVs in ITS). In the subsequent section, the potential strategies for this scenario are described.

In the first possible transition strategy, *Collaboration of an Automated Minibus Provider with a Public MaaS Business Ecosystem*, the hypothesis is that the technology, solution and transition strategies of automated minibuses for public transport must be consistently aligned and integrated with the technology, solution and transition strategies of the public MaaS provider in all aspects of the service portfolio and business model modules. Therefore, the automated minibus strategies must be consequently specified and adapted with the public MaaS providers to ensure that this is a long-term perspective. In conclusion, an alignment with the strategy of the public MaaS business ecosystem orchestrator can be very beneficial in the event of success but can also entail a high risk in case of failure (Antonialli et al., 2021).

The second possible transition strategy describes the *Collaboration of an Automated Minibus Provider with a Private MaaS Business Ecosystem.* For this strategy, the hypothesis is that the technology, solution and transition strategies of automated minibuses must be systematically aligned and integrated with the technology, solution and transition strategies of the private MaaS provider in all aspects of the service portfolio and the business model modules. This illustrates that the solution strategies are consequently specified and adapted to those of the private MaaS provider to ensure that this is a long-term perspective. In summary, aligning the strategy of the private MaaS business ecosystem orchestrator can be very beneficial in case of success, but can also pose a high risk in case of failure (Antonialli et al., 2021).

The third possible transition strategy that is possible is the positioning of an automated minibus provider against other competitor automated minibus providers which means a *Competition within a Public or Private MaaS Business Ecosystem*. The hypothesis here asserts that, 'enhanced automated minibus strategies for USP and technical/business innovations as well as dedicated adaption to the strategies of an Public or Private MaaS Integrator in every facet of the offering portfolio and business model modules are necessary' (Antonialli et al., 2021). This transition strategy concentrates primarily on the USPs in the domains of social and accessibility or in the fields of safety and security as well as on flexible and collaboration-based innovation strategies. Therefore, a strong focus on USP over the relevant competitors within the private and public MaaS business ecosystems as well as a focus on integrators' success factors and an emphasis on the innovation strategy are feasible (Antonialli et al., 2021).

Finally, the fourth possible transition strategy is the development of an own new business ecosystem and thus fosters the *Competition of the Own New (Private) MaaS Business Ecosystem with other (Private and Public) MaaS Business Ecosystems*. In general, the hypothesis for this strategy is that an automated minibus provider which is currently collaborating with other automated minibus providers within the public or private MaaS sub-business ecosystem has sufficient technology and business potential and thus seeks to build his own MaaS business ecosystem. This strategy demonstrates that automated minibus strategies are an essential core for the development of an own MaaS business ecosystem as an integrating network for automated minibuses and other modalities and offerings, like infrastructure, along with innovation strategies to attain USPs. Therefore, this strategy puts emphasis on technology and business innovation to build a designated MaaS business ecosystem and simultaneously a competitive USP strategy (Antonialli et al., 2021).

All five transition strategies aim to realise the desired future scenario AV in MaaS (scenario 3) and represent alternative ways to achieve this.

#### 19.5.3.3 Transition Business Models

Transition business models can be regarded as the conceptional refinement and representation of the business implementation logic of the defined transition strategies which is necessary for implementing the focused scenario 3 (AVs in MaaS). This is precondition and the basis for additional 'intelligence'-based business model updates to achieve the future vision (AVs in ITS).

The following business models for the implementation of AVs in MaaS must be regarded.

In the general description of potential business models for scenario 3, it is assumed that the most likely future will be a customer- and citizen-centric intermodal MaaS ecosystem, where private and public MaaS providers connect different transport options into a seamless travel chain supported by AVs but not consisting of AVs only (Antonialli et al., 2021). In the following sections, the different business models for the scenario AV in MaaS are described in detail.

Firstly, the *partner network* is discussed; to make a business model work, close collaborations with key partners must be established which can serve various functions. In this business model, either a public or a private mobility provider or a consortium of both acts as a MaaS integrator of partners, like service providers, other public and private mobility providers and other IT infrastructure service providers. These can join the MaaS system as contributors and solution providers for the AVs in public transport (Antonialli et al., 2021). A partner network like this therefore combines the experience of many companies, but it also requires close collaboration and trust.

*Collaboration of an Automated Minibus Provider with a Public MaaS Business Ecosystem* can be interesting, since only in public tenders the cooperating companies can get competition. This is very favourable in a niche market or a small market segment giving them a secure position (Antonialli et al., 2021). In this business model, a collaborative strategic approach should be established with mutual alignment and integration of all business modules, where every facet of the offer portfolio is listed. Besides the solution customisation, close cooperation in the delivery and customer module, including joint solution development processes and joint marketing activities, is important for the partnership and business success (Antonialli et al., 2021). The automated minibus providers can deploy their whole solution portfolio to the public MaaS sub-business ecosystem and further reinforce their partner and niche position through specialisation. As a consequence, this is a very appealing business opportunity with long-term prospects (Antonialli et al., 2021).

In the third business model, *Collaboration of an Automated Minibus Provider with a Private MaaS Business Ecosystem*, automated minibus providers establish a close relationship with the private MaaS planner by focussing on complementary offerings or complementing the offerings of other partners through their high-level performance. This business scenario highlights that automated minibus providers can successfully apply their solution portfolio to the private MaaS business ecosystems and augment their partner and niche position through specialising on complementary offerings and high performance (Antonialli et al., 2021). This guiding cooperative and synergistic strategic approach produces a mutual alignment and integration of all business model modules, in all areas of the portfolio of offerings. Beyond solution customisation, close cooperation in the delivery and customer module, including joint solution development processes and joint marketing activities, is pertinent to partnership and business success (Antonialli et al., 2021).

In this paragraph, *Competition of an Automated Minibus Provider within a Public (A) or Private (B) MaaS Business Ecosystem* is evaluated. The providers for automated minibuses can deploy their portfolio to the private and public MaaS business ecosystem but simultaneously participate in competition with other solution providers by specifying USP-focused strategies in the areas of performance leadership, cost leadership etc. or business and technical innovation (Antonialli et al., 2021). This business model is distinctive in its competitive approach to other automated minibus providers, resulting in a strong emphasis on technical/business innovation and USP generation, while tightly aligning business modules in every facet of the offering portfolio. It is marked by the competitive focus, in respect to the innovation pressure, the effort to achieve the unique selling propositions and an emphasis on integration (Antonialli et al., 2021).

In the business model *Competition of the Own New (Private) MaaS Business Ecosystem with other (Private and Public) MaaS Business Ecosystems*, automated minibus providers are in a competitive and challenging situation to preserve their technical focus and ongoing flexibility against competitors. Moreover, this entrepreneurial behaviour is defined by permanent technical and business innovation and the strive for market success and expansion (Antonialli et al., 2021). 'The analysis of this Business Scenario shows that strong innovative and especially USP-driven/competitive Autonomous Minibus transportation offerings have to be developed and provided to customers (Value/Delivery/Customer modules) and at the same time a clear alignment and integration of all modules with those of Public (A) or Private (B) MaaS Business Ecosystems are focused' (Antonialli et al., 2021). Furthermore, the entire MaaS business ecosystem must concentrate on supply, technology, business innovations and USPs that exist in other MaaS. In conclusion, in this business model, the business focus, the competitive focus on new innovations or USPs and the focus on other MaaS must occur at the same time (Antonialli et al., 2021).

For a structured representation of a transition business model for scenario 3 (AVs in MaaS), an adaption of the canvas method for business models[1] shown in Fig. 19.4 as an example is useful for understanding the logic of the business model and deriving refined strategic recommendations for every module of the canvas:

### 19.5.4 Transition Recommendations and Measures (Step 5)

All of the subsequent recommendations relate to the implementation of AV in MaaS/ITS (scenario 3) and have to be operationalised within the previously identified and business models as far as possible.

For further identification and elaboration of transition recommendations and measures, it is inevitable to define these terms precisely to create a clear and common understanding. In this concept we regard that one or more transition recommendations could be related with one or more transition measures (n to m relationship):

[1] https://www.strategyzer.com/canvas

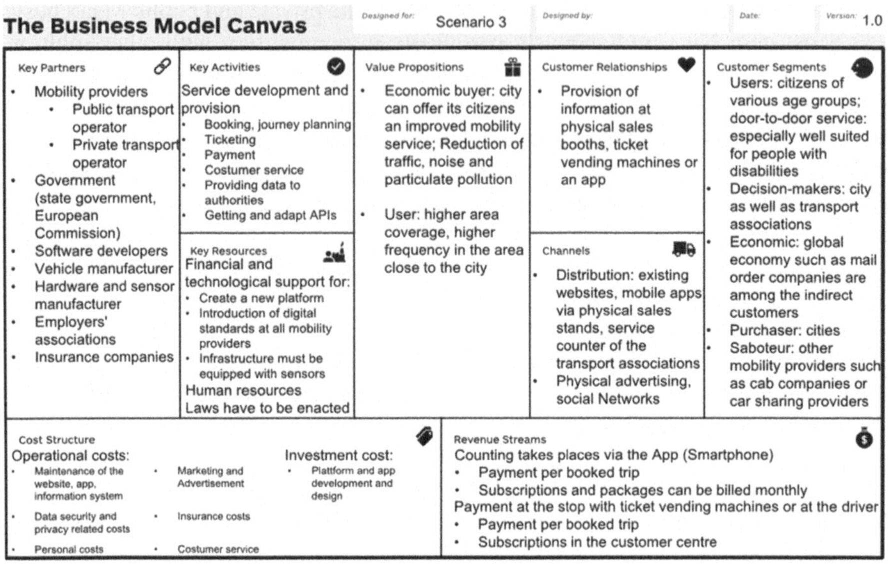


**Fig. 19.4** Exemplary transition business model representation for scenario 3 (AVs in MaaS)

*Transition recommendations* are defined as general (rough) suggestions for transition strategies from experts based on previous elaborations from transition planning modules (results, conclusions documented within the AVENUE work package deliverables) as well as strategic experiences by experts, derived from and aligned with the transition goals, transition strategies and transition business models.

*Transition measures* are operational tasks derived, refined and concretised from the transition recommendations defined above. In this context they can be defined as concretely specified and implementation focused operational tasks described by all relevant organisational attributes (unique identifier, title, description, goal, deliverable, owner, start/end time, resources, dependencies, priority etc.) necessary to start planning and implementation activities.

#### 19.5.4.1 Transition Recommendations

Aiming at a comprehensive and structured identification of transition recommendations based on the discussion of previous elaborations with experts to achieve the AVENUE Vision 2030, it is useful to create a mental framework (grid) where recommendations can be assigned to. An advantage of this framework is to specify recommendations in a rather disjunctive way or to discover new ones during application discussions with experts.

The architecture of this recommendation framework is shown in Fig. 19.5. This two-dimensional grid is compiled by four additive evolution modules of future mobility ('onion model') as vertical dimension (attention: not identical to scenarios in Fig. 19.2) and five PESTLE or SUMP categories as horizontal dimension.

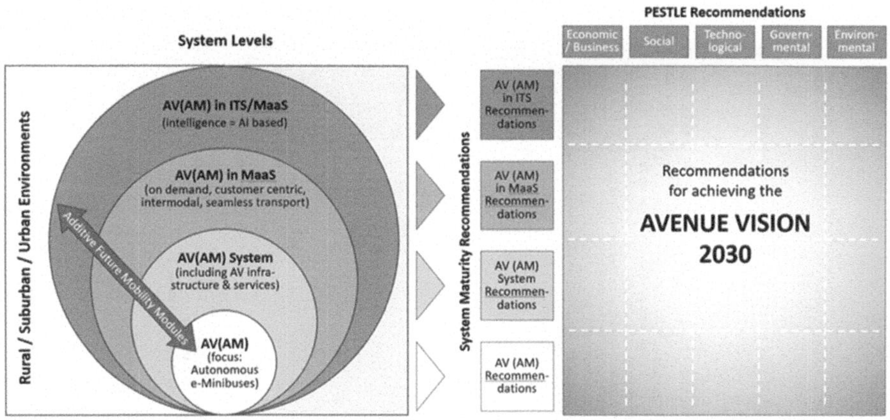


**Fig. 19.5** Framework for (transition) recommendations (Abbreviations: AV: Automated Vehicles; AM: Automated Minibuses; ITS: Intelligent Transport System; MaaS: Mobility-as-a-Service; PESTLE: Political—Economic—Social—Technological—Legal—Environmental)

After a first identification and collection of transition recommendations, they are assigned to the respective fields within the designed framework. This task often requires a split, refinement and further focusing of recommendations together with the chance to identify further valuable recommendations from gaps or white spaces within the methodical grid which is aimed to be comprehensive, well-structured and consistent.

After inserting and updating the framework grid with transition recommendations, this represents a recommendation portfolio basis for prioritisation and scheduling towards a AVENUE Transition Roadmap for achieving the AVENUE Vision 2030.

#### 19.5.4.2 Roadmap of Recommendations for Viable AV in MaaS Business

In the following we provide a summary of the main recommendations within a roadmap (from status quo to a vision for 2030) (see Fig. 19.6) to achieve viable AV in MaaS business (efficiency). Although Sect. 19.5.2 specified different transition strategies and business models within the AV in MaaS deployment area, the following recommendations include the entirety of these strategies and business models. They are designed to be generally applied to develop a viable AV in MaaS business solution and don't detail out different transition strategies.

In the deliverable at hand, this vision is analysed in detail with all the tools of a business model such as the business canvas and the SWOT analysis. However, the costs for this cannot be calculated and presented yet in detail because it is currently a vision that follows the European CCAM vision and will not become reality soon. We anticipate that the large-scale deployment of automated vehicles (AVs) will not become reality in the next 3 years, due to barriers and shortcomings in the technology, high depreciations and legal framework. Too many prerequisites must be

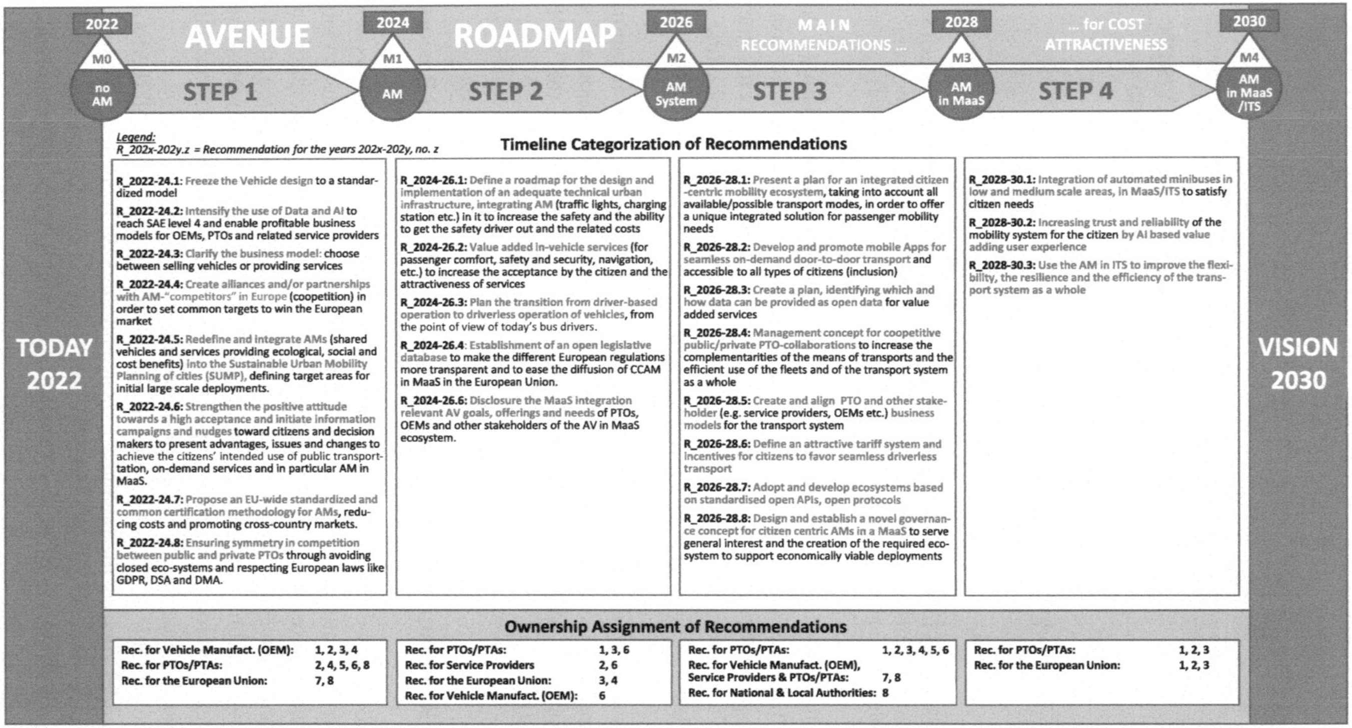


**Fig. 19.6** Roadmap (overview) of the main recommendations for viable AV in MaaS business

fulfilled. It is, for example, common sense that viable AV in MaaS business is only possible if security drivers can be reduced soon in the course of time and replaced by supervisor, but this is currently neither technically nor legally possible at every location in Europe. However, both technology and legal framework are evolving very fast, and we expect that in the next 2–3 years (and at the latest in the time horizon of our Vision 2030), the majority of the issues can be resolved when following the recommendations.

Accordingly, we will present first in summary, then in overview points, a roadmap for achieving this cost-attractive vision. We do not interpret the topic of viable AV in MaaS business in the narrow sense of balance-sheet able costs, as these are currently not yet presentable, but rather in a broader sense including all cost-related factors; we understand our objective beyond the TCO (total cost of ownership) approach not only as a contribution to the consolidation of the total cost of mobility (TCM) but also as a contribution to economic efficiency—cost coverage—and profitability.

The roadmap for viable AV in MaaS business targets all the related actors in the public transportation. Specifically, at the lowest level we have the vehicle manufacturers, who should be able to develop and market at competitive prices large numbers of vehicles, creating profit for their companies. We then have the service provision companies, developing fleet management solutions, who should be able to provide their services to the public transport operators (PTOs) and public transport agencies as commercial for-profit companies. Next, we have the PTOs/PTAs that are in general funded by public funding and have as target to provide high-quality public transportation services, with as low costs as possible for the financing government or municipal authorities, taking into account social and environmental targets defined by the public authorities. Finally, we have the citizens who as users require an affordable public transportation service with a high level of service quality. Our roadmap covers each of the above actors, with different time horizons. However, as we cannot give recommendations to citizens, a few for the EU itself are included below, in addition to the others.

**Roadmap: Step 1–2022–2024**

*R_2022–24.1 (Recommendation for OEM):*

**Freeze the vehicle design to a standardised model**

Today vehicle manufacturers are producing permanent beta-versions of the automated minibuses, with continuous changes and medications. As result each vehicle unit is a prototype almost individually constructed, which results in very high unit costs. The manufacturer should freeze the design and implantation to a standard model and proceed so that they can next advance to chain production, bring the production cost per vehicle to a fraction of today's price and open the way for a large number of orders by the PTOs/PTAs. Currently, there are neither economies of scale nor economies of scope in the market, which are urgently needed for cost-attractiveness.

*R_2022–24.2 (Recommendation for OEM/PTOs/PTAs):*

**Intensify the use of data and AI to reach SAE level 4 and enable profitable business models for OEMs, PTOs and related service providers**

The improvement of road behaviour requires solid automated recognition of different events and situations. This can only be done with the use of AI technologies. However, AI technologies require relevant data, coming from the deployment environments. Today, the vehicles produced in the European Union do not have enough data coming from European streets, thereby hindering the substitution of the safety driver by a back-office supervisor. The collection and analysis of the massively required relevant data are also hindered by the misunderstanding of the GDPR rules and restrictions by the PTOs and PTAs, who simply prohibit the mass collection and analysis of the data collected (or being able to be collected) by the deployed vehicles. In addition, the effort required by the manufactures to analyse the data is beyond their capacity, and the pavement towards profitable business models for PTOs is not realistic. Thus, from one side the public and EU authorities should provide clarifications on the GDPR overreach, reassuring the European OEMs to develop GDPR compliant AI and enable a GDPR performant AI-based ecosystem for EU CCAM mobility. GDPR compliance for AI should become mandatory for non-European competitors to ensure a symmetry in competition with home competitors.

*R_2022–24.3 (Recommendation for OEM):*

**Clarify the business model: choose between selling vehicles and providing services**

Today vehicle manufacturers have an unclear business model that is neither this of a car seller nor of a service provider. The manufacturers try to make their profit from both selling vehicles (hardware) and providing required services (from commissioning to maintenance and supervision). As a result, the overall cost to the PTO/PTA is both high as CAPEX and as OPEX making any large-scale acquisition plan extremely costly. Past experience (in the 1970s–1990s) has shown that equivalent companies (like IBM in the 1970s and 1980s) were able to dominate the market by providing a services model (software rental) and providing the hardware at cost price or even free of cost (but with long-term service and software leasing contract), whereas car companies were making their business by selling hardware only. This means vehicle manufacturers have to identify, analyse and design business strategies/opportunities and models for profitable automated minibus applications, for example, the use of existing infrastructure data to improve the behaviour of the vehicle in the street and its attractiveness.

*R_2022–24.4 (Recommendation for OEM/PTOs/PTAs):*

**Create alliances and/or partnerships with automated minibus: 'competitors' in Europe (coopetition) in order to set common targets to win the European market**

Automated minibuses are manufactured today in many countries around the world, from China to the USA, where the state (China) and private companies (USA) are investing billions per year. In Europe the relevant companies do not have the possibility to raise even a fraction of the USA and Chinese investments. It is more than urgent for the different manufacturers to align their efforts to provide solutions adapted to the European market, before the international players take over the European market offering lower prices. The fragmentation of the European market of vehicle manufacturers and the lack of communication and collaboration do not further allow the creation of a dynamic and sustainable business ecosystem in order to be competitive with non-European manufacturers by offering competitive prices. In order to assure supply chains for vehicle components and to avoid a market consolidation via bankruptcies, vehicle manufacturers should identify, analyse and design partnering strategies/opportunities with service providers and create new partner-oriented business models for a profitable automated minibus system. This could be interesting in particular for open API, open data and open protocols. The recommendations in this deliverable can be implemented independently, but it would be best to be considered together to optimise cost-effectiveness, where each manufacturer can specialise in the areas of its best expertise, thus reducing costs.

*R_2022–24.5 (Recommendation for PTOs/PTAs):*

**Redefine and integrate automated minibuses (shared vehicles and services providing ecological, social and cost benefits) into the sustainable urban mobility planning of cities (SUMP), defining target areas for initial large-scale deployments**

With a laissez-faire strategy, the development from status quo to a mobility system with AVs could end in more individual private transport and weakening of public transport. Considering that the urban public transport mobility strategy implementation can take 10 years from the moment of its definition, PTOs and PTAs must integrate in their development/adaptation urban mobility strategic planning the use and deployment of AVs, even if the technology is not yet to the expected level. With a 10 years lead time to implementation, PTOs/PTAs need to anticipate as early as possible the future deployments and initiate the required studies to evaluate costs, define possible deployment areas and define targeted service levels.

The AVENUE environmental impact assessment shows that if the automated minibuses are well utilised in terms of mileage and regularly used by passengers, they will present great advantages over individual vehicles. Therefore, it is strongly recommended to integrate automated minibuses in the strategic planning of a MaaS/ITS ecosystem to enable positive externalities (through increased citizen choice and inclusiveness, improved fleet efficiency, saving natural resources and energy etc.) and lower external costs (ecological advantage). The governance will decide if the use of automated minibuses is environmentally friendly and sustainable or not. Therefore, a system should be created in which it is standard that vehicles and services are shared, which would also improve the ratio of costs to paying customers.

*R_2022–24.6 (Recommendation for PTOs/PTAs):*

**Strengthen the positive attitude towards a high acceptance, and initiate information campaigns and nudges towards citizens and decision-makers to present advantages, issues and changes to achieve the citizens' intended use of public transportation, on-demand services and in particular automated minibuses in MaaS**

The transition from fixed line and fixed time-table public transportation towards an on-demand and door-to-door public transportation is a major paradigm change for the passengers. If this is not well explained, it could create a negative/rejection attitude of the citizens. An important finding of the AVENUE social impact assessment is that (potential) users present a positive and receptive attitude towards the automated minibuses. Therefore, there is potential to convince others through well-targeted communication campaigns, especially in social media. It is a recognised principle of economics that increasing demand, e.g. through more customers due to a higher acceptance of shared mobility with automated minibuses, could reduce the cost per customer. However, erroneous use of automated minibus-based mobility can drastically increase operational and external costs (e.g. using the automated minibuses for very short distances or as substitute to public mass transport (tramway, trains, etc.), as it is the case when automated minibuses are used like robotaxis) and lower service quality. In this case an information campaign should be started as soon as possible to inform the citizens of the benefits of the new models of transport and the related issues. Create positive incentives towards the use of automated minibuses (nudges).

*R_2022–24.7 (Recommendation for the European Union):*

**Propose an EU-wide standardised and common certification methodology for automated minibuses, reducing costs and promoting cross-country markets**

Throughout Europe each country is setting up its own certification process and rules for automated vehicles. As a result, the same vehicle needs to be homologated under different rules in each deployment country. Although steps have been already made to simplify the process, a European-wide regulation (as is the case with traditional vehicles) will strongly promote the market, reduce the acquisition costs and allow cross-country commerce of AVs and of their components like batteries.

*R_2022–24.8 (Recommendation for the European Union/PTOs/PTAs):*

**Ensuring symmetry in competition between public and private PTOs through avoiding closed ecosystems and respecting European laws like GDPR, DSA and DMA**

We recommend open platforms, open APIs and open data which respect GDPR, DSA and DMA to avoid closed private ecosystems and enable fair competition in the mobility market. To this end, the EU has to create the regulatory and governance conditions for the PTAs to orchestrate a fair and balanced competition on the local, national and EU levels (seamless interoperability) and thus create a European mobility ecosystem which serves the general interest and promotes a European mobility industry ecosystem. The EU has to make laws which anticipate automated minibuses in MaaS/ITS and anticipate the speed of legislative processes. Once closed ecosystems are created with gatekeepers as is the case in several other European industries, it will be too late to regulate the mobility market ex post due to the already created facts and dominant positions.

**Roadmap: Step 2–2024–2026**

*R_2024–26.1 (Recommendation for PTOs/PTAs):*

**Define a roadmap for the design and implementation of an adequate technical urban infrastructure, integrating automated minibuses (traffic lights, charging station etc.) in it to increase the safety and the ability to get the safety driver out and the related costs**

A roadmap for the design and implementation of an adequate technical infrastructure regarding battery charging, communication infrastructure/data exchange, vehicle-related services and maintenance needs to be defined and included in the sustainable urban development planning (SUMP), with an at least 10 years' horizon. The PTAs must be coordinated with urban planning authorities to identify how the city will develop and how the required infrastructure will be installed. This planning will be required in order to secure the required funding and authorisations. The urban planning should take into account the needs of automated minibuses for accurate geolocalisation (centimetric) and identify shadow areas where additional equipment will need to be installed (e.g. RTK (real-time kinematic) positioning, roadside sensorics on traffic lights (vehicle to infrastructure, V2I; vehicle to vehicle, V2V; or vehicle to passenger, V2P)). Furthermore, cybersecurity as well as physical security of the installations should be integrated in the planning, and backup solutions as well as disaster recovery solutions should be anticipated. Automated minibuses in MaaS or better later automated minibuses in ITS should thus integrate the infrastructure as a long-term strategy to create inherent resilience of all the transport system.

*R_2024–26.2 (Recommendation for Service Providers):*

**Value-added in-vehicle services (for passenger comfort, safety and security, navigation etc.) to increase the acceptance by the citizen and the attractiveness of services**

Based on the local needs, regulations and target service quality, a series of passenger in-vehicle services, replacing the driver offered services, must be planned before the deployment of a full-scale, driverless public transportation service. The in-vehicle services should be able to provide easily accessible alternatives to the services offered by the bus driver, preserving the passenger privacy, and enhancing the security of the passengers in the vehicle. A way for the passenger to interact with a human assistant should be available in the vehicle. The services will need to be supported by target information campaign and clearly indicated in the vehicle.

*R_2024–26.3 (Recommendation for PTOs/PTAs/European Union):*

**Plan the transition from driver-based operation to driverless operation of vehicles, from the point of view of today's bus drivers**

The suppression of the bus drivers will be seen as a threat of job losses of the current drivers. A medium- to long-term transition plan should be developed for the gradual reduction of the number of drivers and their conversion to back-office operators and intervention team operators. This plan should consider the natural departure of personnel (e.g. retirement), the lack of bus drivers and the lack of attractivity of the bus driver profession and organise the medium- to long-term re-education of the drivers to the new functions.

*R_2024–26.4 (Recommendation for the European Union):*

**Establishment of an open legislative database to make the different European regulations more transparent and to ease the diffusion of CCAM in MaaS in the European Union**

The fragmentation of European legislation in terms of deployment of CCAM vehicles is an administrative obstacle for pan-European deployment of innovative mobility services, especially for SMEs, who do not have the means to study the differences between EU countries' legislation. A legislative database at European level (similar to the existing platforms in the USA) which brings up-to-date and real-time information about the fast-growing AV legislation on local, state and regional levels is missing, and it should be in place and operational by 2026. This could simplify the planning of all stakeholders, like vehicle manufacturers, deployers of automated minibuses, transport operators, states, municipalities etc., and encourage the diffusion of automated minibus technologies and the improvement of the transport system.

*R_2024–26.5 (Recommendation for PTOs, Vehicle AV Manufacturer (OEMs), AV Service Providers):*

**Disclosure of the MaaS integration-relevant AV goals, offerings and needs of PTOs, OEMs and other stakeholders of the AV in MaaS ecosystem**

Currently minibus manufacturers (OEMs) and AV service manufacturers to realise automation of the vehicles are facing huge troubles with businesses and offerings which are not or not adequately aligned to the needs of PTOs. Thus, OEMs create multiple own AV ecosystems which require a lot of resources but cannot be integrated into a common MaaS ecosystem and into one common citizen-centric MaaS application. As a consequence, these AVs are often applied as robotaxis which are in competition with PTOs and create potentially a weakening of the PTOs and additional traffic and congestion costs particularly in cities. Open standardised API and open data are thus necessary for the AVs to be interoperable in a MaaS ecosystem and enable cooperation with other MaaS partners. Through disclosure of technical specifications and interfaces (APIs) and long-term sourcing and contract strategy from the PTOs, a long-term strategy becomes possible for all the stakeholders of the transport ecosystem of the city. In particular, the OEM strategy and accordingly the financing of the development of AV (e.g. SAE level 4, interoperability etc.) for the PTO of the MaaS can be eased. An alternative would be governmental financing as is the case in China or strong financing companies (like US-GAFAs) which are more interested in capturing value through closed (centralised, non-federated) ecosystems.

**Roadmap: Step 3–2026–2028**

*R_2026–28.1 (Recommendation for PTOs/PTAs):*

**Present a plan for an integrated citizen-centric mobility ecosystem, taking into account all available/possible transport modes, in order to offer a unique integrated solution for passenger mobility needs**

Different mobility modes are becoming available in cities, each with its own targets and different passenger needs. At the same time, many mobile phone applications are appearing, giving access to the different mobility modes. However, today these mobility solution modes represent competing mobility solutions, instead of complementary mobility solutions.

A plan should thus be created for seamless multimodal forms of transport, where automated minibuses play a central role as mobility gap filler, enabling spatial (e.g. area instead of line in particular in time of low demand), temporal (e.g. on-demand) and functional flexibility (door-to-public transportation network, part of the public transport network, door-to-door) similarly attractive to a private car.

*R_2026–28.2 (Recommendation for PTOs/PTAs):*

**Develop and promote mobile apps for seamless on-demand door-to-door transport and accessible to all types of citizens (inclusion)**

Before the deployment of automated minibuses, aggregator apps for seamless and efficient on-demand door-to-door journeys for citizens should be developed and extensively tested, especially for passengers with special needs. All mobility modalities need to be integrated in a single app, increasing thus the mobility offer. The offer could in particular be adapted to all passengers depending on the persona (mobility for work or for leisure), weather (a passenger would probably prefer the automated minibuses in case of rain instead of bicycle) etc. and of the PRM needs and capabilities. In order to provide an integrated citizen-centric mobility ecosystem, the passenger capabilities should be integrated in the app service (e.g. we cannot propose to an 85-year-old person to complete his trip using a bicycle!) which will suggest and reserve the most adapted mobility solution.

*R_2026–28.3 (Recommendation for PTOs/PTAs):*

**Create a plan, identifying which and how data can be provided as open data for value-added services**

The operation of automated minibuses in public transportation will generate large quantities of data which until now were very difficult to collect—from exact passenger trips, to road conditions (congestion, speed etc.), to passenger incidents and vehicle status. These data, if exploited correctly, will allow, e.g. service quality improvement, trip optimisation, energy consumption reduction but also optimisation of automated minibus fleets, infrastructure, city planning etc. However, in order to promote the creation of new services and new mobility models, the data should become available to third parties. A plan must be ready, defining which data can be available (preserving any possible sensible passenger privacy information) and how third parties will be able to access and use them. Detailed specifications and formats should be easily available and published so that the interested parties can start the development of analytical tools and services. This has to be regulated by the ecosystem provided by the PTA in the privacy and security concept.

*R_2026–28.4 (Recommendation for PTOs/PTAs):*

**Management concept for coopetitive public/private PTO collaborations to increase the complementarities of the means of transports and the efficient use of the fleets and of the transport system as a whole**

In the long run, innovations and new mobility services will be developed from different private companies. These innovations will need to be integrated to the overall mobility offer to enable increase in efficiencies and allow a citizen-centric approach. PTAs will thus need to design a management concept for a balanced ecosystem for public/private PTO collaborations to increase the complementarities of the means of transports and the efficient use of the fleets through the use of data and of the transport system as a whole.

*R_2026–28.5 (Recommendation for PTOs/PTAs):*

**Create and align PTO and other stakeholder (e.g. service providers, OEMs etc.) business models for the transport system**

Providing high-quality, low-cost, on-demand, door-to-door public transportation services will eventually have a major impact on the mobility business in the urban environment. Taxi services, private transport services and even delivery services will be highly impacted. The PTAs should clarify the perimeter of the services that will be provided by the automated minibuses, so that they would impact the existing business models of other mobility stakeholders. For example, in order to avoid competition with taxi services (including robotaxis), door-to-door services can be limited to a certain distance or time while imposing no limits to door-to-hub automated minibus-based transport.

*R_2026–28.6 (Recommendation for PTOs/PTAs):*

**Define an attractive tariff system and incentives for citizens to favour seamless driverless transport**

One of the key goals of the deployment of automated minibuses in door-to-door, on-demand urban mobility is to provide citizens an efficient mobility service and 'pull' them to abandon the use of a private car without coercive 'push' policy but with an attractive mobility offer (the so-called pull strategy). However, one of the key elements to this transition is the tariffing model. PTAs and PTOs should therefore create an attractive tariff system together and provide incentives for citizens to utilise the multimodal MaaS system instead of utilising privately owned cars or robotaxis. The proposed tariffs could, e.g. be compatible with the overall cost for using a private car, including parking fees, infrastructure costs, energy consumption, urban toll etc.

*R_2026–28.7 (Recommendation for OEM/Service Provider/PTOs/PTAs):*

**Adopt and develop ecosystems based on standardised open APIs and open protocols**

A key element in the development of interoperability of all the means of transport, their platforms and apps, but also of a competitive market, is the use of standardised interfaces and protocols that open the doors to innovation, marked competition, increased quality and price reductions. In order to avoid a winner-takes-it-all situation, where a major non-European company dominates the market, European manufacturers (OEMs) should propose, national authorities should adopt and PTOs, app providers etc. should use open standards for APIs and protocols.

*R_2026–28.8 (Recommendation for National and Local Authorities):*

**Design and establish a novel governance concept for citizen-centric automated minibuses in a MaaS to serve general interest and the creation of the required ecosystem to support economically viable deployments**

The deployment of CCAM will require the creation of a local, regional or national ecosystem, where key expertise can be found. This will require a dual action from the authorities, in both the legislative and educational domains.

On the level of city or national/regional governments, it will be necessary to design, implement and supervise individual technical certification and licensing concepts for automated minibuses and their integration into MaaS according to respective regulations and selected standards. However, the required expertise, be it technical or regulatory, must be available in the local ecosystem. For this, relevant authorities will need to create technical trainings and integrate at different levels of education the required courses and programmes to create the locally needed expertise.

This implementation will enable economic efficiency and, above all, social and ecological benefits that are only possible through sensitive mobility governance. To do so, a balance between individual and general interest must be found for the stakeholders on the city level.

**Roadmap: Step 4–2028–2030**

*R_2028–30.1 (Recommendation for PTOs/PTAs/European Union):*

**Integration of automated minibuses in low- and medium-scale areas and in MaaS/ITS to satisfy citizen needs**

By 2030, and based on the strategy that was developed earlier, the first medium-scale deployments of AVs should become available in selected targeted areas with low and medium mobility demand, fully integrated to the MaaS services. The deployment should address the needs in rural and suburban areas where public transport is weak and enable flexibility which is nearly as convenient but much cheaper than a private car. The target being the integration of automated minibuses to other means of transport within a MaaS (through an app also adapted for persons with special needs) would increase the attractivity for passengers in terms of time, space, function and usability. Beyond this inclusive approach, starting in rural and suburban areas would further have the advantage to raise the level of education with a lower technical complexity.

*R_2028–30.2 (Recommendation for PTOs/PTAs/European Union):*

**Increasing trust and reliability of the mobility system for the citizen by AI-based value-adding user experience**

As most of the offered mobility services will be eventually based and/or make use of a AI-based real-time self-optimisation, leading to a better information situation and permanently optimised journey for citizens, it is of major importance for the PTA/PTOs to have established, first, a strategy to reply to the questions of the citizens regarding data usage and how decisions are taken and, second, a 'disaster recovery' strategy to reply to incidents and failures of the services, which will eventually increase the positive user experience and trust into the mobility in ITS ecosystems.

*R_2028–30.3 (Recommendation for PTOs/PTAs/European Union):*

**Use the automated minibus in ITS to improve the flexibility, the resilience and the efficiency of the transport system as a whole**

By the application of AI-based real-time self-optimisation of the entire automated minibus in MaaS ecosystem in all facets, the risk of technical failures and downtimes can be significantly reduced, which increases the cost-efficiency at the same time. This increases the attractivity for EU towards a faster market penetration of the automated minibus in MaaS/ITS concept.

#### 19.5.4.3 Transition Measures

When focusing the most challenging transitions towards the Scenario 3: AVs in MaaS and the vision: AVs in ITS, numerous operational measures can be derived from the previously defined transformation recommendations with specific regard to the individual situation of the application field.

As a suggested subset of the any size portfolio of transformation measures, some typical and exemplary implementation measures have been selected, also regarding the fact that there is an n to m relationship between transition recommendations and transition measures:

- *Technological implementation of vehicles and infrastructure*
- By using AI, the insights derived from the journeys can help to optimise the infrastructure. Furthermore, AI can be employed to monitor vehicle utilisation

and subsequently contribute to the timing of the means of transport, which in turn can result in fewer empty journeys. To guarantee optimal communication between the actors in the transport system, it is vital to expand the data network of the infrastructure, vehicles and other actors, with 4G and 5G. This data network is equally required to enable comprehensive function, communication and interaction of the sensors in the vehicles and the infrastructure with the AVs.

- *Software and hardware implementation*
- By integrating AI in the MaaS system, vehicle and user data can be analysed and evaluated. This allows for the recognition of mobility patterns and, as a result, the optimisation of vehicle utilisation and trip adaptation. It can even be used to modify and optimise the infrastructure to the users' mobility needs. This translates to a reduction in journeys without passengers. With the assistance of AI in the MaaS system, route information, such as accidents, disruptions and route closures, can be communicated to the MaaS platform in real time and processed.
- *Regulatory and legal conditions*
- To implement the project successfully and set uniform targets, contractual coordination between all public and private mobility providers and the municipalities is needed. In addition, the adopted laws must be extendable to be applied to new technologies and developments.
- *Client-oriented needs*

- Using AI, the various transport options can be coordinated with each other, leading to schedule optimisation and thus to a reduction in waiting times, enabling seamless transport for the user. On top of that, the additional development of individual mobility in the form of AV on-demand buses promotes the decrease of private vehicle ownership.

### *19.5.5 Design of an Implementation Plan and a Transition Roadmap (Step 6)*

The complete portfolio of transition measures is regarded as the basis for a comprehensive categorisation/structuring, pragmatic characterisation and prioritisation/scheduling in the subsequent step, regarding manifold interdependencies among the defined measures.

- The results of these activities can be displayed within an 'implementation plan' for scenario 3.
- An implementation plan could be structured into implementation categories like business, technical social, environmental and governmental measures. Each measure could be defined by criteria like category (human resources, process, know-how, partners, technology, etc.), measure description, criticality, time frame, cost/budget, responsibility, status, etc.

### 19.5.6 Compilation of a Comprehensive Transition Plan (Step 7)

The final transition planning step of the transition concept (represented in Fig. 19.2) for achieving the future vision (AVs in ITS) is titled as 'transition plan'. It is described as the compilation (comprehensive, systemically structured 'folder' of transition steps and planning results) of all results of the previous transition steps 2–6 for each scenario (mid- or long-term relevance) for the transition from the status quo (step 1) to achieve the future Vision 2030 (step 8).

In analogy to a business or innovation plan, the 'transition plan' is characterised as an orientation guide and thus strategic and operational basis for all stakeholders of the mobility transition. From a scenario-based logic, it also can be divided into partial transition plans of scenario 1 as well as of scenario 2, which are both oriented in a long-term perspective to the future vision of AV in ITS. Each of these scenarios is aiming in a mid-term perspective to prepare and achieve scenario 3, a preliminary evolution stage to achieve AV in ITS as the future vision. Depending on the individual organisation of the urban mobility ecosystem transition as a strategic project, the transition plan can be managed by a dedicated overall or scenario transition-specific transition manager who is responsible for the general design and controlling of the transition plan, the iterative-incremental design and change updates, as well as adequate implementation and adherence of all transition activities to this plan.

## 19.6 Conclusion

The goal of this chapter was to develop an equal and balanced strategic and operational transition concept from the current status quo towards a successfully deployed sustainable urban mobility ecosystem (future vision). Together with all experts mentioned in the introduction, a generic eight-step transition approach following the described criteria was developed: description of the status quo for each identified scenario (step 1); description of a future vision (step 8); description of six transition steps from the status quo to the future vision; identification of success factors and obstacles (step 2); identification and characterisation of stakeholders (step 3); definition of transition goals, strategies and business models (step 4); transition recommendations and measures (step 5); design of an implementation plan and a transition roadmap (step 6); and compilation of a comprehensive transition plan (step 7). In particular, a roadmap with a timeline categorisation of recommendations for the stakeholders was designed to propose how to realise the desired future developed by the team (Vision 2030, Chap. 18). The five dimensions (technical, social, economic, environmental and governance) were used to better understand and underline the interrelationship of the proposed holistic approach.

In terms of methodology, a limitation of the described approach is however that the concrete status quos of the pilot sites were not taken into account in an adequate way. Just the status quos of vehicles (Chaps. 3 and 11) and governance (Chap. 16) were considered. In fact, the status quo as a starting point is different for each pilot site. There are further several transition plans possible to achieve the same vision.

Even the goals derived from this vision must be adapted individually to the specific situation of each pilot site. These weaknesses will be addressed within the subsequent project ULTIMO (2022–2026).

**Acknowledgements** This project has received funding by the European Union's Horizon 2020 research and innovation program under grant agreement No. 769033.

## References

Antonialli, F., Fournier, G., Jaroudi, I., Mira-Bonnardel, S., & Thalhofer, M. (2021). *D8.4 second iteration economic impact*. https://cordis.europa.eu/project/id/769033/results

Araghi, Y., Larco, N., Doll, C., Vonk Noordegraaf, D. M., Bouma, G. M., & Krauss, K. (2020). *Drivers and barriers of mobility-as-a-service in urban areas*. https://www.isi.fraunhofer.de/content/dam/isi/dokumente/ccn/2020/tra2020_drivers_and_barriers_final-after-reviews-19-10-2019.pdf

Deloitte. (2019). *Urbane Mobilität und autonomes Fahren im Jahr 2035*. https://www2.deloitte.com/de/de/pages/trends/urbane-mobilitaet-autonomes-fahren-2035.html

Fournier, G. (2022). *Deliverable 9.3 roadmap for cost-attractiveness*. https://h2020-avenue.eu/public-delivrables/

Fournier, G., Thalhofer, M., Chrétien, P., Boos, A., Korbee, D., Jaroudi, I., Nemoto, E., Naderer, G., Konstantas, D., & Viere, T. (2023). System innovation in transport with automated minibuses and ITS: The citizen centric approach of AVENUE. *Transportation Research Procedia, 72*, 2984–2991. https://doi.org/10.1016/j.trpro.2023.11.845

Khor, N., Arimah, B., Otieno, R., van Oostrum, M., Mutinda, M., & Oginga Martins, J. (2022). *Envisaging the future of cities*. World Cities Report: Vol. 2022. UN-Habitat.

Nemoto, E. H. (2022). *Towards a new sustainable mobility paradigm: The role and impacts of automated minibuses in the mobility transition* [PhD Thesis]. CentraleSupelec, Paris.

Nemoto, E. H., Jaroudi, I., & Fournier, G. (2021). Introducing automated shuttles in the public transport of European cities: The case of the AVENUE project. In E. G. Nathanail, G. Adamos, & I. Karakikes (Vol. 1278). Springer. https://doi.org/10.1007/978-3-030-61075-3_27.

Ruter. (2019). *The Oslo study: How autonomous cars may change transport in cities*. Report.

Trafi online. https://www.trafi.com/

Whim online. How to get there. https://whimapp.com

# Chapter 20
# Conclusion

**Guy Fournier and Dimitri Konstantas**

**Abstract** The AVENUE project, carried out under Horizon 2020, demonstrated remarkable progress in automated public transport by deploying almost SAE Level 4 AVs in six European cities. Despite significant challenges, including COVID-19 restrictions and partner insolvency, the project successfully deployed automated minibuses that carried over 160,000 passengers and travelled a total of 168,000 km. This conclusion highlights AVENUE's achievements in improving urban mobility through on-demand door-to-door services and emphasises the potential of using automated minibuses as game changer towards a more sustainable, efficient and citizen centric mobility which serves the general interest.

AVENUE was one of the biggest European projects with automated vehicles on the road. Despite facing significant challenges, such as the COVID-19 restrictions and the bankruptcy of a key partner, AVENUE made remarkable achievements, including the successful deployment of AVs with "nearly SAE level 4" automation in regular commercial public transportation. These AVs provide on-demand and door-to-door services across four European demonstrator and two replicator cities. Throughout the project, these services amassed 80,000 kilometres travelled and 60,000 passengers and utilised 10 automated minibuses in 4 European demonstrator cities (Geneva (Meyrin, Belle-Idée; Switzerland), Lyon (ParcOL; France), Copenhagen (Nordhavn, Slagelse; Denmark) and Luxembourg (Pfaffenthal, Contern; Luxembourg)) and 2 replicator cities (Uvrier in Switzerland and Esch-sur-Alzette in Luxembourg). Additionally, running projects were incorporated into AVENUE, contributing an additional 82,000 km travelled, 100,000 passengers served and the use of 6 AMs. The term "nearly level 4" AVs indicates that, for technical and legal reasons, a safety operator was required to be present in the AV for supervision and incident handling. Vehicle services that substantially enhance the

G. Fournier (✉)
Pforzheim University, Pforzheim, Germany

CentraleSupélec, LGI, Gif-sur-Yvette, France
e-mail: guy.fournier@hs-pforzheim.de

D. Konstantas
Geneva School of Economics and Management, University of Geneva, Geneva, Switzerland



G. Fournier et al. (eds.), *Automated Vehicles as a Game Changer for Sustainable Mobility*, Contributions to Management Science,
https://doi.org/10.1007/978-3-031-61681-5_20

passenger experience as well as the overall quality and value of the service were introduced, also targeting elderly people, people with disabilities and potentially vulnerable users. Road behaviour, security of the AVs and passengers' safety were crucial points of the AVENUE project.

In the first part of the book, the implementation on the sites and advancements of the vehicle have been documented. Safety and security issues are analysed but also how they can be addressed with, for example, in-vehicles services. Specific requirements for persons with reduced mobility (PRM) are further examined. A comprehensive analysis of stakeholders is finally provided to understand how a collaboration of the AV ecosystem can best provide value and benefits for all the stakeholders and the society.

In Part II, a holistic approach is used to analyse the technical, social, economic, environmental and governance impact assessment of the implementation of AV vehicles. The approach of this in-depth research demonstrates that all the dimensions are interrelated, for example, to show how AVs can enable positive externalities or reduce negative externalities. Technical issues have further business impacts, and governance influences business viability and success for all stakeholders, etc. A comprehensive representative survey depicts a noteworthy goodwill for the innovative services provided with automated minibuses. The aim of the interdisciplinary approach was to enable and objectivise a balance between individual private and general interests as well as between the ecosystem stakeholders in general. The purpose is to maximise the benefits for the society to the best for all the stakeholders (common interest).

As a result, a proposal on how to integrate automated vehicles into a city has been developed in Chap. 18:

- Automated minibus integrated in a MaaS could be a real game changer and enable a sustainable mobility and fair share of the created values between the stakeholders. In this described future, AVs are complementing and fostering other means of transport, in particular mass transport. Rural, remote or suburban areas, for example, could have access to existing mass transport infrastructure and MaaS services and consequently be better deserved. Tangential connection between two suburban areas could be possible. The system can save energy and materials but also space. It provides further temporal flexibility and spatial flexibility nearly similar to private cars and does not compete with public transport. A sustainable and citizen centric mobility is thus possible without a coercive (technology push) transport policy, just in increasing the choice and improving the provided mobility services (demand and attractivity pull). Even privately owned or privately shared cars are part of the model (see Figure 18.2) and justified, for example, in areas with poor infrastructure or where time is a critical issue.
- Integrating AV in an intelligent transport system could further make it possible to use mobility data and artificial intelligence to improve the transport system with fast and slow closed loops. Thus, the recommended approach goes beyond the current product AV innovation and suggests using the automated minibus as a trigger for disruptive innovation of the transport system as a whole. The transport system would become thus ambidextrous in being able to combine antinomic economic goals: efficiency and flexibility. The transport system would also turn more resilient, as in case of failure (incident or accident), an automated minibus could bridge mobility gaps.

A transition planning towards this desirable future has finally been developed in Chap. 19. This is necessary to make live the strategy and deploy it successfully. Recommendations have been formulated within this framework. This transition is necessary as the European AV in MaaS mobility market is currently fragmented and stakeholders in particular OEMs of automated minibuses and the related digital service provider are struggling with the viability of their AV business models (see recommendation R_2024-26.5 in Chap. 19). To create a federated (democratised) and sustainable transport systems which serves the general interest, TPOs and TPAs should shape the market in defining the technical specifications of the vehicles to be integrated in a MaaS and ITS. This makes sense as TPOs and TPAs are already driven by society ("purpose economy") and EU/national regulations to serve the general interest. A disclosure of their procurement goals and strategies would further be necessary to give a long-term perspective to the AV ecosystem and ease their own financing of innovative AV in MaaS/ITS technologies. The emergence of strong OEMs/AV technology provider and their related AV-ecosystems would thus be enabled and supported. This could be a promising alternative to possible strategies in other parts of the world where states or private companies (GAFAM or BATX) or both share the risks with the hidden ulterior motive to capture value for private interest (closed ecosystems focusing on "the winner takes it all") or even for questionable states goals, for example, regarding data privacy.

The AVENUE vision will be implemented and tested in parts within the Horizon Europe ULTIMO project: 15 multivendor vehicles will be implemented each in Oslo (Norway), Herford (Germany) and Geneva (Switzerland) until 2026. The integration of logistics will be tested as well in Geneva. The ULTIMO project will also have to make an update on the proposed vision of Chap. 18. Indeed, in the long run, with the loosing of the boundaries between traditional motorised private and public transport dominance like mentioned in the foreword of FEDRO, the suggested proposal will have to be evaluated again.

**Acknowledgements** This project has received funding by the European Union's Horizon 2020 research and innovation program under grant agreement No. 769033.

SPRINGER NATURE

# GPSR Compliance

*The European Union's (EU) General Product Safety Regulation (GPSR) is a set of rules that requires consumer products to be safe and our obligations to ensure this.*

*If you have any concerns about our products, you can contact us on ProductSafety@springernature.com*

In case Publisher is established outside the EU, the EU authorized representative is:

Springer Nature Customer Service Center GmbH
Europaplatz 3
69115 Heidelberg, Germany

The manufacturer's authorised representative in the EU is Springer Nature Customer Service Centre GmbH, Europaplatz 3, 69115 Heidelberg, Germany. If you have any concerns regarding our products, please contact ProductSafety@springernature.com

Printed and bound by CPI Group (UK) Ltd, Croydon, CR0 4YY
01/10/2026
02225849-0017